Wetlands for Water Pollution Control

Wetlands for Water Pollution Control

Second Edition

Miklas Scholz
The University of Salford, Salford, UK

ELSEVIER

AMSTERDAM • BOSTON • HEIDELBERG • LONDON
NEW YORK • OXFORD • PARIS • SAN DIEGO
SAN FRANCISCO • SINGAPORE • SYDNEY • TOKYO

Elsevier
Radarweg 29, PO Box 211, 1000 AE Amsterdam, Netherlands
The Boulevard, Langford Lane, Kidlington, Oxford OX5 1GB, UK
225 Wyman Street, Waltham, MA 02451, USA

Second edition 2016

Notices
Knowledge and best practice in this field are constantly changing. As new research and experience broaden our understanding, changes in research methods, professional practices, or medical treatment may become necessary.

Practitioners and researchers must always rely on their own experience and knowledge in evaluating and using any information, methods, compounds, or experiments described herein. In using such information or methods they should be mindful of their own safety and the safety of others, including parties for whom they have a professional responsibility.

To the fullest extent of the law, neither the Publisher nor the authors, contributors, or editors assume any liability for any injury and/or damage to persons or property as a matter of products liability, negligence or otherwise, or from any use or operation of any methods, products, instructions, or ideas contained in the material herein.

ISBN: 978-0-444-63607-2

British Library Cataloguing in Publication Data
A catalogue record for this book is available from the British Library

Library of Congress Cataloging-in-Publication Data
A catalog record for this book is available from the Library of Congress

Contents

About the Author xix
Preface xxiii
Acknowledgments and Dedications xxvii
Common Acronyms and Abbreviations xxix

1. Water Quality Standards 1

1.1 Introduction and Historical Aspects 1
1.2 Water Quality Standards and Treatment Objectives 3
1.3 Biochemical Oxygen Demand 5
1.4 Chemical Oxygen Demand 7
1.5 Other Variables Used for the Characterization of Wastewater 7

2. Water Treatment 9

2.1 Sources of Water 9
2.2 Standard Water Treatment 9
2.3 Basic Water Chemistry 10

3. Sewage Treatment 13

3.1 Introduction 13
3.2 Design Flow Rates 13
3.3 Treatment Principles 14
3.4 Engineering Classification of Sewage Treatment Stages 14

4. Stream Pollution and Effluent Standards 17

4.1 Organic Stream Pollution 17
4.2 Prediction of Organic Stream Pollution 18
4.3 Effluent Discharge Standard Principles 19

5. Preliminary Treatment 21

5.1 Introduction 21
5.2 Design of Screening Units 21
5.3 Design Details for Screening Units 23
5.4 Comminutors 23
5.5 Grit Removal 24

6. Primary Treatment 27

6.1 Introduction 27
6.2 Loading Rate Methods 28
6.3 Tank Design 28
6.4 Design Parameters 29
6.4.1 Design Settling Velocity 29
6.4.2 Horizontal Velocity 29
6.4.3 Time Ratio 30
6.5 Economics of Construction 30
6.5.1 Rectangular Settling Tanks 30
6.5.2 Circular Settling Tanks 30
6.6 Design Details 30
6.6.1 Rectangular Settling Tanks 30
6.6.2 Circular Settling Tanks 31
6.7 Hydraulic Losses 31
6.8 General Design Details 32
6.9 Details of Various Types of Sedimentation Tanks 32
6.9.1 Storm Tanks 32
6.9.2 Primary Sedimentation Tanks 33
6.9.3 Secondary Sedimentation 34
6.10 Sedimentation Aids 35

7. Coagulation and Flocculation 37

7.1 Theory of Settling 37
7.2 Classification of Settling Behavior 37
7.3 Ideal Settling 39
7.4 Introduction to Coagulation and Flocculation 39
7.5 Colloidal Suspensions 40
7.6 Coagulation Processes 41
7.7 Coagulation Chemicals 42
7.8 Operation of the Coagulation and Flocculation Process 44
7.9 Rapid Mixing 45
7.10 Flocculation 46

8. Sludge Blanket Clarifiers 47

8.1 Introduction to Sludge Blanket Clarification Systems 47
8.2 Types of Sludge Blanket Clarifier 51
8.3 Plate Settling in Sludge Blanket Clarifiers 53

9. Flotation Systems 55

9.1 Flotation Using Blown Air 55
9.2 Flotation Using Dissolved Air 55
9.3 Flotation Units 57

10. Slow Filtration 61

10.1 Introduction 61
10.2 Slow Sand Filtration 61
10.2.1 Elements of a Slow Sand Filter 61
10.2.2 Mechanisms in a Slow Sand Filter 63
10.3 Algal Actions 66
10.4 Advantages and Disadvantages 67

11. Rapid Filtration 69

11.1 Elements of a Rapid Sand Filter 69
11.2 Sand Bed 70
11.3 Underdrain System 71
11.4 Hydraulics of Filtration 74
11.5 Comparison with Slow Sand Filter 75

12. Biological Treatment 77

12.1 Aerobic Self-Purification 77
12.2 Waste Stabilization Ponds 77
12.2.1 Aerobic Ponds 77
12.2.2 Facultative Ponds 78

13. Biological Filtration 81

13.1 Introduction 81
13.2 Trickling Filter 81
13.3 Basic Ecology 82
13.4 Process Variants 83
13.5 Design of Biological Filters 84

14. Rotating Biological Contactors 87

14.1 Introduction 87
14.2 Principles of Operation 87
14.3 Design and Loading Criteria 87
14.4 Principle Elements 89
14.5 Operational Problems 89

15. Activated Sludge Processes 91

15.1 Background 91
15.2 Activated Sludge Process 91
15.3 Comparison between the Activated Sludge Process, Percolating Filtration, and Wetland System 93
15.4 Activated Sludge Process Types 96
15.4.1 Conventional Complete Mix Activated Sludge Process 96
15.4.2 Series or Plug Flow System 96
15.4.3 Tapered Aeration 96

15.4.4 Step Feed Activated Sludge Process 97
15.4.5 High-Rate Activated Sludge Process 97
15.4.6 Extended Aeration 97
15.4.7 Contact Stabilization 98
15.4.8 Oxidation Ditches 98
15.4.9 Deep Shaft Process 99
15.5 Activated Sludge Process Designs and Kinetics 99
15.5.1 Diffused Air Aeration 99
15.5.2 Mechanical Aerators 100
15.5.3 Process Design 101
15.6 Summary of Key Process Design Criteria 103
15.6.1 Loading Criteria 103
15.6.2 Reactor Types 104
15.6.3 Oxygen Demand 104
15.6.4 Nutrient Requirements 105

16. Iron and Manganese Removal 107

16.1 Introduction 107
16.2 Basic Removal Processes 107
16.3 Advanced Removal Processes 108

17. Water Softening 111

17.1 Introduction 111
17.2 Lime-Soda Softening 112
17.3 Lime Softening 113
17.4 Excess Lime Softening 113
17.5 Lime Recovery 114

18. Water Microbiology 115

18.1 Statistics for Applied Microbiology 115
18.2 Protozoa 116
18.2.1 Trophic Structure 116
18.2.2 Kingdom Protista 117
18.3 Biological Effects of Organic Pollutants 118
18.3.1 Sewage Fungus 118
18.3.2 Saprobic System 119
18.4 Eutrophication and Water Treatment 120
18.5 Protozoology of Treatment Processes 122
18.6 Odor and Toxins of Natural Origin 122
18.7 Public Health Aspects 123
18.7.1 Typical Diseases Related to Waters 123
18.7.2 Invertebrates Found in Main Supplies 126
18.7.3 Monitoring and Prevention of Waterborne Diseases 126

19. Disinfection 129

19.1 Destroying Pathogens and Requirements of a Disinfectant 129
19.2 Traditional Methods of Disinfection 130
19.3 Ozone 130
19.4 Chlorine Dioxide and Chlorine as Disinfectants 131
19.5 Kinetics of Chlorination 133
19.6 Applications of Chlorine 134
19.7 Technology of Chlorine Addition 134
19.8 Advantages and Disadvantages of Chlorine 135

20. Constructed Wetlands 137

20.1 Background 137
20.2 Definitions 137
20.3 Hydrology of Wetlands 138
20.3.1 Hydro-Period and Water Budget 138
20.3.2 Precipitation, Interception, Through-Fall, and Stem-Flow 139
20.4 Wetland Chemistry 140
20.4.1 Oxygen 140
20.4.2 Carbon 142
20.4.3 Nitrogen 142
20.4.4 Phosphorus 144
20.4.5 Sulfur 146
20.5 Wetland Ecosystem Mass Balance 147
20.6 Macrophytes in Wetlands 148
20.6.1 Background 148
20.6.2 Primary Productivity 149
20.6.3 *Phragmites australis* 149
20.6.4 *Typha latifolia* 150
20.7 Physical and Biochemical Parameters 150
20.8 Examples for Natural and Constructed Wetlands 151
20.8.1 Riparian Wetlands 151
20.8.2 Constructed Treatment Wetlands 153
20.8.3 Wetlands for Stormwater Treatment 154

21. Sludge Treatment and Disposal 157

21.1 Introduction 157
21.2 Tests for Dewatering of Sludge 159
21.3 Sludge Treatment and Disposal Objectives and Methods 160
21.4 Treatment Processes 162
21.4.1 Lagoons 162
21.4.2 Aerobic Digestion 162
21.4.3 Other Treatment Methods 163

21.5	**Thickening and Dewatering of Sludges**		163
	21.5.1	Chemical Conditioning	163
	21.5.2	Air Drying	164
	21.5.3	Gravity Thickening	164
	21.5.4	Other Methods	164
21.6	**Partial Disposal**		165
	21.6.1	Incineration	165
	21.6.2	Pyrolysis	166
	21.6.3	Composting	166
21.7	**Land Dumping and Passive Treatment**		167

22. **Wetlands Treating Contaminated Stream Water** 169

22.1	**Introduction**		169
22.2	**Materials and Methods**		170
	22.2.1	Experimental Plan and Limitations	170
	22.2.2	Filter Media Composition	170
	22.2.3	Environmental Conditions and Operation	172
	22.2.4	Analytical Procedures Including Metal Determination	173
	22.2.5	Microbiological Examinations	174
	22.2.6	Statistics	176
22.3	**Results and Discussion**		176
	22.3.1	Comparison of Treatment Efficiency	176
	22.3.2	Water Quality and Macrophytes	178
	22.3.3	Water Quality and Microbiology	182
	22.3.4	Regression and Correlation Analysis as Predictive Tools	184
22.4	**Conclusions**		190

23. **Wetland Systems to Control Roof Runoff** 191

23.1	**Introduction**		191
	23.1.1	Sustainable Roof Runoff Drainage	191
	23.1.2	Case Study: Site Description	192
	23.1.3	Purpose	192
23.2	**Methods**		192
	23.2.1	Design of the Study Site	192
	23.2.2	Engineering Methods	194
	23.2.3	Water Quality Analysis	195
	23.2.4	Control of Algal Growth	196
	23.2.5	System Capacity	197
23.3	**Results and Discussion**		197
	23.3.1	Standard Design Considerations	197
	23.3.2	System Design Comparisons	201
	23.3.3	Water Quality Management	202
	23.3.4	Twenty-Four-Hour Water Quality Monitoring	203
	23.3.5	Aquatic Plant Management	204
23.4	**Conclusions**		206

24. Wetlands Treating Road Runoff 209

24.1 Introduction 209
24.2 Site, Materials, and Methodology 210
24.2.1 Case Study Site 210
24.2.2 Filter Design, Media Composition, and Limitations 211
24.2.3 Environmental Conditions and Operation 212
24.2.4 Metal Nitrates 214
24.2.5 Metal Determinations 214
24.2.6 Organic Strength, Nutrient, and Other Determinations 217
24.3 Experimental Results and Discussion 217
24.3.1 Inflow Water Quality 217
24.3.2 Comparison of Annual Outflow Water Qualities 217
24.3.3 Heavy Metal Removal 223
24.3.4 Link between pH and Treatment of Metals 224
24.3.5 Analysis of Variance and Modeling 224
24.4 Conclusions and Further Work 225

25. Combined Wetland and Below-ground Detention Systems 227

25.1 Introduction 227
25.2 Materials and Methods 228
25.2.1 System Design and Operation 228
25.2.2 Analytical Methods 229
25.3 Results and Discussion 232
25.3.1 Comparison of Costs 232
25.3.2 Inflow Water Quality 232
25.3.3 Comparison of Outflow Water Quality 234
25.3.4 Ecosoil and Turf 236
25.4 Conclusions and Further Research 237

26. Modeling of Constructed Wetland Performance 239

26.1 Introduction 239
26.2 Methodology and Software 241
26.2.1 Experimental Data and Variables 241
26.2.2 K-Nearest Neighbors 242
26.2.3 Support Vector Machine 243
26.2.4 Self-Organizing Map 245
26.3 Results and Discussion 246
26.3.1 Performance Evaluation 246
26.3.2 Correlation Analysis 246
26.3.3 Optimization of Input Variables 246
26.3.4 Comparison of Applications 251
26.4 Conclusions 256

27.	**Infiltration Wetland Systems**		259
27.1	**Introduction**		259
27.2	**Methods**		260
	27.2.1	Design of the Study Site	260
	27.2.2	Hydrological Methods and Water Quality Analysis	261
	27.2.3	Fish Experiment Methodologies	261
27.3	**Results and Discussion**		262
	27.3.1	Design and Operation of Infiltration Ponds	262
	27.3.2	Rainfall, Runoff, and Infiltration Relationships	263
	27.3.3	Water Quality Assessment and Management	263
	27.3.4	Active Control of Algae with Goldfish	267
	27.3.5	Integration of Wetlands into Urban Planning and Development	268
27.4	**Conclusions**		268
28.	**Retrofitting of Sustainable Drainage Systems in the Presence of Vegetation**		271
28.1	**Introduction**		271
	28.1.1	Ecosystem Services and Urbanization	271
	28.1.2	Urban Trees and Linked Ecosystem Services	272
	28.1.3	Example Case Study	275
	28.1.4	Rationale, Aim, and Objectives	276
28.2	**Methodology**		277
	28.2.1	Site Assessment	277
	28.2.2	Ecosystem Services	278
	28.2.3	Traditional "Community and Environment" Variables	279
	28.2.4	Comparison between the Traditional and Revised Methods	280
	28.2.5	Combination of the Traditional and New Approaches	280
	28.2.6	Vegetation Determination	280
28.3	**Results and Discussion**		281
	28.3.1	Site Assessment Outcomes	281
	28.3.2	Strengths and Weaknesses of the Site Assessment	282
	28.3.3	Assessment of Ecosystem Service Variables	282
	28.3.4	Strengths of the New Ecosystem Services Assessment Approach	284
	28.3.5	Limitations of the New Ecosystem Services Assessment Approach	285
	28.3.6	Comparison of Assessment Methods	286
	28.3.7	Vegetation Suitable for Urban Areas	287
28.4	**Conclusions and Recommendations**		293

29. Expert Tool Based on Ecosystem Variables for Retrofitting of Wetland Systems 295

29.1 Introduction 295
29.2 Methodology 297
29.2.1 Site Assessment 297
29.2.2 Assessment of Ecosystem Service Variables 299
29.2.3 SuDS Variable Determination Approaches 300
29.2.4 Uncertainties of the Rapidly Estimated Variables 303
29.2.5 Variability of Estimated Variables and the Learning Process of Estimation 303
29.2.6 Decision Support for Various Professions 304
29.3 Results and Discussion 304
29.3.1 Site Assessment Outcomes and Associated Strengths and Weaknesses 304
29.3.2 Strengths and Limitations of the New Ecosystem Services Assessment Approach 306
29.3.3 Comparison of Assessment Methods 307
29.3.4 Certainty Estimations Based on Drainage Engineering Expert Opinion 309
29.3.5 Variability of Estimated Variables and the Associated Learning Process 309
29.3.6 Decision Support Tool Addressing Different Professional Perspectives 310
29.4 Conclusions and Recommendations 310

30. Sustainable Drainage System Model 313

30.1 Introduction 313
30.1.1 Sustainable Drainage Systems 313
30.1.2 Sustainable Drainage Impact on Water Quantity and Quality 317
30.1.3 Development and Regeneration in Glasgow 318
30.1.4 SuDS in Edinburgh 318
30.1.5 Aims and Objectives 319
30.2 Sites, Methodology, and Modeling 319
30.2.1 Overview of Sites in Glasgow and Edinburgh 319
30.2.2 Decision Support Matrix and Weighting System 319
30.2.3 Decision Support Model 323
30.2.4 Prevalence Rating Approach for Sustainable Drainage Techniques 330
30.2.5 Case-Based Reasoning Model 332
30.3 Results and Discussion 333
30.3.1 Decision Support Model Output 333
30.3.2 Prevalence Rating Approach Analysis 335
30.3.3 Case-Based Reasoning Model Output 337
30.4 Conclusions 337

31. Natural Wetlands Treating Diffuse Pollution 339

31.1 Introduction 339
31.1.1 Background of the Case Study 339
31.1.2 Nutrient Transformations and Removal Processes 340
31.1.3 Aim and Objectives 341
31.2 Materials and Methods 341
31.2.1 Case Study, Sampling, and Data Analysis 341
31.2.2 Ditches and the Channel of the Receiving River 343
31.2.3 Discharge Determination for Open Channels 345
31.2.4 Water Quality Analyses 345
31.2.5 Vegetation Characterization 346
31.3 Results 346
31.3.1 Characteristics of Watercourses in the River Eider Valley 346
31.3.2 Water Quality during Spring and Summer 346
31.3.3 Ditch Vegetation 350
31.4 Discussion 353
31.4.1 Disappearance of Ditches Due to Vegetation Growth 353
31.4.2 Water Quality Variations within the Study Area 354
31.4.3 Vegetation Characterization 355
31.4.4 Hydraulic Changes Due to Summer Flooding 358
31.5 Conclusions and Recommendations 359

32. Integrated Constructed Wetlands for Pollution Control 361

32.1 Introduction 361
32.2 Case Study Sites, Materials, and Methodologies 362
32.2.1 Sites within the Case Study Catchment 362
32.2.2 Sampling and Analytical Methods 362
32.2.3 Deoxyribonucleic Acid Extraction 364
32.2.4 Polymerase Chain Reaction 364
32.2.5 Agarose Gel Electrophoresis 365
32.3 Results and Discussion 365
32.3.1 Wetland Water Quality 365
32.3.2 Flows 368
32.3.3 Design Considerations 368
32.3.4 Nitrogen Transformation Processes 369
32.3.5 Comparison of Ammonia Oxidizers and Denitrifiers 370
32.4 Conclusions and Recommendations 371

33. Infiltration and Contaminant Migration beneath Integrated Constructed Wetlands 373

33.1 Introduction 373
33.2 Case Study, Materials, and Methods 375
33.2.1 Study Site 375
33.2.2 Instrumentation 378
33.2.3 Water Sampling and Analysis 378
33.2.4 Hydraulic Conductivity of Soil Liner 379
33.2.5 Wetland System Water and Hydrological Monitoring 379
33.2.6 Water Quality Analysis 380
33.2.7 Contaminant Loading Rates 380
33.2.8 Statistical Analysis 381
33.3 Results and Discussion 381
33.3.1 Infiltration Rates 381
33.3.2 Hydraulic Conductivities 384
33.3.3 Wetland Water Fluxes 384
33.3.4 Contaminant Concentrations and Input Loading in Wetland Cells 384
33.3.5 Contaminant Concentrations and Input Loading beneath Wetland Cells 385
33.4 Conclusions and Recommendations 388

34. Seasonal Assessment of Vertical-Flow Wetlands Treating Domestic Wastewater 389

34.1 Introduction 389
34.2 Materials and Methods 391
34.2.1 Experimental Set-Up and Operation 391
34.2.2 Water Quality Analysis 392
34.2.3 Clogging Tests 393
34.2.4 Statistical Tests 393
34.3 Results and Discussion 393
34.3.1 Inflow Water Quality Analysis 393
34.3.2 Evaluation of Oxygen Demand Variables 394
34.3.3 Comparison of Nitrogen Compounds 397
34.3.4 Evaluation of Ortho-Phosphate-Phosphorus 398
34.3.5 Comparison of Particles 399
34.3.6 Assessment of Filter Clogging Based on Water Quality Variables 400
34.4 Conclusions and Recommendations 400

35. Recycling of Domestic Wastewater Treated by Vertical-Flow Wetlands for Irrigation 401

35.1 Introduction 401
35.1.1 Background 401

35.1.2 Wastewater Recycling for Crop Irrigation 401
35.1.3 Plant Selection 403
35.1.4 Nutrients and Minerals 404
35.1.5 Motivation and Objectives 405
35.2 Methodology 406
35.2.1 Constructed Wetlands Set-Up and Operation 406
35.2.2 Water Quality Analysis 408
35.2.3 Environmental Monitoring 409
35.2.4 Selected Vegetables 409
35.2.5 Replanting the Vegetables 410
35.2.6 Growth Conditions 411
35.3 Results and Discussion 413
35.3.1 Water Quality Analysis 413
35.3.2 Growth Comparisons 414
35.3.3 Inductively Coupled Plasma Findings 414
35.3.4 Brief Cost—Benefit Analysis and Limitations 415
35.4 Conclusions and Recommendations 416

36. Piggery Wastewater Treatment with Integrated Constructed Wetlands 419

36.1 Introduction 419
36.1.1 Legislative Pressures 419
36.1.2 Traditional Piggery Wastewater Management 420
36.1.3 Alternative Piggery Wastewater Treatment with Wetlands 420
36.2 Site, Materials, and Methods 422
36.2.1 Site Description, System Set-Up, and Test Conditions 422
36.2.2 Analytical Methods 424
36.3 Results and Discussion 424
36.3.1 Raw and Diluted Wastewater 424
36.3.2 Standard Treatment 425
36.3.3 Effluent Recycling Treatment 425
36.3.4 High Nutrient Load Treatment 428
36.3.5 High Flow Rate Treatment 428
36.3.6 Simplified Cost—Benefit Analysis and Limitations 430
36.3.7 Impact on Future Piggery Farming 432
36.4 Conclusions and Recommendations 432

37. Wetland Systems as Part of the Sustainable Flood Retention Basin Concept 433

37.1 Introduction 433
37.1.1 Rationale for Rapid Survey and Classification Method 433
37.1.2 Sustainable Flood Retention Basin Typology 434

37.1.3 Survey Template 438
37.1.4 Manpower and Equipment 438
37.1.5 Rationale and Objectives 439
37.2 Assessment of Classification Variables 440
37.2.1 Engineered 440
37.2.2 Dam Height 440
37.2.3 Dam Length 441
37.2.4 Outlet Arrangement 441
37.2.5 Aquatic Animal Passage 441
37.2.6 Land Animal Passage 442
37.2.7 Flood Plain Elevation 443
37.2.8 Basin Channel Connectivity 443
37.2.9 Wetness 443
37.2.10 Proportion of Flow within the Channel 444
37.2.11 Mean Flooding Depth 444
37.2.12 Typical Wetness Duration 445
37.2.13 Flood Duration 445
37.2.14 Basin Bed Gradient 446
37.2.15 Mean Basin Flood Velocity 446
37.2.16 Wetted Perimeter 446
37.2.17 Maximum Floodwater Volume 446
37.2.18 Floodwater Surface Area 447
37.2.19 Mean Annual Rainfall 447
37.2.20 Drainage 447
37.2.21 Impermeable Soil Proportion 448
37.2.22 Seasonal Influence 448
37.2.23 Altitude 448
37.2.24 Vegetation Cover 449
37.2.25 Algal Cover in Summer 449
37.2.26 Relative Total Pollution 449
37.2.27 Mean Sediment Depth 450
37.2.28 Organic Sediment Proportion 450
37.2.29 Flotsam Cover 451
37.2.30 Catchment Size 452
37.2.31 Urban Catchment Proportion 452
37.2.32 Arable Catchment Proportion 452
37.2.33 Pasture Catchment Proportion 453
37.2.34 Viniculture Catchment Proportion 453
37.2.35 Forest Catchment Proportion 453
37.2.36 Natural Catchment Proportion 453
37.2.37 Groundwater Infiltration 454
37.2.38 Mean Depth of Basin 454
37.2.39 Length of Basin 454
37.2.40 Width of Basin 455
37.3 Conclusions and Recommendations for Further Work 455

38. **Classification of Sustainable Flood Retention Basins** 457

38.1 Introduction to Traditional Classification 457
38.1.1 Flood Risk Management 457
38.1.2 Traditional and Adaptive Structural Measures and Their Classification 458

38.2 Traditional Methodology 459
38.2.1 Identification of Sites and Definitions 459
38.2.2 Recognition of Classification Variables 461
38.2.3 Rationale for the Elimination of Less Relevant Variables 463
38.2.4 Assignment of Basin Types with the Help of Cluster Analyses 464

38.3 Discussion of the Traditional Method 465
38.3.1 Reduction Exercise for the Classification Variables 465
38.3.2 Cluster Analyses 465
38.3.3 Groupings Based on Cluster Analysis and Case Study Application 466

38.4 Introduction to Multilabel Classification 468
38.4.1 Basins and Wetland Systems 468
38.4.2 Environmental Modeling and Multilabel Classification 469

38.5 Data and Methodology 471
38.5.1 Data Acquisition 471
38.5.2 Multilabel Support Vector Machine 472
38.5.3 Multilabel k-Nearest Neighbor 473
38.5.4 Back-Propagation for Multilabel Learning 473
38.5.5 Evaluation Measures 474

38.6 Experimental Results and Discussion 477

38.7 Representative Case Studies 479

38.8 Conclusions and Recommendations 481

References 483
Index 511

About the Author

Prof. Miklas Scholz, Cand Ing, BEng (equiv), PgC, MSc, PhD, CWEM, CEnv, CSci, CEng, FHEA, FIEMA, FCIWEM, FICE, Fellow of IWA, holds the Chair in Civil Engineering at The University of Salford (Figure 1). He is the Head of the Civil Engineering Research Group. Prof. Scholz has shown individual excellence evidenced by world-leading publications, postgraduate supervision, and research impact. His main research areas (Figure 2) in terms of publication output are as follows: treatment wetlands (20%), integrated constructed wetlands (ICW; 15%), sustainable flood retention basins (SFRB; 5%), permeable pavement systems (5%), decision support systems (5%), ponds (5%), and capillary suction time (5%). About 45% and 40% of his research are in water resources management and wastewater treatment, respectively. The remaining 10% is in capillary processes and water treatment.

He has published four books and more than 176 journal articles covering a wide range of topics (Figure 2). Between 2009 and 2015, he topped the publication list in terms of numbers for all members of the staff at The University of Salford. Prof. Scholz's total journal article publications in

FIGURE 1 Miklas Scholz on top of a sustainable flood retention basin near Perth, Scotland, UK. *(Picture taken by Åsa Hedmark.)*

FIGURE 2 Overview of research areas and their corresponding relative importance and linkages between them.

recent years are as follows: 2009, 13 articles; 2010, 19 articles; 2011, 13 articles; 2012, 21 articles; 2013, 17 articles; and 2014, 15 articles.

He publishes regularly in the following journals with high impact factors: *Bioresource Technology, Building and Environment, Construction and Building Materials, Desalination, Ecological Engineering, Environmental Modelling & Software, Environmental Pollution, Industrial & Engineering Chemistry Research, Journal of Chemical Technology and Biotechnology, Journal of Environmental Management, Landscape and Urban Planning, Science of the Total Environment and Water Research.*

Prof. Scholz has total citations of more than 2845 (above 2122 citations since 2010), resulting in an h-index of 28 and an i10-Index of 64. Prof. Scholz is Editor-in-Chief of 13 journals, including the Web of Science-listed journal Water (impact factors for 2014: 1.428). He has membership experience on 35 influential editorial boards. Prof. Scholz was a member of the Institute of Environmental Management and Assessment (IEMA) Council between 2008 and 2015.

Miklas has a currently active (on-going) grant income of usually £270,000. His grant income over any past six years is typically £1,500,000. These figures include research and other grants, as well as consultancy.

His sustainable flood retention basin (SFRB) concept assesses the multi-functionality of all large water bodies, with particular reference to their flood and diffuse pollution control potential. A novel and unbiased classification system allows all stakeholders to clearly define the purpose of a water body that can be classed as an SFRB. Communication among stakeholders regarding the most appropriate management of SFRB is greatly enhanced. Moreover, the SFRB concept addresses the need to assess the flood control potential of all European water bodies as part of new legislation.

His research has led to the incorporation of findings into national and international guidelines on wetland and sustainable drainage systems (SuDS). The greatest impact has been made in the area of integrated constructed wetlands (ICW) in Ireland, Northern Ireland, Scotland, and England. Prof. Scholz contributed to the design guidelines of wetland systems as a research consultant. The guidelines assist designers and managers in all aspects of ICW planning, design, construction, maintenance, and management. Moreover, specific guidelines were written for ICW and used by farmers to treat farmyard runoff in Scotland and Northern Ireland and in Ireland. These guidelines are specifically mentioned in national legislation.

The new guidelines on SFRB and ICW have led to the international uptake of both the SFRB and ICW concepts and the researched hybrid SuDS. This work has particularly benefited the British Isles and Central and Northern Europe. For example, ICW are now being constructed in Belgium, Germany, the United States, and China.

Preface

The first edition of this work, entitled Wetland Systems to Control Urban Runoff, was published by Elsevier in 2006. It follows that the released material is now at least nine years old. This is not a major problem for most of the material, which has a long shelf-life. However, about 30% of the book required updating to make it more relevant for today's market.

This revised edition has both a more detailed and a broader view of the subject area. More detail has been added to some chapters to account for technological advances in treatment units and scientific progress in areas such as molecular microbiology. Furthermore, the subject area has been broadened to account for more multidisciplinary approaches, such as the ecosystem services concept, to solve engineering science challenges with a holistic angle. In order to realize this new approach, both updating and expansion (nine new chapters) of the current content were required. The second edition has therefore been expanded by about 40%, making it more competitive in a market where readers have more choice and flexibility due to advances in technology and the open access policy.

Because the second edition has a much broader focus, it is therefore entitled Wetland Systems to Control Pollution, attracting a wider audience of academics and practitioners. The revised and expanded book covers broad water and environmental engineering aspects relevant for the drainage and treatment of storm water and wastewater, providing a descriptive overview of the complex "black box" treatment systems and general design issues involved.

The fundamental science and engineering principles are explained to address the student and the professional market. Standard and novel design recommendations for, predominately, constructed wetlands and related sustainable drainage systems are provided to account for the interests of professional engineers and environmental scientists. The latest research findings in wastewater treatment and runoff control are discussed to attract academics and senior consultants, who should recommend the proposed textbook to final year and postgraduate students and graduate engineers, respectively.

The revised book deals comprehensively not only with the design, operation, maintenance, and water quality monitoring of traditional and novel wetland systems but also with the analysis of asset performance and modeling of treatment processes and performances of existing infrastructure—predominantly in developed but also in developing countries—and the sustainability and economic issues involved.

The textbook is essential for undergraduate and postgraduate students, lecturers, and researchers in the civil and environmental engineering, environmental science, agriculture, and ecological fields of sustainable water management. It should be used as a reference for the design, operation, and management of wetlands by engineers and scientists working for the water industry, local authorities, nongovernmental organizations, and governmental bodies. Moreover, consulting engineers should be able to apply practical design recommendations and to refer to a large variety of practical international case studies, including large-scale field studies.

The basic scientific principles outlined in the revised edition should be of interest to all concerned with the built environment, including town planners, developers, engineering technicians, agricultural engineers, and public health workers. The book is written for a wide readership, but sufficient hot research topics are also addressed in nine completely new chapters to guarantee a long shelf-life for the book.

Solutions to pressing water quality problems associated with constructed treatment wetlands, integrated constructed wetlands, farm constructed wetlands and stormwater ponds, and other sustainable biological filtration and treatment technologies linked to public health engineering are explained. Case study topics are diverse: wetlands, including natural wetlands and constructed treatment wetlands; sustainable water management, including sustainable drainage systems; and specific applications such as wetlands treating hydrocarbon and piggery wastewater. The research projects are multidisciplinary, holistic, experimental, and modeling-oriented.

The book is predominantly based on experiences gained by the author over the last 14 years. Original material published in articles in more than 170 high-ranking journals and presented in 200 key conference papers has been revisited and analyzed. Experience the author gained as an editorial board member of more than 30 relevant peer-reviewed journals guarantees that the textbook contains sufficient material that fills gaps in knowledge and understanding, and that it documents the latest cutting-edge research in areas such as sustainable drainage.

The book tries to integrate natural and constructed wetlands and sustainable drainage techniques into traditional water and wastewater systems used to treat surface runoff and associated diffuse pollution. Chapters 1–4 introduce water quality management and water and wastewater treatment fundamentals to the inexperienced reader.

Chapters 5–9 review preliminary and predominantly primary treatment units that can be combined with wetland systems. Chapters 10–15 summarize predominantly secondary but also tertiary treatment technologies that can be used in combination with wetland technologies or as alternatives in cases where land availability is restricted due to costs. Usually nonessential traditional technologies are briefly presented in Chapters 16 and 17 for the reason of completeness.

Microbiological and disinfection issues relevant for treatment wetlands are covered in Chapters 18 and 19. Chapter 20 introduces wetland science and biological treatment processes based on microbial biodegradation. Furthermore, examples of different wetland types have been presented for readers new to the subject matter. Chapter 21 highlights sludge treatment and disposal options that should be considered for sludges obtained from wetland systems.

Chapters 22–38 focus predominantly on a wide variety of timely applied research case studies related to constructed wetlands and associated technologies for runoff and diffuse pollution treatment. Moreover, wetlands such as sustainable flood control basins used for both diffuse pollution and flood control purposes are introduced. These chapters are written for professionals and students interested in design, process, and management details.

Miklas Scholz, Salford, October 1, 2015

Acknowledgments and Dedications

I would like to thank all current and previous members of my research groups at The University of Salford, The University of Edinburgh, and the University of Bradford for their research input, and all institutions that provided funding for my research. I am also grateful for the support received from the publishing team at Elsevier.

I would like to dedicate this book to my wider family and friends, who supported me during my studies and career. Particular thanks go to my partner **Åsa Hedmark**, children **Philippa Scholz**, **Jolena Scholz**, **Felix Hedmark,** and **Jamie Hedmark**, twin-sister **Ricarda Lorey** and mother **Gudrun Spieshöfer**.

Common Acronyms and Abbreviations

A	Coefficient *(unknown function of various variables including rainfall intensity and infiltration rate)*
A	Cross-section of flow area (m^2)
A_l	Cross-sectional area of lysimeter (m^2)
AEAICAD	Aesthetic and educational appreciation and inspiration for culture, art, and design (%)
AFTW	Aesthetic flood treatment wetland
ANN	Artificial neural network
ANOVA	Analysis of variance
AS	Activated sludge
ATV-DVWK	*German abbreviation for* German Association for Water, Wastewater and Waste
Avg.	Average (mean)
B	Maximum experimental depth (mm) within the infiltration basin during an individual storm
BC	Biological control (%)
BMP	Best management practice
BMU	Best-matching unit
BOD	Biochemical oxygen demand (mg/l) (*usually five days at* 20°C)
BP	Back-propagation
BP-MLL	Back-propagation for multilabel learning
BRE	British Research Establishment *(company)*
C	Carbon or combined approach or control or chili
C_e	Outflow concentration *(of contaminant in wetland cell)* (g/m)
C_f	Contaminant concentration in infiltration water (g/m^3)
C_0	Inflow concentration *(of contaminant in wetland cell)* (g/m)
CBR	Case-based reasoning
CE	Community and environment approach
CFU	Colony-forming unit
CIRIA	*(British)* Construction Industry Research and Information Association
COD	Chemical oxygen demand (mg/l)
CSS	Carbon sequestration and storage (%)

CST	Capillary suction time (s)
D	Infiltration basin design depth (mm)
DNA	Deoxyribonucleic acid
DO	Dissolved oxygen (mg/l or %)
DWF	Dry weather flow (m^3/s)
E	Global error
EPMSF	Erosion prevention and maintenance of soil fertility (%)
EQS	Environmental quality standard
ES	Ecosystem service approach
ET	Evapotranspiration rate (m/d)
ETAAS	Electrothermal atomic absorption spectrometer
EU	European Union
F	Food (%) or filter
FW	Freshwater (%)
FWS	Free water surface *(flow wetland)*
GAC	Granular activated carbon
GL	Guidance level
GPS	Global positioning system
H_0	Head of water *(in wetland)* (m)
h_{wf}	Average capillary head at the wetting front (m)
Ham.	Hamming
HFR	High flow rate
HFRB	Hydraulic flood retention basin
HNL	High nutrient load
HS	Habitat for species (%)
HSD	Honestly significant difference
I	Hydraulic gradient or infiltration rate *(in wetland cell)* (m/d)
ICP-OES	Inductively coupled plasma optical emission spectrometer
ICW	Integrated constructed wetland
IFRW	Integrated flood retention wetland
IR	*(Empirical)* infiltration rate (m/s)
K	Hydraulic conductivity (m/d)
K	Number of neighbors
K_M	Total roughness
KNN	k-nearest neighbor
L	Loss
L	Depth of wetting front *(beneath ICW cell)* (m) or label *(also l and λ)*
L_0	*(Contaminant)* inflow loading rate (g/m/d)
LCAR	Local climate and air quality regulation (%)
M	Number of instances in a data set
MAC	Maximum admissible concentration
MASE	Mean absolute scaled error
Max	Maximum

MEE	Moderation of extreme events (%)
MGD	Maintenance of genetic diversity (%)
Min	Minimum
MLKNN	Multilabel *k*-nearest neighbor
MLSS	Mixed liquor suspended solids
MLSVM	Multilabel support vector machine
MLVSS	Mixed liquor volatile suspended solids (mg/l)
MR	Medicinal resources (%)
MRP	Molybdate reactive phosphate (mg/l)
N	Number of entries or nitrogen or north
N	Number of instances that are correctly predicted
NFRW	Natural flood retention wetland
nosZ	Nitrous oxide reductase
NTU	Nephelometric turbidity unit *(similar to FTU)*
P	Significance level *(of a test) (also known as p, p-value, or P-value)* or precipitation rate (m/d)
P	Phosphorus (mg/l) or pollination (%) or sweet pepper
PCA	Principal component analysis
PCR	Polymerase chain reaction
PRAST	Prevalence Rating Approach for SuDS Techniques
Pre.	Precision
Q	Volume of water per unit time (m^3/d) or size of the set of labels or hydraulic loading rate (m/d)
Q_f	Daily water volume infiltrating beneath a wetland cell (m^3/d)
Q_0	Inlet wastewater volume flow rate *(in wetland cell)* (m^3/d)
QR	Quantization error
R	*(Mean product moment)* correlation coefficient
R^2	Coefficient of determination
Ran.	Ranking
RBC	Rotating biological contactor
RBF	Radial basis function
Re.	Recall
RM	Raw materials (%)
RMPR	Recreation and mental and physical health (%)
rRNA	Ribosomal ribonucleic acid
SD	Standard deviation
SESP	Spiritual experience and sense of place (%)
SFRB	Sustainable flood retention basin
SFRW	Sustainable flood retention wetland
SOM	Self-organizing map
SRT	Storm runoff treatment (%)
SS	*(Total)* suspended solids (mg/l)
SSSI	Site of special scientific interest
SuDS	Sustainable drainage system

SVM	Support vector machine
T (or t)	Infiltration time (s) or temperature (°C)
TAV	Tourism and area value (%)
TE	Topographic error *(usually in %)*
TFRB	Traditional flood retention basin
TOC	Total organic carbon (mg/l)
TS	Total solids (mg/l)
UK	United Kingdom
U-matrix	Unified distance matrix
USA	United States of America
UV	Ultraviolet *(light)*
W	West
WTW	Wissenschaftlich Technische Werkstätten *(company)*
X	Variable *(here, cost unit)*
X	Domain of instances
x_i	An instance i
Y	Set of labels
y_i	Label i
$\mathbf{Z_1}$	Factor *(defined by the BRE method)*
$\mathbf{Z_2}$	Growth factor *(defined by the BRE method)*
$\boldsymbol{\Gamma}$	Bias parameter of a feed-forward network
$\boldsymbol{\Delta}$	Symmetric difference of two sets

Chapter 1

Water Quality Standards

1.1 INTRODUCTION AND HISTORICAL ASPECTS

Scientific and public interest in water quality is not new. For example, in the United Kingdom (UK), it probably had its origins in the mid-eighteenth century. In 1828, the editor of Hansard, Mr. John Wright, anonymously published a pamphlet attacking the quality of the drinking water in London. This led to the establishment of a Royal Commission, which established the principle that water for human consumption should at all times be "wholesome." The term "wholesome" has been incorporated into virtually every piece of legislation concerned with drinking water ever since.

The first unequivocal demonstration of water-borne transmission of cholera was by Snow in 1854. This stimulated great advances in water treatment practices, in particular the routine application of slow sand filtration and disinfection of public water supplies.

Although the Royal Commission of 1828 was concerned with water quality, it had difficulty in defining it precisely, because there were virtually no analytical techniques available at the time with which to determine either microbial or chemical contamination. Consequently, since that time, there has been a continuing and often fierce debate on what constitutes a suitable quality for human drinking water. Not surprisingly, in the nineteenth and early part of the twentieth centuries, the evaluation was largely based on subjective, usually sensory perception.

Many authorities (e.g., Sir Edwin Chadwick) believed that an atmospheric "miasma" above the water, rather than the water itself, was responsible for disease transmission. As a consequence, great efforts were made to remove the smell, assuming that this would dispel the disease. In 1856, during the "great stink," sheets drenched in chemicals were hung from the windows of the Houses of Parliament to exclude the smell. This action did at least focus the minds of the politicians on the need to take action to improve the quality of London's water supply.

Even today, taste, smell, and appearance (color and turbidity) are considered useful criteria for judging water quality. However, in addition, there are now objective methods for determining the presence and level of many (but by no means all) of the microbial contaminants likely to be present in drinking water.

Wetlands for Water Pollution Control. http://dx.doi.org/10.1016/B978-0-444-63607-2.00001-0

Since the 1960s, the emphasis regarding drinking water quality has shifted from its bacteriological quality to the identification of chemical contaminants. This reflects largely the very considerable success of the water industry in overcoming bacteriological problems, although this victory is not complete (e.g., many viruses and *Cryptosporidium* cause public health concerns).

With the great methodological improvements in analytical chemistry over the past 50 years, it was recognized that water contains trace amounts of several thousand chemicals and that only the limitations of analytical techniques restrict the number of chemicals that can be identified. Many of these chemicals are of natural origin, but pesticides, human and veterinary drugs, industrial and domestic chemicals, and various products arising from the transport and treatment of water are very commonly found, albeit normally at very low concentrations.

In addressing the problem of the contribution of water-borne chemicals to the incidence of human disease, water scientists, whose previous experience has typically been confined to microbiological problems, have tended to focus on acute risks. The absence of detectable short-term adverse effects of drinking water has been taken by many as conclusive evidence that the presence of such chemicals is without risk to humans.

While information on the acute toxicity of a chemical can be very useful in determining the response to an emergency situation such as an accidental spillage or deliberate release of chemicals into a watercourse or even into the water supply, such information is of little use in predicting the effects of daily exposure to a chemical over many years.

However, low levels of chemicals are much more likely to cause chronic (rather than acute) effects to health. Here, direct reliable information is very sparse. Some authorities appear to have accepted the "naïve" assumption that information on the acute effects of a chemical, in either humans or experimental animals, can be used to predict the effects of being exposed over a lifetime. In practice, the chronic effects of a chemical have rarely any resemblance to the acute effects.

An evaluation of health risks associated with drinking water is necessary and timely. If we are to obtain a proper assessment of the health risk that could arise in humans through exposure to chemicals in water over a lifetime, understanding must be developed on the following:

- Identification of the chemicals that are of most concern;
- Data on the effects of long-term exposure in humans and/or animals to each chemical;
- A measure of the extent and form of exposure to each chemical;
- Identification of particularly at-risk groups; and
- The means of establishing how exposure to other chemicals in the water can modify the toxicity.

1.2 WATER QUALITY STANDARDS AND TREATMENT OBJECTIVES

It is commonly agreed that there are three basic objectives of water treatment:

1. Production of water that is safe for human consumption;
2. Production of water that is appealing to the customer; and
3. Production of water treatment facilities that can be constructed and operated at a reasonable cost.

The first of these objectives implies that the water is biologically safe for human consumption. It has already been shown how difficult it is to determine what "safe" actually means in practice. A properly designed plant is not a guarantee of safety, standards will change, and plant management must be flexible to ensure continued compliance.

The second basic objective of water treatment is the production of water that is appealing to the customer. Ideally, appealing water is clear and colorless, pleasant to taste, odorless, and cool. It should be nonstaining, noncorrosive, non-scale-forming, and reasonably soft. The consumer is principally interested in the quality of the water delivered to the tap, not the quality at the treatment plant. Therefore, storage and distribution need to be accomplished without affecting the quality of the water; in other words, distribution systems should be designed and operated to prevent biological growth, corrosion, and contamination.

The third basic objective of water treatment is that it can be accomplished using facilities with reasonable capital and operating costs. Various alternatives in plant design should be evaluated for cost-effectiveness and water quality produced.

The objectives outlined here need to be converted into standards so that proper quality control measures can be used. There are various drinking water standards. The key variables are as follows:

- Organoleptic parameters: color, turbidity, odor, and taste;
- Physical and chemical parameters: temperature, pH, conductivity, dissolved oxygen, dissolved solids, chlorides, sulfate, aluminum, potassium, silica, calcium, magnesium, sodium, alkalinity, hardness, and free carbon dioxide (CO_2);
- Parameters concerning undesirable substances: nitrate, ammonium, total organic carbon (TOC), hydrogen sulfide, phenols, dissolved hydrocarbons, iron, manganese, suspended solids, and chlorinated organic compounds other than pesticides;
- Parameters concerning toxic substances such as arsenic, mercury, lead, and pesticides; and
- Microbiological parameters: total coliforms, fecal coliforms, fecal streptococci, sulfite-reducing clostridium, and total bacterial count.

Standards usually give two values: a guide level (GL) and a maximum admissible concentration (MAC). The GL is the value that is considered satisfactory and constitutes a target value. The MAC is the value that the corresponding concentration in the distributed water must not exceed. Treatment must be provided when the concentration in the raw water exceeds the MAC.

Standards also specify the methods, frequencies, and nature of the analysis. For total hardness and alkalinity, the standards specify minimum values to be respected when water undergoes softening.

Most standards group substances into five categories:

- Microbiological;
- Inorganic with consequences on health;
- Organic with consequences on health;
- Appearance; and
- Radioactive components.

One of the main sources of confusion regarding water standards and their interpretation is the lack of any clear indication as to how the standard was derived. This results in the interpretation of all standards as "health standards" by the public and, subsequently, in the difficulty of assessing what should be done by the water supplier if a threshold is exceeded.

This is particularly true of drinking water quality directives because insufficient explanation of the derivation of the actual numbers is often given. There are even thresholds for variables regarded as toxic that are based on political or other considerations, and they are therefore only loosely based on science (e.g., pesticides). The use of such approaches is acceptable as long as the reasoning behind them is clear to all.

International guidelines are usually intended to enable governments to use them as a basis for standards, taking into account local conditions. They are intended to be protective of public health, and they should be absolutely clear, even down to detailed scientific considerations such as the derivation of uncertainty factors and the rounding of numbers. It is therefore incumbent on the expert groups to justify their thinking and present it openly for all to see. Such a discipline avoids the "fudging" of issues while giving the impression of scientific precision, and it can only be of value in increasing public confidence in the resulting guidelines.

It is clear that, at present, standards for water quality are as follows:

- Loosely based on science (although the situation is improving);
- Not static (the science of monitoring as well as our understanding of the health implications of chronic exposure of many contaminants are improving); and
- Important in the quality control of potable water (for both supplier and consumer).

Concerning the outflow water quality of most wetland systems, standards either are unclear or are currently being developed. The local environment regulator usually sets standards for specific wetland system applications.

1.3 BIOCHEMICAL OXYGEN DEMAND

When wastewater, including urban runoff, is discharged into a watercourse, it exerts a polluting load on that water body. Microorganisms present in the natural water and the wastewater break down (stabilize) the organic matter. In permitting discharges to watercourses, the Environment Agency in the UK, for example, tries to ensure that the conditions are aerobic so that all other life forms in the river (e.g., fish) can continue to survive. The early forms of wastewater treatment developed are aerobic, and so the simplest way of estimating the biodegradability of a wastewater sample is to estimate the amount of oxygen required to stabilize the waste.

To devise an easy and simple method of assessing the oxygen demand, the following constituents of a closed system should be considered:

- Air (in excess);
- A small number of bacteria; and
- A finite amount of substrate (waste representing food).

The following phases of biological growth and decline can be identified in such a system:

- Lag phase: Bacteria are acclimatizing to system conditions, in particular the substrate; very little increase in numbers.
- Log growth: Bacteria are acclimatized; food is not a limiting factor; rapidly increasing population of bacteria.
- Declining growth: Food eventually becomes limiting; declining growth rates.
- Endogenous respiration: As the substrate concentration becomes depleted, competition increases; bacteria start consuming dead bacterial cells and eventually start consuming live cells.

It is a system of this type that is used to assess the oxygen demand of wastes, including organic matter from urban runoff. The test developed from this system is the biochemical oxygen demand (BOD) test.

The BOD test is carried out as follows: a known quantity of a wastewater sample (suitably diluted with prepared water) is placed in a 300-ml BOD bottle. The prepared water is saturated with dissolved oxygen (DO), and nutrients and a buffer are added. The bottles are then sealed airtight. The bottles are subsequently incubated at 20 °C in the dark (Clesceri et al., 1998).

Initially, the bacteria break down the carbon-based molecules. In practice, a second oxygen demand is observed. In the case of raw sewage, this stage usually becomes apparent after approximately 8 days of incubation at 20 °C.

This second stage is due to the oxidation of ammonia present in the waste; this is called nitrification. A large percentage of the nitrogen in the wastewater originates from proteins; the protein molecules are degraded to release ammonia. The oxidation process is described in Eqs. (1.3.1) and (1.3.2):

$$2NH_4^+ + 3O_2 \xrightarrow{Nitrosomonas} 2NO_2^- + 4H^+ + H_2O \qquad (1.3.1)$$

$$2NO_2^- + O_2 \xrightarrow{Nitrobacter} 2NO_3^- \qquad (1.3.2)$$

Nitrification consumes a significant amount of oxygen so that the total demand for nitrification is often comparable with the carbonaceous demand. Nitrification also generates protons (H^+ ions), which increase the acidity (pH) of the waste.

Traditionally, the BOD test is carried out for 5 days; the resulting oxygen demand is referred to as the BOD_5. The BOD is calculated as follows (Eqs. (1.3.3) and (1.3.4)):

$$\text{BOD (mg/l)} = \frac{\text{Initial DO in bottle} - \text{Final DO in bottle}}{\text{Dilution ratio}} \qquad (1.3.3)$$

where:

$$\text{Dilution ratio} = \frac{\text{Volume of wastewater}}{\text{Volume of BOD bottle}} \qquad (1.3.4)$$

In practice, the test is often modified slightly in that a quantity of seed microorganisms are added to the BOD bottle to overcome the initial lag period. In this variant, the BOD is calculated from Eq. (1.3.5):

$$\text{BOD} = \frac{(D_1 - D_2) - f(B_1 - B_2)}{DR} \qquad (1.3.5)$$

where:

D_1 = dissolved oxygen initially in seed and waste bottle;
D_2 = dissolved oxygen at time T in seed and waste bottle;
B_1 = dissolved oxygen initially in seed-only bottle;
B_2 = dissolved oxygen at time T in seed-only bottle;
f = ratio of seed volume in seeded wastewater to seed volume in the BOD test on seed only; and
DR = dilution ratio.

Additional bottles are incubated. These contain only seed microorganisms and dilution water to get the BOD of the seed, which is then removed from the BOD obtained for waste and seed.

However, the BOD test has two major disadvantages: it takes 5 days to obtain the standard test result, and the results can be affected by the process of nitrification (see above). Therefore, a nitrification inhibitor is often used (Chapter 24).

1.4 CHEMICAL OXYGEN DEMAND

The disadvantages of the BOD test have led to the development of a simpler and quicker test. This test is known as the chemical oxygen demand (COD) methodology. In this test, strong chemical reagents are used to oxidize the waste. Potassium dichromate is used in conjunction with boiling concentrated sulfuric acid and a silver catalyst. The waste is refluxed in this mixture for 2 h. The consumption of the chemical oxidant can be related to a corresponding oxygen demand (Clesceri et al., 1998).

The COD test oxidizes material that microorganisms cannot metabolize in 5 days or that are toxic. If the COD is much greater than the BOD in raw wastewater, then the waste is not readily biodegradable, and it may be toxic to the microorganism. If the COD is similar to the BOD, then the waste is readily biodegradable.

1.5 OTHER VARIABLES USED FOR THE CHARACTERIZATION OF WASTEWATER

Most wastewater treatment processes operate best in pH ranges between 6.8 and 7.4; indeed, pH > 10 is likely to kill large numbers of bacteria. Suspended solids (SS) is a measure of the total particulate matter content of wastewater. The nature of the SS is likely to vary considerably depending on the nature of the waste.

The two most important nutrients in wastewater treatment are nitrogen and phosphorus; both are needed for cell growth. Nitrogen (N) is used in protein synthesis (e.g., new cell growth). Phosphorus (P) is used for cell energy storage and is usually present as ortho-phosphate (PO_4).

Organic nitrogen is associated with cell detritus and volatile SS. Free ammoniacal nitrogen (NH_3-N) results from the decay of organic nitrogen. Nitrite–nitrogen (NO_2-N) is formed in the first step in nitrification. Nitrate–nitrogen (NO_3-N) results from the second and final stage in the nitrification process.

For proper microorganism growth, the ratio of C:N:P is important. Carbon (C) is measured by BOD_5. Nitrogen is measured by organic nitrogen and NH_3-N. However, NO_3-N is difficult for microorganisms to use in their growth process. Phosphorus is measured as acid hydrolysable ortho-phosphate (PO_4). To achieve growth, the required minimum values for the C:N:P relationship are 100:5:1.

Chapter 2

Water Treatment

2.1 SOURCES OF WATER

The source of raw water has an enormous influence on the water's chemistry and consequently its treatment. Raw water is commonly abstracted from one of the following four sources:

1. Boreholes extracting groundwater: This water is usually bacteriologically safe as well as aesthetically acceptable. It may require some treatment such as aeration or softening.
2. Rivers: Water can be abstracted at any point along the length of a river. However, the further downstream it is, the more likely the water is to require considerable treatment.
3. Natural lakes: The degree of treatment required for lake water depends on a number of factors such as the catchment use in the immediate vicinity of the lake, the lake's trophic status, and the presence of sewage treatment works.
4. Manmade lakes and reservoirs: These are similar to lakes and rivers, but better managed. The degree of treatment depends on the management of the catchment and upstream catchment usage.

Water for domestic consumption may also come from other sources such as seawater (via desalination) or treated sewage effluents. However, these sources are very rare and therefore beyond the scope of this introductory chapter to water treatment.

2.2 STANDARD WATER TREATMENT

The purpose of screens is simply to remove solid floating objects (e.g., logs and twigs) from the raw water, which may cause damage or blockage in the plant. Sometimes a much finer screening is carried out, called straining. This is usually performed on lake and reservoir water to remove algae.

A coagulant is added to the raw water to destabilize the colloidal material in the water (Chapter 7). Commonly used chemicals are as follows:

- Alum (aluminum sulfate) $Al_2(SO_4)_3 \cdot nH_2O$;
- Ferric chloride $FeCl_3$;
- Ferrous sulfate (copperas) $FeSO_4 \cdot 7H_2O$;

Wetlands for Water Pollution Control. http://dx.doi.org/10.1016/B978-0-444-63607-2.00002-2

- Lime (burnt CaO; slaked $Ca(OH)_2$); and
- Polyelectrolytes (long-chain organic molecules normally used in conjunction with a conventional coagulant).

For the coagulant to function efficiently, it must be rapidly and uniformly mixed through the raw water. This usually takes place in a high-shear (turbulent) environment such as one induced by a hydraulic jump (low-cost option), a pump, a jet mixer, or a propeller mixer.

After the coagulant is uniformly distributed in the water, it requires time to react with the colloid, and then further time (and gentle agitation) to promote the growth (agglomeration) of settleable material (flocks). This is generally accomplished either in a tank with paddles (mechanical mixing) or through a serpentine baffled tank (hydraulic mixing). Once flows of a settleable size have formed, they are removed usually by sedimentation (sometimes by flotation).

In countries such as the UK and Ireland, developments in the 1940s led to the introduction of the sludge blanket clarifier (Chapter 8). This is a single unit that encompasses rapid mixing, flocculation, and settling.

To remove either solids carried over from settling tanks and/or any uncoagulated material (organic or inorganic), a sand bed filter is provided. The water flows downwards through the bed, and the impurities are removed by attachment to the sand grains. The sand grains therefore require periodic cleaning. The frequency of cleaning depends on the type of filter used. The two commonly used types are the following (Chapters 10 and 11):

- Slow sand filter (slow loading rate: approximately one-tenth of that for a rapid gravity sand filter) and
- Rapid gravity sand filter (high loading rate).

Sometimes fluoride is added to the water to reduce the incidence of dental caries. This is a process that provokes public debate. In the United States and UK, chlorine is usually added to the water to disinfect it (Chapter 19). It follows that the water is bacteriologically safe when it leaves the treatment works, and excess chlorine is added to protect the water from contamination during the distribution process.

There are several other commonly used processes. Their usage depends on the nature of the raw water. Air can be introduced to the water to oxidize impurities (e.g., iron, manganese, or chemical compounds affecting the taste of water). pH control is a common process since many of the chemical treatment processes are pH dependent. Softening reduces the hardness and/or alkalinity of water to improve its aesthetic acceptability. This is a complex chemical process depending on the nature of both the anion (HCO_3^-, CO_3^{2-}, or OH^-) and the cation (Ca^{2+} or Mg^{2+}).

2.3 BASIC WATER CHEMISTRY

The most important chemical variables of raw water are usually taken as pH and alkalinity. Alkalinity consists of those chemical species that can neutralize acid. In other words, these species allow the water to resist changes and provide buffering capacity. The major constituents of alkalinity are the hydroxyl (OH^-), carbonate (CO_3^{2-}), and bicarbonate (HCO_3^-) ions. The relative quantities of each are a function of pH.

No significant concentration of hydroxyl ions exists below pH 10, and no significant carbonate concentration can be detected below pH 8.5. For most waters, alkalinity thus consists of the bicarbonate ion. The other species may be formed in the treatment process. The bicarbonate and carbonate ions in the water result from the dissolution of carbonate rocks.

The pH is a measure of the free hydrogen ion concentration in water. Water, and other chemicals in solution, will ionize to a greater or lesser degree. The ionization reaction is given in Eq. (2.3.1).

$$H_2O \Leftrightarrow H^+ + OH^- \tag{2.3.1}$$

In neutral solutions, the [OH^-] activity is equal to the [H^+] activity. Hence, the pH and pOH (a measure of alkalinity) are both equal and have the numerical value of 7. An increase in acidity, for example, leads to higher values of [H^+], thus lowering the pH.

The various chemical reactions that occur in natural waters and in processed water are generally considered to occur in dilute solutions. This permits the use of simplified equilibrium equations in which molar concentrations are considered to be equal to chemical activities.

The assumption of dilute conditions is not always justified, but the error introduced by the simplification is no greater than the error that might be introduced by competing reactions with species that are not normally measured in water treatment.

Concentrations of different chemical species in water may be expressed in moles per liter, in equivalents per liter, or in mass per unit volume (typically, mg/l). The equivalent of a species is its molecular weight divided by the net valence or by the net change in valence in the case of oxidation and reduction reactions.

The number of equivalents per liter (normality) is the concentration divided by the equivalent weight. The number of moles per liter is called the molarity.

Chapter 3

Sewage Treatment

3.1 INTRODUCTION

The waste disposed by domestic households and industry is conveyed to the treatment works by means of pipes (sewers). The arrangement of sewers is known as the sewerage system. Everything that flows in the sewers is sewage. These terms are often confused in practice.

In a traditionally combined sewer, all sewage, both foul and surface water, is conveyed in a single pipe. A foul sewer conveys the "nasties" (i.e., contaminants). A surface water sewer conveys the runoff from roofs and paved areas.

Concerning separate systems, two pipes are laid in the trench for the sewerage system: one for the foul sewer, and the second for the surface water. This book is concerned with the treatment of both wastewater and urban runoff.

The flow in a sewer can be estimated with Eq. (3.1.1). The mean domestic water consumption is typically 140 l/h/day for rural and 230 l/h/day for urban areas.

$$DWF = PQ + I + E \tag{3.1.1}$$

where:

DWF = averaged total flow in 24 h (dry-weather flow) ($Q_T/24$);
P = population;
Q = mean domestic water consumption;
I = rate of infiltration;
E = industrial effluent discharge to the pipe; and
Q_T = total volume of flow in a 24-h period.

3.2 DESIGN FLOW RATES

Normally, at sewage treatment works, flows up to three DWF are given full treatment; >6 DWF (since they are diluted by the surface water) require only preliminary treatment. Flows between three and six DWF are stored temporarily and given full treatment.

However, care needs to be taken in the design of overflow structures, particularly for flows >6 DWF. These must be designed such that the outflow

Wetlands for Water Pollution Control. http://dx.doi.org/10.1016/B978-0-444-63607-2.00003-4

from them has a minimum impact on the receiving water; in particular, care must be taken with the solid material, which occurs in the so-called first foul flush or simply the first flush (i.e., immediately after the rainfall storm commences, accumulated material in the sewer is likely to be flushed out of the system).

3.3 TREATMENT PRINCIPLES

Typically, raw sewage contains 99.9% water and 0.1% solids. The sewage treatment process is fundamentally about separating solids from the water. The treatment of solids and sludge forms an important and costly area of sewage treatment. The impurities in the sewage can be categorized as follows:

- Floating or suspended solids (e.g., paper, rags, grit, and fecal solids);
- Colloidal solids (e.g., organics and microorganisms);
- Dissolved solids (e.g., organics and inorganic salts); and
- Dissolved gases (e.g., hydrogen sulfide and carbon dioxide).

These impurities are removed from the sewage using operations or processes that are physical, chemical, or biological in nature. Physical operations depend on the physical properties of the impurity for efficient removal (e.g., screening, filtration, and sedimentation). Chemical operations depend on the chemical properties of the impurity and use the chemical properties of additives for efficient removal (e.g., coagulation, precipitation, and ion exchange). Biological processes comprise biochemical and/or biological reactions to remove soluble or colloidal organic impurities (e.g., percolating filters and activated sludge).

3.4 ENGINEERING CLASSIFICATION OF SEWAGE TREATMENT STAGES

Wastewater engineers tend to describe the sewage treatment process in terms of the stages of treatment:

- Preliminary treatment (physical): for example, screening and grit removal;
- Primary treatment (physical and/or chemical): for example, sedimentation and flotation;
- Secondary treatment (biological and/or chemical): for example, constructed wetlands, biological filters, and the activated sludge process; and
- Tertiary treatment (physical and/or chemical and/or biological): for example, polishing wetlands, microstraining, grass plots, and lime precipitation.

At the secondary treatment stage, either percolating filters or activated sludge treatment is usually present, but certainly not both in parallel. On occasions, when treating industrial wastes, they may both be used, but always in series. It should be noted that sludge is produced at the majority of the treatment stages. However, in normal practice, the works are organized such that all sludge is collected centrally.

Wetland systems can be designed for each engineering stage and for sludge treatment. However, constructed treatment wetlands (for definitions, refer to Section 20.2) are usually applied for secondary or tertiary treatment stages. Wetlands integrated in sustainable drainage systems (SuDS; previously called sustainable urban drainage systems (SUDS)) are frequently used for preliminary and primary treatment purposes. Urban runoff requires full treatment, which is usually not the case in practice, unless for combined sewer systems and minor storms.

Chapter 4

Stream Pollution and Effluent Standards

4.1 ORGANIC STREAM POLLUTION

Since most effluents, including storm runoff, are discharged to a receiving watercourse, it is important that the concentration of the effluent is such that the receiving water can assimilate the waste and break it down. The receiving water should also remain in a condition appropriate to its use (usually aerobic).

Consider the discharge of an organic effluent into a river. In the receiving water, there are two processes taking place: oxidation of the organic waste and re-aeration, and the introduction of oxygen into the water. Before the effluent is discharged, the river contains dissolved oxygen (DO). The effluent reduces the initial DO concentration progressively to satisfy the BOD.

In the zone of degradation, re-aeration is smaller than the rate of decomposition. In this section, the decomposition of the effluent dominates, and so the DO concentration drops rapidly. The sediment accumulation in the immediate vicinity may be large due to the settling of suspended material in the effluent.

In the zone of active decomposition, re-aeration is approximately equal to the rate of decomposition. The water is likely to contain little diversity of life forms; the bottom sediment may possibly be anaerobic.

In the zone of recovery, the rate of re-aeration is larger than the rate of decomposition. Since the effluent oxygen demand is dropping and the DO deficit is large, then atmospheric oxygen will diffuse into the water body at a greater rate; thus, the DO begins to increase. Nitrification is also likely to begin to take place, and the life forms present in the river increase in diversity.

In the zone of clear water, the DO has now returned to its original value, and the BOD has been virtually eliminated; only a background level remains. However, the river has been permanently changed: there are increased levels of nutrients in the water, and this may lead eventually to eutrophication.

Wetlands for Water Pollution Control. http://dx.doi.org/10.1016/B978-0-444-63607-2.00004-6

4.2 PREDICTION OF ORGANIC STREAM POLLUTION

In the absence of re-aeration, Eq. (4.2.1) should be considered:

$$\frac{dD}{dt} = K_1 L_t \tag{4.2.1}$$

where:

D = dissolved oxygen deficit (DOsat − DOactual);
K_1 = reaction rate coefficient; and
L_t = ultimate BOD remaining at time T.

In water bodies, re-aeration usually occurs as a result of a difference in partial pressures and the turbulence in the river flow. Re-aeration can be expressed with Eq. (4.2.2):

$$\frac{dD}{dt} = -K_2 D \tag{4.2.2}$$

where:

K_2 = re-aeration coefficient.

Note that the negative sign indicates that oxygen transfer tends to reduce the deficit. Combining both decomposition and re-aeration, Eq. (4.2.3) can be obtained.

$$\frac{dD}{dt} = K_1 L_t - K_2 D \tag{4.2.3}$$

This is a simple differential equation that can be easily integrated (with appropriate limits) to yield the Streeter–Phelps equation (Eq. (4.2.4)):

$$D_t = \frac{k_1 L_0}{k_2 - k_1}\left[10^{-k_1 t} - 10^{-k_2 t}\right] + D_0 10^{-k_2 t} \tag{4.2.4}$$

where:

D_0 = dissolved oxygen deficit at time $t = 0$;
L_0 = ultimate BOD initially (i.e., $t = 0$);
$L_t = L_0 10^{-k_1 t}$;
k_1 and k_2 = reaction rate and re-aeration coefficient (base 10), respectively; and
D_t = dissolved oxygen deficit at time t.

The critical point on the DO sag curve is the minimum value of DO. The occurrence of the critical point can be defined with Eqs. (4.2.5)–(4.2.7).

$$k_1 L - k_2 D_c = 0 = \frac{dD}{dt} \tag{4.2.5}$$

$$\therefore D_c = \frac{k_1}{k_2} L_0 10^{-k_1 t_c} \tag{4.2.6}$$

$$t_c = \frac{1}{k_2 - k_1} \log_{10}\left[\frac{k_2}{k_1}\left(1 - \frac{D_0(k_2 - k_1)}{L_0 k_1}\right)\right] \tag{4.2.7}$$

where:

D_c = critical deficit reached at time t_c.

The Streeter–Phelps equation is valid only when no change in the pollution load or dilution occurs. Complex discharge and river problems require a stepwise process to be solved.

4.3 EFFLUENT DISCHARGE STANDARD PRINCIPLES

Water quality standards should achieve the following:

- Safeguard public health;
- Protect water so that it is suitable for abstraction and subsequent use in domestic, agricultural, or industrial circumstances;
- Cater for the needs of commercial, game, and course fisheries; and
- Cater for relevant water-based amenities and recreational requirements.

The limits or standards placed on effluent discharges have been traditionally specified in terms of effluent volume, BOD, and SS. Standards are based traditionally on the fact that a pollution-free stream would have a $BOD_5 = 2$ mg/l, and if the $BOD_5 > 4$ mg/l, the stream may become a nuisance (i.e., occasionally anaerobic). However, these standards are now superseded by various national standards and international directives. Nevertheless, standards for urban water runoff are currently being developed.

Chapter 5

Preliminary Treatment

5.1 INTRODUCTION

Screening is normally the first operation applied at wastewater (sewage) treatment plants, and it is also used just before urban runoff is conveyed into, for example, a wetland system. The general purpose of screens is to remove large objects such as twigs, rags, paper, plastic, metals, and shopping trolleys. These objects, if not removed, may damage the pumping and sludge removal equipment, hang over weirs, and block valves, nozzles, channels, and pipelines, thus creating serious plant operation and maintenance problems.

The second operation constituting preliminary treatment is grit removal. Grit includes sand, dust, and cinders. These are nonputrescible materials with a specific gravity greater than that of organic matter. It is necessary to remove these materials in order to:

- Protect moving mechanical equipment and pumps from unnecessary wear and abrasion;
- Prevent clogging in pipes and heavy deposits in channels;
- Prevent cementing effects on the bottom of sludge digesters and primary sedimentation tanks; and
- Reduce accumulation of inert material in aeration basins and sludge digesters, which would result in the loss of usable volume.

5.2 DESIGN OF SCREENING UNITS

For preliminary treatment, screens normally comprise vertical or inclined bars with openings between 20 and 60 mm for coarse screens and 10 and 20 mm for medium screens. The bars are usually made from steel, and they are between 50 and 75 mm wide and between 10 and 15 mm thick. The spacings used are normally approximately 20 mm wide for mechanically raked screens and between 25 and 40 mm wide for manually cleaned screens.

The hydraulic design of screening units must include the calculation of screen area and the head loss through the screen. The area of submerged screen surface is based on the velocity of flow through the clean openings. The corresponding velocity is 600 mm/s for average flows and 900 mm/s for the maximum rate of flow.

Wetlands for Water Pollution Control. http://dx.doi.org/10.1016/B978-0-444-63607-2.00005-8

The largest area is the controlling area. For small and manually cleaned units, the minimum working width is 450 mm. However, the usual width is between 600 and 900 mm for mechanically cleaned units. In general, the effective screen area should be about two times the cross-sectional area of the incoming pipe, such as a sewer. The head loss through the bars is calculated using Eqs. (5.2.1)–(5.2.3).

$$h_L = \frac{V^2 - V_5^2}{2g} \cdot \frac{1}{0.7} \quad \text{(clean or partly clogged bars)} \tag{5.2.1}$$

$$h_L = \beta\left(\frac{W}{b}\right)^{\frac{4}{3}} h_v \sin\theta \quad \text{(clean screens)} \tag{5.2.2}$$

$$h_L = \frac{1}{2g}\left(\frac{Q}{CA}\right)^2 \quad \text{(common orifice equation used for fine screens)} \tag{5.2.3}$$

where:

h_L = head loss through bars (m);
V = velocity through bars and in the channel upstream of unit (m/s);
W = maximum cross-sectional width of bars facing the direction of flow (m);
b = minimum clear spacing of bars (m);
h_v = velocity head of flow approaching the bars (m);
θ = angle of bars with horizontal (°);
A = effective submerged open area (m^2);
C = coefficient of discharge (i.e., 0.6 for a clean screen); and
β = bar shape factor with different β values for different shapes (sharp-edged rectangular, 2.42; rectangular with semicircular upstream face, 1.83; circular, 1.79; rectangular with semicircular faces upstream and downstream, 1.67; and teardrop shape, 0.76).

The width of the screen channel is usually calculated by applying Eq. (5.2.4):

$$W = \frac{B + b}{b} \cdot \frac{Q}{VD} \tag{5.2.4}$$

where:

W = width of the channel excluding supports (m);
B = width of the screen bars (m);
b = spacing of the screen bars (m);
Q = maximum flow rate (m^3/s);
V = maximum velocity of flow through screens (m/s); and
D = depth of flow at screens (m).

Screens can be either manually cleaned (not popular but common for wetland systems) or mechanically cleaned. In mechanically cleaned screens, cleaning either is continuous or is initiated when the differential head through the screen reaches 150 mm.

5.3 DESIGN DETAILS FOR SCREENING UNITS

Good access is required, because screening units need frequent inspection and maintenance. Ample working space must be provided for screens in deep channels. Protective structures such as guards are traditionally omitted, although trends are changing. Good corrosion protection must be written into the specifications.

Cleaning mechanisms such as front or back cleaning with front or back discharge are usually designed by the manufacturer. Back-cleaning devices are protected from damage by the screen itself. Front discharge by macerated screenings is preferable to back discharge.

The following types of operating controls are used for mechanically cleaned screens, sometimes in combination: manual stop-start, automatic stop-start by time control, high-level alarm, differential head-actuated automatic stop-start, and overload switch and alarm.

5.4 COMMINUTORS

Comminutors are combined screen and macerator units. They consist of an electrically rotated drum with horizontal slots that form a screen. The sewage gravitates from the upstream channel into a spiral flow channel, through the slots and open bottom of the drum, and into the downstream channel via an inverted siphon. Suspended and floating solids are held by the liquid flow against the outside of the drum, and they are macerated by stationary cutters.

It is a great advantage of comminutors that screenings do not have to be removed from the flow. A disadvantage is the tendency for "stringing" or "balling up" of material, and the head loss is higher than that with screens.

These comminutors are designed to take the maximum flow rate. Abnormal overload is diverted through a hand-raked bar screen via an overflow weir. The hydraulic head required depends upon the following:

- Flow rate;
- Machine capacity and other characteristics; and
- Upstream and downstream conduit widths and flow rates.

The presence of the comminutor in the channel will only affect upstream flow depths. The calculation of head losses includes three principle parts:

1. Head required to force the desired flow rates through the comminutor. The differential head varies with the depth of flow, and it may be obtained from the manufacturer's data and literature;

2. Open channel friction losses in the approach and tail channels; and
3. The drop in water surface due to the relative elevations of comminutor, and approach and tail conduits. Careful design can minimize these drops.

Selection of the proper comminutor size and the calculations of the overall head loss involve an iterative approach. In general, a head loss between 150 and 750 mm is involved, depending on the capacity of the unit.

The following are the major areas that must be considered in the detailed design of comminutor units:

- Protection from damage by large floating objects;
- Provision of a flow bypass and emergency manual screening in the event of power failure;
- Standby capacity must be provided;
- Arrangements for isolation of the unit from the flow for maintenance purposes; and
- Provision must be made for drawing the siphon pipe (together with the spiral chamber); this can be obtained prefabricated, thus avoiding complex in situ construction.

5.5 GRIT REMOVAL

These units are used to remove inorganic grit from sewage. The typical organic content of the retained grit is usually <15% of the total. Since inorganic grit has a higher density than organic solids (typically 2.6 times), all grit removal units are velocity-regulating devices that maintain the flow velocity to a critical figure at which only the inorganic grit will settle out. The choice of grit removal methods is dictated by the quantity and quality of the grit, head loss, space requirements, and the type of equipment used in other parts of the plant.

Constant-velocity grit chambers and channels are common in both large and small schemes. The theoretical section for constant velocity at all flows is parabolic. Problems arise at minimum flows because of a high width-to-depth ratio. A trapezoidal section is normally adopted for medium-sized schemes. The side slopes should be a minimum of 45° to the horizontal (preferably 60°).

In the absence of data from settling tests on sewage or storm runoff samples, a velocity of 300 mm/s should be maintained throughout the tank. The velocity must be between 200 and 400 mm/s under all conditions.

The capacity of the channel is normally six DWF. At least two channels should be provided. Standby provision must be made to allow for maintenance and grit removal without decreasing the efficiency of the unit.

Flow regulation is normally undertaken by a standing wave flume. When the velocity of flow is 300 mm/s, the settling velocity is about 30 mm/s. This should be modified by tests in specific case studies.

Theoretically, the length of the channel should be approximately 10 times the depth of flow. However, the length of the channel is usually taken as the depth of flow multiplied by 20 to allow for turbulence effects and varying settling velocities.

Automatic grit removal equipment is available and may include suction or bucket dredgers and conveyor types mounted on rails. Where manual grit removal is employed, the bottom width of the grit channel should not be $<$300 mm.

The other grit removal systems are all proprietary systems. They are usually rectangular or circular tanks with mechanical grit removal and washing equipment, and they are designed to maintain a constant velocity across, throughout, or around the tank.

For example, the Pista Grit Trap has no moving parts in contact with grit. Paddles and peripheral drag keep the horizontal velocity of the sewage reasonably constant throughout the circular tank at varying flow rates. Grit settles in the sump, where it is washed by a counter-flow of air and water under pressure in the sump, before it is air-lifted at regular intervals.

In comparison, spiral-flow aerated tanks are used at larger water treatment works where space is limited. They are provided in duplicate, with each tank or series of tanks capable of treating the maximum rate of flow. The units are approximately rectangular in shape. Air diffusers located at the sides of the tank give vertical movement to the flow. The tanks discharge over weirs or through penstocks. The air induces a spiral motion and consequently pre-aerates the sewage. Grit is removed by gravity, airlift, or pump action.

The design of most spiral-flow aerated tanks is empirical. The forward velocity is usually 0.1 m/s (maximum discharge) with a retention period of $\geq$60 s. The air supply is approximately 0.05 l/s per l/s of sewage.

Chapter 6

Primary Treatment

6.1 INTRODUCTION

A sedimentation tank or at least a silt trap is usually located just before a constructed treatment wetland. Historically, sedimentation tanks have been designed on the basis of three simple principles:

1. Overflow rate;
2. Weir loading rate; and
3. Hydraulic retention time.

However, this has led to an empirical basis for designing efficient settling tanks. The general principles in any sedimentation tank design method are as follows:

1. The objective is to remove settleable solids (SS) and to reduce the SS content of the sewage, thereby reducing the corresponding biochemical oxygen demand (BOD).
2. The principle of gravity settlement in relatively quiescent conditions is employed in the design of all sedimentation tanks.
3. On a typical sewage treatment works, sedimentation units are provided in three stages of treatment: stormwater detention and treatment, primary sedimentation, and final or secondary sedimentation.
4. Sedimentation tanks are designed to operate on a continuous-flow basis. They are usually rectangular or circular in shape and are equipped with mechanical sludge-collecting devices. With the exception of tanks designed for continuous sludge removal (final sedimentation tanks in activated sludge plants), the bottom of sedimentation tanks is essentially flat (up to 15°) and has sludge hoppers with relatively steep sides. Settled sludge on the tank floor is moved by mechanical scrapers into the hoppers for subsequent withdrawal.
5. Imhoff or two-story tanks provide both sedimentation and digestion in a single tank. The use of such devices is limited to small plants serving populations <5000.

Wetlands for Water Pollution Control. http://dx.doi.org/10.1016/B978-0-444-63607-2.00006-X

6.2 LOADING RATE METHODS

Concerning the traditional loading rate methods, primary sedimentation tanks are designed using the data shown below:

- Overflow rate (surface loading rate): 16–32 m^3/m^2/day (normally approximately 24 m^3/m^2/day).
- Length-to-width ratio: between 3:1 and 5:1, with a bottom slope of one in 40.
- Weir loading rate: 450 m^3/m/day (normally approximately 225 m^3/m/day); if >225 m^3/m/day, use a double-sided weir.
- Retention time: 2 h at three dry weather flow (DWF).

Tanks designed using these figures will, typically, give between 30% and 40% BOD removal for domestic sewage. For circular tanks, diameters are usually between 10 and 40 m. The actual depth depends on the type of sludge being removed (i.e., a batch settling test to determine the required depth is necessary).

In general, for mechanically cleaned tanks, the minimum depth is usually between 2.5 and 3.5 m. As the tank depth increases, so does the construction cost, as does the difficulty of keeping the tank in the ground.

Concerning the traditional loading rate methods, secondary tanks are designed using the following data:

- Overflow rate: 32 m^3/m^2/day.
- Side water depth (minimum): 2.1 m (always circular).
- Weir loading rate: 120 m^3/m/day (the lower the better).

Where activated sludge is the secondary treatment process, the actual figures depend on the variant of the process being used, but the major consideration is that there should be a continuous removal of sludge. The tanks are usually designed as settling and thickening tanks.

6.3 TANK DESIGN

Many factors impact on the sedimentation efficiency in a continuous-flow tank, including particle size and density, liquid density, variations in liquid temperature (viscosity), average velocity through the tank, and the influence of the inlet, outlet, and sludge removal arrangements. In a vertical-flow tank, the suspended particles will be eliminated if their settling velocity is equal to the upward velocity of the water. For a horizontal-flow tank, suspended particles are assumed to be retained in the tank, if they can settle through the full depth of the tank in the time it takes for them to pass through the tank.

The design settling velocities are the same for vertical-flow or horizontal-flow tanks and are independent of the tank depth. The flow velocity is sometimes termed the overflow rate or surface loading rate.

The optimal tank design is difficult to achieve in practice due to the following reasons:

- Tanks do not behave in an ideal way;
- The solids in suspension have many different settling velocities;
- The SS content varies from work to work and even from day to day at the same work; and
- Flocculation can occur, either deliberately induced or arising spontaneously from certain conditions in the tank.

As a result of these complications, design methods have been developed to overcome these practical difficulties. The principles used in idealized design methods are as follows:

- Design for a particular settling velocity;
- Take account of high velocities that tend to cause scour near the tank floor; and
- Design for the effect of an imperfect velocity distribution.

6.4 DESIGN PARAMETERS

6.4.1 Design Settling Velocity

Maximum velocities in the tank should not be so great as to prevent the settling of particles, and they should have a design settling velocity based on settling tests for specific samples. The normal design value for the settling velocity of, for example, domestic sewage is 0.339 mm/s. However, note that efficient primary sedimentation is possible even if a surface loading corresponding to an upward velocity >0.339 mm/s is used. Design settling velocities can exceed this figure in specific cases. Many tanks achieve their highest efficiency at too high costs. They could reasonably be loaded at much higher rates.

6.4.2 Horizontal Velocity

The maximum velocity occurring anywhere in the sedimentation tank should not exceed the scour velocity of the previously settled sludge. A theoretical determination of scouring velocities involves understanding of cohesion, friction factor, diameter, and specific gravity of the particles.

Usually, experimental methods are employed on a sludge sample. Generally accepted values for the scour velocity of different sludge types are as follows:

- Settled primary sewage: 31 mm/s.
- Humus sludge: 36 mm/s.
- Activated sludge: 20 mm/s.
- Alum flocculation: 50 mm/s.

6.4.3 Time Ratio

In an ideal tank, the ratio of effective flow-through time to effective settling time would be 1. In practice, the effective flow-through time is reduced because of an uneven distribution of flow throughout the tank; a value >0.95 is unattainable.

In the case of circular tanks, experience shows that the time ratios in existing tanks vary between 0.50 and 0.90 (compared with 0.30 and 0.95 for rectangular tanks). The highest values of both time ratio and sedimentation efficiency have been calculated for tanks with a 10° floor slope, followed by flat-bottomed tanks, and finally by tanks with floor slopes between 20° and 30° (thus to be avoided).

6.5 ECONOMICS OF CONSTRUCTION

6.5.1 Rectangular Settling Tanks

Tanks should be kept short so that depths and costs can be kept down. In some cases, a double row of tanks (back to back) may be convenient to reduce the tank length. A large number of tanks also indicate an increase in capital costs in terms of mechanical and electrical equipment and also increased running costs. The shape of rectangular tanks should be determined on the grounds of cost unless this is contrary to overall plant efficiency.

A numerical addition should be made to the tank depth to allow for sludge storage. The final decision as to the tank breadth depends on the final time ratio, which in turn depends on the inlet arrangements.

6.5.2 Circular Settling Tanks

For tanks constructed below-ground, as the number of tanks increases and the radius decreases, there will be a saving in volume of excavation, since depth is related to the radius. As the number of tanks increases, so does the total perimeter together with the effluent collection channel. Also, for a greater number of tanks, the machinery and support structures, pipework, and maintenance costs rise well above any savings in excavation costs, except in very difficult ground. Therefore, the number of tanks should be as small as the flexibility of the operation allows, and the floor slope should be $\leq 10°$.

6.6 DESIGN DETAILS

6.6.1 Rectangular Settling Tanks

The aim is to feed influent uniformly across the tank. This is achieved by the careful design of inlet pipes, channels, and weirs. Where multiple tanks are employed, it is important to ensure that each tank is fed with the same discharge. This may be achieved by means of a tapered channel.

In order to prevent bottom scour, experience shows that the inflow should be located at about half of the tank depth; also, simple plate baffles have been shown to be very effective when extending to this depth. Sludge hoppers at the inlet end of the tank also help to reduce the maximum velocity near the floor. If these principles are incorporated into the design, the time ratio may be kept at 0.95; otherwise, a proportionate increase in the tank width should be made.

The outlet weir must be level. Castellated or V-notch weirs benefit the efficiency at low flows, but slightly reduce the time ratio at high flows. A shallow scum board should be provided at $\geq$0.25 m from the weir.

6.6.2 Circular Settling Tanks

Research shows that the best time ratio is obtained with a slightly submerged inlet pipe surrounded by a large circular baffle submerged to a depth of 0.25 of the tank depth at the inlet. However, this leads to high floor velocities.

With the pipe submerged to 0.75 of the tank depth and the baffle omitted, the floor velocities can be reduced despite a reduced time ratio value. It is important to ensure equal discharge to each tank where multiple tanks are used.

The baffle or pipe radius significantly affects the time ratio. The small loss of surface area due to the larger inlet is more than offset by the increase of the time ratio.

6.7 HYDRAULIC LOSSES

Hydraulic loss calculations for sedimentation tanks include the following:

- Control chamber losses;
- Losses due to distribution devices;
- Inlet-conduit losses;
- Velocity head losses at inlets;
- Head on outlet weir and free fall below the weir; and
- Outlet channel losses.

The control chamber consists of a small chamber with outlets controlled by sluice gates. Head losses can be calculated for the inlet, changes in flow direction, and outlet. For a system of two or more tank units, equal flow division is necessary. Measuring weirs or flumes are often included. The hydraulic design of conduits and structures downstream of sedimentation tanks should not permit the occurrence of a backwater level above the hydrostatic critical depths at the channel inlets around or across the settling tanks.

6.8 GENERAL DESIGN DETAILS

It is a normal design practice to construct at least two tanks. The standby capacity should be sufficient to allow one tank to be out of operation for maintenance without affecting the overall efficiency. Tanks should be operated in parallel rather than in series because inhomogeneous sludge affects the sludge treatment efficiency.

Tank inlets and outlets must be designed in a satisfactory manner to prevent short circuiting. Inlets should be designed to dissipate the flow velocity in the inlet conduits. This is usually achieved by baffle boards.

Arrangements for scum retention and removal must be made. Scum boards should be no nearer than 250 mm from the outlet weir.

Settled sludge must be regularly removed at intervals sufficient to prevent anaerobic activity. Generally, sludge is removed either manually (small works) or under hydrostatic head. Sludge is much more viscous than normal sewage. Multiplying factors for head losses calculated for clean water and used for sludge of varying moisture content (%) are shown below:

- 98–97%: 2–2.5.
- 96–94%: 4.
- 90–88%: 6–8.
- 86%: 8–10.

Sludge pipework should be ≥150 mm in diameter, and some means of clearing blocked pipes should be provided. Side slopes to sludge hoppers should be ≥60° to the horizontal to prevent sludge deposition and bridging.

6.9 DETAILS OF VARIOUS TYPES OF SEDIMENTATION TANKS

6.9.1 Storm Tanks

The design is based on the total capacity rather than velocity principles. Two or more tanks are provided with a total capacity equal to a DWF of 6 h. This allows 2 h at the maximum flow rate of three DWF. Normally, storm runoff tanks are rectangular with the length between 4 and 5 times the width. Normally, the inlet weirs are fixed at different levels while the outlet weirs are at the same level so that the tanks fill one at a time, with discharge commencing only when all tanks are full.

After a storm, the tank content should be returned to the main inlet of the works or wetland system. The sludge is discharged to the sludge treatment unit. Because the tanks have only intermittent use, the sludge scrapers are generally designed as simply as possible. Traveling bridge units are often too expensive. Usually, the scrapers are of the squeegee type: a cable-hauled boom driven by a trolley-mounted motor.

6.9.2 Primary Sedimentation Tanks

6.9.2.1 Quiescent Tanks

Quiescent tanks work on the principle of "fill, stand, and empty." They are not used on domestic sewage treatment works, but have some applications in the treatment of trade wastes. The tanks are usually rectangular with decanting arms to remove top water when the tanks have stood the required period.

The tanks should be operated continuously, if the inflow is substantial. Manual desludging makes the operation of these tanks costly in labor and does not allow for any flexibility in treatment.

6.9.2.2 Rectangular Horizontal-Flow Tanks

Continuous-flow tanks have detention periods at DWF between 6 and 10 h. The design is based on the criteria given in the method presented in this chapter. Except for the very small schemes, manual desludging should not be used. Because of high labor costs, it is usually economic to provide mechanical desludging on even the smallest units. For mechanically desludged tanks, the floors are nearly flat, and the inlet end has a number of inverted pyramidal-shaped sludge storage hoppers below the general floor level.

Sludge scrapers are usually operated via a power-driven bridge spanning the tank mounted on rails suspending either a vertical or trailing scraper blade on the tank floor. The most economic span is between 7.5 and 10.5 m. The scraper is usually provided with a scum blade. It is possible for one scraper to serve a number of tanks by using a transfer carriage and track. The additional civil engineering work should be costed against the cost of a single scraper spanning two or three tanks. If multiple scraper units are being used, flanges will probably be provided to wall tops; otherwise, intermediate tank walls will be excessively thick.

Another sludge scraper is the flight type, which consists of a number of buoyant flights with blades attached to an endless chain running along the tank sides. The flights move along the tank floor, pushing sludge into hoppers at the inlet, and returning along the surface, where they remove the scum. They are expensive to install and maintain.

6.9.2.3 Imhoff Flow Tanks

Imhoff tanks are designed to provide sedimentation and digestion volume. They may be either round or rectangular (usually the latter). Either design requires deep excavation and a somewhat more costly construction than either horizontal-flow or radial-flow tanks.

Sewage enters these tanks below the top water level but above the maximum sludge level. The flow direction is upward, while settled sludge sinks to the bottom of the tank. The allowable upward velocity determines the surface area.

The side slopes of the cone should be >60° to the horizontal. The inlet and outlet should be designed to produce low velocities. The use is normally confined to small works. The sludge can be drawn off hydrostatically approximately once a week under manual control. The capacity provided within the tanks (excluding sludge capacity) must be between 5 and 15 h of a DWF.

6.9.2.4 Radial-Flow Tanks

Continuous-flow tanks have detention periods at a DWF of between 6 and 10 h. The design is based on the criteria presented above. The effect of the scraper support on the time ratio must be considered (in model tests) after initial sizing of the units. Floor slopes vary between 0° (flat) and 15° (commonly between 7.5° and 10°). V-notch weirs produce a more even flow distribution at low flows, but tend to reduce the efficiency slightly.

Automatic descumming is carried out by the rotating scraper; scum boxes are provided on the tank periphery. Sludge is deflected into a central hopper, which has sufficient capacity for one day. The designs of scrapers are many and varied; all incorporate renewable rubber tips.

Radial-flow tanks cannot be built in batteries, but can be cheaper than rectangular tanks since they are shallower and have less complicated sludge scrapers. If the same tank diameter can be used for final or humus settlement, reusable formwork may further cut costs.

6.9.3 Secondary Sedimentation

Concerning secondary sedimentation tanks, the SS tend to be smaller and less dense, with a lower settling velocity in comparison to the SS in primary sedimentation tanks. Wind can affect the settlement of sludge in secondary tanks, especially if they are large and shallow. The capacity is normally fixed at a DWF between 4 and 6 h. The outlet channels are usually lined with white tiles to assist the visual inspection of the effluent.

There are two classes of secondary sedimentation tanks depending on the biological treatment process. Humus tanks for filtration plants are normally radial-flow tanks, although other types are possible. Sludge removal is usually intermittent. Humus sludge is returned to the feed channel of the primary sedimentation tank and settled with the primary solids since it assists sedimentation and is easier to dewater when mixed with primary sludge.

Concerning final tanks for activated sludge plants, sludge removal must be continuous to feed the aeration units with fresh sludge. Surplus activated sludge is returned to the inlet of the primary tank for the purpose of settlement.

Circular settlement tanks are usually similar to those used for primary and humus sedimentation. Normally, they are flat-bottomed with continuous removal of sludge by airlift or siphon draw-off.

6.10 SEDIMENTATION AIDS

Sewage includes colloidal particles, which may remain suspended indefinitely. However, these very finely divided particles tend to flocculate with the aid of mechanical agitation, aeration, or chemical coagulation. The quality of the final effluent can thus be significantly increased.

Mechanical flocculation is normally achieved in double-zone tanks incorporating an inner mixing zone (paddles rotating at $\leq$450 mm/s) and another conventional settlement zone (e.g., the Dorr Clariflocculator, which gives 20% better effluent than that obtained by plain sedimentation at very little extra cost).

Chemical coagulants in common use are as follows:

- Hydrated lime: $Ca(OH)_2$.
- Aluminum sulfate: $Al_2(SO_4)_3$.
- Copperas: $FeSO_4$.

However, the misuse of chemicals can sterilize sewage or at least slow down the biological degradation rate. Sedimentation aids are usually unnecessary for most wetland systems due to the presence of plants that slow down the flow velocity of the storm runoff.

Chapter 7

Coagulation and Flocculation

7.1 THEORY OF SETTLING

In most wastewater treatment works and wetland systems, sedimentation is the main process used in primary treatment. Primary sedimentation removes between 50% and 70% of the suspended solids (SS), which contain between 25% and 40% of the BOD from the wastewater, including urban runoff.

Sedimentation is normally defined as the removal of solid particles from a suspension by settling under gravity. Clarification is a similar term, which refers specifically to the function of a sedimentation tank in removing suspended matter from the water to give a clarified effluent.

Thickening occurs in sedimentation tanks (called clarifiers in the United States) and is the process whereby settled impurities are concentrated and compacted on the floor of the tank and in the sludge hoppers. The concentrated impurities removed from the bottom of the tank are called sludge, and the impurities floating to the surface of the tanks are called scum.

7.2 CLASSIFICATION OF SETTLING BEHAVIOR

The hydrodynamic problem of settling has been studied by many researchers, and a commonly accepted classification scheme is as follows:

- Class I: unhindered settling of discrete particles;
- Class II: settling of a dilute suspension of flocculent particles;
- Class III: hindered and zone settling; and
- Class IV: compression settling (compaction and consolidation).

Concerning Class I settling, a single particle in a liquid of a lower density should be considered for the purpose of this explanation. The particle will accelerate until a settling (or terminal or limiting) velocity is reached at the point where the gravitational force is balanced by the frictional drag force. The settling velocity can be determined using either Stoke's law (Eq. (7.2.1)) or Newton's law (Eq. (7.2.2)), depending on whether the flow past the particle is laminar or turbulent, respectively.

Wetlands for Water Pollution Control. http://dx.doi.org/10.1016/B978-0-444-63607-2.00007-1

$$U_s = \frac{(S_s - 1)gd^2}{18v} \qquad (7.2.1)$$

where:

U_s = settling velocity;
S_s = specific gravity of the solid matter;
g = gravitational constant;
d = diameter of the solid matter; and
v = kinematic viscosity.

$$U_s = \sqrt{3.3(S_s - 1)gd} \qquad (7.2.2)$$

where:

U_s = settling velocity;
S_s = specific gravity of solid matter;
g = gravitational constant; and
d = diameter of the solid matter.

In many practical cases, even in a relatively dilute suspension, particles coalesce to form particle aggregates (i.e., flocculation resulting in flocs) with increased settling velocities. The extent of such flocculation is a function of many variables, including suspension type and concentration, the prevailing velocity gradient, and time. In wastewater engineering, most sewage, including urban runoff, exhibits Class II settling in primary tanks.

The expected removal of a flocculent suspension in a sedimentation process can be estimated from a laboratory settling test using a water column height equal to that used in the process. At various time intervals, samples are withdrawn from ports at different depths and analyzed for SS. Lines of equal percentage removal are then interpolated between the plotted points. The resulting curves can be used to determine the overall removal of solids for any detention time and depth within the range of the data, bearing in mind that the test conditions are quiescent.

Concerning Class III settling, as the concentration of particles in a suspension is increased, a point is reached where particles are so close together that they no longer settle separately, but the velocity fields of the fluid displaced by adjacent particles overlap. This gives rise to a net upward flow of liquid displaced by settling particles. This results in a reduction of the settling velocity and is called "hindered settling."

For traditional wastewater systems, Class III settling is most commonly exhibited in the sludge blanket clarifier, where the particle concentration is so high that the whole suspension tends to settle as a "blanket" (termed "zone settling"). Furthermore, most urban runoff exhibits Class II or Class IV settling (see above) in a densely planted constructed treatment wetland.

At the bottom of a settling column, as settling continues, a compressed layer of particles begins to form. The particles in this region apparently form a structure in which there is close physical contact between the particles. As the compression layer forms, regions containing successively lower concentrations of solids than those in the compression region extend upward in the cylinder.

Thus, the hindered settling region contains a gradient in solids concentration increasing from that found at the interface of the settling region to that identified in the compression zone. This region of Class IV settling is particularly important when considering the design of thickeners for activated sludge.

7.3 IDEAL SETTLING

The behavior of an ideal sedimentation tank, operating on a continuous basis with a discrete suspension of particles, can be studied as follows:

- Assume quiescent conditions in the settling zone;
- Assume uniform flow across the settling zone;
- Assume uniform solids concentration as the flow enters the settling zone; and
- Assume that solids entering the sludge zone are not resuspended.

Equation (7.3.1) shows how to calculate v, which is termed "overflow rate" and should be considered as a design settling velocity.

$$v = Q/A \tag{7.3.1}$$

where:

v = overflow rate (m/s);
Q = discharge (m^3/s); and
A = tank surface area (m^2).

7.4 INTRODUCTION TO COAGULATION AND FLOCCULATION

When a material is truly dissolved in water, it is dispersed as either molecules or ions. The "particle" sizes of a dissolved material are usually in the range between 2×10^{-4} and 10^{-3} μm. The "particles" cannot settle and cannot be removed by ordinary filtration.

True colloidal suspensions and true solutions are readily distinguished, but there is no sharp line of demarcation. Colloidal particles are defined as those particles in the range between 10^{-3} and 1 μm. In, for example, urban water runoff, colloidal solids usually consist of fine silts, clay, bacteria, color-causing particles, and viruses.

7.5 COLLOIDAL SUSPENSIONS

In water and wastewater treatment, there are generally two types of colloidal systems, both of which have water as the disperse phase: colloidal suspensions (i.e., solids suspended in water) and emulsions (i.e., insoluble liquids such as oils suspended in water). Features of a colloidal suspension are as follows:

- Colloids cannot be removed from a suspension by ordinary filtration. However, they can be removed by ultrafiltration or by dialysis through specialized membranes.
- Colloidal particles are not visible under an ordinary microscope. However, they can be seen as specks of light with a microscope when a beam of light is passed through the suspension. This is caused by the Tyndall effect, which is described as the scattering of light by colloidal particles.
- Brownian motion prevents the settlement of particles under gravity. Some colloids can be removed by centrifugation.
- There is a natural trend for colloids to coagulate and precipitate. Sometimes, this tendency is countered either by mutual repulsion of the particles or by the strong attraction between the particles and the medium in which they are dispersed (usually water). If these effects are strong and coagulation does not occur, the suspension is said to be stable.

When the main factor causing stability of a colloidal suspension is the attraction between particles and water, the colloids are characterized as hydrophilic. When there is no great attraction between particles and water, and stability depends on mutual repulsion, the colloids are said to be hydrophobic. Hydrates of iron and aluminum form hydrophobic colloids in water; stability occurs through mutual electrostatic repulsion (i.e., like charges repel). Proteins, starches, or fats form hydrophobic colloids in water. Stability can be achieved through the attraction between water and particles. Stable hydrophilic colloids are difficult to coagulate.

Colloids have a large surface area per unit volume; for example, if a 10-mm cube was broken into cubical particles with dimensions of 10^{-2} μm, the surface area would be increased by a factor of 10^6 to some 600 m^2. Surface effects are therefore of significance; of these, the following two are important:

1. The tendency for substances to concentrate on surfaces (adsorption); and
2. The tendency for surfaces of substances in contact with water to acquire electrical charges, giving them electro-kinetic properties.

At the surface, electrical charge results from the colloidal material's affinity, from some ions in the water, or from the ionization of some of the atoms (or groups of atoms), which leave the colloid. The surface charge attracts ions carrying a charge of opposite sign and thus creates a cloud of "counter-ions," in which the concentration decreases as the distance from the particle increases.

As an example, clay particles in water are charged because of isomorphous substitution, whereby certain cations of the crystal lattice forming the mineral may have been replaced by cations of a similar size but lower charge. For example, Si^{4+} may be replaced by Al^{3+}, or Al^{3+} by Mg^{2+}. In both cases, the lattice is left with a residual negative charge, which must be balanced by the appropriate number of compensating cations in the dry clay. These may be fairly large ions such as Ca^{2+} or Na^{2+}, which cannot be accommodated in the lattice structure, so these ions are mobile and may diffuse into solution when the clay is immersed in water, resulting in negatively charged clay particles.

For these and other reasons, colloidal particles in water are usually charged. The majority of particles encountered in natural waters are negatively charged. A suspension of colloidal particles as a whole has no net charge, since the surface charge of the particles is exactly balanced by an equivalent number of oppositely charged counter-ions in solution.

Furthermore, the distribution of these counter-ions is not random, since by electrostatic attraction they tend to cluster around charged particles. The combined system of the surface charge of a particle and the associated counter-ions in solution is known as an electrical double layer (stern layer).

In addition to the forces related to the electrical charges, colloidal particles, when close together, are subject to van der Waals forces. These originate in the behavior of electrons, which are part of the atomic or molecular system. The forces of attraction become significant only at short distances (e.g., ≤ 1 μm).

The source of energy may be from either Brownian motion or the relative movement in the water (fluid shear). If the energy of an impact is inadequate, the remaining repulsion will force the particles apart, forming a stable colloidal suspension.

7.6 COAGULATION PROCESSES

It is necessary to reduce or eliminate the energy requirement to destabilize a hydrophobic colloidal suspension to permit coagulation to occur. The van der Waals forces cannot be manipulated, but the electrical forces can. The principle methods are as follows:

- Reduction or neutralization of the charges on the colloid;
- Increase of the density of the counter-ion field, and thus reduction of the range of the repulsive effect (compaction of the double layer); and
- Permanent contact through molecular bridges between particles.

Neutralizing charges on colloids may be accomplished by the addition of either multivalent ions or colloids (or both) that have an opposite charge. These are frequently added as a chemical coagulant. The coagulating power of a chemical rises rapidly with its valence. For example, Al^{3+} and SO_4^{2-} ions are several hundred times more effective than Na^+ and Cl^- ions.

In water, which is almost pure (i.e., it has a low concentration of ions), the cloud of counter-ions is widely dispersed. With an increase in ionic strength of the solution, the counter-ion cloud becomes more concentrated near the colloidal particles, counteracting the charge, and the field of influence of the colloidal charge becomes more restricted. Therefore, an increase in the ionic strength of a colloidal suspension will tend to cause destabilization and consequently coagulation. This phenomenon has been observed in nature where turbid rivers run into brackish estuaries and marine wetlands, and coagulation and sedimentation subsequently occur.

7.7 COAGULATION CHEMICALS

Chemicals used for chemical coagulation of impurities in water treatment should be affordable and not leave any toxic or other undesirable residues in the water. Coagulation in water treatment occurs predominantly by two mechanisms:

- Adsorption of the soluble hydrolysis species on the colloid subsequent and destabilization; and
- Sweep coagulation where the colloid is entrapped within the precipitation product (e.g., aluminum hydroxide).

The specific mechanism occurring is dependent on both the turbidity and alkalinity of the water being treated. The reactions in adsorption and destabilization are extremely fast and occur within 0.1 and 1 s. Sweep coagulation is considerably slower and occurs in the range between 3 and 17 s. The differences between the two types of coagulation in terms of rapid mixing are not commonly delineated in the literature. It is imperative that the coagulants are dispersed in the raw water as rapidly as possible ($\leq$0.1 s) so that the hydrolysis products that develop within 0.01 and 1 s will cause destabilization of the colloid.

The common aluminum salt used is aluminum sulfate (alum). It is available in solid, granular, powdered, or liquid form. The chemical formula is $Al_2(SO_4)_3 \cdot nH_2O$, where n depends on the method used by the manufacturer but is usually in the range between 12 and 16.

When hydrolyzed, alum produces sulfuric acid as well as the hydrate. For example, when forming the simplest hydrate $Al(OH)_3$, the hydrolysis reaction can be described with Eq. (7.7.1):

$$Al_2(SO_4)_3 + 6H_2O \rightarrow Al(OH_3) + 3H_2SO_4 \tag{7.7.1}$$

Therefore, it is regarded as an acid salt, and the water must contain enough alkalinity (natural or added) to react with the acid as it forms to maintain the pH within the desired range for good coagulation and flocculation. It can be shown that 1 mg/l alum is used for 0.5 mg/l alkalinity (as $CaCO_3$), which results in the production of 0.44 mg/l CO_2. After coagulation, taste requirements dictate a residual alkalinity of $\geq$30 mg/l.

Sodium aluminate is an alkaline salt produced by treating aluminum oxide with caustic soda. The hydrolysis reaction is shown in Eq. (7.7.2):

$$NaAlO_2 + 2H_2O \rightarrow Al(OH)_3 + NaOH \qquad (7.7.2)$$

It can be used either in conjunction with alum or on its own for waters, which do not have sufficient natural alkalinity. However, subsequent coagulation may not be as good as when alum is used in combination with an alkali because the divalent sulfate ions introduced with the alum have a favorable influence on coagulation.

Ferric ions can be hydrolyzed and precipitated as ferric hydroxide at pH > 4.5. Poor coagulation occurs in the pH range between 7.0 and 8.5. The ferric ion (like the aluminum ion) hydrolyzes to form hydrates and an acid. Enough alkalinity must be present to combine with the acid and maintain a suitable pH for good coagulation. Unlike $Al(OH)_3$, ferric hydroxide does not redissolve in alkaline solutions, and so there is no particular upper pH limit for ferric coagulation. In addition, the ferric hydroxide floc is usually heavier, and hence faster settling, than the aluminum hydroxide floc.

Considering coagulation of colored acid waters (typically of peat-based catchments), the presence of the divalent sulfate ion assists coagulation to a greater degree than that of the monovalent chloride ion. The commonest ferric salt used in water treatment is ferric chloride ($FeCl_3$), which is very corrosive in the presence of water. Therefore, it requires transportation in rubber-lined tanks or glass containers.

Ferric sulfate ($Fe_2(SO_4)_3$) is normally available as an anhydrous material, making transportation easier. On some occasions, ferrous salts are used. To be effective, the ferrous ion should be oxidized to the ferric form when in solution.

At pH values >8.5, dissolved oxygen (DO) in the water will cause oxidation. However, at lower pH values, chlorine can be used as an oxidizing agent. Ferrous sulfate ($FeSO_4 \cdot 7H_2O$), also known as copperas, is usually available as a granular material. Because it requires a pH of ≥8.5 to be oxidized by DO, it cannot be used on its own in natural waters. When used in conjunction with lime, it is useful in coagulating the precipitate obtained in lime softening of water, and the removal of excessive iron and manganese from waters.

Chlorinated copperas can be used for water with a relatively low pH. It is formed by mixing solutions of copperas and chlorine (Eq. (7.7.3)):

$$6FeSO_4 + 3Cl_2 \rightarrow 2Fe_2(SO_4)_3 + 2FeCl_3 \qquad (7.7.3)$$

Moreover, an excess of chlorine can be used to disinfect water. Copperas can be more readily used than ferric salts because it is not so corrosive and can be handled by ordinary chemical feeding equipment.

Lime is not normally used as a coagulant in water treatment, although it is commonly applied in wastewater treatment. Lime is used in water treatment as a pH control measure for adding alkalinity and in softening waters.

In some waters, coagulation is poor, even with an optimum dose of coagulant. In these cases, a coagulant aid is used. The two most common coagulant aids are clay and polyelectrolytes. Clay (bentonite, Fuller's earth, or other clay) is useful in some waters that are deficient in negatively charged flocs. The clay colloids provide nuclei for the formation and growth of flocs, as well as a medium to add some weight to the particles. In some cases, the adsorptive qualities of the clay may be useful in removing color, odor, and taste.

Polyelectrolytes are long-chain macromolecules that have electrical charges or ionizable groups. Cationic polyelectrolytes are polymers that, when dissolved, produce positively charged ions. They are widely used because the suspended and colloidal solids commonly found in water are generally negatively charged. Cationic polymers can be applied as the primary coagulant or as an aid to conventional coagulants. There are several advantages to using this coagulant aid: the amount of coagulant can be reduced, flocs settle better, there is less sensitivity to pH, and the flocculation of living organisms (bacteria and algae) is improved.

Anionic polyelectrolytes are polymers that dissolve to form negatively charged ions and are used to remove positively charged solids. Anionic polyelectrolytes are used primarily as coagulant aids with aluminum or iron coagulants. The anionic chemicals increase flow size, improve settling, and generally produce stronger flocs. They are not significantly affected by pH, alkalinity, hardness, or turbidity.

Non-ionic polyelectrolytes are polymers that have a balanced or neutral charge, but, upon dissolving, they release both positively and negatively charged ions. Non-ionic polyelectrolytes may be used as coagulants or as coagulant aids. Although they must be added in larger doses than other types, they are less expensive.

Compared with other coagulant aids, the required dosages of polyelectrolytes are very small. The normal dosage range of cationic and anionic polymers is between 0.1 and 1 mg/l. For non-ionic polymers, the range is between 1 and 10 mg/l.

7.8 OPERATION OF THE COAGULATION AND FLOCCULATION PROCESS

There are three fundamental steps in operating the coagulation and flocculation process:

1. Selecting the chemicals;
2. Applying the chemicals; and
3. Monitoring the process efficiency.

The selection of chemical coagulants and aids is a continuing program of trial and evaluation, normally using the jar test (see below). When selecting chemicals, the following characteristics of the raw water to be treated should be measured:

- Temperature;
- pH;
- Alkalinity;
- Turbidity; and
- Color.

The jar test, although still the most widely used coagulation control test, is subjective and depends mostly on the human eye for evaluation and interpretation. The operator should measure pH, turbidity, filterability, and zeta potential to gain further information concerning the coagulation and flocculation process.

However, jar test findings help to determine the type of chemical or chemicals and their corresponding optimum dose to be used. Most jar test results are expressed in mg/l. This unit must be converted to the equivalent full-scale dose in kg/day or m^3/day. The treatment plant operator subsequently applies the chemical to the water by setting the manual or automatic metering of the chemical feed system to the desired dosage rate.

Although the jar test provides a good indication of the expected results, full-scale plant operation may not always match these results. Actual plant performance must be monitored for adequate flash mixing, gentle flocculation, adequate flocculation time, and settled and filtered water quality.

The zeta potential test is increasingly being used at waterworks to help determine the best pH and dosage for cationic polymers and cationic coagulants such as Al^{3+} and Fe^{3+}. The control procedure requires monitoring of the zeta potential of the coagulating water and changing the chemical dosage when the zeta potential varies outside a range known to produce the lowest turbidity. The range, although variable from plant to plant, is between 6 and 10 mV.

7.9 RAPID MIXING

The first contact of a coagulant with water is the most critical period of time in the whole coagulation process. The coagulation reaction occurs quickly, so it is vital that the coagulant and colloidal particles come immediately into contact with each other. After the coagulant has been added, the water should be agitated violently for several seconds to encourage the greatest number of collisions possible with suspended particles.

There are three principle types of mixers used in the rapid mixing stage: mechanical devices, pumps and conduits, and baffled chambers. Because of their positive control features, propeller, impeller, and turbine-type mechanical mixers are widely applied. The detention times in these chambers are very short, usually less than one minute (typically, 30 s). Mechanical mixers can also be mounted directly in the pipeline (in-line mixers). Unlike pump- and conduit-type mixers, the in-line mixers can be adjusted to provide the correct degree of mixing. Since an in-line mixer requires no special tank, it is a low-cost device that is increasing in popularity.

7.10 FLOCCULATION

Once the colloid has been destabilized, the growth of flocs due to the agglomeration of the colloidal particles is promoted and enhanced by gentle mixing. This flocculation process occurs mainly in two stages known as perikinetic and orthokinetic flocculation.

In the first stage (perikinetic flocculation), particles collide and stick together as they move randomly under the influence of Brownian motion. The time taken for particles to grow so large that they are no longer significantly affected by Brownian motion depends on the frequency of collisions. The opportunity for collision is proportional to the concentration of the particles, so perikinetic flocculation is more rapid in concentrated suspensions. The time taken for effective completion of this stage is usually <60.

In the second stage (orthokinetic flocculation), particles are moved together by gentle motion of the water. The rate of flocculation depends on the particle nature, size and concentration, and the velocity shear gradient of the water. Many mathematical models of this process have been developed, although they are all difficult to apply in practice. The rate of flocculation is proportional to:

- The velocity shear gradient;
- The volume of the particle zone of influence; and
- The square of the numerical concentration of particles.

Although the initial rate of floc formation is proportional to the velocity shear gradient, high-velocity gradients can cause large flocs to be ripped apart as a result of internal tension, surface shear stress erosion, or both. The agitation necessary to induce flocculation is created in several ways:

- Hydraulic agitation: jets and baffled channels;
- Air injection: rising bubbles;
- Mechanical agitation: paddle and reel, and turbine and propeller; and
- Solids contact: sludge blanket.

Chapter 8

Sludge Blanket Clarifiers

8.1 INTRODUCTION TO SLUDGE BLANKET CLARIFICATION SYSTEMS

Sludge blanket clarifiers are widely used in water and wastewater treatment, and they can also be applied as a pretreatment step located before a constructed wetland. The sludge blanket effect is obtained simply by allowing chemically pretreated water (i.e., with coagulant added) to flow upward in a clarification unit. As water rises in the clarifier, its velocity progressively decreases until, at a given level, the upflow velocity of the water equals the hindered settling velocity of the floc suspension (e.g., the metal hydroxide precipitate). The flocs are therefore suspended in the water to form the sludge blanket.

The principal elements of the sludge blanket clarification system are as follows:

1. Rapid mixing unit in which the raw water is mixed with coagulants and then subjected to a delay period prior to entering the blanket;
2. Inlet system to the clarifier;
3. Sludge blanket and flocculation zone;
4. Supernatant clear water zone;
5. Excess sludge removal system; and
6. Clarified water collection system.

The addition of chemicals to raw water and their thorough mixing is the first stage in the clarification process. After initial rapid mixing, an additional slower mixing time of a few minutes (i.e., a delay time) is useful to assist perikinetic flocculation (i.e., the formation of primary flocs). This in turn results in improved orthokinetic flocculation.

The nature of the inlet flow to the clarifier is vital to the hydraulic stability of the blanket, the distribution of flow through the blanket, and water quality. It is important to ensure verticality and centering of the inlet. Poor inlet design gives rise to short-circuiting and streaming of the water to one side of the tank. This will disturb the stability of the sludge blanket, with subsequent floc carryover in the effluent. The inlet flow must be dispersed uniformly over the

Wetlands for Water Pollution Control. http://dx.doi.org/10.1016/B978-0-444-63607-2.00008-3

inlet level. A uniform flow distribution to the blanket is vital to ensure its stability.

A zone of high turbulence at the inlet is required to maintain the blanket in a fully fluidized condition. The high energy of the incoming water is reduced in the fluidized bed by the turbulent dissipation of energy and because of the large concentration of solid particles.

A low inlet velocity would be sufficient to prevent sedimentation, whereas an inlet velocity greater than that required would provide excess energy in the incoming water. Excess energy largely increases the instability conditions of the sludge blanket. This demonstrates that even small deviations from the optimum inlet velocity have a large detrimental effect on the performance of the sludge blanket.

Four different inlet distribution systems have been used in practice, namely:

1. Perforated-plate distribution system;
2. Point inlet distribution system;
3. Perforated-pipe distribution system; and
4. Slot distribution system.

The inlet of the perforated-plate distribution system does not produce an adequate uniform flow distribution due to clogging of the space beneath the plate. For this reason, it is not widely used.

The point inlet distribution system tends to cause excess incoming energy, and the resulting blanket tends to be unstable. This excess energy could be reduced by using a multiple-point inlet system, but too many inlets present difficulties in providing a uniform distribution of water.

When using the perforated-pipe distribution system, the energy of the incoming water is reduced substantially and the stability of the blanket increases. However, a large number of pipes should be used to obtain the optimum inlet velocity. In such a pipe system, a uniform flow distribution is difficult to maintain due to clogging of the orifices and pipes by flocs. Pulsation can be used to accomplish a uniform distribution and prevent clogging.

In the slot distribution system, a uniform distribution of coagulated water is provided by means of a narrow slot. For large clarifiers, more than one slot inlet can be used. The dimensions of the inlet slot can easily be designed to obtain the optimum velocity. The adaptability of the size of the inlet slot is one of the greatest advantages of this type of system.

The sludge blanket is a fluidized bed formed by the accumulation of flocs. While the top of the blanket is very often well defined, there is no distinguishable bottom of the blanket. This is because of the varying density of the floc particles forming the blanket. For floc particles to form a blanket, the upflow velocity of the water must be equal to the hindered settling velocity of the floc suspension. In a suspension, the floc particles mutually hinder one another, and the fall velocity of the suspension as a whole is less than that of the individual particles (Eq. (8.1.1)).

$$v_s = v_p(1 - kc)^n \tag{8.1.1}$$

where:

v_s = settling velocity of the suspension;
v_p = settling velocity of an individual floc;
c = concentration of the suspension; and
k and n = constants for a particular suspension.

The specific feature of a fluidized bed in a sludge blanket clarifier is that the suspended particles constituting the sludge blanket are not discrete, uniformly sized particles, but a floc suspension subject to aggregation and breakup. The process of flocculation involves both floc growth and breakup, so even at an equilibrium, there will be a range of floc sizes. The growth and breakup of the flocs depend on the growth properties of the flocs, their shear strength, and the turbulence within the sludge blanket clarifier.

The mechanism of clarification within a sludge blanket tank is complex, involving flocculation, entrapment, and sedimentation. Flocculation is time dependent and greatly influenced by the volumetric concentration of the particles and the velocity gradient in the liquid at the particle surface. With large concentrations of particles associated with sludge blankets, the influence of time and velocity gradients is substantially reduced.

A process of physical entrapment by flocculation and agglomeration, akin to surface capture in deep-bed filtration, occurs throughout the sludge blanket, but the principle mechanism in a sludge blanket is a process of mechanical entrapment and straining, which occurs when rising small particles cannot pass through the voids between the larger particles that comprise the bulk of the sludge blanket.

The separability of the coagulated floc suspension is governed by the physical and chemical properties of the flocs, principally by size and shape, density, shear strength, and surface conditions. For separation processes involved in sludge blanket clarifiers, the aggregational and adhesion properties of the flocs and their resistance to shear forces are of utmost importance. The aggregational and adhesion conditions of the floc suspension are determined by the local forces acting between interacting surfaces and the resistance to shear forces, the chemical nature, and the physical structure of the flocs.

One of the principal hydraulic requirements is that flow across any horizontal layer in the clarifier should be uniform. Nonuniform flow in the blanket causes short-circuiting within the blanket and thus instability. Some sludge blanket clarifier designs have been criticized for the tendency of the blanket to "boil." This can happen, particularly in hot weather or when the tank is subjected to high winds.

In general, it is beneficial to have a deep sludge blanket. Greater blanket depth provides a greater dissipation of turbulence and blanket stability. Blanket stability will enhance blanket concentration, thereby improving clarification by entrapment. Increased depth will also prolong the contact time of the water with the sludge blanket, enhancing chemical and physical reactions.

The concentration of solids within the blanket increases with distance from the bottom; the highest concentration is at the top of the blanket. The size of the flocs in the blanket also enlarges with distance from the inlet. Flocs at the inlet are small due to high turbulence, but since the velocity gradient reduces with movement upward through the blanket, the flocs become larger.

When water exits from the sludge blanket, it enters a clear water zone. This zone serves primarily as a buffer to allow surges in blanket volume to be accommodated. If flocs escape from the surface of the blanket, their free settlement velocity in the clear water zone will exceed the upward velocity, and thus they will return to the blanket. It is essential to have a uniform distribution of flow in the clear water zone. Flow conditions in this zone depend on the depth of the zone, the location of the clear water collection weirs, and the location of the sludge removal devices. The clear water zone must have a minimum depth equal to half the spacing between the water troughs.

Effective removal of excess sludge is very important for the efficient operation of clarifiers. The continuing addition of coagulants to the inflowing water causes an increase of floc volume in the sludge blanket tank. This results in the blanket volume enlarging. It follows that, for steady-state conditions in the clarifier, the rate of sludge removal must balance the rate of sludge formation. The rate of sludge formation equals the dosage rate of a coagulant, if it is assumed that the volume of the natural particles in a suspension of water is insignificant compared with that of the hydrolysis products.

Therefore, it is possible to predict the volume of sludge to be removed, on average, as a function of the coagulant dose. The removal of flocs is usually required with a minimum removal of water. Consequently, it follows that sludge removal should take place where the blanket is most concentrated. Withdrawal of sludge at any other level is inefficient, and withdrawal at the top gives positive control of the sludge blanket level.

An effective method of sludge removal consists of one or more flexible cones suspended in the water and connected to a load cell via a cable. The load cell is sufficiently sensitive that, when the weight of the sludge reaches a preset value, it initiates the opening of the desludging valve on the cone outlet. The removal of excess sludge can also be accomplished by a submerged weir, over which the sludge flows into a sludge concentrator. The weir crest level is set at the upper boundary of the sludge blanket.

The clarified water is collected by a series of overflow weirs. Outlet ducts should not cause biased flow patterns. As is the case for the inlet design, poor outlet design causes short-circuiting and streaming in the blanket, which affect its stability. The collection at the outlet must be uniform over the outlet cross-section. In normal practice, uniformity of weir loadings is accomplished by using castellated weirs. In some instances, it might be necessary to incorporate baffles in the clear water zone to minimize wind-generated surface movement passing down to the blanket and causing disruption.

8.2 TYPES OF SLUDGE BLANKET CLARIFIER

The earliest type of sludge blanket clarifier was the hopper-bottomed tank. This type of tank was first developed during the 1930s and 1940s, and it was used mainly in places where there was a British influence. They are usually square in plan (occasionally circular). The hopper shape in the bottom of the tank assists in the flow distribution. It has been found that square-plan hopper tanks deteriorate in performance at lower upflow rates, in comparison to circular tanks. However, the square plan permits an economical construction.

There has been continuous striving toward simplifying the shape of the tank to reduce construction costs. This is usually achieved by using a flat-bottomed tank. In this type of design, chemically treated water is directed down to the base of the tank through suspended pipe chandeliers before passing upward through the blanket to the surface collection system. The reversal of flow, which takes place at the base of the tank, assists chemical flocculation in the same way as in the hopper-bottomed tank.

The pulsator clarifier represents the industrial application of laboratory observations and theoretical considerations. It permits the use of high upflow rates, which can be up to 8 m/h, depending on the type of suspended matter.

The clarifier consists of a flat-bottomed tank with a series of perforated pipes at its base to distribute the raw water uniformly over the entire bottom of the clarifier. A further set of perforated channels is provided at the top of the clarifier to collect the clarified water evenly and prevent any velocity irregularities in parts of the tank.

Various methods are available for admitting water via the bottom-pipe system at intervals. However, they all require storage of a certain volume of raw water for a certain period of time. The raw water is subsequently introduced into the tank as quickly as possible.

The most economical method of improving this operation is to introduce the raw water into a vacuum chamber from which the air is exhausted by a vacuum pump removing an air flow of about half the maximum water flow to be treated. The operation of the vacuum chamber is coordinated with the bottom-pipe system of the clarifier.

Under these conditions, the level of raw water gradually rises in the vacuum chamber. When it reaches a value between 0.6 and 1.0 m above the level of water in the tank, a contact operates an electrical relay for quick opening of an air inlet valve. Atmospheric pressure is immediately applied to the water stored in the chamber, and the water rushes into the tank at high speed. These appliances are generally set so that the water in the vacuum enters the settling tank in 5–10 s while the filling time for this chamber is between 30 and 40 s. Air is drawn from the vacuum chamber by a fan or electric blower acting as a vacuum pump. The valve connecting the vacuum chamber to the atmosphere is opened and closed when the water reaches the upper or lower level.

The header in the bottom of the settling tank has a large cross-section to reduce head loss. The orifices in all branch pipes are placed so that a uniform sludge blanket forms in the lower half of the settling tank. This blanket is subject to alternating vertical movements and tends to increase in volume because of impurities brought by the raw water and the flocculation reagents added. Its level gradually rises. A particular zone of the tank is used to form hoppers with inclined floors into which the excess sludge discharges and subsequently concentrates. Sludge draw-off is carried out intermittently by the discharge pipes.

The equipment has no mechanical sludge-stirring system likely to break up the flocs already formed. Due to the high concentration in the sludge blanket and its buffer action, faulty adjustment of the treatment dosing rate or a variation in the pH of the raw water has no immediate negative effect. A slow variation in the turbidity of the clarified water is observed, but without any massive loss of the sludge in the tank.

By fitting plate modules in a pulsator above the sludge blanket, the quality of the clarified water can be improved, in comparison with results of a conventional tank. The plates or tubes are usually made of plastic, and they are inclined at about 60° to the horizontal. The floc particles that escape from the sludge blanket settle on the lower surfaces of the plate modules and accumulate there. A thin film develops until cohesion causes the film to slide back into the blanket.

The super pulsator combines the advantages of sludge contact settling, pulsation of the sludge blanket, and plate settling. It has many features in common with the pulsator, from which it has been developed, and it gives improved performance. The principle of raw water feed and distribution at the base of the equipment has been retained.

The flocculated sludge rises vertically in parallel streams crossing the deep zones situated between the bottom distribution pipes and the inclined plates, which are therefore uniformly fed with water. The stilling baffles used in the pulsator can be omitted in the super pulsator.

The flocculated water, distributed evenly by the distribution system, then penetrates the system of parallel plates inclined at 60° to the horizontal and perpendicular to the concentrator. The underside of each plate is fitted with deflectors, which act as supports and create slow eddying movements.

The deflector plates in the sludge blanket enable a high concentration to be maintained, twice that in a pulsator operating at the same rate. The high concentration in the sludge blanket, which can reach 50% by volume, enables the super pulsator to play the role of a filter. This is a major advantage for clarifiers with deep and concentrated sludge blankets.

As with the pulsator, the top surface area of the sludge blanket is limited by the provision made for decanting sludge into the concentrator zone, where no force is exerted by the upward movement of water. The recovery of clarified water is achieved by a system of collectors. The flexibility of operation of the super pulsator permits very quick starting.

8.3 PLATE SETTLING IN SLUDGE BLANKET CLARIFIERS

In a sludge blanket clarifier, the use of plate modules improves the quality of clarified water at the same upward velocity by trapping the residual flocs that escape from the blanket. Conversely, for the same quality of clarified water, the volume of water treated by the equipment can be considerably increased.

If inclined plates are introduced into a sludge blanket maintained in suspension, an accelerated settling process is observed. The sludge settles onto the lower plate, and it is then subjected to a downward flow toward the base of the sludge blanket. At the same time, the water released by this process gathers under the upper plate and tends to rise rapidly toward the top of the tank. Very high settling velocities can be achieved, which can reach 20 m/h, depending on the nature of the water and the type of floc.

Numerous types of plate-type static settling tanks, which are equipped with parallel plates or tubes, exist. Some systems combine a mixing unit, an accelerated flocculator fitted with a super pulsator, deflector plates, and a plate-settling zone. The sludge falls by gravity and is concentrated in a hopper at the bottom of the tank.

Chapter 9

Flotation Systems

9.1 FLOTATION USING BLOWN AIR

Flotation can be spontaneous when the specific mass of particles to be eliminated is less than that of water, or it can be induced by the artificial fixation of bubbles of air or gas onto the particles to be removed, giving them an average specific mass less than that of water. Flotation using blown air is essentially spontaneous flotation improved by blowing bubbles of air (several millimeters in diameter) into the liquid mass. For this purpose, porous devices are used to form the bubbles.

If necessary, where heavily concentrated liquids are involved, medium-sized bubble diffusers are arranged to produce local turbulence designed to encourage the division of the air bubbles. The size of the bubbles must be small enough to enable them to adhere to the particles to be floated.

Two separate zones are generally provided in plants designed to remove light materials (e.g., grease, thick oil, large fibers, and paper): one zone for mixing and emulsifying, and the other, a calmer zone, for flotation. In the emulsion zone, the suspended solids are stirred and intermixed with air. The path of the bubbles is increased by the spiral flow thus created. In the separation and collection zone for floating materials, the cross-flow is very slow, and turbulence is reduced.

9.2 FLOTATION USING DISSOLVED AIR

It is standard practice to reserve the term flotation for processes using very fine air or microbubbles between 40 and 70 μm in diameter, similar to those present in whitewater running from a tap on a high-pressure main. Separation by flotation of solid particles in a liquid suspension obeys the same laws as sedimentation but in a reverse field of force.

There is a variation in the upward flow rate of air bubbles according to their diameter. Bubbles of 20 μm in size have an upward flow rate of several millimeters per second, whereas bubbles of several millimeters in diameter have between 10 and 30 times greater velocities. If, for a given volume, an emulsion is introduced at one point with an outlet provided at the other end, the period of immersion of the air bubbles in the water, and similarly the space

Wetlands for Water Pollution Control. http://dx.doi.org/10.1016/B978-0-444-63607-2.00009-5

filled with bubbles, will be greater, the smaller the upflow rate of the bubbles and their corresponding diameter.

For a flotation unit of a given cross-section, the use of bubbles of several millimeters in diameter will result in an air flow much greater than with microbubbles, if a satisfactory distribution of the bubbles is desired over the cross-section. At the same time, this increase in air flow will set up turbulent currents, disturb satisfactory operation, and induce mechanical mixing.

The bubbles have a flotation effect only to the extent to which they adhere to the particles. This generally assumes that their diameters are less than the diameters of the material or floc in the suspension.

A flotation process using means other than microbubbles can only be used in suspensions containing light and bulky materials whose surface deposits are not disturbed by whirling movements. The application of flotation in the field of wastewater treatment is as follows:

- Separation of flocculated matter in the clarification of surface water (in lieu of settling, generally for which the water is cold, has a low mineral content, and is charged with organic matter);
- Separation of flocculated or nonflocculated oil in wastewater from refineries, airports and steelworks;
- Separation of metallic hydroxides or pigments in the treatment of industrial wastewater; and
- Thickening of activated sludge (or mixed activated sludge and primary sludge) from organic wastewater treatment plants before its application to constructed wetlands functioning, for example, as sludge-drying beds.

The techniques vary according to:

- The method used for bubble formation;
- The method of feeding the flotation unit;
- The shape of the structures; and
- The method of collecting skimming.

The most widely used technique for producing microbubbles is pressurization. The bubbles are obtained by the expansion of a solution enriched with dissolved air at several bars of pressure. The pressurized liquid used is either raw water or recirculated treated water. The rate of flow of the pressurized water is generally only a fraction (10–30% of that to be treated) of the rate of flow of the plant, at pressures between 3 and 8 bar. On average, about 60% of the excess air is allowed to dissolve in relation to the rate of saturation at atmospheric pressure. Hence, the compressed air consumption varies between 15 and 60 l/m^3 of water to be treated.

When the quantity of material to be floated is high and thickening is particularly desired, as in the case of activated sludge, the recycled flow can reach 200% of the nominal flow of the flotation unit. With the use of polyelectrolytes, it is possible to obtain sludge concentrations between 3% and 6%

for specific loadings of 5–13 kg of dry solids/m^2/h at a downward velocity of 2 m/h.

Electroflotation is another technique for which the objective is to produce bubbles of hydrogen and oxygen by electrolysis of water by means of appropriate electrodes. The anodes are highly sensitive to corrosion and the cathodes to scaling by carbonate removal. When the safeguard of anodes is required, protected titanium electrodes may be used, or the periodic reversal of the electrodes may be employed to improve self-cleaning. Preliminary treatment of the water or periodic descaling of the cathodes must be allowed for.

Flotation is often combined with preliminary flocculation. By incorporating a flocculation aid, the floc can be enlarged and the particle surface area increased. This gives improved adhesion of the bubbles and an increase in the upward flow rate of the floc.

The separation or downward flow rate of the water used in flotation units varies according to the nature of the suspensions to be treated and also according to the method of generation and distribution of microbubbles. For a given flotation unit, the downward velocity and concentration of the sludge floated are strongly influenced by the value of the ratio of "quantity of air dissolved" over "quantity of material to be floated." The greater this ratio, the greater is the upward force given to the particles, the higher is the downward velocity, the lower is the sludge density, and the greater is its concentration in dry solids.

This ratio cannot exceed the upward flow of the bubbles. However, extremely fine-sized bubbles are particularly conducive to their distribution over the entire surface. However, this can sometimes limit the traverse speed, and therefore limit the output of water treated by the plant.

9.3 FLOTATION UNITS

Flotation units can be circular or rectangular. Rectangular units are usually employed for the treatment of water and wastewater, as they can be constructed as a monobloc together with the flocculator and filters, so that land requirements are minimal.

With a flotation problem to solve, preliminary tests in the laboratory or a test on a semi-industrial scale is recommended. It is then advisable to proceed with the design calculation stage, which is governed by two main variables: the downward velocity and the quantity of solids to be floated per unit area and time.

A low downward velocity, while reducing the particle flow to the bottom of the tank and theoretically leading to higher separation efficiency, also leads to an increase in the retention time of the floating sludge. This may cause deterioration and splitting up of flocs, resulting in the reduction of the sludge concentration. The specialist must find the best compromise between sludge thickening and separation efficiency. Furthermore, the internal layout has considerable influence over the performance of the flotation unit.

For distribution of air bubbles, circular-shaped flotation units are superior to rectangular units. The distance between the inlet column and the outlet funnel is shorter for the same capacity, and an almost uniform distribution of the bubbles can be maintained over the entire horizontal section of the unit.

The water feed always incorporates a column or chamber with a dual function:

- It brings into mutual contact the water to be treated, which may or may not be flocculated, and the pressurized water. The water should be depressurized immediately upstream of the unit (i.e., bubbles should not be wasted); and
- It disperses the kinetic energy of the raw water and the pressurized water mixture, and it reduces the speed before the process water reaches the actual flotation zone.

The chamber also enables any large bubbles that may have formed upstream to be removed immediately. The points and respective levels of introduction of raw and pressurized water are of major importance in the case of prior flocculation.

Precautions to be taken in calming the mixture, when it is introduced into the flotation zone, vary according to the size, stability, and density of the flocculated or nonflocculated substances to be floated. The flotation of slightly viscous oils should be observed carefully.

The emulsified water is generally introduced in the top half of the unit. Sludge is collected at the open surface. The clear liquor outlet is fitted in the lower third of the unit. The deeper the unit and the greater the quantity of sludge that can be deposited, the further away will be the clear liquor pickup from the bottom. Collection is usually carried out in a peripheral chamber. In certain cases, immersed headers and laterals can be used. The more uniform the distribution of water and microbubbles, the shallower (generally between 2 and 4 m) the flotation unit can be designed.

Several devices require the introduction of air upstream of the pressurization pump. Systems incorporating an air cushion saturation tank are more costly, but their operation is very stable. Pressurized air dispersal devices in the pump discharge pipe have also been proposed. For electroflotation, a bank of electrodes covers the surface area of the tank.

The surface sludge layer can, in certain cases, attain a thickness of many centimeters and be extremely stable (e.g., thickening of activated sludge). In other cases, it is thinner and more fragile (e.g., flotation of flocs of metallic hydroxides or of oils). When the sludge removal is not total, the layer thickens and, with time, acquires a degree of cohesion that facilitates the adhesion of floating particles.

The gradual and regular removal of sludge is an important point. The scraper mechanism must be especially strong if it is fitted in a flotation unit designed for sludge thickening. On circular flotation units, one or more scrapers push the sludge into a radial collection channel with a length equal to half the radius. The access ramp must be constructed in such a way that contact with the scraper is always ensured.

The number of scrapers is governed by the quantity of sludge to be removed, the rapidity with which this removal must be carried out to avoid the risk of de-aeration, and the distance that the sludge can be pushed without breaking up. In rectangular flotation units, the sludge is pushed by a series of scrapers driven by endless chains to a removal channel situated at one end.

Chapter 10

Slow Filtration

10.1 INTRODUCTION

The processes within most filtration systems are similar to those in new wetland systems. Moreover, traditional slow sand filters are operated similar to vertical-flow constructed treatment wetlands.

Filtration is the process by which water is separated from suspended and colloidal impurities. The number of bacteria is reduced, and changes in the chemical characteristics of water are brought about by passing it through a porous substance or filter media. In principle, this porous substance may be any material, including sand, crushed stone, anthracite, glass, cinders, a consolidated layer of porous concrete, stoneware, and plastic.

However, sand is used almost exclusively as the filter material because of its easy availability, relatively low cost, and good performance. On occasions, anthracite has a useful role to play in sand filters. Filtration should not be confused with straining; it is not simply a straining process.

10.2 SLOW SAND FILTRATION

10.2.1 Elements of a Slow Sand Filter

A slow sand filter is essentially a bed of sand through which the pretreated water passes (downward), together with the necessary structures to control the flow and to remove the cleaned water after filtration. The filtration rate varies between 0.1 and 0.4 $m^3/m^2/h$. The filter bed consists of fine grains (effective diameter between 0.15 and 0.35 mm) with a total depth of about 1 m initially. As filtration progresses, the suspended and colloidal matter from the raw water is retained in the very top of the filter bed. The clogging material here may be removed and the filter restored to its original capacity by scraping off this top layer of dirty sand to a depth varying from one to a few centimeters.

The traditional slow sand filter is an open basin (2.5 to 4.0 m deep), usually rectangular in shape, built below finished ground level and varying in area from a few hundred to a few thousand square meters, and sometimes even larger. The box is normally constructed in reinforced or prestressed concrete (traditionally, masonry or earth on puddle clay cores, with sloping sides). This box is filled with a 0.6–1.2 m thick layer of sand on top of which the water to

Wetlands for Water Pollution Control. http://dx.doi.org/10.1016/B978-0-444-63607-2.00010-1

be treated is present to a depth of 1.0–1.5 m. At the lower end, this sand bed is supported by a drainage system, a so-called filter bottom or underdrain, which at the same time allows the passage of the filtered water.

A variety of forms is available for the underdrain, the commonest being the porous concrete filter underdrain. In cold climates, slow sand filters must be covered to prevent freezing in winter. In moderate climates, and particularly in hot climates, covering is sometimes practiced to prevent algal growth in the supernatant water and subsequent rapid clogging of the filter bed, and also to avoid bacteriological and other pollution from entering the filter via, for example, wind and birds. The slow sand filter is not complete without a number of influent and effluent lines, provided with valves and controllers to keep the raw water level, and hence the filtration rate, constant.

During operation, the raw water to be treated enters the filter through a valve, and then it flows down the sand bed, through the underdrainage system, and out through a venturimeter (or other measuring devices) and the regulating valve into a weir chamber. After overflowing the weir, the water is discharged through a further valve (all other valves being closed) toward the clear water well.

With the regulating valve, any rate of filtration, from the minimum to the maximum allowable value, may be obtained; while the weir, on one hand, prevents the filtered water level to drop below the top of the sand bed and, on the other hand, makes the filtration rate independent from uncontrollable variations of the water level in the clear well. At the same time, this weir provides an appreciable amount of aeration (i.e., decreasing carbon dioxide and increasing oxygen concentrations).

Good ventilation must be provided to keep the partial oxygen and carbon dioxide pressures in this chamber as near as possible to those of the outside air. This ventilation action is also required for the filter, allowing air and other gases liberated in the underdrainage system to escape without hindering the movement of water.

During the filtration process, gradual clogging of the pores in the upper part of the filter bed occurs, increasing the resistance to flow. This increase is compensated by opening the regulating valve, keeping the overall loss of head the same and hence the filtration rate constant. When, after an extended period of service, the effluent valve is fully opened, a further increase in filter resistance would result in a lowering of the filtration rate, and the filter must be taken out of service for subsequent cleaning.

If the composition of the raw water is such that a large amount of floating material accumulates at the water surface, this scum must be removed regularly through scum outlets at the four corners of the filters, after the water level has been raised. Scum removal is particularly important near the end of the filter run, so as not to render filter cleaning more difficult than is strictly necessary.

Cleaning of a slow sand filter is commonly executed by scraping off the upper 1–2 cm of sand. However, this requires a dry filter bed with the water level about 0.2 m below the top. In order to drain the filter, the raw water inlet valve is closed, allowing the filter to discharge (e.g., overnight) through further valves to the clear well. The shallow layer of water remaining on the filter bed is removed through another valve connected to a box, of which one wall is built up of stop-logs (usually made of concrete), so that the top of this discharge wall can be kept near the top of the sand bed. The pore water in the upper 0.2 m of the filter bed is finally taken out.

After the cleaning operation has ended, the filter is slowly recharged with filtered water from below to a level of about 0.1 m above the top of the sand bed, allowing air accumulated in the pores of this bed to escape. The remaining depth is filled with raw water from above, taking care not to disturb the surface of the sand bed, because it is protected by a shallow layer of water only. When the raw water level in the filter has regained its normal value, the filter operates at about one-quarter of its normal rate.

In the following 12–24 h, the filtration rate is slowly increased to the required value, after which a sample is taken. If this sample satisfies the quality criteria, the filter is brought back to normal operation. When the water quality is not satisfactory, the filtered water is carried to waste until the water quality is acceptable. After normal cleaning, this ripening process lasts 1–2 days; after repairs such as resanding, it may, however, take longer.

10.2.2 Mechanisms in a Slow Sand Filter

The overall removal of impurities associated with the process of filtration is brought about by a combination of several different processes, the most important of which are sedimentation, mechanical straining, adsorption, and chemical and biological activity.

Mechanical straining is the purifying process that is easiest to understand: removing the particles of suspended matter that are too large to pass through the gaps between the sand grains. As such, it takes place almost entirely at the surface of the filter, where the water enters the pores of the sand bed, and is independent of the filtration rate. Even with a grain size of 0.15 mm, the smallest pores are still a little over 20 μm in diameter and are thus unable to retain either colloidal matter (0.001–1 μm) or bacteria ($\leq$15 μm).

By the twisting movement of the water through the pores of the filter, however, velocity gradients are set up, which cause some flocculation. Some aggregation of the finely divided particulate matter will now occur, and part of the flocs thus created are retained at a greater depth in the sand bed. Moreover, a slow sand filter is not immediately taken into service, but a "ripening" period precedes the filter run. This is similar to a constructed treatment wetland, which has to mature in order to become most efficient.

In this period, coarser suspended solids (SS) caught at or near the surface of the bed form a very fine porous layer, the so-called filterskin or *schmutzdecke* (German for "layer of dirt"), which greatly promotes the straining efficiency and thus reduces effluent turbidity. Straining efficiency is also enhanced with time by deposits on the grains in the filter bed, which constrict the pore openings.

Sedimentation removes particulate suspended matter of finer sizes than the pore openings by precipitation onto the sides of the sand grains. For a normal filter bed (media porosity of approximately 38% and grain size of 0.25 mm), this gives a gross area for sedimentation (i.e., sand grain surface area) of approximately 15,000 m^2/m^3 of filtering material.

Even when only a fraction of this area is effective (facing upward, not in contact with other grains, and not exposed to high-scour velocities), the area for deposition can easily exceed 1000 m^2/m^3 filter volume. The surface loading, as a quotient of the amount of water to be treated, and the area of deposition will now be extremely small, with a filtration rate of 0.2 m/h and $\leq 2 \times 10^{-4}$ m/h.

Sedimentation efficiency is a function of the ratio between surface loading and the settling velocity of the particles. It follows that the settling velocity is larger than the overflow rate, and virtually complete removal by sedimentation may be expected. Small and light particles are only partly removed. Although flocculation will slightly increase the removal efficiency as the water progresses through the bed, truly colloidal matter will not be removed by this mechanism.

Adsorption has many facets, the simplest of which is the collision of suspended particles and sand grains after the former have left a streamline (flowline) by centrifugal force. The impurities will adhere to the "sticky" gelatinous coating formed on the grains by previously deposited bacteria and colloidal matter. Much more important is the active promotion of adsorption by the physical attraction between two particles of matter (van der Waals forces), and especially by the electrostatic attraction between opposite electrical charges (Coulomb forces).

Mass attraction is present everywhere and always, but its magnitude decreases with the sixth power of the distance between particles. For the previous example, 1 m^3 of filter sand has a gross area of 1500 m^2. With a porosity of 38%, the sand contains 0.38 m^3 of water, which corresponds to a thickness of a water film around the grains of 25 μm, if spread over the surface area. This is too large to carry colloidal and molecular dissolved impurities to the sides of the grains. Purification in this way is only possible if other actions (e.g., centrifugal force) have brought the particles to the immediate vicinity of the adsorption surface.

Electrostatic forces are inversely proportional to the second power of the distance, and their influence consequently reaches deeper into the body of the passing liquid. By the nature of its crystalline structure, clean quartz sand has a negative charge, and thus attracts positively charged particles in the form of colloidal matter (e.g., flocs of carbonates, iron hydroxide, and aluminum hydroxide) and cations (e.g., iron, manganese, and aluminum). Colloidal matter of organic origin (e.g., bacteria) mostly carries a negative charge. The organic colloids are consequently not attracted, and when a filter with clean sand is first taken into service, such impurities are not removed.

A slow sand filter needs a ripening period, during which positively charged particles are adsorbed. Positive charges may accumulate on some of the filter grains so that oversaturation occurs, by which the charge on these coated particles reverses and becomes positive. This is similar to the ripening process within a constructed wetland becoming more mature.

After the primary adsorption during the breaking-in period, secondary adsorption is able to remove negatively charged particles, colloidal matter of animal or organic origin, as well as truly dissolved impurities (e.g., anions, including NO_3^- and PO_4^{2-}). When secondary adsorption leads to oversaturation, the overall charge again becomes negative, allowing adsorption of positively charged matter (and so on). Moreover, ions on the sand grains may be dragged away by the flowing liquid, by which again a reversal of charge will occur.

By all the process actions mentioned here, impurities are transferred from the water to the surface of the filter grains. After some time, so much material will have accumulated that, in a unit period of time, the rate of deposition will equal the rate removed by scouring. Indeed, when filters are cleaned regularly by scraping (removal of the upper 1−2 cm), the main body of the filter is left untouched. If no persistent fouling occurs, the filter bed will remain in place virtually for decades. This is due to chemical and microbiological oxidation, which destroys the impurities collected on the surface of the grains.

During the break-in period, bacteria from the raw water are adsorbed on the sand grains, where they multiply selectively, using as food the organic matter deposited on the grains. This food is partly oxidized to provide energy, which the bacteria need for metabolism (dissimilation) and which is partly converted into cell material for growth (assimilation), thus converting dead organic matter into living organic matter.

The dissimilation products are carried downward by the water to be used at greater depths by other bacteria. In this way, the degradable organic matter present in the raw water is gradually broken down (mineralization) into harmless inorganic salts, such as water (H_2O), carbon dioxide (CO_2), nitrate (NO_3^-), and phosphate $\left(PO_4^{2-}\right)$ to be discharged in the finished water. With the limited amount of organic matter supplied by the incoming raw water, only a restricted bacterial population can be maintained, and the growth (assimilation) is therefore accompanied by an equivalent proportion of die-away. This again liberates organic matter, which is broken down as described in this chapter. Thus, all degradable organic matter in the raw water is converted into mineral constituents.

The bacterial activity described is most pronounced in the upper part of the filter bed, and it gradually decreases with depth as food becomes scarcer. With each scraping (cleaning), a portion of the bacteria is removed, and again a ripening period is necessary to bring the population up to the required strength. Below a depth of 30–40 cm (deeper if the filtration rate is higher), bacterial activity is very low. However, chemical reactions occur at greater depths, converting microbiological degradation products such as amino acids into ammonia, nitrites, and nitrates (nitrification).

In order to keep the water quality acceptable, a sufficient oxygen content of the filtered water is required, usually >4 mg/l. This is to prevent local anaerobic conditions. Under anaerobic conditions, the products of the breakdown of organic matter are likely to include hydrogen sulfide, ammonia, manganese, iron (reduced), and various taste- and odor-producing compounds.

The other important variable is temperature. If the temperature drops <2 °C for any long period of time, difficulties such as freezing may arise. The filter must therefore be covered to prevent heat loss.

Bacteria move by themselves to the filter grain surfaces, where their food is concentrated. For intestinal bacteria, the water environment is a decidedly unhealthy place. The temperature is too low, and insufficient organic matter of animal origin is available to suit living requirements.

In the upper part of the filter bed, predatory organisms (e.g., protozoa and lower metazoa) are feeding on bacteria, whereas with increasing depth below the filter bed, surface food becomes so scarce that it results in the starvation of bacteria that venture this far. The microbiological life in a slow sand filter produces various antagonistic actions, such as killing or at least weakening intestinal bacteria with chemical (antibiotic) or biological poisons. The overall effect is a marked decrease in, for example, *Escherichia coli*. As pathogens are less likely to survive in adverse environmental conditions, an even larger drop in their numbers is likely.

10.3 ALGAL ACTIONS

A description of the slow sand filtration process would not be complete without mentioning the role of algae that are present in the supernatant water. A major part of the changes in water quality brought about by slow sand filtration is produced by these algae, which are first carried along by the raw water, and then multiply actively during the often long intervals between successive cleanings. Because algae are autotrophic organisms, they need light for their photosynthetic processes, and hence algae are only present in uncovered filters.

The most important feature of algae is their ability to build cell material from simple minerals with the aid of solar energy. The carbon cycle part of this process is described schematically in Eq. (10.3.1):

$$n(CO_2) + n(H_2O) + \text{energy} \rightarrow (CH_2O)_n + n(O_2) \quad \text{(active growth period)} \tag{10.3.1}$$

For their metabolism, algae need energy, which they produce by oxidizing organic matter. The same reaction occurs when algae die; their cell material is liberated and consumed by, for example, bacteria in the filter bed (Eq. (10.3.2)).

$$(CH_2O)_n + n(O_2) \rightarrow n(CO_2) + n(H_2O) + \text{energy} \tag{10.3.2}$$

Three phases that affect the relative magnitude of algal numbers can be identified: growing, constant, and decreasing populations of algae. During the active growth period, oxygen increases and carbon dioxide decreases. Bicarbonates are broken down into carbonates and carbon dioxide (Eq. (10.3.3)).

$$Ca(HCO_3)_2 \rightarrow CaCO_3 + CO_2 + H_2O \tag{10.3.3}$$

The lowering of the bicarbonate content decreases the temporary hardness (alkalinity); the insoluble carbonates subsequently precipitate, promoting filter clogging. As the growth of algae continues, their volume increases and thereafter hinders water movement, and they need to be removed by cleaning.

Steady-state conditions imply that energy is required. Therefore, the phenomenon is restricted to daylight hours (e.g., oxygen increases during the day and decreases at night). Hence, a diurnal variation in oxygen content and pH is likely to occur. This may affect waters with high iron and manganese contents, resulting in aesthetic problems with the finished water.

In temperate climates, temperature drops in fall cause algae to die. This liberates organic matter, and subsequent biochemical degradation consumes oxygen and produces carbon dioxide. High carbon dioxide content lowers the pH, making water more aggressive (Scholz, 2003). So much organic matter is liberated at this time that some organics may clog the filter and necessitate cleaning.

The final beneficial factor of algal growth requiring consideration is the growth of filamentous species. These form a gelatinous mat on the surface of the filter, removing suspended matter and bacteria by straining and adsorption. Especially, the true water bacteria will multiply in this mat, creating a zooglea or *schmutzdecke* (as discussed in this chapter) of bacterial slime, an adhesive medium for plankton and diatoms. Straining and adsorption efficiencies of the *schmutzdecke* are thus enhanced and less suspended matter will reach the filter, thus prolonging filter runs.

10.4 ADVANTAGES AND DISADVANTAGES

The advantages and disadvantages of traditional slow sand filters are similar to those of constructed treatment wetlands. The advantages of slow sand filtration are summarized below:

- No pretreatment, perhaps other than preliminary sedimentation, is needed;
- No chemicals are required;

- Less skilled supervision is required than for rapid sand filtration;
- The filter has a good bacterial removal efficiency;
- The effluent is less corrosive than that from rapid sand filtration; and
- There is a long period of satisfactory operation between cleanings.

The disadvantages are as follows:

- A large area of land is required, and the structures are expensive (capital cost is high);
- The filter has poor color removal efficiency; and
- Turbidity removal is poor, if the raw water turbidity is >40 number of transfer units (NTU).

Conditions that could favor the installation of slow sand filters:

- The raw water is moderately polluted but has a low turbidity;
- The treatment plant location's chemical supplies are unreliable;
- There is a shortage of skilled labor to operate rapid sand filters (Chapter 11); and
- There is an abundant labor force for construction and cleaning.

Outbreaks of water-borne giardiasis (*Giardia lamblia*) have demonstrated the need for an effective technique for removal of *Giardia* cysts (*Giardia* is a protozoa). These cysts can persist in low-temperature and low-turbidity waters for many months, and they are very resistant to the action of disinfectants. Slow sand filtration, after ripening, has been shown to be highly effective at achieving complete removal of *Giardia* cysts (Chapter 18).

Furthermore, *Cryptosporidium parvum* is another organism that causes acute diarrhea illness in humans. It forms oocysts, which are not affected by chlorine but are, however, removed by slow sand filtration (see also Chapter 18).

Chapter 11

Rapid Filtration

11.1 ELEMENTS OF A RAPID SAND FILTER

A rapid sand filter (also called a rapid gravity sand filter) can be located before a constructed treatment wetland. The mechanisms of filtration in a rapid sand filter are essentially the same as those in a slow sand filter, except that the biological processes are minimized. This results from the much shorter filter run time between cleanings, which prevents the establishment of mature biological growth.

When rapid sand filters were initially introduced, the filtration rates (loading rates) were between 3 and 4 $m^3/m^2/h$. At these relatively high rates of filtration, the following observations were made:

- Coagulation was required in most cases to prevent the impurities from being drawn deep into the filter bed;
- Surface cleaning was no longer adequate, because impurities were drawn deep into the sand bed; and
- Clogging occurred much more rapidly, with filters requiring cleaning at intervals between 2 and 3 days instead of intervals between 30 and 100 days for slow sand filters.

For rapid sand filtration to be acceptable, it was necessary to develop a method of cleaning the full depth of the filter rapidly and economically. The method adopted was to remove the impurities from the sand bed by a reverse flow of water, either preceded or accompanied by some form of agitation to loosen the impurities from the sand grains. Moreover, a rapid sand filter comprises the following construction elements:

- A sand bed in which filtration occurs;
- A support for the sand bed;
- An underdrain system to remove filtered water and to admit backwash water (and air for agitation, if used);
- An inlet for water;
- An outlet for used wash water; and
- Means for controlling the flow through the filter.

In normal operation, the inlet valve and filtered water valve are open, and all other valves are closed. Water enters through the inlet valve, passes down

Wetlands for Water Pollution Control. http://dx.doi.org/10.1016/B978-0-444-63607-2.00011-3

through the sand and subsequently the underdrain system, and exits through the filtered water valve.

To start a backwash cycle, using air scour followed by water scour, the inlet valve is closed, but filtering is allowed to continue for a time to avoid excessive loss of settled water. After a reasonable drawdown period during which some of the settled water in the filter chamber is filtered instead of wasted, the filtered water valve is closed and the waste valve opened, thus dumping into the drain that portion of the settled water that was above the level of the backwash collecting trough.

The air valve is then opened to admit air under pressure into the underdrain system, where it is distributed evenly underneath the filter. The air bubbling upward through the sand bed causes agitation, which loosens the impurities from the sand grains. At the end of the air scour, the air valve is closed and the backwash valve opened to admit water to the underdrain system at a rate sufficient to wash the sand. The water passes upward through the sand and carries with it impurities into the backwash collection troughs, and hence into the drain.

In order to finish the backwash cycle, the backwash and waste valves are closed, and the inlet valve is opened. When the water has reached a satisfactory level in the filter basin, the filtered water valve is again opened, and filtering is resumed.

11.2 SAND BED

The sand bed in a conventional rapid sand filter consists of clean silica sand (at a depth between 0.60 and 0.75 m). The effective size of the sand used is between 0.9 and 1.0 mm. This size range is required, because it is necessary to use a coagulant aid, and water is cold in winter. This arrangement produces a tough floc.

Anthracite is often used as a filter media in the United States. The crushed anthracite has a density of about 1400 kg/m^3 (compared with silica sand, which has a density of 2650 kg/m^3), and therefore it does not require high backwash rates to achieve fluidization of the filter bed.

When a sand bed is washed, there is a tendency for stratification to occur, with the larger particles migrating to the bottom of the bed and the smaller particles migrating to the top. In normal operation, the water is therefore filtered first through the fine sand and later through the coarse sand. The length of a filter run depends on the amount of storage voids available in the sand bed for retention of impurities. If the storage space in the fine sand layer becomes clogged before the storage space in the coarse sand layer (located deeper in the filter) is utilized, the length of the run is likely to be shorter than ideal.

Dual-media filters aim to overcome this problem. If, for example, anthracite with a diameter of 2 mm is placed over sand with a diameter of 0.6 mm, the anthracite, being lighter, will form a surface layer after backwashing so that the water will be filtered through the coarse anthracite and subsequently through the finer sand, and ideal conditions are nearly approached.

Mixed-media filters use three different materials. The bottom layer of the filter consists of fine garnet (with a density of 4200 kg/m^3), the middle layer of silica sand (2650 kg/m^3), and the upper layer of coarse anthracite (1400−1700 kg/m^3). The three materials are not so closely graded that they form separate layers, so mixing takes place, giving a gradual gradation of void sizes from large voids near the surface to fine voids near the bottom. The sand support bed requires a layer of coarse garnet sand to prevent the fine garnet sand from penetrating the gravel bed and distributors.

The sand support commonly consists of layers of gravel placed over the filter compartment floor. The purpose of this support is threefold:

- To provide uniform drainage conditions for removing filtered water from the sand bed;
- To prevent sand from entering the underdrainage system; and
- To assist in the even distribution of backwash water.

In earlier filters, the bottom layer of gravel consisted of material with a diameter between 50 and 70 mm, but the trend nowadays is toward smaller gravel in the bottom layers; a maximum of between 15 and 25 mm is often used. The total thickness of gravel is between 300 and 450 mm for many installations. The uppermost layer of the gravel must prevent the intrusion of the sand; it can consist of coarse sand with an effective diameter between 1 and 2 mm. With mixed-media filters, the uppermost layers of the support consist of (high-density) coarse sand. Much attention should be given to the design of the gravel layers to ensure even backwash flow, effective cleaning of the sand, and freedom from gravel displacement.

11.3 UNDERDRAIN SYSTEM

The underdrain system is hydraulically designed to carry the backwash water. The backwash water flow rates through sand beds in different systems range from about 6 to 16 l/s/m^2, whereas filtering rates are only between 1.3 and 4.1 l/s/m^2. The backwash water and the air for scouring the sand are distributed as evenly as practically possible over the full extent of the sand bed.

Where the distribution nozzles are used as controls, the flow distribution will be uneven unless the combined head loss through the nozzles, gravel bed, and fluidized sand to a depth equal to the original depth of the sand bed is a monotonically increasing function of the upflow rate.

The total head loss in the distribution channel should be $\leq 10\%$ of the nozzle head loss so that the backwash flow rates through the nozzles should not vary by $\geq 5\%$ as a result of friction losses in the underdrain system. Too much coarse gravel in a gravel bed will permit free horizontal movement of backwash water in the space between the nozzles and the sand bed, and therefore will partially negate the distribution value of the nozzles.

Where air scour is used in conjunction with a drainage system that uses nozzles, the nozzles are specially designed to distribute the air evenly. These nozzles are fitted with hollow stems, which project down into the laterals or into the plenum. A small air control orifice is drilled in the stem of each nozzle just below the filter floor. The pressure difference, which can be sustained across this orifice, is dictated by the length of the stem below the hole. The greater the allowable pressure difference, the less is the effect of extraneous disturbing influences on the air flow rate. A hood over the top of the nozzle can help to prevent water from entering the top of the stem and interfering with the air distribution. During the air scour cycle, air is introduced into the underdrain system; it escapes through the nozzles into the filter.

If water and air are used simultaneously for filter cleaning, care must be taken to ensure that the necessary backwash water can be distributed to the various parts of the filter in such a way that the height of the waves in the water surface is small in comparison to the head used for forcing the air through the control orifice (otherwise, uneven air scour will result). This condition can be attained in a plenum distribution system. Particular care should be taken to dissipate the energy of the incoming water, which could cause excessive waves and turbulence.

For the low head loss arrangement, backwash water can freely enter the lower gravel layers. The head loss, which controls the flow distribution, occurs in the upper gravel layers. If air scour is required, air is supplied through a separate layer of perforated air pipes set in one of the gravel layers.

The wash water is forced up through the filter bed and emerges laden with impurities removed from the sand. The troughs to collect the dirty water are placed as low as practically possible, in view of the need to avoid excessive disturbance of the hydraulic flow pattern in the expanded sand bed. The troughs discharge the water into the gullet, which is the main collection channel.

If it is expected that some of the displaced impurities may be too heavy or may settle too rapidly to be carried up into wash water collection troughs, the gullet wall may be used as a weir. The depth of water over the sand during the backwash cycle is kept to a minimum so that the horizontal velocity thus induced may carry the impurities over into the gullet. This is known as a cross-wash system. In order to intensify the horizontal velocity, water is sometimes introduced at the surface of the sand on the side opposite the gullet. Backwash procedures are equally as important as the design of the underdrain, sand support, and sand bed in planning a system, in which the sand is thoroughly and evenly washed, without the occurrence of "mud balls" and gravel movement.

A filter backwash cycle is started in response to one or more of the following conditions: excessive head loss, effluent turbidity, or run time. In most installations, the head loss criterion is the most frequent reason for backwash initiation; the usual head loss limit is set at about 2 m. If the floc in the water is too weak, it sometimes happens that it is not retained in the filter bed until the desired maximum head is attained, and turbidity breakthrough occurs.

In conditions where the pretreatment by clarification is very good, or the follow-through rates are low, the floc trapped in the sand bed may become so well gelled with time that thorough washing would be difficult, if the operator was to wait until the usual maximum head loss was reached. In such a case, the time limitation would control the initiation of backwashing.

Common practice has been to use an air scour method in which air is bubbled through the sand bed at a rate between 5 and 15 $l/s/m^2$ for a period of about 300 s (to let friction dislodge the impurities from the sand grains), followed by a low- to medium-rate water backwash rate at 7–10 $l/s/m^2$ until the water becomes clear after about 600 s.

An alternative is to use a period of combined water and air scour during which the air loosens the impurities, while a low-rate water backwash of about 6 $l/s/m^2$ carries them to the surface and prevents their penetration, which sometimes occurs with air scour only. This period is then followed by medium-rate backwash at a rate between 9 and 12 $l/s/m^2$ to clear out the loosened impurities. Satisfactory combined water and air scour can be achieved only if particular attention has been paid to the design of the water and air distribution systems.

In hot climates, where the raw water may be polluted with nutrients, the growth of algae in the sedimentation and filter basins may be a problem. This can be overcome by chlorination of the water before sedimentation. The viscosity of the warmer water found in hot climates is lower than that in temperate climates, and therefore the backwash rate needed for efficient filter cleaning is greater than that with cooler water.

The flow control for rapid sand filters is achieved in three different ways. Most systems include some means of automatic flow control: control valves operated by signals from level-sensing or flow-sensing elements. Care should be taken to avoid control conditions that lead to controller instability, such as "hunting" caused by continual overcorrection.

Flow control systems in water treatment plants are usually operated hydraulically or pneumatically. The damp conditions often occurring in filter control galleries are not conducive to the reliable long-term operation of electrical equipment.

In a treatment plant equipped with downstream flow control, a flow-measuring device (e.g., venturi, orifice, or weir) is used to provide a signal that operates the flow control valve. The control system adjustment may be set by hand so that the rate of flow is kept close to a predetermined rate. The plant inflow control valve is then operated by a device, which senses the level in the channel feeding the filters. The efficiency of turbidity removal in a filter is greatly reduced by sudden changes in flow. The total head loss across the filter and its control valve is equal to the difference between the water level in the filter chamber and the hydraulic grade level downstream of the control valve.

The flow is controlled by motion of the filter control valve, which automatically adjusts to compensate for changes in the resistance to flow in the filter bed, and, because the filter bed clogs slowly, only a relatively slow movement is required. This allows the valve action to be well damped so that "hunting" will not occur, and sudden flow variations may be avoided.

With this control system, there is a sudden change of the flow through the pretreatment portion of the plant when a filter is taken out of service for backwashing, and an even greater change as the filter chamber is refilled after backwashing. These sudden changes of flow can cause a marked deterioration in the settlement efficiency of any clarifier.

The system of upstream flow control with flow splitting avoids shock loading of the pretreatment units. The plant inflow can either be set to a given flow rate or automatically controlled by the demand for water (perhaps from the level in the filter water storage). The flow to each filter is controlled by a flow-splitting device, through which the flow rate is a function of the level in the distribution channel. In this way, the flow from the pretreatment section is split equally among all operating filters.

A filter that is relatively clean can filter water efficiently at a higher flow rate than one in which clogging with impurities is well advanced. A filter that is starting to pass turbid water can continue to produce acceptable water for longer, if the flow rate is reduced.

For some installations, it would appear that the best use could be made of a filter if a high flow rate could be applied while it is relatively clean, with the rates of flow being gradually reduced as the filter becomes clogged. This is what happens if all filters are subjected to a common head loss. The total head loss through each filter, together with its underdrain and flow-limiting orifice, is equal to the difference in head between the influent header and the common effluent weir.

Filters are sometimes backwashed in a fixed sequence. The time for backwashing the next filter in the sequence is judged from the rising level of the water in all the filter basins. The improvement in performance, coupled with the absence of automatic control equipment, makes this filtration control ideal for use in countries with limited technology.

11.4 HYDRAULICS OF FILTRATION

The hydraulics of filtration are important for design and operation considerations. Accurate prediction of head loss and bed expansion during backwashing is important. Otherwise, filter design may prove to be inaccurate, and the sand in the bed may also be lost.

After the first backwash, rapid (gravity) sand filters will stratify (or were already stratified by design in the case of multimedia filters). Each layer in the filter will consist of differently sized particles.

After a filter has been in operation for a period of time, the head loss builds up as a result of filter clogging. When the head loss reaches an unacceptable level, the filter is stopped, and backwashing is commenced.

Unfortunately, backwashing of vertical-flow constructed wetlands is impossible. However, the degradation of old rhizomes results in hollow passages where liquid can rapidly pass through. This counteracts clogging and makes backwashing unnecessary in a lot of cases.

When a bed of uniform particles is subjected to backwashing, the bed just begins to open when the backwash velocity reaches a critical value. At this stage, the effective weight of the (submerged) particles is exactly balanced by the upward drag on the particles resulting from the upflow velocity. As the velocity increases, the bed opens up further (porosity increases). This increase in velocity does not improve the cleaning action, but it is thought to be important in allowing sufficient open space for trapped suspended matter to be washed away.

For a stratified bed, expansion takes place successively for each media type. The surface strata are expanded at a lower rate of backwash than deeper ones. The bed is fully expanded when the upflow (backwashing) velocity of a filter equals the critical velocity for the largest particles.

11.5 COMPARISON WITH SLOW SAND FILTER

Rapid (gravity) sand filters operate at rates some 10 times those of slow sand filters. It follows that impurities are drawn deep into the bed. Hence, cleaning is automated and hydraulic (i.e., not labor intensive).

The rapid sand filter takes up a relatively small land area (in comparison to a slow sand filter or wetland), the water requires pretreatment, and the operator skill level required is high. It follows that rapid sand filters are particularly suitable for urban treatment plants.

Chapter 12

Biological Treatment

12.1 AEROBIC SELF-PURIFICATION

Essentially, aerobic biological treatment processes in wetland systems depend on supplying colonies of microorganisms with optimum quantities of air (i.e., oxygen) and nutrients to achieve, at a maximum benefit-to-cost ratio, the same reactions that occur in natural self-purification processes. In general, three changes occur during self-purification (aerobic processes only):

- Coagulation of colloidal solids passing through the primary sedimentation stage;
- Oxidation of carbon ($C \rightarrow CO_2$, CO_3, and/or HCO_3), hydrogen ($H \rightarrow H_2O$), nitrogen ($N \rightarrow NH_2$), and phosphorus ($P \rightarrow PO_4$); and
- Nitrification (transformation of NH_3 to NO_2 with the help of *Nitrobacter*, and subsequent transformation of NO_2 to NO_3 fostered by *Nitrosomonas*).

The basic requirements of any aerobic system for successful treatment of organic matter are as follows:

- A community of acclimatized microorganisms;
- Adequate substrate (food) that is waste-free from toxins; and
- A suitable environment (e.g., temperature, light, dissolved oxygen, and pH).

The basic biological treatment processes are as follows:

- Waste stabilization ponds and constructed wetland systems (a mixture of catabolic and anabolic processes);
- Trickling (or percolating) filter systems (primarily catabolic); and
- Activated sludge systems (primarily anabolic).

12.2 WASTE STABILIZATION PONDS

12.2.1 Aerobic Ponds

In their simplest form, an aerobic stabilization pond is a large and shallow excavation in the ground. The treatment of the waste is by natural processes involving the use of both bacteria and algae. These ponds can be aerobic (Scholz, 2003), anaerobic, or facultative (both aerobic and anaerobic). The

Wetlands for Water Pollution Control. http://dx.doi.org/10.1016/B978-0-444-63607-2.00012-5

oxygen required in aerobic and facultative ponds is supplied by algae (through respiration during sunshine) and by diffusion from the atmosphere to the bulk liquid. Mixing in the ponds is usually achieved by wind, although some ponds are mechanically mixed.

In aerobic ponds, oxygen is supplied by natural surface re-aeration and by algal photosynthesis. Except for the algal population, the biological community (predominantly bacteria, protozoa, and metazoa) in stabilization ponds is similar to that in an activated sludge system. The oxygen released by the algae through the process of photosynthesis is used by bacteria in the aerobic degradation of the organic matter. The nutrients and carbon dioxide released in this degradation process are (in turn) used by the algae.

Higher animals such as rotifers and protozoa are also present in the pond. Their main function is to predate on bacteria and, to a lesser extent, on algae. Therefore, higher animals control the suspended solids (SS) concentration in the effluent. This process is also called polishing.

The presence of particular bacteria, algae, and animals depends on variables such as the organic loading rate (kg BOD m^3/day), degree of pond mixing, pH, nutrients, sunlight, and temperature. Temperature has a profound effect on the operation of aerobic ponds, particularly in regions with cold winters.

12.2.2 Facultative Ponds

Ponds in which the stabilization of wastes is brought about by a combination of aerobic, anaerobic, and facultative bacteria are known as facultative stabilization ponds. Three zones exist in such ponds, and they can be described as follows:

- A surface zone where aerobic bacteria and algae exist in a symbiotic relationship;
- An intermediate zone that is partly aerobic and partly anaerobic (usually on a time basis) in which the decomposition of organic matter is carried out by facultative bacteria; and
- An anaerobic bottom zone in which accumulated solids are decomposed by anaerobic bacteria.

Conventional facultative ponds are excavations filled with screened and, in some cases, comminuted raw wastewater or primary effluent. They are approximately 1 m deep. Large and heavy solids settle to form an anaerobic sludge layer on the bottom of the pond. Soluble and colloidal organic materials are oxidized by aerobic and facultative bacteria using the oxygen produced by the algae growing abundantly near the surface (within the photic depth). The carbon dioxide produced in the organic oxidation process serves as the carbon source for the algae. Anaerobic breakdown of the solids in the sludge layer results in the production of dissolved organics and gases such as carbon dioxide (CO_2), hydrogen sulfide (H_2S), and methane (CH_4) which are either oxidized by the aerobic bacteria or vented to the atmosphere.

In practice, oxygen is maintained in the upper layer of facultative ponds by the presence of algae and by surface re-aeration. The biological community in the upper (aerobic) layer is similar to that of an aerobic pond. The microorganisms in the middle and bottom layers are facultative or strict anaerobes (usually, these are only found in the bottom layer).

Respiration (by bacteria) also occurs in the presence of sunlight. However, the net reaction in the pond is the production of oxygen. Because algae use carbon dioxide in photosynthetic activity, high-pH conditions ($pH > 10$) can occur, particularly in wastewater with low alkalinity (buffering capacity). The simplified reactions for both photosynthesis and respiration are presented in Eqs. (12.2.2.1) and (12.2.2.2):

$$\text{Photosynthesis: } CO_2 + 2H_2O \xrightarrow{\text{in presence of light}} (CH_2O) + O_2 + H_2O \tag{12.2.2.1}$$

$$\text{Respiration: } CH_2O + O_2 \rightarrow CO_2 + H_2O \tag{12.2.2.2}$$

Facultative pond systems are natural systems, and as such they require a minimum energy input in the form of electricity; they also require large land areas and are difficult to control, particularly in temperate climates. These factors have led to the development of engineered systems, such as constructed wetlands, percolating filters, and activated sludge processes.

Chapter 13

Biological Filtration

13.1 INTRODUCTION

The discussion in this chapter is limited to aerobic processes. However, it is important to remember that anaerobic biological filtration processes are used, particularly to treat strong industrial wastes, where the bulk of the biological oxygen demand (BOD) is present as soluble BOD.

Aerobic biological filtration processes have been around since the 1900s, although little was understood about their mode of operation until the early 1920s. Scientific and rational design methods for these units are still not in common use.

13.2 TRICKLING FILTER

The trickling (or percolating) filter, also called a bacteria bed, consists of a bed of suitable coarse porous media on which a biological film or slime grows. Hence, the generic name for such types of processes is a fixed-film system. The film consists mainly of bacteria upon which higher organisms graze. This grazing process helps to keep the film active.

Settled sewage is periodically distributed over the surface of the medium, and, as it flows down through the bed, the fine suspended and dissolved organic matter is adsorbed by bacteria. Oxygen to sustain aerobic biological oxidation is provided by air, which circulates through the bed via vents at the base of the filter.

Clogging of the interstices within the filter, sometimes termed ponding, is caused by excessive, unchecked, and uncontrolled biological growth. This is normally prevented by the film or biomass being sheared off by the flow of wastewater. This material, which constitutes the net increase in biomass in the system and which would otherwise contribute to high BOD and SS concentrations in the effluent, is removed in secondary sedimentation tanks, called humus tanks.

It is difficult to obtain an adequate measure of the active mass of biological solids in a trickling filter. However, the total surface area of the medium gives some indication of the possible area available for growth and thus biomass. Indeed, media manufacturers often quote the available surface area on the

Wetlands for Water Pollution Control. http://dx.doi.org/10.1016/B978-0-444-63607-2.00013-7

media. However, the actual thickness of the biofilm and the percentage of its active bacteria proportion cannot be practically determined without disturbing or even destroying at least some parts of the system.

Knowledge of the available surface area is often of little use to design engineers. Therefore, it is normal practice to use the volume of media as the measure of microorganism activity in a biological filter; thus, loading rates to biological filters are expressed in organic (kg BOD/m^3 of media/day) and/or volumetric (m^3/m^2 of surface area/day) units. The volumetric loading rate is important since it affects the distribution of the flow over the surface of the medium and hence the quality of the contact between the applied organic matter and the active biomass. Trickling filters can be compared to some types of constructed wetlands. However, they are not planted with macrophytes, but contain more voids to foster air entrainment.

13.3 BASIC ECOLOGY

Trickling filters are well known for their ecological diversity of life forms participating in the wastewater's stabilization. These include prokaryotic and eukaryotic organisms as well as higher life forms such as rotifers, nematodes, annelid worms, snails, and many insect larvae.

The bacteria are active in the uptake and degradation of soluble organic matter. Nitrifying bacteria convert ammonia to nitrate. In a low-rate trickling filter, there is a high nitrifier population and the effluent is well nitrified. In a high-rate filter, there is more "sloughing" of the biomass due to higher fluid shear, and so little or no nitrification takes place.

Fungi, which are also to be found in trickling filters, are active in the biofilm and are actively involved in waste stabilization. These organisms tend to dominate at lower pH values, which are usually associated with industrial waste treatment. Algae are once again active in the biofilm and produce oxygen during daylight hours. This helps to keep the uppermost portion of the biofilm aerobic. Both the fungi and algae are important components of the biofilm in trickling filters.

Protozoa are unicellular prokaryotic organisms that feed on the bacteria within the biofilm. The continuous removal of bacteria by protozoa helps to maintain an active bacterial population and thus maintains a high decomposition rate.

Rotifers are also present in the biofilm. Once again, these organisms predate on bacteria, algae, and fungi. Most rotifers are taken to be indicative of a high degree of treatment efficiency and, when present, serve to reduce the effluent SS.

The major macroinvertebrates present in trickling filters are insect larvae (e.g., chironomids). These feed on the biofilm and help to control its thickness, thus avoiding clogging of the pores in the filter by microbial extrapolymeric substances. The larvae develop into adult insects (filter flies) in 2–3 weeks.

These can be a nuisance to plant operators and local residents, particularly in summer. Numbers as high as $3 \times 10^5/m^2$ have been reported. Insects are controlled by increasing the frequency of wetting of the filter surface (larvae only emerge on dry filters), by the use of insecticides, or by biological control using *Bacillus thuringiensis* var. *israelensis*. This pathogen contains a toxin that, when ingested by the insect, causes its death.

Other important controls on the ecology of the trickling filter include cold temperature and direct application of toxins. These slow down or stop predator activity and thus increase the likelihood of clogging. This in turn may adversely affect the performance of the humus tank, particularly in spring when predator activity usually resumes after winter. Excess biofilm from the spring sloughing is likely to overload the humus tank and causes high SS to be discharged in the effluent. Nitrification is also reduced in winter as a result of the low average temperature.

13.4 PROCESS VARIANTS

Various factors affect the efficiency of the aerobic biological filtration process, including the organic loading rate; the volumetric or hydraulic loading rate; the depth of media; the type, size, and shape of media; and the total surface area of media available in the filter. A number of process variants have been developed over the years to deal with specific waste types and specific circumstances.

Standard rate filters have low hydraulic and organic loading rates, usually without recirculation (e.g., the effluent is applied to the filter, passes through, goes to the humus tank, and is then discharged). This type of filter will give an effluent with BOD < 20 mg/l and SS < 30 mg/l. Moreover, the effluent is well nitrified from a domestic sewage feedstock. Single-pass units (i.e., through the filter once) can be fed by gravity without any source of external power. Typical loading rates are between 0.07 and 0.10 kg BOD/m^3/day, and volumetric loading rates are between 0.25 and 1.2 m^3/m^2/day.

High-rate filters have loading rates that are significantly higher than those of the low-rate filters. Recirculation is usually included, so pumping is normally necessary. Recirculation is a process whereby a proportion of the effluent is recycled to the influent of the filter and mixed with the incoming effluent, thus increasing the volume to be treated but decreasing its BOD.

Traditionally, rocks were used as filter media (aggregates). Nowadays, it is much more likely that the media consist of a plastic packing material. Organic rates are variable, and volumetric rates are up to 3.5 m^3/m^2/day.

High-rate filters are sometimes operated as part of a two-stage biological treatment process for strong effluents. In these cases, the organic loading rate can be up to 2 kg BOD/m^3/day with volumetric loadings between 10 and 20 m^3/m^2/day and with BOD removals between 60% and 65%. This type of filter is usually referred to as a roughing filter.

Alternating double filtration uses two filters in series. The effluent from the first filter passes to a humus tank from which the settled effluent is dosed onto a second filter and subsequently passes to a second humus tank. The first filter is subjected to a high organic loading rate, and so quickly accumulates a heavy biomass growth, while the second filter is lightly loaded. The filter sequence is altered on a weekly or monthly basis, depending on the degree of ponding present on the first filter.

When the flow pattern is changed, the previously lightly loaded filter begins to develop a greater biomass, while the previously heavily loaded filter sheds biomass because the loading rate is now insufficient to sustain a similar growth rate. Effluents with BOD < 20 mg/l and SS < 30 mg/l are usually produced, although extensive pumping is required, and the design of the humus tanks must account for the increased solids loading.

13.5 DESIGN OF BIOLOGICAL FILTERS

A biological filter is suitable as a complete secondary treatment unit or as a roughing process. Although a biological filter has a lower power requirement and thus a lower operational cost than the activated sludge process, it has a relatively high capital cost (particularly the land requirement). This had made it unpopular for large treatment works. Moreover, biological filters can create a fly nuisance during summer. However, they do produce a well-nitrified effluent. The main aspects to be considered in designing these units are as follows:

- Pretreatment requirements;
- Dosing system for applying effluent to the filter;
- Filter bed, volume, and type;
- Dimensions of filter media, including plastic material;
- Underdrain system;
- Ventilation system; and
- Humus tanks.

Pretreatment is required to remove all material that is likely to clog the nozzles on the distribution arms. Preliminary and primary treatment are both used to achieve this. Moreover, they reduce the organic loading rate to the units.

Dosing equipment distributes the settled sewage regularly and evenly over the surface of the filter bed. A circular filter has a multiple-arm rotary distributor supported on a central bearing, which incorporates the inlet. Rotation of the distributor is induced by the jet reaction of the flow from the nozzles along one side of the distributor arm. Occasionally, these arms require a brake to be fitted to slow down the speed of rotation. With a gravity-fed system, a dosing siphon and chamber are provided to hold back the flow until a sufficient volume is stored to ensure effective operation.

Filter beds are usually circular, although rectangular beds have been used in an effort to reduce land requirements. The media used in the filter must provide a suitable environment for the growth of the biofilm and additionally be inert to chemical attack.

The filter media for standard-rate filters are usually gravel, crushed rock, blast furnace slag, or other granular material. It is usual to limit the minimum size of particles to between 40 and 50 mm, with typical sizes between 50 and 100 mm. The filter depth is of the order of 2 m. However, the actual volume of media required can be calculated.

The underdrain collects the effluent and usually consists of an impervious base beneath the perforated filter floor, graded toward one or more collecting channels. The capacity of the underdrain should be large enough to accommodate the maximum flow rate while maintaining an adequate air space.

This is normally achieved by natural air circulation, induced by the heating or cooling effect of the applied wastewater. Sufficient openings should be provided at the base of the filter to allow air to flow into or out of the space under the suspended filter floor. In temperate climates, adequate ventilation can be assumed by providing plentiful air inlets around the base of the filter. Care must be taken in cold climates to minimize heat loss, and thus prevent low-temperature inhibition of biological activity. In hot and dry climates, plant operators must prevent the air from drying out the biofilm, if the flow rate is low.

Furthermore, sloughing of biomass tends to be a seasonal phenomenon, occurring predominantly in spring in, for example, the United Kingdom. The corresponding humus sludge is difficult to dewater. However, constructed wetlands can be used for sludge dewatering in warm countries.

Chapter 14

Rotating Biological Contactors

14.1 INTRODUCTION

Rotating biological contactors (RBC) were first installed in West Germany in the 1960s and later were introduced to other countries such as the United Kingdom (UK) and the United States of America (USA). Approximately 70% of RBC systems are removing carbonaceous BOD only and 25% are used for combined carbonaceous BOD removal and nitrification, with the remainder being used for nitrification of secondary effluents.

RBC consist of a series of closely spaced circular disks of polystyrene or polyvinyl chloride (PVC). The disks are submerged in wastewater and rotated slowly through it. Constructed wetlands can be used to polish the effluent of RBC.

14.2 PRINCIPLES OF OPERATION

In operation, biomass (slime or film) grows on the surface of the disks and eventually covers the entire wetted surface. The rotation of the disks alternately contacts the biomass with organic matter in the wastewater and then with the atmosphere for the adsorption of oxygen.

The disk rotation affects oxygen transfer and maintains the biomass in an aerobic condition. The rotation is also the mechanism for removing excess solids from the disks by the shearing forces it creates as the disk passes through the wastewater. The sloughed or sheared biomass is then maintained in suspension so that it can be removed from the unit to the sedimentation tanks. Some RBC are designed to allow the sloughed solids to settle to the bottom of the RBC tank and break down, with the sludge being periodically removed.

14.3 DESIGN AND LOADING CRITERIA

Although a few theoretical loading performance models have been developed, the usefulness of these models is still questioned by design engineers. As a consequence, the design of RBC is largely based on the use of simple design parameters. These design parameters are derived from the experience of the operation of full-scale plants.

Wetlands for Water Pollution Control. http://dx.doi.org/10.1016/B978-0-444-63607-2.00014-9

When RBC systems were originally introduced, the process design was based on hydraulic loading expressed in m^3/m^2/day to achieve the required removal efficiency. The design approach has shifted, first to the use of total BOD per unit of surface area (kg total BOD/m^2/day) and subsequently to soluble BOD per unit of surface area per day (soluble BOD/m^2/day) or, in the case of design for nitrification, to kg NH_3/m^2/day.

Poor performance has been observed where systems are overloaded, resulting in low DO concentrations in the units, H_2S odors, and poor removal rates (particularly in the first stage). Typical design loading rates are shown in Table 14.3.1.

The design parameters are for mixtures of domestic wastewater with minor amounts of industrial (nontoxic) wastewater. Selection of a loading value within the noted range is made based on effluent requirements, temperature range, degree of uncertainty as to waste load, and expected competency of the operating staff. The total media area is normally sized based on the mean annual design year conditions, unless data on significant loading variations throughout the year are available. After the total surface area required has been calculated, the design must be checked to avoid exceeding the oxygen transfer capacity of the first-stage unit.

TABLE 14.3.1 Traditional Design Data for Rotating Biological Contactors at a Wastewater Temperature of 13 °C

Item	Secondary	Combined	Separate Nitrification
Hydraulic loading (m^3/m^2/day)	0.08–0.16	0.003–0.08	0.04–0.1
Organic loading (kg soluble BOD_3/m^2/day)	0.0037–0.01	0.0025–0.0075	0.005–0.015
Organic loading (kg total BOD_3/m^2/day)	0.01–0.0175	0.0075–0.015	–
Maximum load on the first stage (kg soluble BOD_5/m^2/day)	0.02–0.03	0.02–0.03	–
Maximum load on the first stage (kg total BOD_5/m^2/day)	0.04–0.06	0.04–0.06	–
NH_3 loading (kg NH_3/m^2/day)	–	0.00075–0.0015	0.001–0.004
Hydraulic retention time θ (day)	0.7–1.5	1.5–4.0	1.2–2.9
Effluent BOD_5 (mg/l)	15–30	7–15	7–15
Effluent NH_3 (mg/l)	–	<2	1–2

Note: BOD_5 = 5-day at 20 °C N-allylthiourea biochemical oxygen demand; NH_3 = Nitrate–nitrogen.

14.4 PRINCIPLE ELEMENTS

The principle elements of an RBC are as follows: shaft, media, drive system, tankage, enclosure, and settling tank. A shaft is used to support and rotate the disks. In practice, the shaft should not exceed 8.2 m, and it should have a maximum of 7.6 m occupied by the media. The minimum length occupied by the media is usually about 1.5 m. The major problem with shafts in RBC is shear failure due to the overload from excessive media growths.

Many different media have been used in RBC. The first plants used timber disks, although these have fallen out of fashion. The disk media are commonly manufactured from high-density polyethylene, PVC, or polystyrene, depending upon the manufacturer and the specification by the client.

Most RBC are driven by a direct mechanical drive. Air drive units are also available. They comprise a deep plastic cup attached to the perimeter of the media disk and are driven by compressed air. Variable speed control is essential on all drive units.

The capacity of the tankage (i.e., space occupied by wastewater in the RBC) has been optimized at 0.0049 m^3/m^2 of media. This results in a hydraulic retention time of 1.44 h at a hydraulic loading rate of 0.08 m^3/m^2/day. A typical side water depth of 1.5 m is usually sufficient to give a 40% disk submergence.

Usually, RBC are covered with a fiberglass or reinforced plastic shell over each separate drive shaft assembly. This gives protection to the biofilm and plant in adverse weather conditions.

Settling tanks are not essential to the proper operation of a RBC. If they are provided, they should be designed as humus tanks.

14.5 OPERATIONAL PROBLEMS

Many of the early RBC units had operational problems. These were mainly shaft failures, media breakage, bearing failures, and odor problems. Shaft failures have been the most serious equipment problem because of the loss of a process unit from service and the damage to a portion of the media. The causes of shaft breakage may be attributed to inadequate structural design, metal fatigue, and excessive biomass accumulation on the media.

Media breakage has been caused by exposure to heat, organic solvents, or ultraviolet radiation or by inadequate design of the media support systems. Bearing failures have been attributed to inadequate lubrication and poor maintenance. Odor problems are frequently caused by excessive organic loading rates, particularly in the first stage.

Modifications to the equipment have been made to mitigate against many of these problems and to simplify maintenance. When designing or specifying a RBC unit, it is essential to consult the manufacturer's literature.

Chapter 15

Activated Sludge Processes

15.1 BACKGROUND

The activated sludge (AS) process is currently the most widely used biological wastewater treatment process in the developed world. Since its conception in the late nineteenth century and subsequent development into a full-scale process in 1913 by Arden and Lockett at the Davyhulme sewage treatment works in Manchester, United Kingdom, the basic process has been widely adopted and further developed, giving it a unique flexibility of operation.

Constructed treatment wetlands have the potential to replace costly AS processes, particularly in warm developing countries. Moreover, wetland systems are often used to polish the effluent from AS systems throughout the world.

15.2 ACTIVATED SLUDGE PROCESS

The AS process consists of two separate phases: aeration and sludge settlement. This unit is usually operated with no settlement allowed in the aeration tank, and a completely separate settlement tank with continuous sludge removal and return to the aeration tank is therefore operated after the aeration tank. In the first phase, wastewater (from the primary settlement tanks) is added to the aeration tank containing a mixed microbial population. Air or sometimes pure oxygen is added either by surface agitation or via diffusers using compressed air. The aeration has a dual function:

1. To supply oxygen, for respiration, to the aerobic microorganisms in the reactor; and
2. To maintain the microbial flocs in a continuous state of agitated suspension, ensuring maximum contact between the surface of the flocs and the wastewater.

The continuous mixing action is important not only to ensure that an adequate food supply gets to the microorganisms but also to support a maximum oxygen concentration gradient to enhance mass transfer and to help disperse the metabolic waste products from within the flocs.

Wetlands for Water Pollution Control. http://dx.doi.org/10.1016/B978-0-444-63607-2.00015-0

As the settled wastewater enters the aeration tank, it displaces the mixed liquor (i.e., the mixture of wastewater and microbial mass) into a sedimentation tank. This is the second stage, where the flocculated biomass settles rapidly out of suspension to form sludge, and where the clarified effluent, which is virtually free from solids, is discharged as the final effluent.

In the conventional AS process, between 0.5 and 0.8 kg dry weight of sludge is produced for every kilogram of BOD_5 removed. The sludge is rather like weak slurry containing between 0.5% and 2.0% dry solids, and so it can be easily pumped. As the solids content increases, the viscosity rapidly becomes greater, although under normal operating conditions AS is difficult to concentrate to >4% dry solids by gravity alone.

Most of the AS is returned to the aeration tank to act as an inoculum of microorganisms, ensuring that there is an adequate microbial population to fully oxidase the wastewater during its retention period within the aeration tank. The excess sludge (8−10% of that produced daily) requires treatment before disposal.

The most important function in the AS process is the flocculent nature of the microbial biomass. Not only do the flocs have to be efficient in the adsorption and subsequent absorption of the organic matter in the wastewater, but also they have to be rapidly and effectively separated from the treated effluent within the sedimentation tank. Any change in the operation of the reactor will lead to changes in the nature of the flocs, which can adversely affect the overall process in a number of ways; most notably, poor settlement (bulking) can result in turbid effluents and the loss of microbial biomass.

Although some variants of the AS process are used to treat sewage that has only been screened and degritted, the majority of AS processes use settled sewage as the feedstock. The sludge produced from the process should not be confused with primary sludge (which contains coarse organic and inorganic solids), as it is composed entirely of microbial biomass and adsorbed particulate matter.

The main components of all AS systems are as follows:

- Reactor: This can be a tank, lagoon, or ditch. The main design criteria for a reactor are that its content can be adequately mixed and aerated. The reactor is also known as the aeration tank or basin.
- Activated sludge (AS): This is the microbial biomass within the reactor, which comprises bacteria and other microfauna and microflora. The sludge is a flocculant suspension of these organisms and is often referred to as the mixed liquor. The normal concentration is estimated as the mixed liquor suspended solids (MLSS; usually between 2000 and 5000 mg/l).
- Aeration and mixing system: Aeration and mixing of the AS and the incoming wastewater are essential. While these processes can be undertaken independently in separate tanks, they are normally combined with each other using a single system. Either surface aeration or diffused air is used.

- Sedimentation tank: Final settlement (or clarification) of the AS displaced from the aeration tank by incoming wastewater is required. This separates the microbial biomass from the treated effluent.
- Returned sludge: Some of the settled AS in the sedimentation tank is recycled back to the reactor to maintain the microbial population at a required concentration to ensure continuation of treatment.

Ideally, the AS process should be operated as close to a food-limited condition as possible to encourage endogenous respiration. This is where each microorganism is utilizing its own cellular content, thus reducing the quantity of biomass produced. During the endogenous respiration phase, the respiration rate will fall to a minimum value, which is sufficient for cell maintenance only. However, under normal operating conditions, the growth of the microbial population and the accumulation of nonbiodegradable solids result in an increase of the amount of AS produced.

The two principle removal mechanisms in the AS process are as follows:

- Assimilation: Utilization of waste to create new biomass. Colloidal and soluble BOD is transformed into biomass, which is then settled out.
- Mineralization: Oxidation (degradation) of waste to inert end products, which are either vented to the atmosphere or are left in solution in the effluent.

Activated sludge plants can be operated to favor either of these two processes. In plants dominated by assimilation, there is rapid removal of the BOD with a correspondingly high production of sludge, which means that sludge treatment costs are high. Plants that operate in a way that favors mineralization require long aeration times. Thus, the operating costs increase due to increased air or oxygen requirements. This is offset by the reduced production of sludge and correspondingly low sludge treatment costs.

Whichever method is favored, the wastewater itself must contain adequate nutrients for biological growth. This nutrient requirement is usually expressed in terms of the carbon, nitrogen, and phosphorus ratio (Eq. (15.2.1)). Note that the 5-day BOD is used as a measure for the carbon content.

$$\text{5-day BOD}:N:P = 100:5:1 \qquad (15.2.1)$$

15.3 COMPARISON BETWEEN THE ACTIVATED SLUDGE PROCESS, PERCOLATING FILTRATION, AND WETLAND SYSTEM

The relative merits and demerits of the AS process, percolating filtration (Chapter 13), and wetland system (Chapter 20) parameters are summarized in Table 15.3.1.

TABLE 15.3.1 Comparison of the Activated Sludge (AS) Process, Percolating Filtration, and Wetland System Parameters

	AS Process	Percolating Filtration	Wetland System
Capital cost	Low	High	Very low
Area of land	Low; advantageous where land availability is restricted or expensive	Large; 10 times more area required	Large; at least 10 times more area required
Operating cost	High	Low	Very low
Influence of weather	Works well in wet weather; slightly worse in dry weather; less affected by low winter temperatures	Works well in summer but possible ponding in winter	Independent of weather, if the wetland is sufficiently large
Technical control	High; the microbial activity can be closely controlled; requires skilled and continuous operation	Little possible; does not require continuous or skilled operation	Little possible; dces not require continuous or skilled operation
Nature of wastewater	Sensitive to toxic shocks and changes in loading; trade wastewaters can lead to bulking problems	Strong wastewaters acceptable; able to withstand changes in loading and toxic discharges	Strong wastewaters acceptable; able to withstand changes in loading and toxic discharges

Hydrostatic head	Small; low pumping requirement; suitable for a site where the available hydraulic head is limited	Large; site must provide natural hydraulic head, otherwise pumping is required	Small; suitable for a site where the available hydraulic head is limited
Nuisance	Low odor and no fly problems; noise may be a problem in both urban and rural areas	Moderate odor; severe fly problem in summer; quiet	No unacceptable nuisance due to an ecological equilibrium
Final effluent quality	Poor nitrification but low in suspended solids, except when separation problems occur	Highly nitrified; relatively high load of suspended solids	High-quality effluent unless the system is overloaded
Secondary sludge	Large volume; high water content; difficult to dewater; less stabilized	Small volume; less water; highly stabilized	Virtually no sludge production (except for detritus)
Energy requirement	High; required for aeration, mixing and maintaining sludge flocs in suspension, and recycling sludge	Low; natural ventilation; gravitational flow	Virtually no energy required
Synthetic detergents	Possible foaming; especially with diffusers	Little or no foam	No problem
Robustness	Not very robust; high degree of maintenance on motors; not possible to operate without power supply	Very sturdy; low maintenance; possible to operate without power	Very sturdy; low degree of maintenance; possible to operate without power

15.4 ACTIVATED SLUDGE PROCESS TYPES

15.4.1 Conventional Complete Mix Activated Sludge Process

Since its introduction, the AS process has undergone many variations and adaptations. In some cases, these adaptations have resulted from basic research into the principles of the process, but more often they have arisen as empirical solutions to particular problems in plant operation.

Concerning the conventional AS process, also called complete mix, the effluent from primary sedimentation tanks is aerated in the aeration tank. This aeration usually lasts 6–12 h at 3 DWF. The MLSS are kept fairly high in order to increase the efficiency of the biological reactions. Typical MLSS are between 3.0 and 3.5×10^{-3} mg/l. The organic loading rate for these units is between 3.0 and 3.5 kg BOD_5/m^3/day. The air supplied to the aeration unit is in the order of 6 m^3/m^3 sewage. This type of plant does not produce a nitrified effluent.

The effluent quality of the treated sewage after 2 h of sedimentation is as follows: BOD_5 between 10 and 15 mg/l, and SS of 20 mg/l. The aeration in the aeration tank is achieved at a reasonably uniform rate over the length of the tank. This leads to a possible oxygen shortage at the inlet. Reactions may therefore be oxygen limited. Hence, the process efficiency may be reduced. The conventional system provides good buffering against shock loads and toxic conditions and is thus particularly suitable for the treatment of industrial effluents.

15.4.2 Series or Plug Flow System

In this arrangement, the waste to be treated and the return AS are introduced at one end of a channel, in which a number of aerators are located, and treated mixed liquor is withdrawn at the other end. With mechanical aerators, there may be as many as 10 aerators in a line, but it is usual to have more lines with three or four aerators per line. Compared to a single line, this provides a higher initial degree of aeration to the untreated waste. Domestic sewage is commonly treated by this method as long as there are no inhibitory substances present. One advantage of this arrangement is that a progressively reducing substrate level produces sludge with improved settling characteristics.

15.4.3 Tapered Aeration

This is a modification of the plug flow system. The process seeks to adjust the rate of oxygen supply throughout the length of the tank to the rate of oxygen demand. Hence, a higher proportion of the total air supply is introduced at the inlet end, and the rate of supply is tapered (reduced) toward the tank outlet. The rate at the outlet is normally set to a value approaching that of endogenous respiration. This system is particularly suited to the treatment of strong, readily biodegradable wastes.

15.4.4 Step Feed Activated Sludge Process

This is another variation of the plug flow process, which attempts to equalize the oxygen supply and demand. In this case, instead of varying the rate of oxygen supply along the aeration tank, partial equalization of the demand is achieved by adding the waste to the aeration tank at various points along the length of the tank. There is no apparent advantage for this system concerning the treatment of domestic sewage.

15.4.5 High-Rate Activated Sludge Process

This is sometimes called the modified aeration process. This type of plant operates at very much higher food-to-microorganism ratios than the conventional process:

- Conventional: 0.2–0.4 kg BOD_5/kg mixed liquor volatile suspended solids (MLVSS)/day; and
- High rate: 1.5–5.0 kg BOD_5 kg/MLVSS/day.

This high loading rate is achieved by a reduced hydraulic retention time (about 2 h) and a lower MLSS (about 1000 mg/l). The air supply is 3 m^3/m^3, and BOD_5 reductions between 60% and 70% are possible. The process results in more of the incoming organic matter being synthesized to sludge organisms. Therefore, the total oxygen requirements are somewhat less than in conventional AS processes, but the rate of oxygen demand is higher per unit of MLVSS.

15.4.6 Extended Aeration

This variant is characterized by a very low food-to-microorganism ratio, low net sludge yield, higher hydraulic retention times, and higher MLSS. This makes the process suitable for small isolated communities.

The disadvantages of aeration systems for small communities are as follows:

- The effect of high hydraulic loading, causing loss of AS solids;
- Widely fluctuating BOD loads; and
- The uneconomic demands of conventional aeration systems for regular attention (e.g., surplus AS management).

In the extended aeration plant, primary sedimentation can be omitted so that only sludge from the wastage line is to be disposed of. It follows that this process is ideal for rural locations. The extended aeration system provides an aeration tank capacity of >24 h DWF, which is a large buffer for high instantaneous loadings. Nevertheless, the final settlement tanks should be designed for the treatment of peak loads.

The loading rate applied to these systems is between 0.24 and 0.32 kg $BOD_5/m^3/day$. The hydraulic retention time is between 24 and 36 h. This should produce an effluent of typically 40 mg/l BOD_5 and 50 mg/l SS.

One advantage of the extended aeration system is that the aeration process is taken well into the endogenous respiration stage. Consequently, the sludge yield is minimized. The sludge age is ≥10 days, which means that the effluent is at least partially nitrified.

15.4.7 Contact Stabilization

Contact stabilization tries to exploit the rapid reduction in BOD of the untreated waste by biosorption and bioflocculation with the return AS. In this system, the untreated waste and return AS are aerated in a contact stage with a retention between 0.5 and 3 h at DWF. The sludge, separated by settlement, is passed to a stabilization tank, where it is re-aerated for a period (usually <6 h) based on the AS return flow to complete the oxidation of the adsorbed BOD.

The degree of purification obtained in the contact zone depends on the nature of the waste, the contact time, and the MLSS concentration. As the proportion of BOD in true solution increases, the contact period must be increased until the process may become conventional. An effluent with a BOD of <20 mg/l and SS of <30 mg/l can be achieved for domestic sewage with this treatment process.

The advantage of this system is that the sludge re-aeration is done in a small tank. Hence, the tank volume and capital costs are low. Large volumes of sludge are produced in this process, and aerobic digestion is therefore sometimes adopted to reduce the yield. The typical design criteria are as follows:

- The BOD loading should be ≤480 mg/l for the combined contact and re-aeration stages; and
- The capacity of the contact stage based on DWF should be between 0.5 and 2 h.

15.4.8 Oxidation Ditches

Oxidation ditches are a relatively low-cost adaptation of the extended aeration process, which is especially suitable for small installations. Originally developed in Holland, they are sometimes called Pasveer Oxidation Ditches. The oxidation ditch can be operated continuously or intermittently. In continuous-flow plants, the aeration tank consists of a racetrack-shaped aeration basin and horizontal aerators. Sewage flows continuously into one part of the ditch, where it is mixed with the AS and circulated around the ditch by the action of the aerators. The velocity is such that the sewage and SS are kept in suspension.

In intermittent-flow plants, both biological oxidation and settling take place in the ditch. Effluent discharge is intermittent, and so the water level varies throughout the cycle of operation. This means that the aerator must be able to move vertically to maintain the correct level of immersion.

Sewage is continuously admitted to the ditch, while effluent is withdrawn by a decanting bell during a 0.5-h period at the end of every cycle. A typical cycle takes 4 h; in other words, 2.5-h aeration, 1-h settlement, and then 0.5-h effluent removal (then repeat).

The circulation velocity should be around 0.3 m/s. The organic loading rate is between 0.12 and 0.18 kg BOD_5/m^3/day with a food-to-microorganism ratio of 0.03 kg BOD/kg MLVSS/day. The hydraulic retention time is about 24 h. With the low food-to-microorganism ratio, a highly nitrified effluent can be produced.

15.4.9 Deep Shaft Process

The final variant was developed from research into methods of improving and enhancing oxygen transfer in the AS process. This can be achieved by an increase in the oxygen partial pressure and by an increase in the total pressure applied to the system. In the deep shaft process, a circular shaft is drilled to a depth between 60 and 150 m, and it is lined and fitted with a second liner, concentric with the first but suspended clear of the floor of the shaft.

A circulation of flow is induced into the system by the injection of air or oxygen into the outer annulus. Once the circulating flow pattern is developed, air is injected into the down-flowing liquid in the center tube. Air is then carried downward toward the bottom of the shaft, where it is subjected to a steady increase in pressure and an enhanced rate of solution into the surrounding liquid. Oxygen transfer rates are correspondingly enhanced so that such a system is able to sustain a higher rate of aerobic biological activity than is normal for AS plants.

It is important to realize that there is always a trade-off in the AS system between oxidation to minimize sludge production and keeping oxidation at a minimum but still stabilizing the waste. All variants operate at different levels of sludge production and various degrees of oxidation.

15.5 ACTIVATED SLUDGE PROCESS DESIGNS AND KINETICS

15.5.1 Diffused Air Aeration

The purposes of aerating wastewater are:

- To introduce air or pure oxygen into the wastewater with submerged diffusers or other aeration devices and
- To agitate the wastewater mechanically, so as to promote solution of air from the atmosphere.

A diffused air system consists of diffusers submerged in the wastewater, header pipe air mains, and the blowers through which the air passes. In the past, the various diffusion devices have been classified as either fine-bubble or coarse-bubble aeration systems, with the connotation that fine bubbles were more efficient in transferring oxygen.

The current preference is to categorize the diffused aeration systems by the physical characteristics of the equipment. The following three categories have been defined:

1. Porous or fine-pore diffusers;
2. Nonporous diffusers; and
3. Other diffusion devices, such as jet aerators, aspirating aerators, and U-tube aerators.

Porous diffusers are made in many shapes, the most common being plates, domes, discs, and tubes. Plate-type diffusers are costly to install and difficult to maintain, while domes and discs are easy to remove and can be made from ceramics or porous plastic. It is essential that the air supplied to these devices is clean and free from dust, which might clog the diffusers.

Nonporous diffusers are available in a variety of types. The fixed and valve orifice diffusers produce larger bubbles than porous diffusers. Consequently, they have lower aeration efficiency, but the advantages of lower cost, less maintenance, and the absence of stringent air purity requirements offset the slightly lower efficiency.

In the static tube aerator, air is introduced at the bottom of a circular tube, which can vary in height between 0.5 and 1.25 m. Internally, the tubes are fitted with alternately placed deflection plates to increase the contact of the air with the wastewater. Mixing is accomplished, because the tube aerator acts as an airlift pump.

Other types of diffuser are available on the market. Jet aeration, for example, combines liquid pumping with air diffusion. The pumping system recirculates liquid in the aeration basin and ejects it with compressed air through a nozzle assembly. This system is useful for deep tanks (6–7 m). Aspirating aeration consists of a motor-driven aspirator pump. The pump draws in air through a hollow tube and injects it underwater, where both high velocity and propeller action create turbulence, which diffuses the air bubbles. U-tube aeration is used in the deep shaft process. Typical transfer efficiencies for these devices range between 9% and 40%.

15.5.2 Mechanical Aerators

This class of aerators is divided into two groups: aerators with a vertical axis and aerators with a horizontal axis. Both groups can be further classified as surface or submerged aerators. Surface aerators with a vertical axis are designed to induce either updraft or downdraft flows through pumping action.

High-speed aerators are used in ponds and lagoons. The water level varies so that a fixed support is impractical. Therefore, nearly always floating devices are used in practice.

Low-speed aerators are used in the AS process, usually on a fixed platform. They are designed to throw wastewater into the atmosphere as small droplets. Sometimes, they are fitted with a draft tube to improve efficiency.

Submerged devices with a vertical axis rely on violent agitation of the surface and subsequent air entrainment to achieve enhanced oxygen transfer. These are not popular devices, although they have been used in pure-oxygen systems.

Mechanical aerators with a horizontal axis can also be subclassified into surface and submerged devices. The surface aerator types are based on the so-called Kessener brush aerators, a device normally used in oxidation ditches (Section 15.4.8). The brush type aerator has a horizontal axis with bristles mounted just above the water surface. The bristles are then submerged, and the cylinder is rotated rapidly, spraying wastewater across the tank, promoting circulation, and entraining air.

Submerged aerators with a horizontal axis are similar to the surface-type aerators, except that they use disks or paddles attached to the rotating shaft to agitate the fluid. The disk aerators have been used previously in oxidation ditches. The disks are submerged to approximately between 0.125 and 0.375 of their diameters. Recesses on the disks introduce entrapped air beneath the surface as the disks turn. Typical efficiency figures for mechanical aerators are in the range between 8% and 15%.

15.5.3 Process Design

An empirically developed relationship between biological growth and substrate utilization, which is commonly used in biological systems stabilizing organic and inorganic waste, is expressed in Eq. (15.5.3.1).

$$\text{Net growth rate of microorganisms} = \text{birthrate} - \text{deathrate}. \qquad (15.5.3.1)$$

The food-to-microorganism ratio is known as the process loading factor, the substrate removal rate, and the specific utilization. The fractional growth, the microorganism growth rate, and the food-to-microorganism growth rate are directly related with each other. However, the following assumptions have to be made:

- All the necessary nutrients for growth are present;
- Temperature and pH are regulated to achieve the optimum growth rate; and
- The equations apply only to that portion of the waste that is soluble or biodegradable.

The four principle types of reactors (i.e., tank systems) for biological wastewater treatment are classified according to their hydraulic characteristics as batch, plug, complete mix (continuous but stirred flow), and arbitrary flow.

In plug flow, fluid particles pass through the tank and are discharged in the same sequence in which they enter. The particles retain their identity, and they remain in the tank for a time equal to the theoretical detention time. Complete mixing occurs when particles entering the tank are immediately dispersed throughout the tank. The particles leave the tank in proportion to their statistical population. The batch reactor is characterized by the fact that flow neither enters nor leaves the reactor on a continuous basis. Arbitrary flow represents any degree between partial mixing and plug flow and is difficult to describe mathematically. Hence, ideal plug flow or complete mix flow models are usually assumed.

The reactor is completely mixed, and there are no organisms in the influent. The hydraulic retention time and the mean cell residence time are the same. The mass balance for the microorganisms in the reactor is as follows: the "rate of change of organism concentration in the reactor" is equal to the net "rate of organism growth in the reactor" subtracted by the "rate of organism outflow from the reactor." The microorganism concentration is a function of the mean cell residence time.

The complete mix (no recycle) model can be used to simulate conventional anaerobic treatment systems (usually sludge digestion) and some modified AS processes, oxidation ponds, and lagoons, provided an allowance is made for settling where appropriate. The treatment efficiency of a process with a high cell residence time can be limited by the depletion of nutrients, oxygen transfer difficulties, and problems with the mixing of the large microorganism mass resulting from extended cell residence times. It is assumed that the reactor content is completely mixed and that there are no microorganisms in the waste effluent. Since such a system involves a settling unit, further simplifying assumptions must be made:

- Waste stabilization occurs only in the reactor; and
- The volume used to derive the mean cell residence time only includes the reactor volume.

For this system, the mass balance will be as follows: the "rate of change of organism concentration in the reactor" is equal to the "net rate of organism growth in the reactor" subtracted by the "rate of organism outflow from the reactor." There is not a need to determine the amount of biological solids in the system or the amount of food utilized. A specified percentage of the cell mass in the system must be wasted each day to control the growth rate of microorganisms and hence the degree of waste stabilization.

Unlike the complete mix (recycle) scheme, in most biological treatment processes, cell wastage occurs from the sludge recycle line (surplus AS). Both the mixed liquor and return sludge microorganism concentration should be known, if cells are wasted from the sludge recycle line.

In a plug flow model, all particles entering the reactor will remain there for an equal amount of time. Thus, simplifying assumptions can be made.

For example, the influent microorganism concentration is approximately equal to the effluent microorganism concentration. Furthermore, the excess microorganisms are wasted from the reactor outflow and not from the recycle line.

Theoretically, the plug flow recycle system is more efficient at waste stabilization than is the complete mix recycle system. In practice, however, longitudinal dispersion prevents a true plug flow regime. The plug flow system is also more susceptible to shock loads (see below). Therefore, the aeration tank is normally divided into a series of complete mix reactors (i.e., a combination of plug flow with complete mix) to give better treatment together with improved resistance to shock loads.

15.6 SUMMARY OF KEY PROCESS DESIGN CRITERIA

15.6.1 Loading Criteria

The factors to be considered in the design of AS systems are as follows:

- Loading criteria;
- Selection of the reactor type;
- Sludge production;
- Oxygen requirements and transfer;
- Nutrient requirements;
- Environmental requirements;
- Solid–liquid separation; and
- Effluent characteristics.

The following parameters are used to control and design the AS process:

1. Food-to-microorganism ratio; and
2. Mean cell residence time.

Generally, the food-to-microorganism ratio is between 0.2 and 0.5. Values for the mean cell residence time are between 7 and 15 days and result in a stable and high-quality effluent. The corresponding sludge should be easy to dewater.

Empirical relationships based on detention time and volume have also been used. Detention times may be based on the waste flow disregarding recirculation (sewage detention period) or on the mixed liquor flow inducing recirculation (mixed liquor detention period).

This empirical design method is based on the consideration of the maximum organic loading in terms of kg BOD/m^3/day, together with the specification of the detention time. The method ignores the concentration of the mixed liquor, the food-to-microorganism ratio, and the mean cell residence time. However, it has the merit of requiring a minimum aeration tank volume that should be adequate for satisfactory treatment, assuming proper selection of other design criteria.

15.6.2 Reactor Types

Initial construction, operation, and maintenance costs will affect the reactor selection process in practice. Operational factors that must be considered include the following:

- The reaction kinetics governing the treatment process;
- Oxygen transfer requirements;
- The nature of the wastewater to be treated; and
- Local environmental conditions.

For first-order substrate removal kinetics, the total volume required for a series of complete mix reactors (i.e., four or more) is considerably less than that required for a single complete mix reactor. Indeed, the volume differential becomes more pronounced as the removal efficiency increases and also as the order of substrate removal kinetics increases. In practice, however, neither a plug flow nor a complete mix reactor functions as assumed in theory.

With conventional plug flow aeration systems, insufficient oxygen is supplied to meet the oxygen requirements of the input and the reactor. Modifications to improve the oxygen transfer route can be summarized as follows:

- Tapered aeration, where the air supply is matched to the oxygen demand;
- Step aeration, where the incoming waste and the return solids are distributed along the length of the reactor; and
- A complete mix AS process, where the aeration matches the oxygen demand.

In a complete mix reactor, the incoming waste is uniformly dispersed. As a result, it can withstand shock loads (i.e., discharges of a sudden and/or toxic nature) better than a plug flow reactor. Thus, the complete mix reactor is the preferred system for industrial wastewater treatment.

Theoretically, the most important environmental variable is temperature. However, the conventional AS process is not significantly influenced by temperature in practice. It should be noted that AS processes are generally less sensitive to low temperatures than trickling filters (Chapter 13).

The quantity of sludge produced is vital to the design of the sludge-handling and disposal facilities. The volumetric sludge production depends on the volume of the reactor and the efficiency of the final settling tank.

15.6.3 Oxygen Demand

The oxygen requirement is calculated from the BOD (Section 1.3) of the wastewater and from the quantity of cells wasted from the system per day. Assuming that the entire BOD is converted to end products, the total oxygen demand can be computed by converting BOD_5 to BOD_{ult} and subtracting the BOD_{ult} of the cells wasted from the system. Using the molecular weight relationships for typical waste ($C_5H_7NO_2$) and the associated oxygen requirement, it follows that the oxygen demand (kg/day) can be estimated by the food utilized per day reduced by 1.42 times the organisms wasted per day.

If the oxygen transfer efficiency of the aeration system is known, the air requirement can be determined. The minimum dissolved oxygen concentration throughout the reactor should be between 1 and 2 mg/l. Generally, an aeration system should be capable of delivering 150% of the normal air requirement.

15.6.4 Nutrient Requirements

Adequate nutrients must be available if the treatment plant is to function properly. The principle nutrients for all biological treatment systems, including constructed wetlands, are nitrogen and phosphorus. Assuming a cell composition of $C_5H_7NO_2$, then 12.4% nitrogen by weight of the mass of organisms produced each day is required. The corresponding phosphorus demand is about 0.2 of the nitrogen demand. In general, wastewater contains all nutrients required for cell growth. However, industrial wastewaters may require additional nutrients for optimum biodegradation.

Chapter 16

Iron and Manganese Removal

16.1 INTRODUCTION

Selecting an appropriate process sequence for the removal of iron and manganese, for example, from water supplies, is made complex by the variety of reactions that can occur with these elements. The choice depends on the presence or absence of manganese, carbon dioxide, organic acids, turbidity, dissolved oxygen, hardness, and bicarbonates. It also depends on the binding of iron and manganese to organic matter (e.g., peat) and also on the pH. In general, pilot plant studies are essential to optimize the processes.

Iron and manganese have to be removed from some waters for various reasons:

- They cause corrosion and pipe blockages (directly by precipitation or indirectly by creating favorable growth conditions);
- They negatively affect the appearance of water;
- They impart a metallic taste to water;
- They cause laundering difficulties; and
- They pose a potential pollution problem to the receiving watercourse.

Particular problems may be experienced with groundwater that has a high iron and manganese content. These precipitate on contact with air and turn dark brown or black, creating obvious aesthetic problems.

16.2 BASIC REMOVAL PROCESSES

Iron alone in groundwater that contains little or no organic material can be removed by simple aeration, followed by detention, sedimentation, and filtration processes. However, these processes are usually carried out in towers containing slats or trays, which either are seeded with catalytic material such as pyrolusite (MnO_2) or are permitted to accumulate deposits of iron(III) oxide (Fe_2O_3). The latter deposits also serve to catalyze the oxidation of iron.

The detention and sedimentation basin common to such systems is unlikely to accumulate much in the way of solids. Its purpose is primarily to provide sufficient time for the oxidation reaction to reach completion and for iron to come out of solution. The precipitated iron is removed by filtration. A number

Wetlands for Water Pollution Control. http://dx.doi.org/10.1016/B978-0-444-63607-2.00016-2

of manufacturers provide packaged iron removal systems, which incorporate an aeration tower, a small detention tank, and a filter in a single unit. This type of system can be completely satisfactory for small flows with the characteristics described in this chapter.

If both iron and manganese are present or if water contains organic material such as humic or fulvic acid, aeration is sufficiently rapid only if it is catalyzed by pyrolusite or by accumulation of oxidation products (Fe_2O_3 and manganese dioxide (MnO_2)) on a porous bed such as coke. Simple aeration will not provide oxidation and subsequent precipitation within a reasonable time, although elevation of the pH will increase the precipitation rate substantially. Manganese is much more slowly oxidized than iron. In fact, the rate is negligible at pH levels <9. Organic material interferes with the removal of iron and manganese by peptization (i.e., by complexing with the metal without complete neutralization of the molecular charge, by forming soluble complexes with both the reduced and oxidized metals, by reduction of the oxidized metals, or by a combination of these processes).

The application of strong oxidizing agents such as chlorine, ozone, chlorine dioxide, or potassium permanganate can serve to oxidize iron and manganese more rapidly and can also oxidize or destroy the organic materials present. Such oxidizing agents have the potential of forming trihalomethanes from humic and fulvic acid and thus must be applied with care. Permanganate in this application will function adequately at neutral pH. The other oxidants require pH levels of about 9.5 for adequate manganese removal. Iron is oxidized satisfactorily at neutral pH, but its oxidation is enhanced under more alkaline conditions, for example at pH values between 8.0 and 8.5, and followed by the removal of the oxidized metals via rapid filtration (Chapter 11).

16.3 ADVANCED REMOVAL PROCESSES

Catalytic action may be required for the oxidation of manganese. This occurs in beds of either pyrolusite ore or zeolite regenerated with potassium permanganate. In some cases, potassium permanganate is fed continuously into the water before it enters the filter. A bed of coke coated with deposits of ferric hydroxide can also act as a catalyst. Lime softening increases the pH. Preliminary aeration removes carbon dioxide and results in the conversion of iron to its ferric state and subsequently its precipitation as ferric hydroxide.

Ion exchange using either zeolites or other resins can remove ferrous and manganous ions, if they are associated with bicarbonate. These conditions occur most frequently with oxygen-deficient groundwater. Care must be taken to avoid aeration or oxidation of such waters, as this would produce Fe^{3+} and Mn^{4+} ions, which will precipitate and clog the ion exchange bed. This type of process is practicable when used as an incidental step to softening only and when combined iron and manganese is less than 0.3 times the hardness.

Sequestration of iron and manganese may be satisfactory in some circumstances. This involves the use of complex molecules, which encase (or combine with) the iron and manganese ions so that they, although still in solution, can no longer take part in other reactions. The usual sequestering agents are polyphosphates or organic compounds such as sodium hexametaphosphate (e.g., marketed as Calgon). Subsequent heating of the water may destroy the effectiveness of this treatment, so it must be used with care.

Most wetland systems can also be used to remove iron and manganese. The main processes are oxidation, detention, sedimentation, and predominantly filtration.

Chapter 17

Water Softening

17.1 INTRODUCTION

Softening is practiced when the hardness of the outflow, for example from a wetland system, is not acceptable. The degree of softening depends upon the use to which the water will be put. For domestic water supply, for example, hardness creates nuisance scums, wastes soap when used for laundry or ablution, and deposits scale in hot-water services. Softening is seldom applied to municipal water supplies if the hardness is ≤ 150 mg/l as calcium carbonate ($CaCO_3$). A reasonable target after treatment is a hardness of approximately 80 mg/l as $CaCO_3$.

For general treatment of municipal supplies, a process of precipitating the calcium and magnesium hardness by adding lime ($Ca(OH)_2$) and/or soda (caustic soda ($NaOH$) or soda ash (Na_2CO_3)) is usually adequate.

Hardness is caused by calcium and magnesium ions in the water. The presence of bicarbonate ions affects the dose of chemicals required to precipitate the calcium and magnesium, and therefore hardness is classified in two ways:

- As calcium or magnesium hardness and
- As carbonate or noncarbonate hardness.

Carbonate hardness is caused by calcium or magnesium ions associated with bicarbonate ions, and noncarbonate hardness is caused by calcium or magnesium ions associated with chloride, sulfate, or other ions.

The chemical reactions involved in water softening are as follows (Eqs. (17.1.1) and (17.1.2)):

$$CaO + H_2O \rightarrow Ca(OH)_2 \quad (17.1.1)$$

$$CO_2 + Ca(OH)_2 \rightarrow \underline{CaCO_3} + H_2O \quad (17.1.2)$$

For carbonate hardness with calcium (Eq. (17.1.3)):

$$Ca(HCO_3)_2 + Ca(OH)_2 \rightarrow \underline{2CaCO_3} + 2H_2O \quad (17.1.3)$$

For noncarbonate hardness with calcium (Eq. (17.1.4)):

$$CaSO_4 + Na_2CO_3 \rightarrow \underline{CaCO_3} + Na_2SO_4 \quad (17.1.4)$$

Wetlands for Water Pollution Control. http://dx.doi.org/10.1016/B978-0-444-63607-2.00017-4

For carbonate hardness with magnesium (Eq. (17.1.5)):

$$Mg(HCO_3)_2 + 2Ca(OH)_2 \rightarrow \underline{Mg(OH)_2} + \underline{2CaCO_3} + 2H_2O \quad (17.1.5)$$

For noncarbonate hardness with magnesium (Eq. (17.1.6)):

$$MgSO_4 + Ca(OH)_2 + Na_2CO_3 \rightarrow \underline{Mg(OH)_2} + \underline{CaCO_3} + Na_2SO_4 \quad (17.1.6)$$

For excess lime removal (Eq. (17.1.7)):

$$Ca(OH)_2 + Na_2CO_3 \rightarrow \underline{CaCO_3} + 2NaOH \quad (17.1.7)$$

For carbon dioxide removal (Eq. (17.1.8)):

$$CO_2 + Ca(OH)_2 \rightarrow \underline{CaCO_3} + H_2O \quad (17.1.8)$$

Equations (17.1.1−17.1.8) show important reactions that can take place in the softening process. Inspection of these chemical equations shows that the required dose may be estimated as follows:

$$\begin{aligned} \text{Equivalents of lime} &= \text{equivalents of carbonate hardness} \\ &+ \text{equivalents of magnesium hardness} \\ &+ \text{equivalents of carbon dioxide} \\ &+ \text{excess lime to raise the pH} \\ \text{Equivalents of soda} &= \text{equivalents of noncarbonate hardness} \\ &+ \text{equivalents of excess of lime} \end{aligned}$$

17.2 LIME-SODA SOFTENING

The quantities of soda ash (Na_2CO_3) and lime ($Ca(OH)_2$) are estimated from a chemical analysis of the water and are subsequently added to the influent stream. Sometimes, a small amount of iron salts or alum is added as a flocculation aid. The dosed water passes to a sludge blanket clarifier (or a similar treatment unit), where the chemicals precipitate out and are subsequently removed. Lime is a cheap form of alkali, which will raise the pH to between 9.5 and 10.0 so that bicarbonates will be converted to carbonates and the calcium precipitates as calcium carbonate.

A further increase of the pH will cause the magnesium hardness to precipitate as magnesium hydroxide. This is achieved by adding an excess of lime over that indicated as needed by the chemical equations and sufficient soda to precipitate the calcium leaving sodium hydroxide in the water. The amount of excess lime added is often between 40 and 65 mg/l as $CaCO_3$, depending on the desired efficiency of magnesium removal. As the pH is increased, any carbon dioxide (CO_2) present will be converted to carbonate and will be available for the precipitation of calcium. Water with a high CO_2 concentration can be aerated before lime treatment to reduce the chemical requirements by eliminating some of the CO_2.

Lime-soda softening is used for all forms of calcium hardness. By adding soda ash, noncarbonate hardness is converted to $CaCO_3$, which will then precipitate. As much soda ash is added as there is noncarbonate hardness associated with calcium.

Excess lime-soda softening is used for all forms of magnesium hardness. The chemistry is complex, and the required dose rate is best determined experimentally.

17.3 LIME SOFTENING

Lime softening is used for calcium hardness of the carbonate form. The addition of lime (equivalent to the amount of bicarbonate present) will form the insoluble calcium carbonate. The solubility of $CaCO_3$ at normal temperatures is about 20 mg/l. Considering the limited contact time available in a standard plant, a residual of about 40 mg/l can be measured. Thus, the water must be stabilized prior to filtration. The water leaving has a fairly high pH (9.5–11). If it were applied to the filter, calcium carbonate would precipitate in the filter and cement the sand grains together.

The usual method of reducing the pH and bringing the water to a stable condition in which it will not precipitate calcium carbonate is to add CO_2, usually by burning gas, oil, or coke. The products of combustion are bubbled through the water in a recarbonation tank. If too little CO_2 dissolves, the tendency to precipitate $CaCO_3$ is increased. In contrast, if too much CO_2 dissolves, the water can become corrosive.

In cases where the target hardness and raw water quality are suitable, split treatment can eliminate the need for recarbonation, resulting in the saving of chemicals. However, extra tanks increase capital costs. In split treatment, one stream is dosed with lime and softened by the excess lime process. Some of the calcium and most of the magnesium in that stream are removed. The flows are then recombined and dosed with soda ash, and, if necessary, passed to a second clarifier. The caustic alkalinity of the water, which has been treated with excess lime, reacts with the bicarbonates in the untreated stream to form carbonates. These carbonates react with the remaining calcium to form $CaCO_3$, most of which will precipitate.

Split treatment removes virtually all the magnesium from the first stream, but practically none from the second. Recarbonation can be expressed with Eq. (17.3.1):

$$CaCO_3 + CO_2 + H_2O \rightarrow Ca(HCO_3)_2 \tag{17.3.1}$$

17.4 EXCESS LIME SOFTENING

Excess lime softening is used for the removal of magnesium carbonate hardness. The other methods discussed in this chapter are not effective for the removal of magnesium, since magnesium carbonate is soluble (Eq. (17.4.1)).

$$Mg(HCO_3)_2 + Ca(OH)_2 \rightarrow \underline{CaCO_3} + MgCO_3 + H_2O \quad (17.4.1)$$

However, at about pH 11, the following process occurs (Eq. (17.4.2)):

$$MgCO_3 + Ca(OH)_2 \rightarrow \underline{Mg(OH)_2} + \underline{CaCO_3} \quad (17.4.2)$$

The practical solubility of magnesium hydroxide ($Mg(OH)_2$) is approximately 10 mg/l. For excess lime softening, it is necessary to add calcium hydroxide ($Ca(OH)_2$) equivalent to bicarbonate (HCO_3^-), $Ca(OH)_2$ equivalent to Mg^{2+}, and 35 mg/l excess $Ca(OH)_2$ to raise the pH to 11.

Carbonation is necessary to remove the excess lime and to reduce the pH after treatment. The high pH also produces disinfection as a by-product, and so lower chlorine doses may be necessary after softening, for example in a water treatment works.

17.5 LIME RECOVERY

All forms of precipitation softening produce considerable volumes of sludge. Lime recovery is possible by calcining $CaCO_3$ sludge and subsequently slaking calcium oxide (CaO) with water as shown in Eqs. (17.5.1) and (17.5.2). In this way, more lime than is required in the plant is produced, and the surplus may be sold. This solves the sludge disposal problem at the same time.

$$CaCO_3 \rightarrow CaO + CO_2 (CO_2 \text{ can be used for re-carbonation}) \quad (17.5.1)$$

$$CaO + H_2O \rightarrow Ca(OH)_2 \quad (17.5.2)$$

Chapter 18

Water Microbiology

18.1 STATISTICS FOR APPLIED MICROBIOLOGY

The quantitative and qualitative assessment of wetland system microbial dynamics is important when estimating the treatment performance and assessing public health issues. There are three common ways to estimate the average of a population; averages can be described as mean, median, and mode values. The mean is equivalent to the average. The median is the middle observation in a set of observations that have been ranked in magnitude. The (crude) mode is the class in a frequency distribution that contains more observations than any other.

In a symmetrical distribution, mean, median, and mode are equivalent. In contrast, the quantitative relationship in a skewed distribution is as follows: mean > median > mode. Skewed distributions may require data transformation in order to perform advanced statistical tests on the observations. The estimate of the standard deviation for a population is defined in Eq. (18.1.1).

$$s = \sqrt{\frac{\sum (x - \overline{x})^2}{n - 1}} \tag{18.1.1}$$

where:

s = population;
x = the value of an observation;
$\overline{x}$ = the sample mean; and
n = the number of observations.

The standard deviation is the square root of the variance, which describes the variability of a sample. A logarithmic transformation (x is replaced by $\log_{10} x$), in order to convert skewed into symmetric data, is recommended if the variance of a sample of count data is larger than the mean. In contrast, the square root transformation (x is replaced by $\sqrt{x}$) is appropriate when the variance of a sample of count data is close to the mean. Other data transformation methods, including the arcsine transformation, are less popular in praxis.

The number of species present in a sample is a very simple measure of the diversity. However, there are a number of other indices that take into

Wetlands for Water Pollution Control. http://dx.doi.org/10.1016/B978-0-444-63607-2.00018-6

account how evenly the total number of individuals in a sample is apportioned between each species. The fundamental diversity index can be expressed as a ratio of the number of species and the number of individuals. Furthermore, the Shannon–Weaver Diversity Index, a popular (often applied) but controversial (difficult to interpret) population diversity index, is defined in Eq. (18.1.2).

$$H = -\sum^{S} p_i \ln p_i \tag{18.1.2}$$

where:

H = Shannon–Weaver Diversity Index;
S = total number of species; and
p_i = proportion of a particular species in a sample.

For practical purposes, a diversity index can be regarded as a number on an ordinal scale. It is sensible to use diversity indices only to compare "like with like": for example, protozoa species A counts with protozoa species B counts, and not protozoa species A counts with nematode species A counts. Comparisons between diversity indices are valid only if the samples from which the indices are calculated have approximately the same number of observations.

18.2 PROTOZOA

18.2.1 Trophic Structure

Protozoa have key roles in both the autotrophic and heterotrophic food chains of wetland systems (Scholz and Martin, 1998a,b):

18.2.1.1 Autotrophic Food Chain (Fueled Principally by Sunlight)

18.2.1.1.1 Producers

Photosynthesis: algae, floating plants, attached forms, and rooted plants.
Chemosynthesis: bacteria.

Primary consumers
Grazers (herbivores): protozoa, rotifers, crustaceans, and other herbivores.

Secondary consumers
Carnivorous: small fish.

Tertiary consumers
Carnivorous: large fish.

18.2.1.2 Heterotrophic Food Chain (Fueled by Organic Matter)

Particle producers, including bacteria.
Detritus, including dead organisms, excretions, and feces.

18.2.1.2.1 Decomposers (Including Bacteria and Fungi)

Primary consumers
Protozoa

Secondary consumers
Rotifers

Tertiary consumers
Fish

18.2.2 Kingdom Protista

Living things can be divided into five kingdoms: Monera, Protista, Plantae, Fungi, and Animalia. Protista can be subdivided as follows:

18.2.2.1 Phylum I: Sarcomastigophora

Flagella and/or pseudopodia present; spores not produced.

18.2.2.1.1 Subphylum 1: Mastigophora

Flagella typically present; division by longitudinal binary fission.

Class A: Phytomastigophorea
Plant-like flagellates typically with chloroplasts; if missing, relationship to pigmented forms clearly evident (e.g., *Euglena* and *Peranema*).

Class B: Zoomastigophorea
Animal-like flagellates; chloroplasts absent; one to many flagella; ameboid forms; with or without flagella; many parasitic forms (e.g., *Bodo* and *Giardia*).

18.2.2.1.2 Subphylum 2: Opalinata

Binary fission takes place between rows of cilia, which cover the entire body in oblique rows; two or many monomorphic nuclei; all parasitic; none associated with polluted waters.

18.2.2.1.3 Subphylum 3: Sarcodina

Pseudopodia typical; flagella restricted to developmental stages when present.

Superclass A: Rhizopoda
Locomotion by pseudopodia or by protoplasmic flow without production of discrete pseudopodia. It includes the naked (*Acanthamoeba*, *Amoeba*, *Entamoeba*, and *Naegleria*) and testate amebae (*Arcella* and *Euglypha*), the foraminifera, and the slime-mold amebae.

Superclass B: Actinopoda
Spherical; typically planktonic; axopodia with delicate internal microtubular skeleton; some naked, others with tests of chitin, silica, or strontium sulfate (e.g., *Actinophyrys*).

18.2.2.2 Phylum II: Apicomplexa

Apical complex visible with an electron microscope; all species are parasitic.

Class A: Perkinsea
No sexual reproduction; incomplete cone; none in polluted waters.

Class B: Sporozoea
Sexual and asexual reproduction typical; oocysts generally contain invective sporozoites, which result from sporogony. Locomotion of mature organisms by gliding. It includes the gregarines and coccidia (e.g., *Cryptosporidium*).

18.2.2.3 Phylum III: Microspora

Unicellular spores each with an imperforated wall, containing one uninucleate or dinucleate sporoplasm; always with polar tube and cap; obligatory intracellular parasites in nearly all major animal groups.

18.2.2.4 Phylum IV: Ciliophora

Cilia or compound ciliary organelles present in at least one stage of the life cycle; two types of nucleus present; sexual and asexual reproduction; plane of division transverse across ciliary rows.

Class A: Kinetofragminophorea
Oral cilia slightly distinct from body ciliature; cytostome often apical or midventral on surface of body; body ciliation commonly uniform (e.g., *Acineta*, *Amphileptus*, *Chilodonella*, *Colpoda*, and *Litonotus*).

Class B: Oligophymenophorea
Oral apparatus, at least partially in buccal cavity, generally well defined, although absent in one group; oral ciliation clearly distinct from body cilia; cytostome usually ventral at or near anterior end at bottom of buccal cavity (e.g., *Carchesium*, *Colpidium*, *Glaucoma*, *Paramecium*, *Uronema*, and *Vorticella*).

Class C: Polyhymenophorea
Dominated by well-developed, conspicuous adoral zone of numerous buccal or peristomial ciliary organelles; cytostome at bottom of buccal cavity (e.g., *Aspidisca*, *Euplotes*, and *Metopus*).

18.3 BIOLOGICAL EFFECTS OF ORGANIC POLLUTANTS

18.3.1 Sewage Fungus

Sewage fungus is present in organically polluted waters commonly found at the inlet of a wetland system. Most samples of sewage fungus contain protozoa (mainly flagellates and ciliates). Colorless microflagellates, including *Bodo caudatis*, are the most common flagellated protozoa, although a number of species of *Euglena* and its colorless relative *Peranema trichophorum* are sometimes also observed.

The most important ciliates are *Colpidium colpoda* and *Chilodonella cucullulus*, which swim or glide between and along the filamentous material.

Some of the ciliates, including *Chilodonella*, are known to feed on filamentous organisms. Most of the others feed on suspended bacterial populations. Some ciliates (e.g., *Litonotus* and *Trachelophyllum*) are carnivorous, and prey upon other ciliates. Furthermore, diatoms are frequently present in slimes (*Navicula, Fragilaria,* and *Synedra*). Organisms most frequently found in sewage fungus growth are listed below.

Bacteria
Zoogloeal bacteria, *Sphaerotilus natans*, *Flavobacterium* spp., *Beggiatoa alba*, *Flexibacterium* sp., and *Sirillum* sp.

Blue-Green "Algae"
Oscillatoria spp. and *Phormidium* spp.

Fungi
Geotrichum candidum, *Leptomitus lacteus*, and *Fusarium aquaeductum.*

Algae
Stigeoclonium tenue and *Ulothrix* spp.

Diatoms
Navicula spp., *Fragilaria* spp., and *Synedra* spp.

Protozoa

Ciliated protozoa
Hemiophrys fusidens, *Litonotus fasciola*, *Trachelophyllum pusillum*, *Chlodonella cucullulus*, *Chilodonella uncinata*, *Colpidium campylum*, *Colpidium colpoda*, *Glaucoma scintillans*, *Cinetochilum margaritaceum*, *Paramecium trichium*, *Carchesium polypinium*, *Vorticella convallaria*, *Aspidisca cicada*, *Aspidisca lynceus*, and *Tachysoma pellionella.*

Other protozoa
Colorless microflagellates, thecate amebae, and naked amebae.

18.3.2 Saprobic System

A simple saprobic system can be applied for most treatment wetland systems to indicate the different zones of polluted waters.

Polysaprobic zone
High concentration of biodegradable organic compounds; low dissolved oxygen (DO) concentration; high concentrations of ammonia and hydrogen sulfide; high BOD; low microorganism diversity; and bacteria, "sewage fungus" community, turbificid worms, and chironomid larvae are present.

Mesosaprobic zones

α-Mesosaprobic zone: less organic material present; preponderance of amino acids rather than proteins; ammonia concentration has been reduced, but nitrates have appeared; blanket weed (*Cladophora*) increases, but other polysaprobic organisms decrease; slow recovery of DO.

β-Mesosaprobic zone: recovery of DO and increase of population diversity; declining blanket weed.

Oligosaprobic zone
Self-purification complete; organic matter has been degraded and mineralization completed; great diversity of species but reduced numbers of bacteria.

Flagellates and bacterivorous protozoa are more frequent in the polysaprobic to mesosaprobic zones, while carnivorous ciliates are more frequent in the oligosaprobic zones. The following sequence of saprobic zones and corresponding protozoa can be identified in rivers receiving sewage:

Polysaprobic zone (p): *Bodo*, *Colpidium*, *Glaucoma*, and *Paramecium*.
α-Mesosaprobic zone (m or α): *Paramecium*, *Carchesium*, and *Vorticella*.
β-Mesosaprobic zone (m or β): *Vorticella*, *Chilodonella*, *Spirostomum*, *Didnium*, and *Coleps*.
Oligosaprobic zone (o): *Didinium*, *Coleps*, and *Litonotus*.

The five major objections related to the limitations of the use of saprobic systems are as follows:

1. Highly trained personnel are required to identify microorganisms;
2. Incomplete knowledge about microorganisms and their ecology and potential harm to humans;
3. Samples from aquatic habitats have to be statistically representative;
4. A saprobic system generalizes pollution of watercourses; and
5. The original saprobic system is limited to pollution caused by sewage discharge to rivers.

18.4 EUTROPHICATION AND WATER TREATMENT

The input water of water treatment works is often polluted with large numbers of algae, in particular diatoms, during spring. Table 18.4.1 gives an example of different levels of pollution by algae for lake and reservoir water.

The effects of eutrophication on the receiving ecosystems can be summarized as follows:

- Species diversity often decreases, and the dominant biota subsequently change;
- Plant and animal biomass increases;
- Turbidity increases;
- The rate of sedimentation increases, shortening the life span of the wetland; and
- Anoxic conditions may develop.

TABLE 18.4.1 Characteristic Algal Associations of Oligotrophic and Eutrophic Lakes and Reservoirs

State	Algal Group	Examples
Oligotrophic	Picoplankton (often small cyanobacteria)	*Synechococcus*
	Desmid plankton	*Staurdesmus* and *Staurastrum*
	Chrysophycean plankton	*Dinobryon*
Mesotrophic	Diatom plankton	*Cyclotella* and *Tabellaria*
	Dinoflagellate plankton	*Peridinum* and *Ceratium*
	Chlorococcal plankton	*Oocystis* and *Eudorina*
Eutrophic	Diatom plankton	*Asterionella, Fragilaria, Stephanodiscus,* and *Melosira*
	Dinoflagellate plankton	*Peridinium* and *Glenodinium*
	Chlorococcal plankton	*Scenedesmus*
	Cyanobacterial plankton	*Alphanizomenon, Anabaena,* and *Microcystis*

The problems to humans associated with the effects of eutrophication on ecosystems are summarized as follows:

- The treatment of water may be difficult, and the supply may have an unacceptable taste and odor;
- The water may be injurious to health;
- The amenity value of the water may decrease;
- Increased vegetation may impede water flow and navigation; and
- Commercially important species such as salmonids and coregonids (whitefish) may disappear.

Typical trophic classification system data are summarized in Table 18.4.2. Most constructed treatment wetlands are eutrophic or hypertrophic.

In trickling filters, suspended solids including plankton are partly consumed directly by macro-invertebrates, and suspended bacteria by protozoa and rotifers. Annelids, collembola, and the majority of fly larvae are important in preventing fixed film filters from being blocked during warm temperatures.

TABLE 18.4.2 Trophic Classification System Variables

Trophic Category	TP	Mean Chl	Maximum Chl	Mean Secchi	Minimum Secchi
Ultra-oligotrophic	<4.0	<1.0	<2.5	>12.0	>6.0
Oligotrophic	<10.0	<2.5	<8.0	>6.0	>3.0
Mesotrophic	10–35	2.5–8.0	8–25	6–3	3.0–1.5
Eutrophic	35–100	8–25	25–75	3.0–1.5	1.5–0.7
Hypertrophic	>100	>25	>75	<1.5	<0.7

TP = Mean annual total phosphorus concentration (μg/l); mean Chl = mean annual chlorophyll a concentration in surface waters (μg/l); maximum Chl = peak annual chlorophyll a concentration in surface waters (μg/l); mean Secchi = mean annual Secchi depth transparency (m); and minimum Secchi = minimum annual Secchi depth transparency (m).

18.5 PROTOZOOLOGY OF TREATMENT PROCESSES

Protozoa are plentiful in constructed treatment wetlands, AS, and other polluted mixed liquors (up to about 50,000 cells per ml). Protozoa species and their abundances can be related to water and wastewater treatment processes, and the corresponding water quality.

The spatial distribution of protozoa in a filter depends on the level of saprobity. It is possible to monitor the filter effluent quality by interpreting protozoa species and levels in various parts of the filter. The temporal succession of protozoa groups within polluted waters that become gradually cleaner is as follows: flagellates → free-swimming ciliates → crawling ciliates → attached ciliates (Scholz and Martin, 1998a,b).

Protozoa may be used as indicator organisms to predict treatment plant efficiencies in terms of BOD effluent concentrations (Scholz and Martin, 1998a,b). Under protozoa-free conditions, plants produce effluent of low quality with respect to BOD, permanganate value, organic nitrogen, suspended solids, viable bacteria, chemical oxygen demand, and optical density.

18.6 ODOR AND TOXINS OF NATURAL ORIGIN

There are four main groups of algae that are able to cause odor problems: blue-green algae, diatoms, green algae, and yellow-brown algae. The important genera of blue-green algae (*Cyanophyceae*) are *Anabaena*, *Anacystis*, *Aphanizomenon*, *Gomphosphaeria*, and *Oscillatoria*. These generally produce grassy odors that intensify into piggy, almost septic odors as the cells disintegrate. Some algae can also produce toxic substances.

The most important genera of diatoms (*Bacillariophyceae*) are *Asterionella*, *Fragillaria*, *Melosira*, *Tabellaria*, and *Synedra* (Scholz and Martin, 1998b). They secrete oils that impart a fishy or aromatic odor. When moderate densities of *Asterionella* and *Tabellaria* are present, then geranium odors are produced.

Green algae (*Chlorophyceae*) can impart a fishy or grassy odor. Important genera include *Volvox* and *Staurastrum*.

Yellow-brown algae (*Chrysophyceae*) can produce pungent odors. *Uroglenopsis* can produce a very strong fishy odor, whereas *Synura* produces a cucumber odor at low, a fishy odor at moderate, and a piggy odor at high numbers.

Blue-green algae are responsible for some toxins in drinking waters. Despite their name, these algae are actually a group of bacteria that are capable of photosynthesis. It follows that blue-green algae are also called cyanobacteria.

There are no water quality standards set for algal toxins. However, water companies monitor toxic algal blooms regularly. Most algal blooms can be prevented by adding copper sulfate to the untreated water. Activated carbon treatment helps to oxidize algal toxins (Scholz and Martin, 1998a).

18.7 PUBLIC HEALTH ASPECTS

18.7.1 Typical Diseases Related to Waters

Wetland systems might harbor organisms that are potentially pathogenic. Protozoa are known to feed upon pathogenic bacteria, including those that cause diseases (e.g., diphtheria, cholera, typhus, and streptococcal infections) and fecal bacteria (e.g., *Escherichia coli*). Waterborne infectious diseases are associated with certain types of organisms:

- Waterborne virus infections are epidemic viral gastroenteritis, hepatitis A, pharyngoconjunctival fever, poliomyelitis, and traveler's diarrhea.
- Some bacteria may cause diseases such as cholera, enteropathogenic diarrhea, inclusion conjunctivitis, leptospirosis, melioidosis, paratyphoid fever, salmonellosis, traveler's diarrhea, shigellosis, tularemia, typhoid fever, *Vibrio parahemolyticus*, food poisoning, and yersiniosis.
- Certain species of protozoa are related to the following diseases: amebiasis, balantidiasis, giardiasis, and primary amebic meningoencephalitis.
- The presence of certain nematodes in (particularly drinking) water may cause angiostrongyliasis, anisakiasis, dracontiasis, gnathostomiasis, hepatitic capillariasis, and intestinal capillariasis.
- Chlonorchiasis, echinostomiasis, fascioliasis, fasciolopsiasis, and heterophyidiasis could be caused by certain trematodes.
- Some cestodes are associated with hydatidosis.

Infections related to water may be classified into four main groups:

1. Waterborne disease: A pathogen is transmitted by ingestion of contaminated water (e.g., cholera and typhoid fever).
2. Water-washed disease: These include fecally and orally spread diseases or diseases spread from one person to another that are facilitated by a lack of an adequate supply of water for washing (e.g., many diarrheal diseases as well as diseases of the eyes and skin).
3. Water-based infections: These diseases are caused by pathogenic organisms, which spend part of their life cycle in aquatic organisms (e.g., schistosomes and other trematode parasites).
4. Water-related disease: Insect vectors breeding in water (e.g., malaria) and filariasis arthropods are responsible for water-related illnesses (e.g., yellow fever).

Some typical diseases and/or symptoms associated with bacteria, viruses, and protozoa are summarized in Table 18.7.1.1. Infection is defined as the end result measured for each organism (except *Shigella dysenteriae*). It is assumed that the population is equally susceptible to a single exposure, the latter being defined as the consumption of 2 l of water per day.

TABLE 18.7.1.1 Bacterial, Viral, and Protozoan Diseases and/or Symptoms Generally Transmitted by Contaminated Water

		Well-documented Incubation Time (days)	
Agent	**Disease or Symptoms**	**Minimum**	**Maximum**
Bacteria			
Shigella spp.	Shigellosis; bacterial dysentery	1	7
Salmonella spp.	Gastroenteritis		
S. typhimurium	Salmonellosis	0.25	3
S. typhi	Typhoid fewer	1	3
S. paratyphi	Paratyphoid fewer		
Escherichia coli	Gastroenteritis; diarrhea	0.5	3
Leptospira icterohaemorrhagia	Weil's oil disease		
Campylobacter spp.	Gastroenteritis; intestinal infections	1	7

TABLE 18.7.1.1 Bacterial, Viral, and Protozoan Diseases and/or Symptoms Generally Transmitted by Contaminated Water—cont'd

		Well-documented Incubation Time (days)	
Agent	**Disease or Symptoms**	**Minimum**	**Maximum**
Francisella tularensi	Chills, fever, weakness		
Mycobacterium	Tuberculosis		
Vibrio cholera	Cholera; gastroenteritis	1	3
Viruses			
Hepatitis A	Hepatitis	15	45
Norwalk-like agent	Gastroenteritis	1	7
Virus-like particles	Gastroenteritis	1	7
Enteroviruses	Many diseases, including poliomyelitis, respiratory diseases, meningitis, and infectious hepatitis		
Rotavirus	Gastroenteritis; diarrhea; enteritis	1	2
Protozoa			
Giardia lamblia	Giardiasis; diarrhea; malabsorption	7	10
Naegleria fowleri	Amebic meningoencephalitis		
Entamoeba histolytica	Amebiasis; amebic dysentery	14	28
Cryptosporidium spp.	Cryptosporidiosis; diarrhea	5	10
Helminths			
Diphyllobothrium latum	Tapeworm infection		
Taenia saginata	Tapeworm infection		
Schistosoma spp.	Bilharzia		
Clonorchis sinensis	Trematode infection		
Dracunculus medinensis	Guinea worm		

18.7.2 Invertebrates Found in Main Supplies

The main groups of animals found in water distribution pipes, for example on the British Isles, are described as follows:

- *Isopoda* (e.g., *Asellus aquaticus*)
- *Amphipoda* (e.g., *Gammarus pulex*)
- *Insecta* (e.g., *Chironomid* larvae). *Chironomid* larvae are very common in water distribution systems. They can be between 5 and 25 mm in length.
- *Oligochaeta* or segment worms (e.g., *Nais communis*)
- *Hirudinea* or leeches (e.g., *Erpobdella octoculata*). Leeches are very seldom present in water mains.
- *Mollusca* or snails (e.g., *Potamopyrgus jenkinsi*). They are harmless to consumers in temperate climates.
- *Nematoda*, roundworms, or eelworms. They are the most common larger microorganisms in water supply systems. Large numbers of bacteria and an elevated BOD concentration can be correlated with the presence of nematodes.
- Smaller *Crustacea* (e.g., *Cyclops* and *Chydorus*). They are common in water mains.
- Terrestrial animals (e.g., slugs and earthworms). They die rapidly in waters.

The presence of some invertebrates correlates well with high counts of certain bacteria in potable water distribution networks (e.g., *Nematoda* correlate well with *Pseudomonas cepacia*). Furthermore, fly larvae are associated with *Aeromonas hydrophila*, *P. cepacia*, and *Pseudomonas fluorescens*, and copepods with *Flavobacterium meningosepticum*, *Moraxella* spp., *Pasteurella* spp., *Pseudomonas diminuta*, and *Pseudomonas paucimobilis*.

Finally, amphipods are associated with many bacteria genera and species in water distribution systems, including *Acinetobacter* spp., *Achromobacter xylosoxidans*, *Bacillus* spp., *Chromobacter violaceum*, *Moraxella* spp., *Pasteurella* spp., *Pseudomonas fluorescens*, *Pseudomonas maltophilia*, *Pseudomonas paucimobilis*, *Pseudomonas vesicularis*, *Serratia* spp., and *Staphyloccocus* spp.

18.7.3 Monitoring and Prevention of Waterborne Diseases

Relevant criteria for potential indicator microorganisms of fecal contamination are as follows:

- The microorganisms should be a member of the normal intestinal flora of healthy people.
- They should be exclusively intestinal in habitat and therefore exclusively fecal in origin if found outside the intestine.
- Ideally, they should only be found in humans.
- They should be present when fecal pathogens are present and only when fecal pathogens are expected to be present.

- They should be present in greater numbers than the pathogen that they are intended to indicate.
- An indicator microorganism should be unable to grow outside the intestine with a death rate slightly less than the pathogenic organism.
- They should be resistant to natural environmental conditions and to water and wastewater treatment processes in a manner equal to or greater than the pathogens of interest.
- They should be easy to isolate, identify, and enumerate.
- They should preferably be nonpathogenic.

Traditionally, the following maximum admissible conditions apply for most finished waters:

- Zero total coliforms per 100 ml sample; where sufficient numbers of samples are examined, then a 95% consistent result is acceptable.
- Zero fecal coliforms (*E. coli*) per 100 ml.
- Zero fecal streptococci per 100 ml sample.
- Less than one sulfite-reducing clostridia per 20 ml sample.
- There should be no significant increase in the total bacteria colony counts above background levels, although a guide value of <10 per ml at 22 °C and <100 per ml at 37 °C has been set.

Furthermore, guidelines require that there is no significant increase from background levels of heterotrophic bacteria in either tap or bottled waters. Heterotrophic plate counts represent the aerobic and facultative anaerobic bacteria that derive their carbon and energy from organic compounds. This group includes Gram-negative bacteria belonging to the following genera: *Pseudomonas*, *Aeromonas*, *Klebsiella*, *Flavobacterium*, *Enterobacter*, *Citrobacter*, *Serratia*, *Acinetobacter*, *Proteus*, *Alcaligenes*, and *Moraxella*.

Changes in heterotrophic bacteria counts indicate a variation in water quality and the potential for pathogenic survival and regrowth. They are also applied to monitor disinfection efficiencies. Changes in the pattern of colony counts are more significant than single numerical counts for the interpretation of bacterial activity within a treatment process.

Heterotrophic plate counts are normally done by using the spread plate count method using yeast extract agar (incubated at 22 °C for 72 h and 37 °C for 24 h). Results are expressed as colony-forming units. Counts at 37 °C are particularly useful as they can provide a rapid indication for possible contamination of water supplies.

Rapid sand filters (Chapter 11) remove 99.99% of *Entamoeba histolytica* cysts. However, these figures cannot be expected for poor conditions of coagulation. In contrast, diatomaceous earth filters can essentially remove all cysts. As with the control of giardiasis, the usual recommended procedure for the safe control of amebiasis is to use some form of wetland filtration with prior chemical coagulation, followed by chlorination.

Both physical and biological potable water filtration systems (Chapter 13) are effective in cyst removal. When rapid physical filtration systems are used, it is essential that filters are adequately maintained and operated, preferably with a pretreatment of chemical coagulation and sedimentation procedures.

Pathogenic bacteria have been associated with protozoa. Under specific environmental conditions, some protozoa can be related indirectly to certain diseases. Certain pathogenic bacteria can live in the cytoplasm of free-living and nonpathogenic protozoa. For example, the ciliates *Tetrahymena* and *Cyclidium* can act as hosts to *Legionellae*.

Chapter 19

Disinfection

19.1 DESTROYING PATHOGENS AND REQUIREMENTS OF A DISINFECTANT

Disinfection of the outflow water from wetland systems might be required for particular applications such as recycling of treated water for irrigation purposes. Disinfection is carried out to destroy the microbiological agents that cause disease. It differs from sterilization, which implies the destruction of all microorganisms. Water supplies are usually disinfected. The widespread adoption of disinfection is a major factor in reducing waterborne diseases, and this has been interpreted as the major single factor in increasing average human life expectancy.

Chlorination, usually performed as the final treatment process, is the most common means of disinfecting predominantly drinking water. Other processes, natural and artificial, aid in the destruction or removal of pathogens.

Most pathogens are accustomed to the temperatures and conditions in the bodies of humans and animals; they do not survive outside such environments. Nonetheless, significant numbers can survive in potable water. Some pathogens, particularly certain viruses and those organisms that form cysts, can survive for long periods even under the most adverse conditions. Because such organisms also tend to be resistant to chlorine doses normally used in water treatment, chlorination alone cannot always ensure safe drinking water.

Storing water for extended periods of time in open tanks or reservoirs prior to treatment can accomplish some destruction of pathogens through sedimentation and natural death of the organisms. Significant pathogen removal occurs in the water and wastewater treatment processes.

A disinfectant must be able to destroy particular pathogens at the concentrations likely to occur, and it should be effective in the normal range of environmental conditions. Disinfectants, which require extremes of temperature or pH or which are effective only for low-turbidity waters, are unlikely to be suitable for large-scale use.

While the disinfectant should destroy pathogens, it must not be toxic to humans or other higher animals (e.g., fish) in a receiving watercourse. Ideally, some residual disinfecting capacity should be provided for a water supply network to provide protection against reinfection while the water is in the

Wetlands for Water Pollution Control. http://dx.doi.org/10.1016/B978-0-444-63607-2.00019-8

distribution system. The residual, which passes to the consumer, should be neither unpalatable nor obnoxious.

A disinfectant should be safe and easy to handle, both during storage and addition to the process water. The other major consideration is cost, which is particularly important for municipal plants where large volumes are commonly disinfected continuously.

19.2 TRADITIONAL METHODS OF DISINFECTION

Although chlorination is the most common technique for disinfection, other methods are available and may be useful in some situations. The three general types of disinfection commonly available are as follows:

- Heat treatment;
- Radiation treatment; and
- Chemical application.

Heat was probably the first method used for disinfection, and it is still an excellent method in emergencies for small quantities of water. Boiling water for 15–20 min (depending on altitude) destroys pathogens. Pasteurization, which is essentially heating to a temperature ≥60 °C, is effective in destroying many pathogens as well. Moreover, autoclaving, where the temperature is ≥100 °C (typically, 121 °C) at increased pressure, can also be used for the disinfection of small-scale wastewater treatment plants.

Heat is uneconomical for large-scale use, primarily because of the high cost of energy required. However, it is acceptable and commonly used in emergency situations. The water contains no residual, and care must be taken to prevent recontamination before use.

Concerning radiation treatment, sunlight (high-intensity ultraviolet radiation) is a natural means of disinfection. It works by breaking down long-chain organic molecules and subsequently kills pathogens. Solar drinking water disinfection is popular in some developing countries on a small-scale.

Since ultraviolet (UV) light is readily absorbed by water and "filtered" by turbidity, the normal penetration depth of radiation is less than 120 mm. This makes the process useful for small-scale disinfection only. No residual is provided, so care against reinfection must be exercised.

UV disinfection of water is growing in its usage. In both the United States and Australia, UV light is used to disinfect recycled water prior to its use for irrigation or landscaping purposes. The technology is now at a stage where the lamps are self-cleansing, and the power and number of tubes needed for a given log kill can be predicted.

19.3 OZONE

Ozone (O_3) is an allotrope of oxygen. It is formed when a high-voltage arc passes through the air between two electrodes. It is also formed photochemically in the atmosphere, and it is one of the constituents of smog. Ozone is a bluish and toxic gas with a pungent odor. It is considered hazardous to health at a concentration of 0.25 mg/l in air (by volume) and is extremely hazardous at levels of $\geq$1 mg/l. It is corrosive in nature, so only resistant materials like stainless steel and polyvinyl chloride (PVC) can be used to handle it. Ozone is unstable because it breaks down to give molecular oxygen. Its low solubility and instability require that it is to be generated on site and introduced into the water as fine bubbles.

A retention time between 5 and 10 min is usually sufficient to destroy pathogens. Ozone is an effective method of destroying pathogens. It is a strong oxidant and is effective for removing taste, odor, iron, manganese, and color constituents from waters. Its ability to completely oxidize organic materials minimizes the organic content of water and so inhibits the growth of lower life forms in water. It is not known to produce dangerous or objectionable compounds as chlorine does when it yields chlorophenols and trihalomethanes.

The only product from the addition of excess ozone is the breakdown product oxygen, and so nothing foreign is added. The breakdown of ozone does mean that there is no lasting residual disinfectant in the water. If a residual is required, it may be necessary to add chlorine or chlorine dioxide (Section 19.4) to act as a residual in the distribution system, while ozone is used as the primary disinfectant and for its other effects such as taste and odor suppression.

Ozone acts through the destruction of all protoplasm. Disinfection by ozone depends mainly upon the concentration of the gas added. Once the concentration of ozone falls below a critical value for each pathogen, its effectiveness as a disinfectant is significantly reduced. This behavior differs from that of chlorine.

Ozone is usually more expensive than chlorine; it requires a high energy input during production (approximately 10 to 15 times that of chlorine). It is only slightly soluble in water and leaves no residual. The difficulties associated with the production of toxic or noxious compounds by reaction with chlorine have led to a renewed interest in ozone as a disinfectant for water and effluents.

19.4 CHLORINE DIOXIDE AND CHLORINE AS DISINFECTANTS

Chlorine dioxide (ClO_2) is an unstable gas that can be generated by the addition of acid or chlorine to sodium chlorine. In water, it reacts to produce two unstable acids, chorus acid and chlorite acid, which can act as disinfectants. Chlorine dioxide is a highly reactive gas, and particular care is required to ensure its safe handling. Its instability requires it to be generated on site and used immediately.

Although chlorine dioxide is relatively expensive as a disinfectant, its strong oxidizing characteristics make it valuable for the control of taste, odor, iron, manganese, and color. Chlorine dioxide is not affected by ammonia compounds. It oxidizes compounds to form tasteless products and provides a persistent residual in the distribution system.

Chlorine is available in gaseous, liquid, and solid forms. Under normal conditions, chlorine is a green-yellow and corrosive gas with a density 2.5 times that of air. It can be liquefied under relatively small pressure (3.7 atm). The gas is very soluble in water and is a potent disinfectant even at low concentrations. When dissolved in water, chlorine forms two acids by reaction with molecules of water, as shown in Eq. (19.4.1).

$$
\begin{array}{ccccc}
 & & & & \text{(free available chlorine)} \\
Cl_2 + H_2O & \rightarrow & HCl & + & HOCl \\
 & pH = 9 & \uparrow\downarrow & pH = 5 & \uparrow\downarrow \\
 & & H^+ \; Cl^- & & H^+ \; ClO^- \\
 & & & & \text{(free residuals)}
\end{array}
\tag{19.4.1}
$$

Hydrochloric acid (HCl) is a common mineral acid and plays no significant role in disinfecting water. Hypochlorus acid (HOCl) is a disinfecting agent and is referred to as free available chlorine (HOCl). Both acids can dissociate to produce their respective anions and protons. For hypochlorus acid, this dissociation is important since the hypochlorite ion is not a disinfectant. The hyperchlorus acid is dissociated by 50% at pH 7.5. To ensure effective disinfection, the pH must be acidic so that at least 90% is in the undissociated and active disinfectant form (at pH 5, it is completely undissociated).

Chlorine rapidly penetrates microbial cells and subsequently kills the corresponding microorganisms through cell lyses. However, the effectiveness of chlorine is influenced greatly by the physical and chemical characteristics of the waste. The presence of SS or the clustering of microorganisms may protect pathogens and so reduce the disinfecting ability. Chlorine is a strong oxidizing agent, and any reducing agents such as nitrites, ferrous ions, and hydrogen sulfite rapidly react with it, reducing the concentration available to destroy pathogens.

Ammonia is present in both water and effluents. It reacts with chlorine in water to form sequentially monochloramine, dichloramine, and trichloramine. The reactions can be considered as successive replacements of hydrogen atoms in ammonia by chlorine (Eqs. (19.4.2)–(19.4.4)).

$$NH_3 + HOCl \rightarrow H_2O + NH_2Cl \quad \text{(monochloramine)} \tag{19.4.2}$$

$$NH_2Cl + HOCl \rightarrow H_2O + NHCl_2 \quad \text{(dichloramine)} \tag{19.4.3}$$

$$NHCl_2 + HOCl \rightarrow H_2O + NCl_3 \quad \text{(trichloramine)} \tag{19.4.4}$$

In addition, several other reactions may occur. The monochloramine and dichloramine may react further to give nitrogen gas (N_2) and/or nitrous oxide (N_2O). The proportions of any of these products depend upon the reaction conditions, notably concentrations, temperature, and pH.

Chlorine reacts rapidly with any reducing agents in water. This reduces chlorine to chloride, which is not a disinfectant. The residual chlorine is low, and there will be no disinfecting properties for a small chlorine dose.

The addition of more chlorine leads to complete oxidation. Monochloramines and dichloramines are formed by reaction with ammonia. The chloramines are disinfected and are detected as chlorine residual. However, they are much less powerful disinfectants than free chlorine. If they are present in high concentrations or if a long reaction time is permitted, these chloramines will destroy pathogens. Monochloramine and dichloramine are referred to as combined available chlorine. The addition of further chlorine produces an approximately proportional increase of combined chlorine residual.

Even further addition of chlorine produces trichloramine, nitrogen, nitrous oxide, and other products, which are not disinfectants. Therefore, the addition of further chlorine reduces the available chlorine and hence the ability of the solution to destroy pathogens.

On further addition of chlorine, these reactions are complete, and the ammonia is absolutely oxidized. Any subsequent addition of chlorine will remain as HOCl, which will act as a strong chlorine residual if the pH is adequate. This point is often referred to as breakpoint chlorination and is commonly carried out beyond the breakpoint to ensure free chlorine residual. The chlorine dose required to reach the breakpoint is several times the ammonia concentration.

The chlorine–ammonia reaction (described above) should be regarded as a simplified representation. The free and combined residual concentrations depend upon the reaction rate and boundary conditions for any particular sample.

19.5 KINETICS OF CHLORINATION

The ability of a reagent to destroy pathogens is related to the concentration of the disinfectant and the contact time between the disinfectant and pathogens. The distribution of microorganisms is controlled by processes of diffusion.

The rate of kill depends upon the number of microorganisms that were originally present. If the microorganisms all possess the same resistance, the kill follows an exponential pattern. A complete kill is not feasible. The efficiency of disinfection is reported in terms of the ratio of microorganisms killed to the number of microorganisms originally present (e.g., 99%, 99.9%, or 99.99% kills, sometimes also referred to as two, three, or four log kills).

The kill rate depends upon such factors as the penetration of the cell wall, the time to penetrate vital centers within the cell, and the distribution of the disinfectant and microorganisms. Therefore, each microorganism species is likely to have a different sensitivity to each disinfectant.

For disinfectants such as chlorine, the contact time is usually the most important single factor to ensure adequate pathogen destruction. A contact time of 0.5 h (or up to 1 h) with a residual afterward of between 0.2 and 0.5 mg/l is the norm in practice.

19.6 APPLICATIONS OF CHLORINE

Chlorine can be used directly as a disinfectant, in which case the active reagent is hypochlorus acid. Alternatively, combined chlorine residuals can be formed, and they can also act as disinfectants.

Typical contact times for hypochlorus acid in waters of satisfactory purity (e.g., post filtration) are around 0.5 h. Chloramines are weaker disinfectants and require contact times of the order of 1 h.

When chlorine is added to water, the residual is a function of both the dose and the time elapsed. The necessary conditions for an adequate kill of pathogens are incorporated in the dosing requirements (e.g., the total free residual after x h shall be not less than y mg/l).

In addition to disinfection, chlorine has some further benefits for the water supply industry. It is a strong oxidant, capable of oxidizing some materials in the water. Color, taste, and odor in water supplies are often of biological origin, and they are caused by the presence of organic molecules. Iron and manganese salts may be present, particularly in groundwater. They are usually in their reduced forms, and they precipitate out in the distribution system. Chlorine will oxidize some of these materials, reducing color, taste, and odor, and converting ferrous to ferric ions, which can be removed as a precipitate.

The presence of phenolic compounds in water supplies is a particular problem. Phenols (C_6H_5OH) react with chlorine to form chlorinated phenols, which have a very penetrating taste and odor. Conventional chlorination will produce chlorophenols and is therefore unsuitable if phenols are present. The use of chloramines sometimes avoids this problem.

The other use of chlorine in water treatment is as a prechlorination stage prior to rapid sand filtration (Chapter 11). This reduces the bacterial load on filters and enables longer filter runs before backwashing is needed. It provides additional safety in that a lower number of pathogens will be present for final disinfection, and microbial growth within the treatment works is also controlled.

19.7 TECHNOLOGY OF CHLORINE ADDITION

Chlorine may be purchased as solid calcium hypochlorite, liquid sodium hypochlorite, or liquid chlorine. Liquefied chlorine, available in drums or cylinders, is used at large treatment plants. The rate of removal of chlorine from a drum or cylinder is a function of temperature. The vapor pressure of chlorine is temperature dependent, and, as liquid chlorine vaporizes, latent heat for vaporization must be supplied. The temperature of a cylinder will be less than the average air temperature as the latent heat is absorbed. If chlorine is withdrawn too rapidly from a cylinder or drum, frosting or excessive sweating due to condensation occurs. In very damp climates, the rates of removal should be reduced.

Therefore, with high air temperature, it is possible to withdraw chlorine relatively quickly, whereas for low air temperature, slow withdrawal is needed. For example, it may be possible to use an entire cylinder in 4 days at an air temperature of 21 °C, while 30 days would be needed at 2 °C. At high temperatures, frothing of gas (while boiling) causes a carryover of impurities. Condensation of chlorine in gas lines should be avoided. A pressure-reducing valve attached near the cylinder or drum should keep the pressure in the pipes below the vapor pressure associated with the pipe temperature.

Chlorine is dissolved in a small part of the flow to be disinfected. This produces a strong solution of hypochlorus acid, which can then be added to the main stream to provide the required chlorine dose. The mechanism for dosing chlorine into the service water to produce strong hypochlorus acid requires careful handling and control.

The chlorinator includes a pressure-reducing valve, vacuum, pressure-relief valve, and antisuction valve. The high solubility of chlorine in water, combined with the hydraulic design of the injector, ensures that chlorine is drawn into the water under vacuum. This reduces the possibility of release of chlorine to the atmosphere if the equipment develops a leak. The hypochlorus acid solution can be released into a bulk water stream by turbulent diffusion. For example, at a Reynolds number of $\geq$4000, the chlorine will be dispersed well within approximately 10 standard pipe diameters. The chlorine can be dispersed at points of hydraulic discontinuity (e.g., near a submerged weir).

In practice, it is most convenient to control chlorination by adjustment of the chlorine concentration. Other possible variables such as contact time and form of chlorine residual are fixed by the design of the plant and the quality of the corresponding water. The control vacuum is varied manually or automatically on the basis of the measured flow rate and the determination of the chlorine residual.

19.8 ADVANTAGES AND DISADVANTAGES OF CHLORINE

Chlorine is the most widely used chemical disinfectant. It has found widespread use not only for large and small water supply systems but also for the treatment of swimming pool waters and effluent disinfection. Chlorine is a powerful disinfectant at low concentrations. Its ability to destroy pathogens without adversely affecting the future use of water is an advantage. Analytical methods have been developed that provide a simple, direct, and accurate measurement of chlorine residuals.

Chlorine is a dangerous chemical that requires careful handling. However, the widespread and long-standing use of chlorine-based products has resulted in the establishment of well-documented controls for safe operation. Moreover, it can be produced in large quantities relatively cheaply.

There are some disadvantages in the use of chlorine. Its reaction with phenolic compounds can make it unsuitable in some circumstances. Both free chlorine and chloramines have adverse effects upon some higher animals (notably fish). Chlorination has been shown to produce trihalomethane compounds such as chloroform (CH_3Cl) by reaction with organic matter in the water. These compounds are known as carcinogens. Chlorine is not effective in removing certain tastes, odors, and colors from water.

Chapter 20

Constructed Wetlands

20.1 BACKGROUND

Wetlands have been recognized as a natural resource throughout human history. Their importance in their natural state is appreciated by such people as the Marsh Arabs around the confluence of the Tigris and Euphrates rivers in southern Iraq, as well as in managed forms (e.g., rice paddies, particularly in Southeast Asia) (Mitsch and Gosselink, 2000).

The water purification capability of wetlands is being recognized as an attractive option in wastewater treatment. For example, the UK Environment Agency spends significant amounts of money on reed bed schemes in England and Wales. Such systems are designed, for example, to clean up mine water from collieries on which the constructed wetlands and associated community parks are being built.

Reed beds provide a useful complement to traditional sewage treatment systems. They are often a cheap alternative to expensive wastewater treatment technologies such as trickling filters and activated sludge processes (Cooper et al., 1996; Kadlec and Knight, 1996). Vertical-flow and horizontal-flow wetlands based on soil, sand, and/or gravel are used to treat domestic and industrial wastewater (Sun et al., 1999). They are also applied for passive treatment of diffuse pollution, including mine drainage (Mungur et al., 1997), as well as urban and motorway runoff after storm events (McNeill and Olley, 1998).

Furthermore, wetlands serve as a wildlife conservation resource and can also be seen as natural recreational areas for the local community (Hawke and José, 1996). The functions of macrophytes within constructed wetlands have been reviewed previously. *Phragmites* spp., *Typha* spp., and other macrophytes typical for swamps are widely used in Europe and North America (Brix, 1999).

20.2 DEFINITIONS

Defining wetlands has long been a problematic task, partly due to the diversity of environments, which are permanently or seasonally influenced by water, but also due to the specific requirements of the diverse groups of people involved with the study and management of these habitats. The Ramsar Convention, which brought wetlands to the attention of the international community,

Wetlands for Water Pollution Control. http://dx.doi.org/10.1016/B978-0-444-63607-2.00020-4

proposed the following definition (Convention on Wetlands of International Importance Especially as Waterfowl Habitat, 1971): "Wetlands are areas of marsh, fen, peatland, or water, whether natural or artificial, permanent or temporary, with water that is static or flowing, fresh, brackish, or salt, including areas of marine water, the depth of which at low tide does not exceed 6 m."

Another, more succinct, definition is as follows (Smith, 1980): "Wetlands are a half-way world between terrestrial and aquatic ecosystems and exhibit some of the characteristics of each." This complements the Ramsar description since wetlands are at the interface between water and land. This concept is particularly important in areas where wetlands may only be "wet" for relatively short periods of time during a year, such as in areas of the tropics with marked wet and dry seasons.

These definitions put emphasis on the ecological importance of wetlands. However, the natural water purification processes occurring within these systems have become increasingly relevant to those people involved with the practical use of constructed or even seminatural wetlands for water and wastewater treatment. There is no single accepted ecological definition of wetlands.

Wetlands are characterized by the following (United States Army Corps of Engineers, 2000):

- The presence of water;
- Unique soils that differ from upland soils; and
- The presence of vegetation adapted to saturated conditions.

Whichever definition is adopted, it can be seen that wetlands encompass a wide range of hydrological and ecological types, from high-altitude river sources (contradiction to United States Army Corps of Engineers (2000)) to shallow coastal regions, in each case being affected by prevailing climatic conditions. For the purpose of this chapter, however, the main emphasis will be upon constructed wetlands in temperate and oceanic climates.

20.3 HYDROLOGY OF WETLANDS

20.3.1 Hydro-Period and Water Budget

The biotic status of a wetland is intrinsically linked to the hydrological factors by which it is affected. These affect the nutrient availability as well as physicochemical variables such as soil and water pH and anaerobiosis within soils. In turn, biotic processes will have an impact upon the hydrological conditions of a wetland.

Water is the hallmark of wetlands. Therefore, it is not surprising that the input and output of water (i.e., water budget) of these systems determine the biochemical processes occurring within them. The net result of the water budget (hydro-period) may show great seasonal variations, but ultimately it delineates wetlands from terrestrial and fully aquatic ecosystems.

From an ecological point of view as well as an engineering one, the importance of hydrology cannot be overstated, as it defines the species diversity, productivity, and nutrient cycling of specific wetlands. That is to say, hydrological conditions must be considered if one is interested in the species richness of flora and fauna or if the interest lies in utilizing wetlands for pollution control.

The stability of particular wetlands is directly related to their hydro-period; that is, the seasonal shift in surface and subsurface water levels. The terms flood duration and flood frequency refer to wetlands that are not permanently flooded and give some indication of the time period involved in which the effects of inundation and soil saturation will be most pronounced.

Of particular relevance to riparian wetlands is the concept of flooding pulses as described by Junk et al. (1989). These pulses cause the greatest difference in high and low water levels and benefit wetlands by the inflow of nutrients and washing out of waste matter. These sudden and high volumes of water can be observed on a periodic or seasonal basis. It is particularly important to appreciate this natural fluctuation and its effects since wetland managers often attempt to control the level by which waters rise and fall. Such manipulation might be due to the overemphasis placed on water and its role in the life cycles of wetland flora and fauna, without considering the fact that such species have evolved in such an unstable environment (Fredrickson and Reid, 1990).

The balance between the input and output of water within a wetland is called the water budget. This budget is summarized by Eq. (20.3.1.1), where the volumetric units could be cubic meters:

$$\Delta V/\Delta t = P_n + S_i + G_i - ET - S_o - G_o \pm T \qquad (20.3.1.1)$$

where:

V = volume of water storage within a wetland;
$\Delta V/\Delta t$ = change in volume of water storage in a wetland per unit time (t);
P_n = net precipitation;
S_i = surface inflows, including flooded streams;
G_i = groundwater inflows;
ET = evapotranspiration;
S_o = surface outflows;
G_o = groundwater outflows; and
T = tidal inflow (+) or outflow (−).

20.3.2 Precipitation, Interception, Through-Fall, and Stem-Flow

In general terms, wetlands are most widespread in those parts of the world where precipitation exceeds water loss through evapotranspiration and surface runoff. The contribution of precipitation to the hydrology of a wetland is influenced by a number of factors. Precipitation such as rain and snow often passes through a canopy of vegetation before it becomes part of the wetland. The volume of water retained by this canopy is termed interception. Variables such as precipitation intensity and vegetation type will affect interception, for which median values of several studies have been calculated as 13% for deciduous forests and 28% for coniferous woodland (Dunne and Leopold, 1978).

The precipitation that remains to reach the wetland is termed the through-fall. This is added to the stem-flow, which is the water running down vegetation stems and trunks and is generally considered a minor component of a wetland water budget, such as 3% of through-fall in cypress dome wetlands in Florida (Heimburg, 1984). Thus, through-fall and stem-flow form Eq. (20.3.2.1), where the volumetric units could be cubic meters; this is the most commonly used precipitation equation for wetlands:

$$P_n = TF + SF \tag{20.3.2.1}$$

where:

P_n = net precipitation;
TF = through-fall; and
SF = stem-flow.

20.4 WETLAND CHEMISTRY

20.4.1 Oxygen

Because wetlands are associated with waterlogged soils, the concentration of oxygen within sediments and the overlying water is of critical importance. The rate of oxygen diffusion into water and sediment is slow, and this (coupled with microbial and animal respiration) leads to near-anaerobic sediments within many wetlands (Moss, 1998). These conditions favor rapid peat buildup, since decomposition rates and inorganic content of soils are low. Furthermore, the lack of oxygen in such conditions affects the aerobic respiration of plant roots and influences plant nutrient availability. Wetland plants have consequently evolved to be able to exist in anaerobic soils.

While the deeper sediments are generally anoxic, a thin layer of oxidized soil usually exists at the soil–water interface. The oxidized layer is important, since it permits the oxidized forms of prevailing ions to exist. This is in contrast to the reduced forms occurring at deeper levels of soil. The state of reduction or oxidation of iron, manganese, nitrogen, and phosphorus ions determines their role in nutrient availability and also toxicity. The presence of oxidized ferric iron (Fe^{3+}) gives the overlying wetland soil a brown coloration, whereas reduced sediments have undergone glaying, a process by which ferrous iron (Fe^{2+}) gives the sediment a blue-gray tint.

Therefore, the level of reduction of wetland soils is important in understanding the chemical processes that occur most likely in sediment and influence the corresponding above water column. The most practical way to determine the reduction state is by measuring the redox potential, also called the oxidation–reduction potential, of the saturated soil or water. The redox potential quantitatively determines whether a soil or water sample is associated with a reducing or oxidizing environment. Reduction is the release of oxygen and the gain of an electron (or hydrogen), whereas oxidation is the reverse (i.e., the gain of oxygen and loss of an electron). This is shown by Eq. (20.4.1.1) and explained in detail by Mitsch and Gosselink (2000).

$$E_H = E^0 + 2.3[RT/nF]\ \log[\{ox\}/\{red\}] \qquad (20.4.1.1)$$

where:

E_H = redox potential (hydrogen ion scale);
E^0 = potential of reference (mV);
R = gas constant = 81.987 cal/deg/mol;
T = temperature (°C);
n = number of moles of electrons transferred; and
F = Faraday constant = 23,061 cal/mole/volt.

Oxidation (and therefore decomposition) of organic matter (very reduced material) occurs in the presence of any electron acceptor, particularly oxygen; although NO^{3-}, Mn^{2+}, Fe^{3+}, and SO_4^{2-} are also commonly involved in oxidation, the rate will be slower in comparison with oxygen. A redox potential range between +400 mV and +700 mV is typical for environmental conditions associated with free dissolved oxygen. Below +400 mV, the oxygen concentration will begin to diminish and wetland conditions will become increasingly more reduced (>−400 mV).

Redox potentials are affected by pH and temperature, which influence the range at which particular reactions occur. The following thresholds are therefore not definitive:

- Once wetland soils become anaerobic, the primary reaction at approximately +250 mV is the reduction of nitrate (NO_3^-) to nitrite (NO_2^-), and finally to nitrous oxide (N_2O) or free nitrogen gas (N_2).
- At about +225 mV, manganese is reduced to manganous compounds. Under further reduced conditions, ferric iron becomes ferrous iron between approximately +100 mV and −100 mV, and sulfates become sulfides between approximately −100 and −200 mV.
- Under the most reduced conditions (<−200 mV) the organic matter itself and/or carbon dioxide will become the terminal electron acceptor. This results in the formation of low-molecular-weight organic compounds and methane gas ($CH_4 \uparrow$).

20.4.2 Carbon

Organic matter within wetlands is usually degraded by aerobic respiration or anaerobic processes (e.g., fermentation and methanogenesis). Anaerobic degradation of organic matter is less efficient than decomposition occurring under aerobic conditions.

Fermentation is the result of organic matter acting as the terminal electron acceptor (instead of oxygen, as in aerobic respiration). This process forms low-molecular-weight acids (e.g., lactic acid), alcohols (e.g., ethanol), and carbon dioxide. Therefore, fermentation is often central in providing further biodegradable substrates for other anaerobic organisms in waterlogged sediments.

The sulfur cycle is linked with the oxidation of organic carbon in some wetlands, particularly in sulfur-rich coastal systems. Low-molecular-weight organic compounds that result from fermentation (e.g., ethanol) are utilized as organic substrates by sulfur-reducing bacteria during the conversion of sulfate to sulfide (Mitsch and Gosselink, 2000).

Previous work suggests that methanogenesis is the principal carbon pathway in freshwater. Between 30% and 50% of the total benthic carbon flux has been attributed to methanogenesis (Boon and Mitchell, 1995).

20.4.3 Nitrogen

The prevalence of anoxic conditions in most wetlands has led to them playing a particularly important role in the release of gaseous nitrogen from the lithosphere and hydrosphere to the atmosphere through denitrification (Mitsch and Gosselink, 2000). However, the various oxidation states of nitrogen within wetlands are also important to the biogeochemistry of these environments.

Nitrates are important terminal electron acceptors after oxygen, making them relevant in the process of oxidation of organic matter. The transformation of nitrogen within wetlands is strongly associated with bacterial action. The activity of particular bacterial groups is dependent on whether the corresponding zone within a wetland is aerobic or anaerobic.

Within flooded wetland soils, mineralized nitrogen occurs primarily as ammonium (NH_4^+), which is formed through ammonification, the process by which organically bound nitrogen is converted to ammonium under aerobic or anaerobic conditions. Soil-bound ammonium can be absorbed through rhizome and root systems of macrophytes and be reconverted to organic matter, a process that can also be performed by anaerobic microorganisms.

The oxidized top layer present in many wetland sediments is crucial in preventing the excessive buildup of ammonium. A concentration gradient will be established between the high concentration of ammonium in the lower reduced sediments and the low concentration in the oxidized top layer. This may cause a passive flow of ammonium from the anaerobic to the aerobic layer, where microbiological processes convert the ion into further forms of nitrogen.

Within the aerobic sediment layer, nitrification of ammonium, firstly to nitrite (NO_2^-) and subsequently to nitrate (NO_3^-), is shown in Eqs. (20.4.3.1) and (20.4.3.2), preceded by the genus of bacteria predominantly involved in each process step. Nitrification may also take place in the oxidized rhizosphere of wetland plants.

$$\textit{Nitrosomonas}: \quad 2NH_4^+ + 3O_2 \rightarrow 2NO_2^- + 2H_2O + 4H^+ + \text{energy} \tag{20.4.3.1}$$

$$\textit{Nitrobacter}: \quad 2NO_2^- + O_2 \rightarrow 2NO_3^- + \text{energy} \tag{20.4.3.2}$$

A study in southern California indicated that denitrification was the most likely pathway for nitrate loss from experimental macrocosms and larger constructed wetlands (Bachand and Horne, 1999). Very high rates of nitrate–nitrogen removal were reported (2800 mg N/m^2/day). Furthermore, nitrate removal from inflow wastewater is generally lower in constructed wetlands compared to natural systems (Spieles and Mitsch, 2000). There is considerable interest in enhancing bacterial denitrification in constructed wetlands in order to reduce the level of eutrophication in receiving waters such as rivers and lakes (Bachand and Horne, 1999).

An investigation into the seasonal variation of nitrate removal showed maximum efficiency to occur during summer. This study also indicated a seasonal relationship in the pattern of nitrate retention, in which nitrate assimilation and denitrification are temperature dependent (Spieles and Mitsch, 2000).

Further evidence supporting the importance of denitrification was presented by Lund et al. (2000). The proportion of nitrogen removed by denitrification from a wetland in southern California was estimated by analyzing the increase in the proportion of the nitrogen isotope ^{15}N found in the outflow water. This method is based on the tendency of the lighter isotope ^{14}N to be favored by the biochemical thermodynamics of denitrification, thus reducing its proportion in water flowing out of wetlands in which denitrification is prevalent. Denitrification seems to be the favored pathway of nitrate loss from a treatment wetland, as this permanently removes nitrogen from the system, compared to sequestration within algal and macrophyte biomass.

In some wetlands, nitrogen may be derived through nitrogen fixation. In the presence of the enzyme nitrogenase, nitrogen gas is converted to organic nitrogen by organisms such as aerobic or anaerobic bacteria and cyanobacteria (blue-green algae). Wetland nitrogen fixation can occur in the anaerobic or aerobic soil layer, overlying water, in the rhizosphere of plant roots, and on leaf or stem surfaces. Cyanobacteria may contribute significantly to nitrogen fixation.

In northern bogs, which are often too acidic for large bacterial populations, nitrogen fixation by cyanobacteria is particularly important (Etherington, 1983). However, it should be noted that while cyanobacteria are adaptable organisms, they are affected by environmental stresses. For example, cyanobacteria are particularly susceptible to ultraviolet radiation, whereby their nitrogen metabolism (along with other functions) is impaired (Donkor and Häder, 1996).

20.4.4 Phosphorus

In wetland soils, phosphorus occurs as soluble or insoluble and organic or inorganic complexes. The phosphorus cycle is sedimentary rather than gaseous (as with nitrogen) and predominantly forms complexes within organic matter in peatlands or inorganic sediments in mineral soil wetlands. Over 90% of the phosphorus load in streams and rivers may be present in particulate inorganic form (Overbeck, 1988).

Soluble reactive phosphorus is the analytical term given to biologically available ortho-phosphate, which is the primary inorganic form. The availability of phosphorus to plants and microconsumers is limited due to the following main effects:

- Under aerobic conditions, insoluble phosphates are precipitated with ferric iron, calcium, and aluminum.
- Phosphates are adsorbed onto clay particles, organic peat, and ferric and aluminum hydroxides and oxides.
- Phosphorus is bound up in organic matter through incorporation into bacteria, algae, and vascular macrophytes.

There are three general conclusions about the tendency of phosphorus to precipitate with selected ions (Reddy et al., 1999):

1. In acid soils, phosphorus is fixed as aluminum and iron phosphates.
2. In alkaline soils, phosphorus is bound by calcium and magnesium.
3. The bioavailability of phosphorus is greatest at neutral to slightly acid pH.

The phosphorus availability is altered under anaerobic wetland soil conditions. The reducing conditions that are typical of flooded soils do not directly affect phosphorus. However, the association of phosphorus with other elements that undergo reduction has an indirect effect upon the phosphorus in the environment. For example, as ferric iron is reduced to the more soluble ferrous form, phosphorus as ferric phosphate is released into solution (Faulkner and Richardson, 1989; Gambrell and Patrick, 1978). Phosphorus may also be released into solution by a pH change brought about by organic, nitric, or sulfuric acids produced by chemosynthetic bacteria. Phosphorus sorption to clay particles is greatest under strongly acidic to slightly acidic conditions (Stumm and Morgan, 1996).

Great temporal variability in phosphorus concentrations of wetland influent in Ohio has been reported (Nairn and Mitsch, 2000). However, no seasonal pattern in phosphorus concentration was observed. This was explained by precipitation events and river flow conditions. Dissolved reactive phosphorus levels peaked during floods and on isolated occasions in late fall. Furthermore, sedimentation of suspended solids appears to be important in phosphorus retention within wetlands (Fennessy et al., 1994).

The physical, chemical, and biological characteristics of a wetland system affect the solubility and reactivity of different forms of phosphorus. Phosphate solubility is regulated by temperature (Holdren and Armstrong, 1980), pH (Mayer and Kramer, 1986), redox potential (Moore and Reddy, 1994), the interstitial soluble phosphorus level (Kamp-Nielson, 1974), and microbial activity (Gächter and Meyer, 1993; Gächter et al., 1988).

Where agricultural land has been converted to wetlands, there can be a tendency toward the solubilization of residual fertilizer phosphorus, which results in a rise of the soluble phosphorus concentration in floodwater. This effect can be reduced by physicochemical amendment, that is, by applying chemicals such as alum and calcium carbonate to stabilize the phosphorus in the sediment of these new wetlands (Ann et al., 1999a,b).

The redox potential has a significant impact on dissolved reactive phosphorus of chemically amended soils (Ann et al., 1999a,b). The redox potential can alter with fluctuating water table levels and hydraulic loading rates. Dissolved phosphorus concentrations are relatively high under reduced conditions, and they decrease with increasing redox potential. Iron compounds (e.g., $FeCl_3$) are particularly sensitive to the redox potential, resulting in the chemical amendment of wetland soils. Furthermore, alum and calcium carbonate are suitable to bind phosphorus even during fluctuating redox potentials.

Macrophytes assimilate phosphorus predominantly from deep sediments, thereby acting as nutrient pumps (Carignan and Kaill, 1980; Mitsch and Gosselink, 2000; Smith and Adams, 1986). The most important phosphorus retention pathway in wetlands is via physical sedimentation (Wang and Mitsch, 2000).

Model simulations on constructed wetlands in northeastern Illinois (USA) showed an increase in total phosphorus in the water column in the presence of macrophytes mainly during the nongrowing period, with little effect during the growing season. Most phosphorus taken from sediments by macrophytes is reincorporated into the sediment as dead plant material and therefore remains in the wetland indefinitely. Macrophytes can be harvested as a means to enhance phosphorus removal in wetlands. By harvesting macrophytes at the end of the growing season, phosphorus can be removed from the internal nutrient cycle within wetlands (Wang and Mitsch, 2000).

Moreover, the model showed a phosphorus removal potential of three-quarters of that of the phosphorus inflow. Therefore, harvesting would reduce phosphorus levels in upper sediment layers and drive phosphorus movement into deeper layers, particularly the root zone. In deep layers of sediment, the phosphorus sorption capacity increases along with a lower desorption rate (Wang and Mitsch, 2000).

20.4.5 Sulfur

In wetlands, sulfur is transformed by microbiological processes and occurs in several oxidation stages. Reduction may occur if the redox potential is between −100 and −200 mV. Sulfides provide the characteristic "bad egg" odor of wetland soils.

Assimilatory sulfate reduction is accomplished by obligate anaerobes such as *Desulfovibrio* spp. Bacteria may use sulfates as terminal electron acceptors (Eq. (20.4.5.1)) in anaerobic respiration at a wide pH range, but usage is highest around neutral pH (Mitsch and Gosselink, 2000).

$$4H_2 + SO_4^{2-} \rightarrow H_2S\uparrow + 2H_2O + 2OH^- \qquad (20.4.5.1)$$

The greatest loss of sulfur from freshwater wetland systems to the atmosphere is via hydrogen sulfide ($H_2S\uparrow$). In oceans, however, this is through the production of dimethyl-sulfide from decomposing phytoplankton (Schlesinger, 1991).

Oxidation of sulfides to elemental sulfur and sulfates can occur in the aerobic layer of some soils and is carried out by chemoautotrophic (e.g., *Thiobacillus* spp.) and photosynthetic microorganisms. *Thiobacillus* spp. may gain energy from the oxidation of hydrogen sulfide to sulfur, and further, by certain other species of the genus, from sulfur to sulfate (Eqs. (20.4.5.2) and (20.4.5.3)).

$$2H_2S + O_2 \rightarrow 2S + 2H_2O + \text{energy} \qquad (20.4.5.2)$$

$$2S + 3O_2 + 2H_2O \rightarrow 2H_2SO_4 + \text{energy} \qquad (20.4.5.3)$$

In the presence of light, photosynthetic bacteria, such as the purple sulfur bacteria of salt marshes and mud flats, produce organic matter as indicated in Eq. (20.4.5.4). This is similar to the familiar photosynthesis process, except that hydrogen sulfide is used as the electron donor instead of water.

$$CO_2 + H_2S + \text{light} \rightarrow CH_2O + S \qquad (20.4.5.4)$$

Direct toxicity of free sulfide in contact with plant roots has been noted. There is a reduced toxicity and availability of sulfur for plant growth if it precipitates with trace metals. For example, the immobilization of zinc and copper by sulfide precipitation is well known.

The input of sulfates to freshwater wetlands, in the form of Aeolian dust or as anthropogenic acid rain, can be significant. Sulfate deposited on wetland soils may undergo dissimulator sulfate reduction by reaction with organic substrates (Eq. (20.4.5.5)).

$$2CH_2O + SO_4^{2-} + H^+ \rightarrow 2CO_2 + HS^- + 2H_2O \qquad (20.4.5.5)$$

Protons consumed during this reaction (Eq. (20.4.5.5)) generate alkalinity. This is illustrated by the increase in pH with depth in wetland sediments (Morgan and Mandernack, 1996). It has been suggested that this alkalinity effect can act as a buffer in acid rain-affected lakes and streams (Rudd et al., 1986; Spratt and Morgan, 1990).

The sulfur cycle can vary greatly within different zones of a particular wetland. The stable isotope $\delta^{34}S$ within peat, the $^{35}SO_4^{2-}:Cl^-$ ratio, and the stable isotopes $\delta^{18}O$ and $\delta^{34}S$ of sulfate within different waters were analyzed by Mandernack et al. (2000). The variability in the sulfur cycle within the watershed can affect the distribution of reduced sulfur stored in soil. This change in local sulfur availability can have marked effects upon stream water over short distances.

The estimation of generated alkalinity may be complicated due to the potential problems associated with the use of the $^{35}SO_4^{2-}:Cl^-$ ratio and/or the $\delta^{34}S$ value, which are used to estimate the net sulfur retention. These problems may exist because ester sulfate pools can be a source of sulfate availability for sulfate reduction, and as a $\delta^{34}S$ sulfate buffer within stream water.

20.5 WETLAND ECOSYSTEM MASS BALANCE

The general mass balance for a wetland system, in terms of chemical pathways, uses the following main pathways: inflows, intrasystem cycling, and outflows. The inflows are mainly through hydrologic pathways such as precipitation, (particularly urban) surface runoff and groundwater. The photosynthetic fixation of both atmospheric carbon and nitrogen comprises important biological pathways. Intrasystem cycling is the movement of chemicals in standing stocks within wetlands, such as litter production and remineralization. Translocation of minerals within plants is an example of the physical movement of chemicals. Outflows involve hydrologic pathways but also include the loss of chemicals to deeper sediment layers, beyond the influence of internal cycling (although the depth at which this threshold occurs is not certain). Furthermore, the nitrogen cycle plays an important role in outflows, such as nitrogen gas lost as a result of denitrification. However, respiratory loss of carbon is also an important biotic outflow.

There is great variation in the chemical balance from one wetland to another, but the following generalizations can be made:

- Wetlands act as sources, sinks, or transformers of chemicals depending on wetland type, hydrological conditions, and length of time the wetland has received chemical inputs. As sinks, the long-term sustainability of this function is associated with hydrologic and geomorphic conditions as well as the spatial and temporal distribution of chemicals within wetlands.
- Particularly in temperate climates, seasonal variation in nutrient uptake and release is expected. Chemical retention will be greatest in the growing seasons (spring and summer) due to higher rates of microbial activity and macrophyte productivity.
- The ecosystems connected to wetlands affect and are affected by the adjacent wetland. Upstream ecosystems are sources of chemicals, while

those downstream may benefit from the export of certain nutrients or the retention of particular chemicals.
- Nutrient cycling in wetlands differs from that in terrestrial and aquatic systems. More nutrients are associated with wetland sediments than with most terrestrial soils, whereas benthic aquatic systems have autotrophic activity, which relies more on nutrients in the water column than in the sediments.
- The ability of wetland systems to remove anthropogenic waste is not limitless.

Equation (20.5.1) indicates a general mass balance for a pollutant within a treatment wetland. Within this equation, transformations and accretion are long-term sustainable removal processes, whereas storage does not result in long-term average removal but can lessen or accentuate the cyclic activity.

$$\begin{aligned}\text{In} - \text{Out} = {} & \text{Transformation} + \text{Accretion} + \text{Biomass Storage} \\ & + \text{Water/Soil Storage}\end{aligned} \tag{20.5.1}$$

20.6 MACROPHYTES IN WETLANDS

20.6.1 Background

Wetland plants are often central to wastewater treatment wetlands. The following requirements of plants should be considered for use in such systems (Tanner, 1996):

- Ecological adaptability (no disease or weed risk to the surrounding natural ecosystems);
- Tolerance of local conditions in terms of climate, pests, and diseases;
- Tolerance of pollutants and hypertrophic waterlogged conditions;
- Ready propagation and rapid establishment, spread, and growth; and
- High pollutant removal capacity through direct assimilation or indirect enhancement of nitrification, denitrification, and other microbial processes.

Interest in macrophyte systems for sewage treatment by the UK water industry dates back to 1985 (Parr, 1990). The ability of different macrophyte species and their assemblages within systems to most efficiently treat wastewater has been examined previously (Kuehn and Moore, 1995). The dominant species of macrophyte varies from locality to locality. The number of genera (e.g., *Phragmites* spp., *Typha* spp., and *Scirpus* spp.) common to all temperate locations is great.

The improvement of water quality with respect to key water quality variables, including BOD, COD, total SS, nitrates, and phosphates, has been studied by Turner (1995). Relatively little work has been conducted on the enteric bacteria removal capability of macrophyte systems (Perkins and Hunter, 2000).

20.6.2 Primary Productivity

There have been many studies to determine the primary productivity of wetland macrophytes, although estimates have generally tended to be fairly high (Mitsch and Gosselink, 2000). The estimated dry mass production for *Phragmites australis* (Cav.) Trin. ex Steud. (common reed) is between 1000 and 6000 g/m^2/year in the Czech Republic (Kvet and Husak, 1978), between 2040 and 2210 g/m^2/year for *Typha latifolia* L. (cattail) in Oregon, USA (McNaughton, 1966), and 943 g/m^2/year for *Scirpus fluviatilis* (Torr.) A. Gray [JPM][H&C] (river bulrush) in Iowa, USA (van der Valk and Davis, 1978).

Little of aquatic plant biomass is consumed as live tissue; rather, it enters the pool of particulate organic matter following tissue death. The breakdown of this material is consequently an important process in wetlands and other shallow aquatic habitats (Gessner, 2000). Litter breakdown has been studied along with intensive work on *P. australis*, one of the most widespread aquatic macrophytes (Wrubleski et al., 1997).

There has been an emphasis on studying the breakdown of aquatic macrophytes in such a way that most closely resembles natural plant death and decomposition, principally by not removing plant tissue from macrophyte stands. Many species of freshwater plants exhibit the so-called standing-dead decay, which describes the observation of leaves remaining attached to their stems after senescence and death (Kuehn et al., 1999). Different fractions (leaf blades, leaf sheaths, and culms) of *P. australis* differ greatly in structure and chemical composition and may exhibit different breakdown rates, patterns, and nutrient dynamics (Gessner, 2000).

20.6.3 *Phragmites australis*

Phragmites australis (Cav.) Trin. ex Steud. (common reed, formerly known as *Phragmites communis* (Norfolk reed)) is a member of the large family Poaceae (roughly 8000 species within 785 genera). Common reed occurs throughout Europe to 70° North and is distributed worldwide. It may be found in permanently flooded soils of still or slowly flowing water. This emergent plant is usually firmly rooted in wet sediment but may form lightly anchored rafts of "hover reed."

It tends to be replaced by other species at drier sites. The density of this macrophyte is reduced by grazing (e.g., consumption by waterfowl) and may then be replaced by other emergent species such as *Phalaris arundinacea* L. (reed canary grass).

Phragmites australis is a perennial, and its shoots emerge in spring. Hard frost kills these shoots, illustrating the tendency for reduced vigor toward the northern end of its distribution. The hollow stems of the dead shoots in winter are important in transporting oxygen to the relatively deep rhizosphere (Brix, 1989).

Reproduction in closed stands of this species is mainly by vegetative spread, although seed germination enables the colonization of open habitats. Detached shoots often survive and regenerate away from the main stand (Preston and Croft, 1997).

Common reed is most frequently found in nutrient-rich sites, and it is absent from the most oligotrophic zones. However, the stems of this species may be weakened by nitrogen-rich water and are subsequently more prone to wind and wave damage, leading to an apparent reduction in density of this species in Norfolk (England) and elsewhere in Europe (Boar et al., 1989; Ostendorp, 1989).

20.6.4 *Typha latifolia*

Typha latifolia L. (cattail, reedmace, or bulrush) is a species belonging to the small family Typhaceae. This species is widespread in temperate parts of the northern hemisphere but extends to South Africa, Madagascar, Central America, and the West Indies and has been naturalized in Australia (Preston and Croft, 1997). *Typha latifolia* is typically found in shallow water or on exposed mud at the edge of lakes, ponds, canals, and ditches and less frequently near fast-flowing water. This species rarely grows at water depths >0.3 m, where it is frequently replaced by *P. australis*.

Reedmace is a shallow-rooted perennial producing shoots throughout the growing season, which subsequently die in fall. Colonies of this species expand by rhizomatous growth at rates of 4 m/year, whereas detached portions of rhizome may float and establish new colonies. In contrast, colony growth by seeds is less likely. Seeds require moisture, light, and relatively high temperatures to germinate, although this may occur in anaerobic conditions. Where light intensity is low, germination is stimulated by temperature fluctuation (Hammer, 1989).

20.7 PHYSICAL AND BIOCHEMICAL PARAMETERS

The key physicochemical parameters relevant for wetland systems include the BOD, turbidity, and the redox potential. The BOD is an empirical test to determine the molecular oxygen used during a specified incubation period (usually 5 days) for the biochemical degradation of organic matter (carbonaceous demand) and the oxygen used to oxidize inorganic matter (e.g., sulfides and ferrous iron). An extended test (up to 25 days) may also measure the amount of oxygen used to oxidize reduced forms of nitrogen (nitrogenous demand), unless this is prevented by an inhibitor chemical (Scholz, 2004a). Inhibiting the nitrogenous oxygen demand is recommended for secondary effluent samples (Clesceri et al., 1998).

Upper BOD limits of around 3 mg/l for salmonid rivers and about 6 mg/l for coarse fisheries are traditionally set by fishery agencies. A river is deemed polluted if the BOD exceeds 5 mg/l. Municipal wastewater values are usually between approximately 150 and 1000 mg/l (Kiely, 1997).

Turbidity is a measure of the cloudiness of water, caused predominantly by suspended material such as clay, silt, organic and inorganic matter, plankton and other microscopic organisms, and scattering and absorbing light. Turbidity in wetlands and lakes is often due to colloidal or fine suspensions, whereas in fast-flowing waters the particles are larger and turbid conditions are prevalent predominantly during floods (Kiely, 1997).

The redox potential is another key parameter for monitoring wetlands. The reactivities and mobilities of elements such as iron, sulfur, nitrogen, carbon, and a number of metallic elements depend strongly on the redox potential conditions. Reactions involving electrons and protons are pH- and redox-potential-dependent. Chemical reactions in aqueous media can often be characterized by pH and the redox potential together with the activity of dissolved chemical species. The redox potential is a measure of intensity and does not represent the capacity of the system for oxidation or reduction (American Public Health Association, 1995). The interpretation of the redox potential values measured in the field is limited by a number of factors, including irreversible reactions, electrode poisoning, and multiple redox couples.

20.8 EXAMPLES FOR NATURAL AND CONSTRUCTED WETLANDS

20.8.1 Riparian Wetlands

Riparian wetlands are ecosystems under the influence of adjacent streams or rivers (Scholz and Trepel, 2004). A succinct definition is as follows (Gregory et al., 1991): "Riparian zones are the interface between terrestrial and aquatic ecosystems. As ecotones, they encompass sharp gradients of environmental factors, ecological processes and plant communities. Riparian zones are not easily delineated, but are composed of mosaics of landforms, communities and environments within the larger landscape." There are four main reasons as to why periodic flooding, which is typical of riparian wetlands, contributes to the observed higher productivity compared to adjacent upland ecosystems:

1. There is an adequate water supply for vegetation.
2. Nutrients are supplied and coupled with a favorable change in soil chemistry (e.g., nitrification, sulfate reduction, and nutrient mineralization).
3. In comparison to stagnant water conditions, a more oxygenated root zone follows flooding.
4. Waste products (e.g., carbon dioxide and methane) are removed by periodic flushing.

Nutrient cycles in riparian wetlands can be described as follows:

- Nutrient cycles are "open" due to the effect of river flooding, runoff from upslope environments, or both (depending on season and inflow stream or river type).
- Riparian forests have a great effect on the biotic interactions within intrasystem nutrient cycles. The seasonal pattern of growth and decay often matches available nutrients.
- Water in contact with the forest floor leads to important nutrient transformations. Therefore, riparian wetlands can act as sinks for nutrients.
- Riparian wetlands have often appeared to be nutrient transformers, changing a net input of inorganic nutrients to a net output of their corresponding organic forms.

The nitrogen cycle within a temperate stream—floodplain environment is of particular interest to ecological engineers. During winter, flooding contributes to the accumulation of dissolved and particulate organic nitrogen that is not assimilated by the trees due to their dormancy. This fraction of nitrogen is retained by filamentous algae and through immobilization by detritivores on the forest floor.

As the water of a wetland gets warmer in spring, nitrogen is released by decomposition and by shading of the filamentous algae by the developing tree canopy. Nitrate is then assumed to be immobilized in the decaying litter and gradually made available to plants. As vegetation increases, the plants take up more nitrogen, and water levels fall subsequently due to evapotranspiration. Ammonification and nitrification rates increase with exposure of the sediments to the atmosphere. Nitrates produced during nitrification are lost when denitrification becomes prevalent as flooding later in the year creates anaerobic conditions.

In terms of reducing the effects of eutrophication on open water by urban runoff, the use of riparian buffer zones, particularly of *Alnus incana* (Gray Alder) and *Salix* spp. (Willow) in conjunction with perennial grasses, has been recommended (Mander et al., 1995). Riparian zones are also termed riparian forest buffer systems (Lowrance et al., 1979). Such zones were found to reduce the nutrient flux into streams.

The role of riparian ecosystems in nutrient transformations is specifically important in relation to the production of the greenhouse gas nitrous oxide (N_2O). Due to the inflow of excess agricultural nitrogen into wetland systems, the riparian zones in particular are likely hot spots for nitrous oxide production (Groffman, 2000).

The control of nonpoint source pollution can be successfully achieved by riparian forest buffers in some agricultural watersheds and, most effectively, if excess precipitation moves across, in, or near the root zone of the riparian forest buffers. For example, between 50% and 90% retention of total nitrate loading in both shallow groundwater and sediment subject to surface runoff within the Chesapeake Bay watershed (USA) was observed. In comparison, phosphorus retention was found to be generally less than nitrate retention (Lowrance et al., 1979).

20.8.2 Constructed Treatment Wetlands

Natural wetlands usually improve the quality of water passing through them, acting effectively as ecosystem filters. In comparison, most constructed wetlands are artificially created wetlands used to treat water pollution in its variety of forms. Therefore, they fall into the category of constructed treatment wetlands. Treatment wetlands are solar-powered ecosystems. Solar radiation varies diurnally, as well as on an annual basis (Kadlec, 1999).

Constructed wetlands have the purpose to remove bacteria, enteric viruses, SS, BOD, nitrogen (predominantly as ammonia and nitrate), metals, and phosphorus (Pinney et al., 2000). Two general types of constructed wetlands are usually commissioned in practice: surface-flow (i.e., horizontal-flow) and subsurface-flow (e.g., vertical-flow). Surface-flow constructed wetlands most closely mimic natural environments and are usually more suitable for wetland species because of permanent standing water. In subsurface-flow wetlands, water passes laterally through a porous medium (usually sand and gravel) with a limited number of macrophyte species. These systems often have no standing water.

Constructed treatment wetlands can be built at, above, or below the existing land surface, if an external water source is supplied (e.g., wastewater). The grading of a particular wetland in relation to the appropriate elevation is important for the optimal use of the wetland area in terms of water distribution. Soil type and groundwater level must also be considered if long-term water shortage is to be avoided. Liners can prevent excessive desiccation, particularly where soils have a high permeability (e.g., sand and gravel) or where there is limited or periodic flow.

Rooting substrate is also an important consideration for the most vigorous growth of macrophytes. A loamy or sandy topsoil layer between 0.2 and 0.3 m in depth is ideal for most wetland macrophyte species in a surface-flow wetland. A subsurface-flow wetland will require coarser material such as gravel and/or coarse sand (Kadlec and Knight, 1996).

Furthermore, the use of flue-gas-desulfurization by-products from coal-fired electric power plants in wetland liner material was researched (Ahn et al., 2001). These by-products are usually sent to landfill sites. This is now recognized as an increasingly unsuitable and impractical waste disposal method. Although the study was short (2 years), no detrimental impact on macrophyte biomass production was reported. Moreover, flue-gas-desulfurization material may be a good liner and substrate for phosphorus retention in constructed wetlands.

The following conclusions with implications for treatment wetland design were made by Pinney et al. (2000):

- High levels of dissolved organic carbon may enter water supplies where soil aquifer treatment is used for groundwater recharge, as the influent for this method is likely to come from long-hydraulic-retention-time wetlands. Consequently, there is a greater potential for the formation of disinfection by-products.
- A shorter hydraulic retention time will result in less dissolved organic carbon leaching from plant material compared to a longer hydraulic retention time in a wetland.
- Dissolved organic carbon leaching is likely to be most significant in wetlands designed for the removal of ammonia, which requires a long hydraulic retention time.

20.8.3 Wetlands for Stormwater Treatment

Most constructed wetlands in the USA and Europe are soil- or gravel-based horizontal-flow systems planted with *T. latifolia* and/or *P. australis*. They are used to treat urban runoff and domestic and industrial wastewater (Cooper et al., 1996; Kadlec and Knight, 1996; Scholz, 2003; Scholz et al., 2005), and they have also been applied for passive treatment of mine wastewater drainage (Mays and Edwards, 2001; Mungur et al., 1997).

Storm runoff from urban areas has been recognized as a major contributor to pollution of the corresponding receiving urban watercourses. The principal pollutants in urban runoff are BOD, SS, heavy metals, deicing salts, hydrocarbons, and fecal coliforms (Scholz and Martin, 1998a; Scholz, 2004b).

Although various conventional methods have been applied to treat stormwater, most technologies are not cost-effective or are too complex. Constructed wetlands integrated into a best management practice concept are a sustainable means of treating stormwater and prove to be more economic (e.g., construction and maintenance) and energy efficient than traditional centralized treatment systems (Kadlec et al., 2000; Scholz et al., 2005). Furthermore, wetlands enhance biodiversity and are less susceptible to variations of loading rates (Cooper et al., 1996; Scholz and Trepel, 2004).

Contrary to standard domestic wastewater treatment technologies, stormwater (e.g., gully pot liquor) treatment systems have to be robust to highly variable flow rates and water quality variations. The stormwater quality depends on the load of pollutants present on the road and the corresponding dilution by each storm event (Scholz, 2003; Scholz and Trepel, 2004).

In contrast to standard horizontal-flow constructed treatment wetlands, vertical-flow wetlands are flat and intermittently flooded and drained, allowing air to refill the soil pores within the bed (Cooper et al., 1996; Gervin and Brix, 2001; Green et al., 1998). While it has been recognized that vertical-flow constructed wetlands have usually higher removal efficiencies with respect to organic pollutants and nutrients in comparison to horizontal-flow wetlands, denitrification is less efficient in vertical-flow systems (Luederits et al., 2001). When the wetland is dry, oxygen (as part of the air) can enter the top layer of debris and sand. The following incoming flow of runoff will absorb the gas and transport it to the anaerobic bottom of the wetland.

Furthermore, aquatic plants such as macrophytes transport oxygen to the rhizosphere. However, this natural process of oxygen enrichment is not as effective as the previously explained engineering method (Kadlec and Knight, 1996; Karathanasis et al., 2003).

Heavy metals within stormwater are associated with fuel additives, car body corrosion, and tire and brake wear. Common metal pollutants from cars include copper, nickel, lead, zinc, chromium, and cadmium. Freshwater quality standards are most likely to be exceeded by copper (Cooper et al., 1996; Scholz et al., 2002; Tchobanoglous et al., 2003).

Metals occur in soluble, colloidal, or particulate forms. Heavy metals are most bioavailable when they are soluble, in either ionic or weakly complexed form (Cooper et al., 1996; Cheng et al., 2002; Wood and Shelley, 1999).

There have been many studies on the specific filter media within constructed wetlands to treat heavy metals economically. Media used include limestone, lignite, activated carbon (Scholz and Martin, 1998a), peat, and leaves. Metal bioavailability and reduction are controlled by chemical processes including acid volatile sulfide formation and organic carbon binding and sorption in reduced sediments of constructed wetlands (Kadlec, 2002; Obarska-Pempkowiak and Klimkowska, 1999; Wood and Shelley, 1999). It follows that metals usually accumulate in the top layer (fine aggregates, sediment, and litter) of vertical-flow and near the inlet of horizontal-flow constructed treatment wetlands (Cheng et al., 2002; Scholz and Xu, 2002; Vymazal and Krasa, 2003).

Physical and chemical properties of the wetland soil and aggregates affecting metal mobilization include particle size distribution (texture), redox potential, pH, organic matter, salinity, and the presence of inorganic matter such as sulfides and carbonates (Backstrom et al., 2004). The cation exchange capacity of maturing wetland soils and sediments tends to increase as texture becomes finer because more negatively charged binding sites are available. Organic matter has a relatively high proportion of negatively charged binding sites. Salinity and pH can influence the effectiveness of the cation exchange capacity of soils and sediments because the negatively charged binding sites will be occupied by a high number of sodium or hydrogen cations (Knight et al., 1999).

Sulfides and carbonates may combine with metals to form relatively insoluble compounds. Particularly the formation of metal sulfide compounds may provide long-term heavy metal removal because these sulfides will remain permanently in the wetland sediments as long as they are not reoxidized (Cooper et al., 1996; Kadlec and Knight, 1996).

Chapter 21

Sludge Treatment and Disposal

21.1 INTRODUCTION

Wetland systems produce considerable amounts of sludge and sediment. It can be said that the main objective in wastewater treatment with treatment wetlands is to convert a water-based pollution problem to one that concerns solid waste management. The treatment and disposal of sewage solids (sludge) is a major component in the overall treatment costs and should therefore be considered. The treatment and disposal of sludge has historically been an area that has attracted little attention. However, as overall costs continue to rise and the available sites for disposal decrease, sludge treatment has begun to become an area of interest for environmental engineers.

The main objective of all forms of wastewater treatment is the production of an effluent suitable for disposal into the environment or for some form of reuse. It follows that the impurities present in the wastewater must be either transformed into innocuous end products or effectively separated from the effluent stream. Impurities are either drawn off as a side stream to the main flow or converted into gaseous products. Treatment and disposal of side streams is an essential part of the overall treatment process, and they frequently contribute significantly to the total cost of the treatment.

In wastewater treatment works, the main side stream products (apart from screenings and grit) are the various forms of sludge from the underflows of sedimentation tanks. Treatment and disposal of these sludges is dependent on the volume and characteristics of the sludges.

The simplest classification of wastewater sludges is based on the process from which they are produced. Raw or primary sludge is drawn from the primary sedimentation tanks. It contains readily settleable matter from wastewater, has a high organic content (mainly fecal matter and food scraps), and is therefore highly putrescible. In its fresh state, raw sludge is gray in color and has a heavy fecal odor. Both color and odor intensify on prolonged storage under anoxic conditions, leading rapidly to the onset of putrefaction and very unpleasant odors. This is often evident in small works, when sludge is drawn from the sedimentation tank into an open pit for transfer to a digestion tank.

Wetlands for Water Pollution Control. http://dx.doi.org/10.1016/B978-0-444-63607-2.00021-6

Humus sludge comprises the underflow from humus tanks, which are usually located after trickling filter units. It consists of biological solids sloughed off or scoured from the surface of the filter media and thus represents the net growth of biomass in the filters. The sloughing rate varies throughout the year but it is usually most intense during spring. Being largely organic in content, humus sludge exhibits problems similar to those of primary sludge under anoxic or anaerobic conditions.

Excess activated sludge (AS; Chapter 15) is usually drawn off the return AS stream from the underflow of the secondary sedimentation tanks in the AS treatment process. It consists of light flocculent biological solids with a significant demand for oxygen, which is largely owing to the respiration of the sludge microorganisms. The mass of excess sludge produced each day represents the net growth in biomass and is related to the loading rate of the process. Typically, AS has much higher moisture content than other organic wastewater sludges, and therefore it presents greater dewatering problems.

Chemical sludge is produced by processes involving chemical coagulation or chemical precipitation. In conventional wastewater treatment, such processes are seldom used. Nevertheless, they appear to offer the most economical means of removing phosphorus from wastewater and for the treatment of intermittent flows such as combined sewage overflows. Such sludges comprise mainly the reaction products of the added chemicals and the impurities to be removed. When chemicals are applied to raw sewage, the sludge produced is a mixture of chemical sludge and organic raw sludge. If the chemical is dosed following full secondary treatment, there will be relatively little organic contamination of the sludge.

Digested sludge is the product of either aerobic or anaerobic digestion and is a well-stabilized material capable of being dewatered on open drying beds without severe odor problems. Well-stabilized anaerobically digested sludge has a black appearance with a tarry odor; it is nonputrescible and is therefore no longer attractive to flies and other insects.

The rational selection of sludge treatment and disposal units requires the definition of both the volume and the characteristics of the sludge to be treated. The operations involved can be categorized under the general headings of conditioning, thickening, stabilization, and dewatering, with the associated processes of storage, transport, and ultimate disposal.

The overall aim of municipal wastewater treatment is to concentrate solids, which comprise $\leq 0.05\%$ of the mass of the raw wastewater, in sludges having gross solids contents generally in the range of $1-10\%$.

The total solids residue is the variable that measures the gross solids content. It is determined by evaporating the water content of a measured amount of sludge, weighing the residue, and expressing it as a percentage of the original wet sludge weight. The sludge moisture content is an alternative variable, commonly quoted as a measure of the gross sludge composition.

TABLE 21.1.1 Moisture and Organic Content of Sludges

Sludge Type	Moisture Content (% by Weight)	Organic Content (% Dry Weight)
Primary sludge	93–97	48–80
Activated sludge	98–99.5	65–75
Digested sludge	96–99	30–60
Humus sludge	94–99	65–75
Sludge associated with high-rate plastic media	92–97	65–75

The volatile solids content is measured as the weight loss on ignition of the dried sludge solids obtained from the total solids (TS) residue test at a standard temperature (usually 600 °C). It is a measure of the organic content of the sludge. It is therefore related to the possible reduction in the sludge mass by incineration processes, and it gives an indication of the degree of stabilization, which could be achieved by biological treatment processes. The volatile solids content is usually quoted as a percentage of the TS residue. The solids content remaining after ignition (ash) is termed the fixed residue and defines the weight of inorganic matter in the sludge (and thus the minimum weight of solids) that would remain for ultimate disposal after incineration.

A more detailed chemical analysis of the solids component is required where disposal methods such as composting and land application are contemplated. Such analyses should involve the determination of nutrients such as carbon, nitrogen, phosphorus, and potassium.

The most important characteristics of sludges are their organic and moisture contents. A high organic content in an unstabilized sludge indicates that the sludge will continue to degrade and is therefore likely to present health, odor, and rodent problems. Table 21.1.1 summarizes the typical moisture and organic content ranges of several sludges.

21.2 TESTS FOR DEWATERING OF SLUDGE

Since sludges commonly contain 1–10% solids (by weight), their major component is therefore water. Moreover, since sludge solids (called biosolids in the United States to highlight their recycling value as a semi-natural resource) are of similar density to water, the water content accounts for most of the volume of wet sludges. The sludge moisture content is therefore the single variable that has the greatest effect on the volume of sludge to be processed at a given plant.

Furthermore, removing water from sludge with low solids content affords a considerable reduction in volume. Doubling the solids content from 1% to 2% usually halves the volume of the wet sludge.

Wherever sludges have to be disposed of in restricted land areas or transported over long distances for ultimate disposal, some form of volume reduction is usually necessary. Sludge dewatering is an effective method of volume reduction in such cases. It is also an essential pretreatment method, where incineration is required.

Dewatering processes in common use, such as pressure filters, vacuum filters, and centrifuges, require for their design some measure of the sludge dewatering characteristics. Two "competing" methods are used to measure the ease of dewatering: specific resistance (to filtration) and capillary suction time (CST). The latter is most commonly used as a rapid and mobile test and can be measured by CST measurement equipment manufactured by Triton Electronics Ltd (England).

21.3 SLUDGE TREATMENT AND DISPOSAL OBJECTIVES AND METHODS

The main objectives of sludge treatment and disposal are as follows:

- Stabilization of the organic matter contained in the sludge;
- Reduction in the volume of sludge for disposal by removing some of the water;
- Destruction of pathogens;
- Collection of by-products, which may be used or sold to offset some of the costs of sludge treatment; and
- Disposal of the sludge in a safe and aesthetically acceptable manner.

Unfortunately, the recycling of by-products (see above) is an "ideal scenario" rarely achieved in practice, except in the case of methane gas, which is produced in anaerobic digestion. The methane is often collected and used as a fuel to provide heat for controlling the temperature of the digesting sludge and, occasionally, for driving dual fuel engines, which may be used to generate power for the treatment plant. The production of compost and the use of sludges for agricultural purposes can be viewed as waste recycling.

Sludge treatment and disposal at any particular location may comprise any or all of the following steps:

1. Concentration: reduction in the volume of sludge to be treated by encouraging the sludge to compact to higher solids content.
2. Treatment to stabilize organic matter: destruction of pathogens and/or yield of by-products.
3. Dewatering and drying: removal of water, thus reducing the sludge volume. Sludges with $\leq$80% moisture content are usually spadeable.

TABLE 21.3.1 Treatment and Disposal Options for Wastewater Sludges

Thickening	Stabilization	Dewatering	Partial Disposal	Ultimate Disposal
Gravity	Anaerobic digestion	Drying bed	Incineration	Sanitary landfill
Flotation	Aerobic digestion	Filter press	Pyrolysis	Cropland
Centrifuge	Lagooning	Centrifuge	Wet air oxidation	Ocean
Elutriation	Heat treatment	Vacuum filter, belt press, and lagooning	Composting	
Constructed wetland (drying bed)	Constructed wetland	Constructed wetland		

4. Disposal: the only places where sludges can be disposed of are into the air, onto land, or into water. Whether or not the impact on the receiving environment is legally, aesthetically, and ecologically acceptable depends on both the degree of treatment provided and the method of dispersing the sludge into the environment.

In comparing the various alternative methods of sludge treatment and disposal, it is important to consider all steps necessary for the final disposal of the treated sludge. An overview of the majority of available processes is shown in Table 21.3.1. Not all of these methods or types of methods would be practiced at any one site, and not all of the methods are necessarily compatible or legal. For each site, the range of options requires consideration.

The most commonly used method of sludge stabilization is anaerobic digestion. Digested sludges are suitable for direct disposal, or they can be further treated and subsequently handled without odor or hygiene problems. The principles of anaerobic digestion are set out in earlier chapters and should be incorporated within a sludge treatment scheme. The sludge produced is amenable to dewatering. It has a reduced and stabilized organic content, and it is hygienic. However, sludge still requires final disposal.

Digested sludge is often further treated and dewatered in drying beds or in lagoons. Mechanical dewatering further reduces sludge volumes and is a necessity for incineration. Ultimate disposal of any residue or sludge is to land or sea, though some national and international guidelines have stopped or restricted these options. If nondigested sludges are being handled, particular attention has to be paid to odor, insect and rodent infestation, and hygiene. Subsequent sections deal with particular processes in greater detail.

21.4 TREATMENT PROCESSES

21.4.1 Lagoons

The most common treatment process is anaerobic digestion, and the digested sludge can be further treated in a lagoon. Aerobic digestion is employed particularly at small treatment plants. This process includes a mixture of cold digestion, air drying, and gravity thickening. Sludge lagoons are usually designed on a simple "volume per capita" basis (0.2–0.5 m^3/cap) for an expected service between 7 and 15 years before desludging is required. A water depth between 3 and 5 m is provided, and a freeboard of $\geq$1 m is required.

Internal slopes at the sides should have a width-to-height ratio of 3:1. Moreover, the sides should be lined with impervious material above, and to about 1 m below the top water level in the lagoon. Inlet arrangements should ensure even distribution of the digested sludge flow over the whole area of the lagoon floor. An outlet weir should be provided to draw off displaced water, which should then be returned to the treatment plant inlet. It is common practice to construct two lagoons; one is filled first and can be emptied while the second is filling up. The lagoons should not pollute the groundwater. Therefore, they require lining in some locations.

The sludges removed from a lagoon vary in solids content. Depending on the influent sludge, the lower compacted layers can be of 20% solids content, while surface layers contain only a few percent solids. The sludges removed from the lagoon require final disposal.

21.4.2 Aerobic Digestion

Sludges can be stabilized by an aerobic process. They are maintained in an aerobic condition by external aeration, through which organic material is oxidized. This process is widely used for sludges from small extended-aeration AS plants (Chapter 15).

An aerobic digester has modest construction requirements when compared to an anaerobic digester because it does not require gas collection. The supernatant is well oxidized so that any return flows do not impose heavy loads on the rest of the treatment plant. The loss of methane, if compared with anaerobic digestion, is not always significant when the capital investment and running costs for using methane are taken into account.

21.4.3 Other Treatment Methods

Several other treatment methods are used in restricted circumstances. These include freezing and thawing of sludge to improve its dewatering characteristics. A low-pressure coolant such as butane can be employed, but it tends to dissolve greases and oils from the sludge, reducing its useful life. The fragile nature of the frozen and thawed sludge particles, together with cost, has prevented widespread use of this technique. Where freezing and thawing can be provided naturally (e.g., storage through several cold winters), the method is very effective.

Heat treatment processes such as composting can also be considered as sludge treatment processes but are usually considered to be partial disposal methods. Other processes such as ultrasonic treatment have only restricted use owing to operational problems and relatively high costs.

21.5 THICKENING AND DEWATERING OF SLUDGES

21.5.1 Chemical Conditioning

The difference between thickening and dewatering is arbitrary. Thickening can be considered to produce sludge of $\leq$10% solids content and tends to precede a sludge treatment process. Dewatering produces a sludge of $>$10% solids content and is used after a sludge treatment process. Similar techniques can be used for thickening and dewatering. A centrifuge operated with a high throughput produces a thickened sludge, but it can also be used to produce sludges of 20% solids content. Both thickening and dewatering methods can be improved by the addition of coagulants or conditioning agents.

The addition of certain chemicals increases the rate of water loss by sludge. Chemical conditioning is sometimes used in conjunction with drying beds and gravity thickening and is a prerequisite for other processes such as filter presses, vacuum filters, and centrifuges. The common conditioners are iron salts ($FeCl_3$, $FeSO_4Cl$, and $Fe(SO_4)_3$), alum ($Al_2(SO_4)\cdot 18H_2O$), limes (CaO and $Ca(OH)_2$), and polyelectrolytes. All these conditioners act by binding sludge particles together and permitting water to run out. Combinations of inorganic salts and polyelectrolytes can also be effective.

Lime is used to condition primary and digested sludges prior to filter pressing or vacuum filtration; it is added as 10–25% of the dry sludge solids. Applying lime to primary sludges suppresses odor formation. Iron salts can be expensive and difficult to obtain. Alum is also used prior to vacuum and pressure filtration and drying beds. Polyelectrolytes appear to make the sludge floc stronger and less prone to collapse during mechanical treatment. Polyelectrolytes are widely used in association with centrifugation.

21.5.2 Air Drying

Sludges can be air-dried in drained drying beds, but the constraints imposed by odor control limit this method to digested sludges. Rectangular beds are usually constructed in brick or concrete with a suitable underdrain system overlaid with sand, clinker, or both materials. The beds should have a gradient of 1 in 200 to facilitate drainage. Design should promote access by foot or mechanical equipment and should provide adequate underflow. At least four drying beds, preferably more, are required to allow a suitable drying interval between sludge applications to each bed.

Sludge is run to a depth of approximately 300 mm. After a few days, excess liquor can be removed and returned to the treatment plant. The sludge can be lifted at approximately 70–75% moisture content when it is spadeable or preferably at 0–55% moisture content, when it appears to be solid.

The size of a drying bed depends upon the local climate and sludge characteristics since water is lost by evaporation and percolation. Design figures between 0.1 and 0.5 m^2/cap provide a suitable guide.

Moreover, air drying can be combined with constructed wetland drying beds. However, this is only economically feasible if evapotranspiration rates are high. This technique is therefore limited to warm climates.

21.5.3 Gravity Thickening

Waste sludge can be thickened by allowing it to resettle. The general principles of settling apply to this method, with particles undergoing hindered settlement while the bottom sludges are compressed.

Surface loadings for this process are usually between 300 and 500 m^3/m^2/day. Primary sludges can be thickened at 150 kg/m^2/day and give thickened sludge of 10% solids content. Activated sludge loads are approximately 25 kg/m^2/day and subsequently yield sludges with 25% solids. This process is a useful pretreatment method prior to mechanical methods of dewatering. The effectiveness of gravity thickening can be increased by the addition of chemical conditioners. Care has to be taken during the operation of gravity thickening to minimize odors. If an undigested sludge is held for several hours, it becomes anaerobic and hence very offensive.

21.5.4 Other Methods

Washing sludge with water or effluent is known as elutriation. It removes ammonia compounds, which would otherwise interfere with coagulants, as well as very fine suspended matter, which is difficult to dewater. Elutriation is most effective if carried out as several washings, and it is often used for primary and digested sludges prior to chemical conditioning and filtration.

Pressure filtration is a batch process in which conditioned sludge is pumped slowly with increasing pressure between filter plates, which support cloths to retain the solids and permit the passage of liquid. The pressure applied reaches approximately 700 kPa, the resultant cake is between 20 and 40 mm thick, and 40% dry solids are achieved. The process is intermittent. Modern presses have minimized the labor requirements for desludging.

In a vacuum filtration process, a continuous-filter septum is partially immersed in the conditioned sludge and is slowly rotated through it. A vacuum of approximately 85 kPa is applied inside the drum, and so liquid (i.e., water) is drawn into the drum. A cake of dewatered sludge accumulates on the outside of the drum to a thickness of approximately 50 mm. The cake is removed by scrapers. By this method, primary and digested sludges can be concentrated to 32% solids and activated sludges (Chapter 15) to 25% solids.

With the centrifuge method, conditioned sludge is added to a rapidly (>0 rev/s) rotating bowl. The rotation separates water and sludge. Sludge solids concentrate at the outer side of the bowl, where they are removed by a screw conveyor or scroll. The sludge solids are removed at one end of the bowl and the liquid at the opposite end. The method is compact but requires careful control of the process variables. Centrifuges will concentrate sludges to 20% solids, but they are more commonly employed to achieve solids concentrations of approximately 10%.

Several other methods can be applied to dewater sludges:

- Air flotation, which uses fine air bubbles to carry sludge solids to the surface.
- Filter belts, which compress sludge between two endless belts.
- Heat treatment processes, in which the sludge is heated to a high temperature (180 °C–250 °C) under pressure.
- Wetland systems, which are used to dry sludge via high evapotranspiration rates (efficient in warm climates only).
- Several hybrid methods that include various combinations of the methods covered above.

21.6 PARTIAL DISPOSAL

21.6.1 Incineration

Sludges can be burned to produce ash, which contains very little water and small amounts of organic matter. The sludge is therefore reduced to the nonvolatile fraction. A sludge containing 30% solids, of which 50% are volatile, would reduce to approximately 15% of the original wet sludge volume. Stringent controls to prevent secondary pollution are required on any gaseous or liquid discharge from incinerators.

Sludge solids have calorific values similar to conventional fossil fuels (e.g., coal and oil: 20,000–50,000 kJ/kg). Thus, dry sludges can be burned with no additional fuel consumption. In practice, the fuel value is reduced considerably by the moisture in the sludge so that effective dewatering is necessary prior to incineration. Designs are usually based on the production of sufficient heat to evaporate the associated water from sludge. Sludge can be incinerated with municipal refuse.

The two types of furnace employed are the multiple hearths and the fluidized bed. Multiple hearth furnaces consist of a series of floors in a cylindrical tower; the cake is introduced at the top, and it gradually falls to the lower floors. The material is moved over floors by rabble arms. The major combustion occurs at lower levels, and heat from these levels dries out the sludge at the upper levels. In comparison, fluidized bed furnaces have a cylindrical chamber, which contains approximately 1 m of sand on a heat-resistant steel grid. The bed is fluidized by the injection of compressed air, and as sludge is injected into the sand under pressure through an air-cooled lance, water evaporates and the organic material burns.

21.6.2 Pyrolysis

Pyrolysis is combustion under conditions of reduced oxygen; complete oxidation is not possible. The partially combusted gaseous products from the low temperature-regulated (600 °C) pyrolysis unit is passed to the high temperature-regulated (2000 °C) furnace, where complete oxidation is carried out. This sequence reduces the formation of large solids (clinker) during combustion and ensures that the gases released to the atmosphere are fully oxidized. As with incineration, the low energy value of wet sludge makes it necessary to dewater the sludge before pyrolysis and, usually, to provide additional fuel.

An economical source of fuel is provided by municipal refuse, which can be classified to separate noncombustible materials (such as metals and glass) from the combustible fraction (such as paper and plastics). Shredded municipal refuse gives a high-energy material, often referred to as refuse-derived fuel. Pyrolysis of municipal refuse-derived fuel and dewatered municipal wastewater sludge can therefore provide a system that uses the properties of these two waste products and facilitates their disposal.

21.6.3 Composting

Dewatered sludge can be composted with household refuse to produce agricultural or horticultural compost. Nondegradable materials such as rags, metals, and glass are removed from the refuse before composting. This method produces a commercial product that has been well accepted in many communities. The process appears to be most suitable for communities in which the industrial input into garbage and wastewater is not significant.

In composting, approximately between 20% and 30% of the volatile solids are converted to carbon dioxide and water. As the organic matter in the sludge decomposes, the compost heats to temperatures in the pasteurization range between 50 °C and 70 °C, and enteric pathogenic organisms are destroyed.

The two most commonly used composting methods are the aerated static pile and the windrow process. The aerated static pile system consists of a grid of aeration or exhaust pipes, over which a mixture of dewatered sludge and bulking agent is placed.

In a typical static pile, the bulking agent consists of wood chips, which are mixed with the dewatered sludge. Material is composted for 21–28 days and is typically cured for >30 days. Pile heights are between 2 and 2.5 m. A layer of screened compost is often placed on the top of the pile for insulation. Disposable corrugated plastic drainage pipe are commonly used for air supply. Each pile is recommended to have an individual blower for more effective aeration control. Screening of the cured compost is usually done to reduce the quantity of the end product requiring ultimate disposal and to recover the bulking agent. For improved odor and process control, many new facilities cover or enclose all or significant portions of the system.

In the windrow process, the mixing and screening operations are similar to those for the aerated static pile system. Windrows are constructed with a height between 1 and 2 m and with corresponding measurements at the base between 2.0 and 4.3 m. The rows are turned and mixed periodically during the composting period. Under typical operational conditions, the windrows are turned a minimum of five times while the temperature is maintained at or above 55 °C. Turning of the windrows is often accompanied by the release of offensive odors. The composting period is between 21 and 28 days. Specialized equipment has been developed to mix the sludge with bulking agents and to turn the composting windrows. Some windrow operations are covered or enclosed (similar to aerated static piles).

21.7 LAND DUMPING AND PASSIVE TREATMENT

Sludges can be disposed of at a sanitary landfill site, provided that this does not cause contamination of surface waters or groundwaters. The high water content of sludges usually makes it uneconomical to transport wet sludge to a landfill site, so sludges are dewatered before disposal. At small works, the sludge can be buried at or near the works. The disposal practice is labor-intensive in that a trench is dug, sludge is added, and the area is covered again with topsoil.

Wastewater sludges contain some chemical compounds required for plant growth. Digested sludge contains nitrogen, phosphorus, and potassium in the dry sludge solids and can therefore be considered as a cheap fertilizer. The sludge does contain a high organic content, which is valuable in maintaining a solid structure. Sludge can therefore also be regarded as a solid conditioner. Some sludges are often marketed as biosolids in countries such as the United States.

Sludge also contains pathogenic organisms and can be a source of odors. Therefore, it is most common to recycle sludge only after digestion, and in some cases pasteurization or disinfection is required before the sludge can be used on agricultural land. The requirements for the pretreatment of sludge before land application also depend on the crop to be grown.

For agricultural use of sludge, approximately 1 ha is required for a population of 1000. Most sludge is produced in large urban areas, and the logistics of transporting it to agricultural land can pose problems. Additionally, there is usually a significant industrial component in municipal wastewaters. Toxic metals may concentrate in the sludge and can restrict its land application value. Sustainable land treatment can be regarded as recycling of organic materials back into the food chain.

Chapter 22

Wetlands Treating Contaminated Stream Water

22.1 INTRODUCTION

The functions of macrophytes within constructed wetlands have been reviewed in detail elsewhere (Brix, 1999; Kadlec and Knight, 1996). Macrophytes enrich their substrates with oxygen, reduce filter erosion, and modify the texture as well as hydraulic conductivity of the soil media by growth of plant roots and rhizomes. Plant structures initially serve as pathways for gaseous transport into and out of the sediments. Roots propagate through the soil, pushing into pore space and moving soil grains. This is possible in fine-grained media—either soils or fine gravels. Roots typically die with a turnover of a few years, leaving a fibrous mat of root litter. In some cases, the decaying rhizomes offer new pore space for water and gas movement. The root mat (live and dead) provides strength and stability to organic soils. *Phragmites*, *Typha*, *Juncus*, and other swamp plants are widely used within treatment wetlands in Europe and North America (Kadlec and Knight, 1996; Mungur et al., 1997).

Metals occur in soluble or particulate forms. Heavy metals such as lead and copper are most bioavailable when they are soluble, as either ionic or weakly complex forms (Cooper et al., 1996). Many trace metals are commonly found in sulfide ores; e.g., galena (PbS) and chalcopyrite ($CuFeS_2$). Metal concentrations are often high in mining wastes with associated phytotoxic effects. Lead concentrations above 20,000 mg/kg are not uncommon (Ross, 1994).

Plant root absorption of heavy metals can be of either passive (nonmetabolic) or active (metabolic) uptake in nature. Nonmetabolic uptake diffuses ions in the soil solution into the outer root structures (endodermis). In comparison, active uptake takes place against an ionic concentration gradient but requires metabolic energy to proceed (Alloway, 1995). For example, the uptake of lead is commonly considered to be passive, while copper is thought to be taken up either by active or by a combination of both active and passive processes (Alloway, 1995).

A diverse microbial community, including a variety of bacteria, fungi, algae, and protozoa, is present in both the aerobic and anaerobic zones of

Wetlands for Water Pollution Control. http://dx.doi.org/10.1016/B978-0-444-63607-2.00022-8

wetlands (Kadlec and Knight, 1996). Moreover, most wetland species have a symbiotic relationship of their roots with specialized fungi, known as mycorrhizae (fungus roots). In the presence of mycorrhizae, an increased uptake of copper has been observed in soils low in nutrients (Wild, 1993).

The aim of this experimental study is to characterize the function of biomass within mature experimental wetland filters with different design characteristics, including the selection of appropriate macrophytes and aggregates. The removal efficiencies of mature experimental constructed wetlands treating urban water receiving high loads of lead and copper will be assessed. An overall water quality assessment focusing on physical and chemical variables as well as bio-indicators will be undertaken. Moreover, statistical relationships between water quality variables will be identified to reduce sample costs and effort.

22.2 MATERIALS AND METHODS

22.2.1 Experimental Plan and Limitations

The experimental plan focuses on the water quality investigation of six different experimental wetland filters receiving urban water contaminated with lead and copper. The metal concentrations were increased after 1 year of operation to assess the associated impact on the outflow water quality and biomass within the filters.

This experimental case study was based on applied engineering science principles; therefore, no replicate filters were operated. Moreover, the operation of two or three replicates for each filter (up to 24 filters in total) would have been impractical and unrealistic on a small pilot plant scale like the one presented in this chapter.

Results are not represented in terms of mass balance equations because most experimental data vary greatly with season. Moreover, the focus of this study is on the water quality that is traditionally not expressed in terms of mass balance but with concentration units.

22.2.2 Filter Media Composition

Wetland habitats were simulated on a laboratory scale with experimental filters of the following dimensions: height (58 cm), upper diameter (40 cm), and lower diameter (32 cm). The empty filter volume was 59 l. Six packing arrangements of filter media and plant roots were used in six vertical-flow wetland filters (Table 22.2.2.1).

Filter 1 is similar to a slow sand filter (Chapter 10). In comparison, Filters 2 and 3 are typical reed bed filters. However, Filters 4–6 are predominantly of academic interest because they contain different absorption media.

Additional adsorption media were used because they are associated with high heavy metal removal potentials (Scholz and Martin, 1998a,b). The objective was to study the influence of different compositions of aggregates on the overall treatment efficiency.

TABLE 22.2.2.1 Packing Order Arrangements of Vertical-Flow Filters Simulating Wetlands

Height (cm)	Filter 1	Filter 2	Filter 3	Filter 4	Filter 5	Filter 6
49–58 (top)	Water	Water	Water	Water	Water	Water
47–48 (top)	Sand	Sand + P	Sand + P + T	GAC + P + T	GAC + P + T	Charcoal + P + T + F
41–46 (top)	Sand	Sand + P	Sand + P + T	GAC + P + T	GAC + P + T	GAC + P + T
37–40 (top)	Sand	Sand + P	Sand + P + T	Sand + P + T	Filtralite + P + T	Filtralite + P + T
35–36 (top)	Sand	Sand	Sand + T	Sand + T	Filtralite + T	Filtralite + T
33–34 (top)	Gravel	Gravel	Gravel + T	Sand + T	Sand + T	Sand + T
25–32 (top)	Gravel	Gravel	Gravel	Sand	Sand	Sand
21–24 (top)	Gravel	Gravel	Gravel	Gravel	Sand	Sand
0–20 (bottom)	Gravel	Gravel	Gravel	Gravel	Gravel	Gravel

GAC, granular activated carbon; P, *Phragmites australis* (Cav.) Trin. Ex Steud (Common reed); T, *Typha latifolia* L. (Cattail); F, fertilizer.

22.2.3 Environmental Conditions and Operation

The experiment was performed in the center of a large unheated laboratory hall (windows at two sides and at the top) that was subject to direct and indirect variable natural (day-length) and artificial light. The system was operated in batch flow mode to save pumping costs. The inflow water was taken from the Westbrook (a local brick-lined urban stream in Bradford, UK), which is polluted by urban runoff and sewage from combined sewer overflows during and after storms. The pH of the raw stream water was slightly acidic. The experiment was split into two operational stages, both of which lasted 1 year each in order to assess the system performance for all seasons: from January 21, 2000, to January 20, 2001, and from January 21, 2001, to January 20, 2002.

Lead and copper were selected as pollutants because they are both not easily bioavailable. Lead and copper sulfate were added to the inflow water to give total concentrations of dissolved lead and copper of approximately 1.30 and 0.98 mg/l, respectively, for the first year, and 2.98 and 1.93 mg/l, respectively, for the second year. The magnitude of lead and copper values is comparable to figures reported for (pH-adjusted) mine wastewater and urban water heavily contaminated with heavy metals (Cooper et al., 1996; Hammer, 1989; Kadlec and Knight, 1996).

Natural background concentrations of the Westbrook water (before contamination with heavy metals in the laboratory) were approximately 0.02 mg/l for lead and 0.06 mg/l for copper. These background concentrations are very low in comparison to the values measured after artificial contamination.

The calculated theoretical hydraulic load per filter was approximately 1.8 cm/day for the first stage and 4.2 cm/day for the second stage. The hydraulic load is based on the mean inflow volume (m^3) of water that flows through the cross-section (m^2) of a filter per day. Approximately 40% of the filter water was drained and replaced with 6 liters on Mondays, Wednesdays, and Fridays for the first stage and 7 liters contaminated inflow water on each working day for the second stage of the experiment.

Calculated mass load rates for lead and copper were 24.3 and 18.0 $mg/m^2/day$, respectively, for the first stage, and 118.8 and 77.0 $mg/m^2/day$, respectively, for the second stage of the experiment. The mass load rate is based on the metal inflow concentrations (mg/m^3) associated with the mean daily inflow volume (m^3) distributed over the cross-section (m^2) of each wetland filter. After 1 year of operation, high metal loads were applied to stress the biomass even further and to explore the limits of adequate system performance.

The water level was lowered carefully within 2 min until only 0.5 cm of water remained on top of the submerged litter. The purpose of this procedure was not to destroy the anoxic biofilm structure. The inflow was distributed slowly and evenly into the *schmutzdecke* (i.e., a layer of dirt containing suspended solids and biomass) and litter zone of each filter (Chapter 10). The effluent point was located at the center of the bottom of the filter.

The balanced controlled-release N—P—K fertilizer Miracle-Gro (Scotts Miracle-Gro Company, Marysville, OH, USA) was added to the top of Filter 6 to stimulate the growth of macrophytes. Approximately 50 g of the controlled-release Miracle-Gro pellets was added in January, April, July, and October. The applied fertilizer load was low according to the manufacturer's guidelines. The initial load was therefore doubled after 1 year of operation to enhance further macrophyte and biomass growth in Filter 6 only.

22.2.4 Analytical Procedures Including Metal Determination

For heavy metal determination, sample water (5 cm^3) from the inflow (before adding metals) and outflow, and from the water layer on top of the litter zone (sample point selected at random), was directly analyzed with a Palintest Scanning Analyzer SA-1000 approximately twice a month. Precalibrated electrodes (SE-1), prepared by the manufacturer, were used. Mean values based on at least two readings per sample were calculated. Sample water dilutions were made, if appropriate.

In October 2001, metal concentrations within the *schmutzdecke*, litter, macrophytes, and liquid samples were analyzed by atomic absorption spectroscopy in order to increase sample accuracy. Samples were taken randomly, and two or three sample replicates were considered. A Perkin Elmer 1100 series Atomic Absorption Spectrophotometer was used for metal determination with an accuracy of 0.19 mg/l at 283.3 nm for lead and 0.032 mg/l at 324.9 nm for copper (Perkin Elmer, 1982).

Macrophyte and aggregate (including microbial biomass and litter) samples were taken randomly. Different sample types for macrophytes (leaf, stems, and rhizome) and depth profiles for aggregates were analyzed for lead and copper concentrations to determine the effective heavy metal load removal capacity from the contaminated waters by the separate components of the constructed wetland filters.

Water samples were analyzed directly, but solid materials such as rhizomes were dissolved before the corresponding solution was analyzed. Nitric acid was considered to be suitable for flame and nonflame atomization. In comparison, sulfuric acid can interfere with flame atomization. Perchloric acid and hydrochloric acid were not used for samples containing lead because volatile chlorides could be formed. These chlorides can be lost before flame atomization and subsequent analysis (Tan, 1996).

It was necessary to use an oxidizing acid such as nitric acid (alternatively, perchloric acid) in combination with hydrofluoric acid because decomposed organic material (e.g., aquatic plant litter) may contain sulfides (Alloway, 1995). Due to health and safety considerations regarding the use of hydrofluoric acid and the requirement for lead determination, a mixed acid digestion of concentrated nitric acid (10 ml) and sulfuric acid (1 ml) was used to maximize biomass dissolution with minimal risk to health, safety, and experimental accuracy.

During the atomic absorption analysis, the spectrophotometer was calibrated using 0.0, 0.5, 1.0, 1.5, 2.0, 5.0, 7.5, 10.0, 15.0, and 20.0 ppm (i.e., mg/l for liquid samples) standards for lead analysis and 0.00, 0.25, 0.50, 0.75, 1.00, 1.50, 2.00, 4.00, 5.00, and 8.00 ppm standards for copper analysis. The calibration curves were calculated by least-squares linear regression analysis (coefficient of determination $R^2 > 0.97$). The metal concentrations for the samples were then calculated with the help of the corresponding calibration curves.

A Hanna HI 8519N microprocessor benchtop pH meter (Orion Model 420 used as a replacement after September 14, 2001), a Hanna HI 9142 portable waterproof dissolved oxygen (DO) meter, a Hanna C 102 turbidity meter, and a Hanna HI 9033 conductivity meter were used to determine pH, DO, turbidity, and conductivity, respectively. An ORP HI 98,201 redox meter with a platinum tip electrode HI 73,201 was applied.

Measurement units used are listed in Table 22.2.4.1. Samples were analyzed on Mondays, Wednesdays, and Fridays of most working weeks. No replicates were taken. All analytical procedures were performed according to standard methods (Clesceri et al., 1998), except counting of microorganisms (explained below).

22.2.5 Microbiological Examinations

Samples were taken twice a week for analysis of microorganisms. Five water samples (5 ml each) were taken randomly from the inflow water tank, the outflow water tank, and the top layer of filter media (mixed with plant litter). The samples from each sample set were mixed with each other to reduce sample bias and to increase sample accuracy and validity of the subsequent analysis of subsamples. The sample regime is based on the assumption that all water sources have a homogeneous water quality at the time of sampling. Immediately after mixing, microorganisms including protozoa and algae groups were counted in a first subsample (two replicates) of each filter by using a Fuchs–Rosenthal Hemacytometer (counting chamber: $0.20 \times 0.25 \times 0.25$ mm) and a Wang Biomedical Research Microscope 6000 (bright field and phase-contrast).

During counting, one microorganism counted represented 313 organisms per milliliter. Protozoa were identified based on their mode of movement (i.e., life action; e.g., movement with flagella or cilia) and morphology according to the methodology applied for the general subdivision of protozoa in British sewage treatment processes (Scholz and Martin, 1998a,b). The research was focused on microbiological parameters that can be interpreted as bio-indicators for water quality (e.g., ciliated protozoa) or as essential to assess filter clogging (e.g., diatoms).

TABLE 22.2.4.1 **First (January 21, 2000–January 20, 2001) and Second (January 21, 2001–January 20, 2002) Year of the Experiment: Inflow Westbrook Water Quality Parameter after Contamination with Lead and Copper Sulfate**

Variable	Unit	Number	Mean	Median	Minimum	Maximum	SD
First Year							
Lead	mg/l	23	1.30	1.28	≈1.0	≈1.5	n/a
Copper	mg/l	23	0.98	0.99	≈0.5	≈1.5	n/a
BOD	mg/l	127	2.7	1.9	0.1	11.3	2.4
SS	mg/l	77	15.8	12.0	<0.1	609.0	15.4
Total solids	mg/l	96	470	472	114	1048	147
Turbidity	NTU	91	8.9	6.8	0.6	49.5	8.0
DO	mg/l	139	8.2	8.1	5.8	10.4	0.6
pH	–	135	7.69	7.71	6.96	8.07	0.20
Redox potential	mV	51	284	302	76	423	70
Conductivity	S	74	674	631	271	997	186
T (at source)	°C	114	11	12	7	26	6
T (laboratory)	°C	115	20	20	10	28	3
Second Year							
Lead	mg/l	18	2.98	2.98	2.5	3.5	n/a
Copper	mg/l	18	1.93	1.93	1.5	2.5	n/a
BOD	mg/l	139	2.8	2.1	<0.1	10.0	2.45
SS	mg/l	77	19.4	18.0	<0.1	242.0	12.1
Total solids	mg/l	56	404	430	20	846	170
Turbidity	NTU	104	11.6	8.2	<0.01	80.0	12.5
DO	mg/l	140	8.2	8.3	3.9	10.2	0.9
pH	–	123	7.47	7.48	6.50	8.0	0.26
Redox potential	mV	105	195	195	98	310	42

Continued

TABLE 22.2.4.1 First (January 21, 2000–January 20, 2001) and Second (January 21, 2001–January 20, 2002) Year of the Experiment: Inflow Westbrook Water Quality Parameter after Contamination with Lead and Copper Sulfate—cont'd

Variable	Unit	Number	Mean	Median	Minimum	Maximum	SD
Second Year							
Conductivity	S	90	657	675	143	836	118
T (at source)	°C	163	9	10	2	24	6
T (laboratory)	°C	147	18	18	11	26	3

SD, standard deviation; BOD, 5-day biochemical oxygen demand; SS, suspended solids; DO, dissolved oxygen; n/a, not available (the SD could not be estimated accurately, because most estimated sample values were below or above the detection limit of the mostly diluted sample); T, temperature.

22.2.6 Statistics

Microsoft Excel correlation analysis, coefficient of determination, and trendline-fitting tools were used for practical statistical data analysis. Furthermore, all second-order polynomials were fitted by calculating the least-squares fits through points representing corresponding pairs.

22.3 RESULTS AND DISCUSSION

22.3.1 Comparison of Treatment Efficiency

Table 22.2.4.1 summarize the inflow water quality after the addition of dissolved lead and copper sulfate to natural Westbrook water. The standard deviations for all inflow variables (except for DO; the stream is fast-flowing) were elevated (Table 22.2.4.1) due to natural and artificial water quality variations of the Westbrook. The water quality of the stream in terms of BOD and SS was sometimes inadequate due to irregular sewage input via a combined sewer overflow structure located upstream of the abstraction site. The correlation between SS and turbidity was low because of the presence of large inorganic solids (mostly fine sand).

After 2 years of operation, the overall performance in terms of copper, lead, BOD, and turbidity removal was high for most filters (Table 22.3.1.1) if compared to published data for similar reed beds (Hammer, 1989; Kadlec and Knight, 1996). However, the reduction in SS was negligible. An analysis of variance (Fowler and Cohen, 1998) indicated that reductions of lead and copper were similar ($P < 0.05$).

TABLE 22.3.1.1 First (January 21, 2000–January 20, 2001) and Second (January 21, 2001–January 20, 2002) Year of the Experiment: Relative Reduction (%) of Outflow Variables

	Change (%) per Wetland Filter[a]					
Variables	1	2	3	4	5	6
First Year						
Lead[b]	>98	>98	>98	>99	>99	>99
Copper[b]	>83	>89	>90	>93	>92	>93
5-day biochemical oxygen demand	77	79	70	75	80	72
Suspended solids	50	37	40	46	48	25
Total solids	4	−7	−7	−7	−4	−24
Turbidity	94	93	88	87	96	94
Dissolved oxygen	54	50	69	68	61	68
pH	1	7	8	7	6	8
Redox potential	−2	−7	14	10	6	3
Conductivity	−2	−15	−18	−20	−15	−36
Second Year						
Lead[b]	>100	>99	>99	>99	>100	>100
Copper[b]	>96	>93	>96	>97	>93	>97
5-day biochemical oxygen demand	74	77	70	72	76	62
Suspended solids	53	44	49	39	36	24
Total solids	5	−3	2	−1	−4	−27
Turbidity	87	86	86	87	78	86
Dissolved oxygen	53	49	72	74	65	75
pH	0	7	7	7	7	11
Redox potential	−4	−11	−3	24	15	4
Conductivity	1	−5	−2	−6	−5	−25

[a]*Change (%) = (in − out) × (%)/in, where in = inflow value and out = outflow value.*
[b]*Some outflow concentrations were below the detection limit.*

It has been suggested that mature biomass, in contrast to aggregates with high adsorption capacities, is responsible for the high overall filtration performances (Cooper et al., 1996; Scholz and Martin, 1998a,b). However, it is difficult to classify objectively a biological system as mature. Nevertheless, there was no breakthrough of metals after approximately 9 weeks. Furthermore, a stable biological ecosystem in terms of the abundance of protozoa and crustaceans has been noted.

22.3.2 Water Quality and Macrophytes

The filters containing macrophytes (in particular, *Typha latifolia*; Filters 3–6) contributed artificially to the inflow BOD. Approximately 75% of *T. latifolia* (infested by aphids) and 30% of *Phragmites australis* died during late fall. Dead plant material made a fair contribution to the BOD measured at the top of Filters 3–6. The 5-day BOD resulting from plant decay was greatest for filters containing *T. latifolia*. The addition of fertilizer (Filter 6 only) increased the degradation rate of dead plant matter and the breakthrough of TS. Furthermore, DO was reduced.

A positive correlation was expected between decaying macrophytes (expressed as an increase in BOD) and conductivity. However, the presence of macrophytes did not increase conductivity (Figure 22.3.2.1); for example, Filters 1 and 2 had similar conductivity values and temporal distributions.

Due to the photosynthesis of green algae, Filter 1 had the highest daily observed DO concentrations (at the top of the filter), which were even higher than the ones for the inflow water for most days in spring and summer. Concerning Filters 3–6, the top and outflow DO concentrations were almost

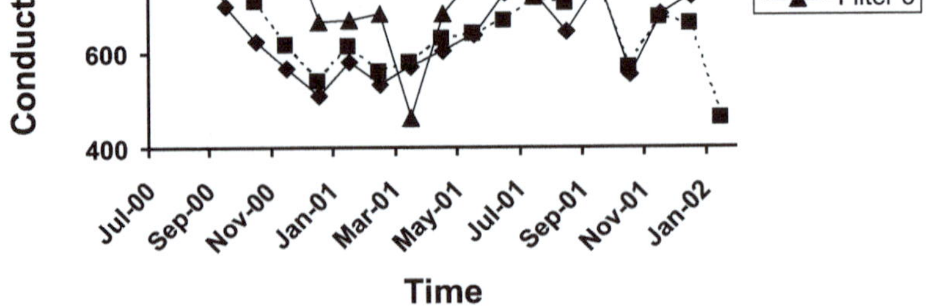

FIGURE 22.3.2.1 Monthly average conductivity values (July 1, 2000–January 20, 2002) for the inflow, Filter 1 (outflow), and Filter 6 (outflow).

always <4.0 mg/l. However, the outflow DO values were slightly higher than the top concentrations.

The differences in top DO can be explained as follows: decaying plants in the litter zone indirectly supplied ample dissolved organic matter for the active biomass. It follows that microbial respiration reduced the DO concentrations within the litter zones of Filters 2–6. Another reason for the differences in DO is the presence of green algae, which were thriving in Filter 1 during summer.

In contrast, green algae decreased in Filters 2–6 due to the shading effect from the emergent plants. As a result of the photosynthetic oxygen production by green algae, the DO and pH concentrations for Filter 1 were higher than the concentrations in Filters 2–6. Therefore, algal growth did not compensate for the oxygen utilized by microbial respiration. Furthermore, dead plants floating on the water surface of Filters 2–6 impeded oxygen dissolution in the open water zones by reducing the areas of water–oxygen interface. Moreover, the decay of macrophytes reduced the pH as indicated in Figure 22.3.2.2.

In comparison to the inflow, concentrations of lead and copper were low within macrophytes and aggregates (Table 22.3.2.1). Because macrophytes contain low concentrations of metals at the end of the growing season, harvesting of macrophytes (i.e., mature leaves and stems only) leads only to a small reduction of the overall metal concentration within the filters. In contrast, harvesting macrophytes during the growing season, where metals are accumulated in the new stems and fresh leaves, might be more promising but will damage the ecosystem considerably.

Relatively high concentrations of copper were present in the top aggregate layer and the litter zone (44–48 cm) of the reed bed filters. In contrast, relatively high lead concentrations occurred throughout the top part

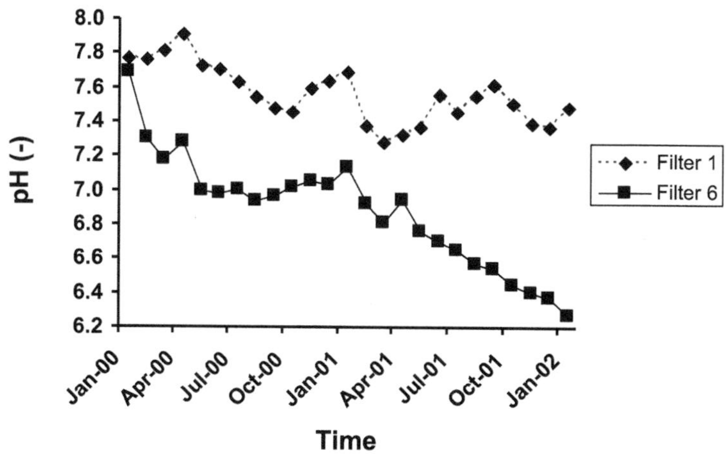

FIGURE 22.3.2.2 Monthly average pH values (July 1, 2000–January 20, 2002) for Filter 1 (outflow) and Filter 6 (outflow).

TABLE 22.3.2.1 Atomic Absorption Spectroscopy Results: Mean Concentrations (mg/kg) of Lead and Copper for Decomposed Macrophyte Material and Biomass Located within the Aggregates of Different Filter Depths (December 2001). Values Are Based on the Mean of Either Two or Three Replicates. Metal Concentrations Exceeding the Inflow Concentrations Are Highlighted (i.e., Numbers in Italic)

	Filter 1		Filter 2		Filter 3	
Variable	**Pb**	**Cu**	**Pb**	**Cu**	**Pb**	**Cu**
Inflow water	2.98	1.93	2.98	1.93	2.98	1.93
Phragmites (leaf)	n/a	n/a	0.37	0.47	0.28	0.38
Phragmites (mature stem)	n/a	n/a	0.28	0.47	0.24	0.48
Phragmites (new stem)	n/a	n/a	0.36	0.91	0.20	0.51
Phragmites (rhizome)	n/a	n/a	0.63	0.74	*0.79*	1.03
Typha (leaf)	n/a	n/a	n/a	n/a	*0.17*	1.07
Typha (stem)	n/a	n/a	n/a	n/a	0.36	0.39
Layer height: 44–48 cm	*0.16*	*0.83*	0.28	*0.76*	0.32	*0.38*
Layer height: 40–44 cm	2.17	*0.98*	0.55	0.57	0.32	0.59
Layer height: 36–40 cm	0.67	0.78	*0.22*	0.80	*0.74*	0.12
Layer height: 32–36 cm	2.02	*0.36*	1.01	1.82	1.32	*0.99*
Layer height: 28–32 cm	1.00	1.82	0.90	1.76	1.04	1.84
Layer height: 24–28 cm	0.85	1.74	0.89	1.76	1.09	1.87
Fertilizer	n/a	n/a	n/a	n/a	n/a	n/a
Charcoal	n/a	n/a	n/a	n/a	n/a	n/a
GAC	n/a	n/a	n/a	n/a	n/a	n/a
Filtralite	n/a	n/a	n/a	n/a	n/a	n/a
Sand	0.41	0.10	0.41	0.10	0.41	0.10
	Filter 4		**Filter 5**		**Filter 6**	
Variable	**Pb**	**Cu**	**Pb**	**Cu**	**Pb**	**Cu**
Inflow water	2.98	1.93	2.98	1.93	2.98	1.93
Phragmites (leaf)	0.22	0.39	0.38	0.61	0.55	0.50
Phragmites (mature stem)	0.40	0.47	0.33	0.46	0.38	0.48
Phragmites (new stem)	0.12	0.17	0.29	1.31	2.91	*0.06*
Phragmites (rhizome)	0.12	0.19	0.59	0.72	0.29	0.38

TABLE 22.3.2.1 Atomic Absorption Spectroscopy Results: Mean Concentrations (mg/kg) of Lead and Copper for Decomposed Macrophyte Material and Biomass Located within the Aggregates of Different Filter Depths (December 2001). Values Are Based on the Mean of Either Two or Three Replicates. Metal Concentrations Exceeding the Inflow Concentrations Are Highlighted (i.e., Numbers in Italic)—cont'd

	Filter 4		Filter 5		Filter 6	
Variable	Pb	Cu	Pb	Cu	Pb	Cu
Typha (leaf)	0.18	0.42	0.31	0.50	0.12	0.72
Typha (stem)	0.31	0.63	0.14	0.62	0.20	0.60
Layer height: 44–48 cm	0.71	*0.28*	1.91	*0.22*	*0.35*	*0.67*
Layer height: 40–44 cm	0.53	0.80	0.38	1.00	0.44	0.63
Layer height: 36–40 cm	*0.97*	0.18	*0.23*	0.16	0.33	0.32
Layer height: 32–36 cm	2.78	*0.76*	1.10	1.87	1.26	*0.96*
Layer height: 28–32 cm	0.95	1.79	0.92	1.78	0.88	1.76
Layer height: 24–28 cm	1.08	1.86	0.90	1.77	0.88	1.76
Fertilizer	n/a	n/a	n/a	n/a	0.33	0.12
Charcoal	n/a	n/a	n/a	n/a	0.43	0.08
GAC	0.16	0.06	0.16	0.06	0.16	0.06
Filtralite	n/a	n/a	0.16	0.21	0.16	0.22
Sand	0.41	0.10	0.41	0.10	0.41	0.10

Pb, lead; Cu, copper; n/a, not applicable; GAC, granular activated carbon.

(32–48 cm) of the filter beds (Table 22.3.2.1). Much higher metal concentrations were expected for the top of the filters compared to the inflow concentrations. A realistic assumption is that all filters have a high remaining filtration capacity and that particulate and colloidal metals integrated into the biomass and absorbed by aggregates are difficult to detect fully with the atomic absorption methodology presented in this chapter. Lead (and, to a lesser extent, copper) forms also inert particulate matter that breaks slowly through the filters. Some aggregate layers help to slow down this process not only by physical straining but also by complex filtration processes.

Biological processes within the process are considered to be important, but their investigation was beyond the scope of this experiment. For example, precipitation of metal oxides and sulfates may be caused by the presence of metal-oxidizing bacteria in the aerobic zone and sulfate-reducing bacteria in the anaerobic zone, respectively (IWA, 2000). However, lead and copper are not easily biologically available to aquatic plants and microbes. Furthermore, a relatively high pH value reduced the risk of metals going back into solution (Table 22.2.4.1).

22.3.3 Water Quality and Microbiology

Microorganisms can be used as bio-indicators to determine the water quality. Several groups of protozoa, particularly ciliates (Figure 22.3.3.1), are good indicators for the water quality within specific BOD ranges and in relation to physical and chemical conditions (Scholz and Martin, 1998a,b).

The abundance of microorganisms during the setting-up period was low within the *schmutzdecke* and/or litter zones of the filters. After about 2 months, protozoa increased in numbers, and their community was relatively diverse during spring and summer, but this diversity decreased slightly during fall. In comparison to the first stage, ciliated protozoa declined sharply in numbers during the second stage, indicating adverse environmental conditions for active biomass (Figure 22.3.3.1).

High numbers of flagellated protozoa (>2500 counts/ml) were observed throughout the study (60% of all experimental days) for Filters 1–3. High populations of flagellated protozoa within a biological system indicate a stress situation such as a system setup or toxic shock. In this study, the tidal flow (i.e., the filter was temporary flooded) operation and the addition of heavy metals to the stream water resulted in a stress situation for the biota within filters containing no adsorption filter media.

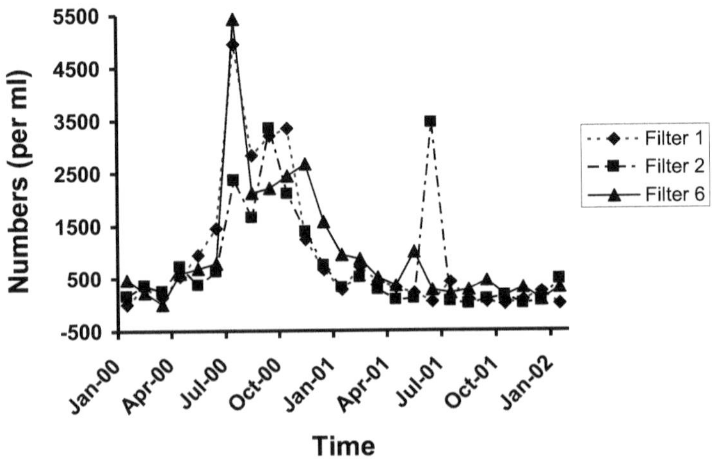

FIGURE 22.3.3.1 Monthly average ciliated protozoa numbers (January 21, 2000–January 20, 2002) for the top water of Filters 1, 2, and 6. Ciliates include free-swimming, crawling, and stalked protozoa.

The total abundance of green algae in Filters 3–6 was high in spring (>6000 counts/ml) but had decreased by late summer (<500 counts/ml). In contrast, the green algal abundances in Filters 1 and 2 increased in spring and were high in fall (data not shown). In comparison, diatom numbers were high throughout the experiment but increased sharply during the second stage because the theoretical flow rate was increased after the second year. This increase resulted in a rapid accumulation of diatoms on top of the filter (Figure 22.3.3.2). Diatoms contributed to the filter-clogging process by reducing the hydraulic conductivity.

During the first year of operation, a diverse invertebrate community, which included rotifers, crustaceans (Figure 22.3.3.3), and segmented worms, formed the top of the food chain in the liquid phase of the litter zone of all filters. In comparison to Filters 2–6, Filter 1 contained fewer invertebrate numbers and groups because of low food availability (low BOD) and unsuitable habitat. The absence of dead leaves and stems within the main water body of Filter 1 resulted in a reduced opportunity for invertebrates to hide and/or attach themselves to detritus. Furthermore, zooplankton decreased rapidly during the second year of operation, indicating a deterioration of the water quality (Figure 22.3.3.3).

Figure 22.3.3.3 is based on monthly average mean values. Some individual counts for Filter 6 are high (e.g., 640 numbers on June 23, 2000). Therefore, monthly mean values are distorted. It follows that crustaceans are not suitable bio-indicators if sampled infrequently. High numbers of crustaceans subsequently lead to lower numbers of protozoa due to predation.

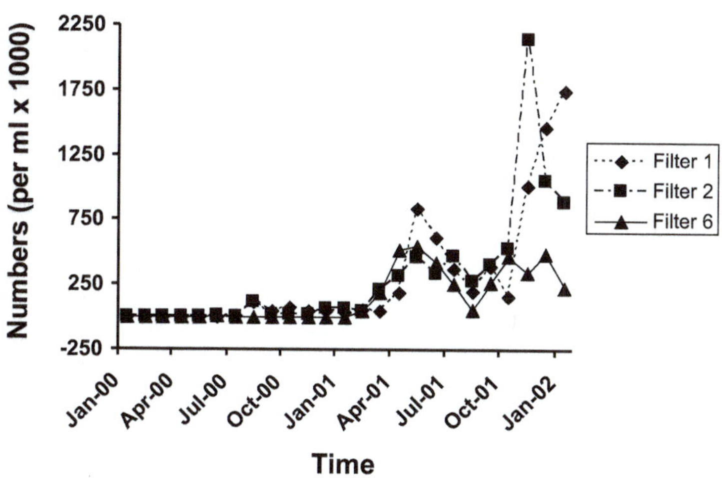

FIGURE 22.3.3.2 Monthly average total diatom numbers (January 21, 2000–January 20, 2002) for the top water of Filters 1, 2, and 6. Total diatoms include the genera *Diatoma*, *Nitzschia*, and *Navicula*.

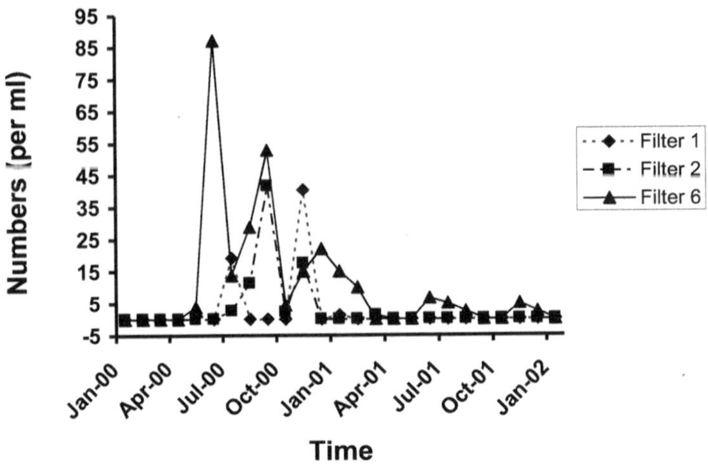

FIGURE 22.3.3.3 Monthly average crustacean numbers (January 21, 2000–January 20, 2002) for the top water of Filters 1, 2, and 6. Crustaceans include water fleas and copepods.

Invertebrates are cost-efficient and long-term water quality indicators, and they are significant within the food chain of a wetland. They help to stabilize the ecological equilibrium within the treatment process and are a major food source for higher organisms.

22.3.4 Regression and Correlation Analysis as Predictive Tools

Tables 22.3.3.1 and 22.3.4.1 summarize findings of a correlation analysis. Only important correlations with more than 20 corresponding pairs of daily data and a correlation coefficient $R > 0.59$ (threshold for a strong correlation according to Fowler and Cohen (1998)) are shown. Small data sets were ignored because they increase the likelihood of a spurious correlation arising by chance or sampling error. One variable is usually quick and inexpensive to measure or determine on a frequent monitoring routine. The other variable could consequently be estimated through the first variable.

However, a correlation does not necessarily imply a direct causal relationship. Only those correlations are listed that are both empirically related to each other and useful for researchers interested in continuous monitoring techniques. Correlations may be strong only under specific environmental conditions and are therefore not frequently used for occasional water quality monitoring by institutions such as environment agencies.

The range of variables presented is restricted to those that are key to traditional and biological water quality analysis. Furthermore, findings for different seasons are not presented because of the great volume of associated additional information.

TABLE 22.3.3.1 Correlation Analysis for Variables Representing Either the Top of the Water Body (Top) or the Outflow Water (Out) for the First Year of the Experiment (January 21, 2000–January 20, 2001)

Filter	First Variable	Second Variable	Pairs	*R*
1	Total heterotrophic bacteria (top)	Biochemical oxygen demand (top)	23	+0.70
2	Dissolved oxygen (top)	pH (top)	39	+0.66
2	Total diatoms (top)	Flagellated protozoa (top)	109	+0.77
3	Conductivity (top)	Dissolved oxygen (top)	24	−0.68
5	Dissolved oxygen (top)	Biochemical oxygen demand (top)	34	−0.71
5	Conductivity (top)	Dissolved oxygen (top)	24	−0.71
1	Total solids (out)	Conductivity (out)	34	−0.63
2	Conductivity (out)	Dissolved oxygen (out)	71	+0.73
2	Total solids (out)	Conductivity (out)	35	+0.77
3	Total solids (out)	Conductivity (out)	35	+0.84
3	Redox potential (out)	Conductivity (out)	37	+0.67
3	Total solids (out)	Redox potential (out)	30	+0.61
4	Redox potential (out)	Conductivity (out)	38	+0.82
4	Total solids (out)	Conductivity (out)	35	+0.77
5	Total solids (out)	Conductivity (out)	35	+0.89
5	Redox potential (out)	Conductivity (out)	38	+0.84
6	Redox potential (out)	Conductivity (out)	38	+0.74
1	Conductivity (out)	Total green algae (top)	21	+0.80
2	Actinopoda (top)	Suspended solids (out)	25	+0.70
2	Conductivity (out)	Total diatoms (top)	20	−0.64
3	Dissolved oxygen (top)	pH (out)	39	+0.61
4	Dissolved oxygen (top)	Conductivity (out)	23	−0.70
5	Dissolved oxygen (top)	Conductivity (out)	24	−0.82
6	Conductivity (out)	Total green algae (top)	21	−0.67
6	Biochemical oxygen demand (out)	Conductivity (top)	21	+0.60

R, correlation coefficient; out, outflow water; top, top of the water body.

TABLE 22.3.4.1 Correlation Analysis between Inside and Outside Temperature as Well as between Temperature and Other Variables for the First Year of the Experiment (January 21, 2000–January 20, 2001)

Filter	First Variable	Second Variable	Pairs	*R*
–	Temperature (inside)	Temperature (outside)	113	+0.00
1	Temperature (outside)	Conductivity (out)	74	+0.77
2	Temperature (outside)	Conductivity (top)	24	+0.68
2	Temperature (outside)	Conductivity (out)	72	+0.80
3	Temperature (outside)	Conductivity (out)	24	+0.86
3	Temperature (outside)	Dissolved oxygen (top)	30	−0.79
3	Temperature (inside)	Dissolved oxygen (top)	32	−0.64
4	Temperature (outside)	Biochemical oxygen demand (top)	26	+0.66
4	Temperature (outside)	Conductivity (out)	76	+0.87
4	Temperature (outside)	Conductivity (top)	24	+0.68
4	Temperature (inside)	Conductivity (out)	74	+0.64
4	Temperature (outside)	Dissolved oxygen (top)	30	−0.61
4	Temperature (outside)	Redox potential (out)	47	+0.60
4	Temperature (inside)	pH (top)	32	−0.63
5	Temperature (outside)	Dissolved oxygen (top)	28	−0.86
5	Temperature (outside)	Conductivity (top)	24	+0.73
5	Temperature (outside)	Conductivity (out)	73	+0.86
5	Temperature (inside)	Dissolved oxygen (top)	32	−0.71
5	Temperature (inside)	Conductivity (out)	74	+0.63
6	Temperature (outside)	Conductivity (out)	76	+0.84
6	Temperature (inside)	Conductivity (out)	74	+0.65
6	Temperature (outside)	Suspended solids (out)	61	+0.63

R, correlation coefficient; out, outflow water; top, top of the water body.

Strong positive correlations between conductivity and other variables, including temperature, TS, DO, and the redox potential, were identified (Tables 22.3.3.1 and 22.3.4.1). Similar findings for reed bed treatment systems have been reported elsewhere (Hammer, 1989; Kadlec and Knight, 1995). There is a strong negative correlation between conductivity and DO. However, the correlation between outflow conductivity and outflow DO is positive. This can be explained by an increase in conductivity during the biological filtration process in summer and a strong decrease of DO throughout the year. Biological activity can also explain the strong correlation between total heterotrophic bacteria and BOD and a consequent decrease of available DO. All correlations between groups of microorganisms were positive. However, microorganism groups within the litter zones of Filters 2–6 were negatively correlated with conductivity, despite an increase of both conductivity and biomass activity during summer (Table 22.3.3.1).

Automated online water quality monitoring is important if outflow water quality needs to be controlled (Kadlec and Knight, 1996). Online measurements of variables such as temperature and conductivity are more cost-efficient, easier to perform, and more accurate than online measurements of variables such as pH (electrode requiring frequent maintenance) and suspended solids. Moreover, some other variables such as flagellated protozoa, total green algae, and total diatoms cannot be measured online at all. Therefore, it will reduce sampling effort and cost if the sampling frequencies of these microbiological water quality indicators can be reduced by estimating missing data with variables such as conductivity, which correlates well with these variables (Table 22.3.3.1).

Table 22.3.4.1 indicates very consistent results: correlations between temperature and conductivity, BOD, redox potential, and SS are all positive. In contrast, conductivity correlates negatively with DO and pH.

Table 22.3.4.1 summarizes r^2 for all strong correlations with >30 corresponding pairs of daily data and a correlation coefficient $R > 0.69$. Small data sets were ignored because they increase the likelihood of a spurious correlation arising by chance or sampling error. Since a correlation does not necessarily imply a direct causal relationship, only those correlation examples are listed in Table 22.3.4.1 that are both empirically related to each other and useful for researchers interested in continuous monitoring techniques.

In Table 22.3.4.2, the standard second-order polynomial Eq. (22.3.4.1) was fitted in Microsoft Excel for data ranges determined by given minimum and maximum values of y and x.

$$y = ax^2 + bx + c \qquad (22.3.4.1)$$

For a specific filter, variable y may be estimated with variable x for a given range of x defined by x_{min} and x_{max}. For example, some expensive and time-consuming variables such as BOD can be predicted with less expensive ones like temperature. This reduces considerably the sampling effort and costs for water quality monitoring.

Figure 22.3.4.1 shows one example of how outflow conductivity (expensive and difficult to measure online) can be predicted with outside temperature (inexpensive and easy to record online). The associated absolute mean error for outflow conductivity regarding Filter 4 was 8%. The corresponding standard deviation was 7%.

TABLE 22.3.4.2 Coefficient of Determination (r^2). Equation (22.3.4.1), $y = ax^2 + bx + c$, Was Fitted for Experimental First-Year Data Only (January 21, 2000–January 20, 2001). For a Specific Filter, Variable y Can Be Predicted with Variable x for a Given Range of x Defined by X_{min} and X_{max}

Number	Filter	Y	X	A	B	C
1	2	FP_{top}	TD_{top}	0.000	0.484	3500.90
2	5	BOD_{top}	DO_{top}	0.292	2.670	7.11
3	2	DO_{out}	$Cond_{out}$	6.339	17.194	1029.80
4	2	$Cond_{out}$	TS_{out}	0.002	0.364	516.04
5	3	$Cond_{out}$	TS_{out}	0.002	0.814	632.09
6	4	$Cond_{out}$	RP_{out}	0.001	1.168	450.81
7	4	$Cond_{out}$	TS_{out}	0.003	1.802	919.72
8	5	$Cond_{out}$	TS_{out}	0.002	3.900	610.89
9	5	$Cond_{out}$	RP_{out}	0.005	4.812	106.49
10	6	$Cond_{out}$	RP_{out}	0.011	8.692	588.52
11	–	$T_{outside}$	T_{inside}	0.004	1.798	22.26
12	1	$Cond_{out}$	$T_{outside}$	−0.063	14.944	541.91
13	2	$Cond_{out}$	$T_{outside}$	0.038	19.352	576.16
14	3	DO_{top}	$T_{outside}$	0.001	0.201	5.25
15	4	$Cond_{out}$	$T_{outside}$	0.225	18.841	586.54
16	5	$Cond_{out}$	$T_{outside}$	0.278	17.627	561.79
17	5	DO_{top}	T_{inside}	0.015	0.885	14.30
18	6	$Cond_{out}$	$T_{Outside}$	0.072	26.084	646.52
Number	**Filter**	r^2	y_{min}	y_{max}	x_{min}	x_{max}
1	2	0.62	0	180,288	0	420,772
2	5	0.55	1.4	10.2	0	5.6
3	2	0.53	1.7	9.0	481	991
4	2	0.61	472	991	266.0	654.8
5	3	0.78	463	1035	148.0	680.5
6	4	0.67	526	1025	89	383
7	4	0.65	480	1025	264	664.9
8	5	0.79	462	990	374.0	651.6

TABLE 22.3.4.2 Coefficient of Determination (r^2). Equation (22.3.4.1), $y = ax^2 + bx + c$, Was Fitted for Experimental First-Year Data Only (January 21, 2000–January 20, 2001). For a Specific Filter, Variable y Can Be Predicted with Variable x for a Given Range of x Defined by X_{min} and X_{max}—cont'd

Number	Filter	r^2	y_{min}	y_{max}	x_{min}	x_{max}
9	5	0.72	508	990	148	382
10	6	0.58	557	1163	172	394
11	–	0.64	7	26	10.0	27.8
12	1	0.59	445	845	−7	19.2
13	2	0.65	472	991	−7	19.2
14	3	0.63	0.0	5.6	2.3	18.1
15	4	0.75	480	1025	−7	19.2
16	5	0.75	462	990	−7	19.2
17	5	0.52	0.1	4.8	12.9	26.2
18	6	0.71	516	1188	−7	19.2

FP, flagellated protozoa; TD, total diatoms; BOD, 5-day biochemical oxygen demand; DO, dissolved oxygen; Cond, conductivity; TS, total solids; RP, redox potential; T, temperature; top, top of the water body; out, outflow water.

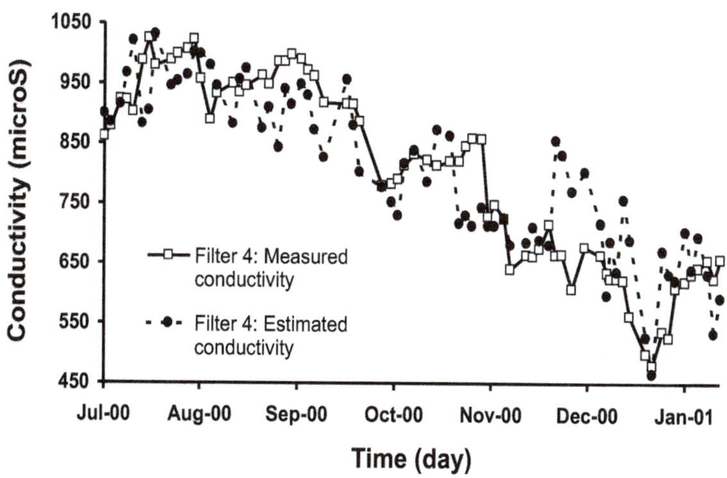

FIGURE 22.3.4.1 Measured and predicted outflow conductivity for Filter 4.

22.4 CONCLUSIONS

Adsorption filter media (as replacements for aggregates such as sand and gravel) did not significantly reduce lead and copper concentrations after the establishment of an active biomass (2 years old). The overall filtration performance was acceptable and similar for all experimental filters after biomass maturation. It follows that the biomass (not the aggregates and macrophytes as expected) within each filter is predominantly responsible for the removal of lead and copper. This will have implications on the future design and operation of constructed wetlands. Costs can be saved on filter material and aquatic plants.

Lead and copper accumulated predominantly within the litter zone and the rhizomes of the macrophytes (practically not accessible). It follows that costs could be saved by constructing shallow filter beds. Furthermore, only a limited load of heavy metals could be removed from the constructed reed beds by harvesting macrophyte stems and leaves. The costs associated with harvesting do not currently justify the benefit of metal removal.

The dissolved fraction of metals within the litter zone led to a reduction in bioactivity (e.g., ciliated protozoa and zooplankton) during the second spring and summer of the experiment. The reduction of these bio-indicators reflected the accumulation of toxic metals within sediments and the food chain. This has a negative environmental impact on the wildlife associated with wetlands. Furthermore, diatom numbers increased during the second stage, resulting in potential filter-clogging problems.

Fertilizer use in one of the filters led to a deterioration of the water quality, particularly with respect to SS and pH. Additional nutrients resulted in an artificial increase of the biodegradation rate, which is not desirable in the case of treatment wetlands for metal removal. Thus, costs can be saved by not applying fertilizers.

Strong positive correlations between conductivity and other variables, including temperature, were identified. Some expensive and time-consuming (i.e., high sampling effort) variables such as BOD, DO, and conductivity can be predicted with less expensive ones such as temperature that are easy to measure. This would result in savings of costs and sampling effort for future constructed wetland water quality monitoring schemes.

Chapter 23

Wetland Systems to Control Roof Runoff

23.1 INTRODUCTION

23.1.1 Sustainable Roof Runoff Drainage

Considering that sustainable drainage system (SuDS) technology is a novel research area, definitions for the terminology used in this chapter are required. Reduction of the rate of flow through a system, which has the effect of reducing the peak flow and increasing the duration of a flow event, is defined as attenuation. A pond is specifically referred to as a wet pond if it is a permanently wet depression designed to retain stormwater for several days and to permit settling of suspended solids (SS). It follows that an attenuation pond combines the meaning of both attenuation and pond. In comparison, an infiltration basin is a dry basin designed to promote infiltration of surface water into the ground. If vegetated, a small infiltration basin is often referred to as a dry pond.

Conventional stormwater and urban drainage systems are designed to dispose of rainfall runoff as quickly as possible. This results in "end of pipe" solutions that often involve the provision of large interceptor and relief sewers, huge storage tanks in downstream locations, and centralized wastewater treatment facilities (Abbott and Comino-Mateos, 2001; Butler and Davies, 2000; Ellis et al., 2002).

In contrast, combined attenuation pond and infiltration basin systems can be applied as cost-effective end of pipe drainage solutions for local source control; e.g., collection of roof drainage. It is often possible to divert all roof drainage for infiltration or storage and for subsequent recycling. As runoff from roofs is a major contributor to the quantity of surface water requiring disposal, this is a particularly beneficial approach, where suitable ground conditions prevail (Butler and Davies, 2000).

Furthermore, roof runoff is usually considered to be cleaner than road runoff. However, diffuse pollution can have a significant impact on the water quality of any receiving watercourse (Ellis et al., 2002).

Wetlands for Water Pollution Control. http://dx.doi.org/10.1016/B978-0-444-63607-2.00023-X

23.1.2 Case Study: Site Description

A domestic property in Sandy Lane (Bradford, West Yorkshire, England) was selected for this pilot plant case study. The study area is located approximately 1.8° west of Greenwich and 53.8° north of the Equator. The surface water (subject to disposal) came from the roofs of a single-family house and a tandem (double) garage.

Concrete roof tiles were used for the house, and the garage roof was covered with a layer of gravel. The roof materials had not been cleaned for more than 5 years prior to the study (house owners, personal communications). One can assume that dirty roofs could be a source of inorganic and organic pollution, and it is likely that the roof material supports micro-habitats that include algae. However, a detailed discussion of these challenges is beyond the scope of this chapter.

23.1.3 Purpose

The purpose of this chapter is not only to research a case study of a SuDS designed according to best management practice (Martin et al., 2000) but also to address the following key objectives to assess the potential for scaled-up systems:

- Identification of technical constraints associated with the design, operation, and maintenance;
- Identification of the infiltration characteristics of stormwater into the vegetated infiltration basin;
- Suggestion of water quality monitoring and management strategies, including algal control techniques, to increase public acceptance of SuDS; and
- Assessment of the water treatment potential of stormwater ponds for roof runoff.

23.2 METHODS

23.2.1 Design of the Study Site

The pilot plant was designed considering SuDS guidelines of the Building Research Establishment (BRE, 1991), Construction Industry Research and Information Association (CIRIA) (Bettes, 1996; Martin et al., 2000), and German Association for Water, Wastewater, and Waste (ATV-DVWK, 2002). The British system design allows flooding to occur only once within 10 years, which is defined as the return period. However, the German guideline recommends a design allowing for a 5-year storm only (ATV-DVWK, 2002).

The system was based on a combined attenuation pond and infiltration basin design (Figure 23.2.1.1). Runoff from the roofs of one house (tandem

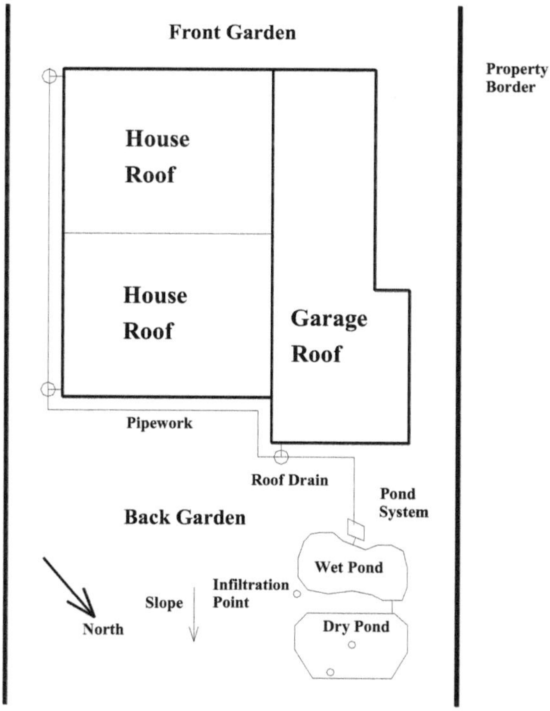

FIGURE 23.2.1.1 Drawing of the case study site showing roof areas, pipework, and the combined attenuation wet and dry pond system. Runoff flows from the roof areas into the drainage pipe network, which conveys the water first into the silt trap, then into the wet pond, and finally into the dry pond.

garage attached) was drained directly into a silt trap, which fulfilled the purpose of a small sedimentation tank.

The surface areas of both house roofs measured 29 m^2 each. The angle between each roof and the ceiling of the house was 23°. Therefore, the total theoretical horizontal area of the house roofs was 53 m^2. The roof area of the tandem garage covered 33 m^2. It follows that the total theoretical horizontal area to be drained was 86 m^2.

The distances between the nearest building (garage) and the attenuation pond and infiltration basin were 1.5 and 5.0 m, respectively (Figure 23.2.1.1). The total length of the horizontal plastic pipework close to the ground (mean angle of 2°) was 19.6 m. Guttering and down pipes were not included in this sum. The inside diameter of all pipes was 6.5 cm. The pipeline layout could have been optimized, but the SuDS was retrofitted in this case study, and it was therefore the aim to recycle as much of the old pipework as possible.

In the original pipeline layout dated 1972, rainwater drained into the public sewer. However, in April 2001, this layout was changed to feed a semi-natural attenuation pond structure (Figure 23.2.1.1) with rainwater. The storage water was predominantly used for watering garden plants in summer, but there is a much greater potential for other usage, such as recycling of stormwater to flush toilets (Butler and Davies, 2000). If the attenuation pond structure overflows, the water is transferred to a vegetated infiltration basin structure (Figure 23.2.1.1).

The traditional drains were sealed off temporarily, although the house owners preferred a system that would allow rainwater diversion into the main sewer in cases of exceptional rainfall or failure of the infiltration basin. However, neither of these events occurred during the duration of this experiment.

The maximum horizontal dimensions (length × width) of the silt trap, attenuation pond, and infiltration basin were 0.7 × 0.4 m, 3.2 × 1.7 m, and 3.7 × 2.5 m, respectively (Figure 23.2.1.1). The maximum depths of the lined attenuation pond and the vegetated infiltration basin were approximately 40 cm each. All water level measurements were taken daily at a reference level point that was part of the attenuation pond outflow structure (Figure 23.2.1.1).

The total area of the experimental attenuation pond was 5.5 m^2 when completely filled with water (Figure 23.2.1.1). The slope ratios of the infiltration basin were 1:1.6 (toward the attenuation pond in the west), 1:1.9 (toward north and south), and 1:2.4 (toward the lower garden in the east).

The attenuation pond and infiltration basin were planted with common aquatic plants, which were collected from local natural habitats. The dominant macrophytes of the attenuation pond included common reed (*Phragmites australis*), reedmace (*Typha latifolia*), and yellow iris (*Iris pseudochorus*). water starwort (*Callitriche stagnalis*) and frogbit (*Hydrocharis morsus-ranae*) were common floating aquatic plants. Canadian waterweed (*Elodea canadensis*) was the dominant submerged plant. Common reed and reedmace (both deep-rooting) were also planted in the infiltration basin to enhance infiltration, although this expected purpose has been disputed by some findings reviewed by Brix (1997). This combination of plants was chosen because it is typical for similar watercourses around Bradford.

Two koi carp (*Cyprinus carpio*; introduced during pond construction in April 2001), at least 25 European three-spined sticklebacks (*Gasterosteus aculeatus*; introduced in September 2001), and the common grass frog (*Rana temporaria*; approximately 20 adults and at least 2500 tadpoles during spring and summer) were at the top of the food chain in the attenuation pond.

23.2.2 Engineering Methods

Two simple rain gauges were used (consisting of a measurement cylinder fed through a funnel with a diameter of 9 cm) to give a more representative estimate of rainfall. Experimental rainfall data were compared with official data (Bradford measurement station) supplied by the Meteorological Office (2002).

The rain gauges were located between the attenuation pond and the infiltration basin (Figure 23.2.1.1).

A cost-effective, nondestructive test on a small scale was required to assess the stormwater (roof runoff) infiltration properties into the infiltration basin. This information was important for the estimation of the infiltration rates required for the design calculations (see above). Therefore, a simple infiltration device specifically designed to promote the passage of gravity-fed tap water into the natural ground was used.

Infiltration rates were determined every two weeks or monthly by measuring the actual infiltration time of 200 ml tap water through the first 4 cm of a 24-cm-long drainage pipe (diameter of 6.5 cm). The pipe was buried vertically in the ground at a depth of approximately 20 cm. The infiltration test was carried out at three infiltration point stations in parallel under varying geological, biochemical, and, particularly, meteorological conditions. The infiltration stations are indicated by small circles in Figure 23.2.1.1.

23.2.3 Water Quality Analysis

Daily or weekly water quality sampling schemes were applied. Daily sampling and subsequent analysis took place at approximately 6:00 AM and/or 6:00 PM. All analytical procedures to determine the water quality were performed according to American standard methods (Clesceri et al., 1998) that also outline the corresponding water quality criteria. Water samples were tested for temperature (air and water), biochemical oxygen demand (BOD), SS, total solids (TS), conductivity, turbidity, dissolved oxygen (DO), and pH. The Hanna instruments HI 9033 conductivity meter, C 102 turbidity meter, HI 9142 DO meter, and HI 8519N pH meter were used throughout the study.

Oxidized aqueous nitrogen was determined as the sum of nitrate and nitrite. Nitrate was reduced to nitrite by cadmium and determined as an azo dye at 540 nm (using a Peristorp Analytical EnviroFlow 3000 flow injection analyzer) following diazotization with sulfanilamide and subsequent coupling with N-1-naphthyl ethylenediamine dihydrochloride. This technique also simultaneously measured nitrite. Furthermore, aqueous ammonia reacted with hypochlorite and salicylate ions in solution in the presence of sodium nitrosopentacyanoferrate (nitroprusside). Aqueous phosphate reacted with acidic molybdate to form a phosphomolybdenum blue complex. The associated colored complexes were measured spectrophotometrically at 655 and 882 nm, respectively, using a Bran and Luebbe Autoanalyzer (Model AAIII). All analyses for nutrients were carried out in triplicate.

Vegetative, sediment, and soil samples were dried at 105 °C overnight in a drying oven (UM500, Memmert) prior to being ashed at 400 °C for 12 h in a muffle furnace (ELF 11/14, Carbolite). Ashed samples (0.2–0.6 g) were then digested under reflux in aqua regia for 2 h, cooled, filtered through Whatman No. 5412 filter paper, and diluted to 100 ml with de-ionized water ready for analysis.

An inductively coupled plasma optical emission spectrometer (ICP-OES; TJA IRIS instrument, manufactured by ThermoElemental, USA) was used at 1350 W with coolant, auxiliary, and nebulizer argon gas flows of 15, 0.5, and 0.7 ml/min, respectively. A pump flow rate of 1 ml/min was used to screen for total concentrations of analyte elements in filtered (0.45 μm) water samples and digests.

Multi-element calibration standards in the concentration range between 0.1 and 10 mg/l were used, and the emission intensities were measured at appropriate wavelengths. For all elements, analytical precision in terms of the relative standard deviation was typically between 5% and 10% for individual aliquots. Three replicates for each sample were analyzed.

An electrothermal atomic absorption spectrometer (ETAAS; Varian SpectrAA 400, manufactured by Varian Inc., Australia) with auto-sampler and powered by a GTA-96 graphite tube atomizer was used to analyze some of the water samples for their zinc content. A 20 μl injection volume was used for samples and standards in notched GTA partition tubes (coated). Nitrogen was applied as the carrier gas. The char temperature was 300 °C with a ramp rate of 10 °C/s and a hold time of 3 s. For atomization, the temperature was set to 1900 °C with ramp and hold times of 1 and 2 s, respectively. Precision in terms of the relative standard deviation was typically $<5\%$ for triplicate injections. Three sample replicates were taken from prefiltered (Whatman 0.45 μm cellulose nitrate membrane filters) subsamples.

23.2.4 Control of Algal Growth

Visual algal cover estimations were undertaken using a 0.2×0.2 m reference metal grid located on top of the attenuation pond during the experiment that could also be used as a permanent safety barrier. Microorganisms including algae such as *Microspora* spp. (dominates "blanket weed") were counted using a Sedgewick Rafter Cell S50 (counting chamber: $1 \times 50 \times 20$ mm) and a Wang Biomedical Research Microscope 6000 (bright field and phase-contrast).

Excess algae and decomposing litter were occasionally removed from the attenuation pond and infiltration basin. Wet algae and litter were weighed after the biomass had drained for 2 min by applying maximum pressure with both hands to small portions of the organic waste. Blanket weed control with barley straw bales (a commercial product called Frogmat) was also practiced.

The straw bales were subdivided into smaller bales before application. The new bales (approximately 0.2 kg straw each) were located at eight sites (approximately 1.6 kg straw in total) near the attenuation pond margins, and remained there almost fully submerged between August 31 and October 31, 2001. This procedure was repeated between February 22 and May 18, 2002.

Between March 15 and August 12, 2002, an experimental granular activated carbon (GAC) filter cleaned the stormwater before it reached the silt trap. Approximately 1.4 kg GAC (Grade 207 EA, US mesh: 12×40) was used to test the purification potential with respect to nutrients and trace elements that might contribute to algal growth.

23.2.5 System Capacity

The silt trap and attenuation pond had a combined total effective volume of approximately 1.9 m^3 during storm events. The rainwater drained into the silt trap with a maximum capacity of 0.1 m^3. Suspended solids (e.g., weathered building materials, decayed leaves, bird droppings, and particles from atmospheric pollution) from the roofs settled predominantly in the silt trap.

Water from the silt trap overflowed into the attenuation pond (volume of approximately 1.7 m^3), which also served the purpose of a storage pond. If the attenuation pond was full, water flowed into the infiltration basin, which accommodated a maximum volume of approximately 1.8 m^3 during heavy storm events before it overflowed (Figure 23.2.1.1).

23.3 RESULTS AND DISCUSSION

23.3.1 Standard Design Considerations

Tables 23.3.1.1–23.3.1.3 summarize the calculation procedures for the BRE, CIRIA, and ATV-DVWK design guidelines (ATV-DVWK, 2002; Bettes, 1996;

TABLE 23.3.1.1 Calculation of the Required Infiltration Basin Storage Volume (Italic Number) Using the Building Research Establishment Guidelines (BRE, 1991)

Duration	min	10	15	30	60	120
Factor Z_1	–	0.46	0.56	0.75	1.00	1.30
Design rainfall M5-D (5-year return period)	mm	9.2	11.2	15.0	20.0	26.0
Growth factor Z_2	–	1.22	1.22	1.24	1.24	1.24
Design rainfall M10-D (10-year return period; calculated from M5 rainfall)	mm	11.2	13.7	18.6	24.8	32.2
Inflow (10-year return period)	m^3	0.97	1.18	1.60	2.13	2.77
Outflow drainage	m^3	0.12	0.18	0.36	0.72	1.44
Required storage (10-year return); M10-D	m^3	0.85	1.00	1.24	*1.41*	1.33

Assumptions: ratio = 0.24 (ratio between the M5-60 storm and the M5-2D rainfall); impermeable area = 86 m^2; M5-60 = 20 mm; infiltration rate = 0.0001 m/s; perimeter = 10 m; depth = 0.4 m; surface area to 50% effective depth (excluding base) = 2 m^2.

TABLE 23.3.1.2 Calculation of the Required Infiltration Basin Storage Height (Italic Number) Using the Construction Industry Research and Information Guidelines (Bettes, 1996)

Duration	min	10	15	30	60	120
Intensity of storm	mm/h	67.3	54.6	37.2	24.8	16.1
Maximum height of water to be stored	m	0.154	0.184	*0.213*	0.210	0.142

Assumptions: infiltration rate = 0.0001 m/s; safety factor = 2 (area to be drained <100 m^2; minor inconveniences, if the system fails); area of base = 4.95 m^2; perimeter = 10 m; impermeable area = 86 m^2; void ratio of the soak-away fill material = 1 (represents air).

TABLE 23.3.1.3 Calculation of the Required Infiltration Basin Storage Volume (Italic Number) Using the German Association for Water, Wastewater, and Waste Guidelines (ATV-DVWK, 2002)

Duration	min	10	15	30	60	120
Runoff for each design storm	l/(s × h)	186.9	151.7	103.3	68.9	44.7
Volume to be stored	m^3	0.85	0.99	1.19	*1.27*	0.96

Assumptions: impermeable area = 86 m^2; surface factor = 0.9 (considering roof angle and building material); infiltration area = 4.95 m^2; infiltration rate = 0.0001 m/s; safety factor = 1.1 (minor inconvenience, if the system fails).

BRE, 1991; Martin et al., 2000). Findings are based on the design assumption that flooding for the case study site might statistically occur only once within 10 years (defined as the return period). The estimated mean infiltration rate applied for all methods was 10^{-4} m/s (Tables 23.3.1.1–23.3.1.4).

The critical storm durations for the BRE, CIRIA, and ATV-DVWK design calculations were 1.0, 0.5, and 1 h, respectively. The associated maximum infiltration basin height requirements were 28, 21, and 26 cm in this order. Furthermore, the maximum infiltration storage volume requirements were 1.41, 1.04, and 1.27 m^3, respectively.

Water level fluctuations within the attenuation pond and infiltration basin are indicated in Figure 23.3.1.1. Maximum infiltration basin height requirements calculated according to BRE (1991), CIRIA (Bettes, 1996), and ATV-DVWK (2002) guidelines were not sufficient for the period of the experiment. Figure 23.3.1.1 indicates that the wetland system would have failed if the recommended design depths for the infiltration basin had been applied.

TABLE 23.3.1.4 Summary Statistics: Water Quality of the Attenuation Pond and Infiltration Rates for the Infiltration Basin (April 26, 2001–August 12, 2002)

Number	Variable	Unit	Time	Sample Number	Mean
1	Air temperature	°C	PM[i]	345	12.2
2	Water temperature	°C	PM[i]	325	11.1
3	BOD_5[a]	mg/l	AM[j]	57	4.3
4	Suspended solids	mg/l	AM[j]	40	46.8
5	Total solids	mg/l	AM[j]	40	193.3
6	Conductivity	μS	PM[i]	281	39.8
7	Turbidity	NTU[h]	AM[j]	39	2.8
8	Dissolved oxygen	mg/l	PM[i]	319	9.3
9	pH	—	PM[i]	263	7.77
10	Algal cover	%	PM[i]	306	44
11	IR (station 1)[b]	10^{-7} m/s	PM[i]	23	8
12	IR (station 2)[b]	10^{-7} m/s	PM[i]	23	158
13	IR (station 3)[b]	10^{-7} m/s	PM[i]	24	9946

Number	1st Summer[c] Mean	Fall[d] Mean	Winter[e] Mean	Spring Mean[f]	2nd Summer Mean[g]
1	16.9	8.9	6.5	13.2	17.1
2	15.4	8.6	9.1	11.9	15.3
3	4.2	3.1	5.6	6.7	2.5
4	132.5	51.3	3.8	54.8	3.4
5	238.7	293.8	104.7	151.5	91.7
6	75.1	37.5	33.0	37.7	34.2
7	4.3	2.7	2.2	2.0	1.5
8	11.5	4.8	10.1	12.4	7.3
9	8.36	7.27	7.86	8.69	6.84
10	61	50	36	39	38
11	3	23	1	2	5
12	4270	95	42	36	580
13	8152	4778	1286	6342	31,841

[a]BOD_5 = 5-day biochemical oxygen demand at 20 °C.
[b]IR = infiltration rate; these were calculated for the soil layers located 20 cm below the bottom of the infiltration basin center (Station 1), the middle of the slope of the infiltration basin (Station 2), and the lawn (Station 3); see also Figure 23.2.1.1.
[c]Summer: 06/21/01–9/21/01.
[d]Fall: 09/22/01–12/20/01.
[e]Winter: 12/21/01–03/19/02.
[f]Spring: 03/20/02–06/20/02.
[g]Summer: 06/21/02–08/12/02.
[h]NTU = nephelometric turbidity unit.
[i]PM = afternoon.
[j]AM = morning.

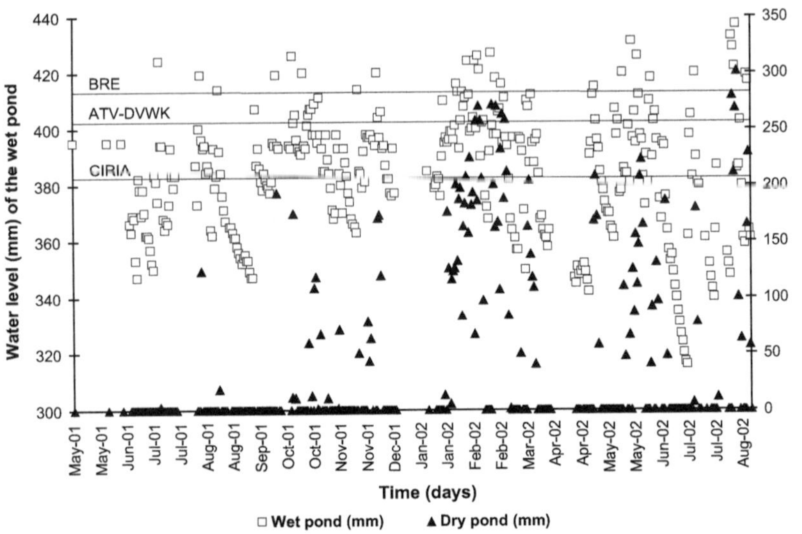

FIGURE 23.3.1.1 Maximum daily water level fluctuations within the wet pond and dry pond between May 13, 2001, and August 12, 2002. Maximum infiltration basin height requirements calculated according to Building Research Establishment (BRE, 1991), Construction Industry Research and Information Association (CIRIA) (Bettes, 1996; Martin et al., 2000), and the German Association for Water, Wastewater, and Waste (ATV-DVWK, 2002) guidelines are indicated by horizontal lines.

Equation (23.3.1.1) indicates the mathematical relationship between the infiltration basin design depth D (mm) and the infiltration time T (h). The corresponding R for the function is 0.92.

$$D = -a + \ln(T) + b \tag{23.3.1.1}$$

where:

$a = 2.54 \times b^{0.51}$, $R = 0.58$ (in general);
$a = 0.59 \times b^{0.83}$, $R = 0.89$ (for $b < 190$ mm);
$a = 0.015 \times b^{1.51}$, $R = 0.77$ (for $b \geq 190$ mm during spring, summer, and fall); and
$a = 0.001 \times b^{1.94}$, $R = 0.75$ (for $b \geq 190$ mm during winter).

Figure 23.3.1.2 indicates the mathematical relationship between the terms a and b, which are both based on logarithmic trendline functions for 36 recorded storms (between three and eight depth measurements each) between May 2001 and August 2002. Term a is a coefficient, whereas b represents the maximum depth (mm) within the infiltration basin during an individual storm. The corresponding R accounts for differences in infiltration behavior for systems that are either shallower or deeper than approximately 185 mm. The mathematical relationships summarized in Eq. (23.3.1.1) and

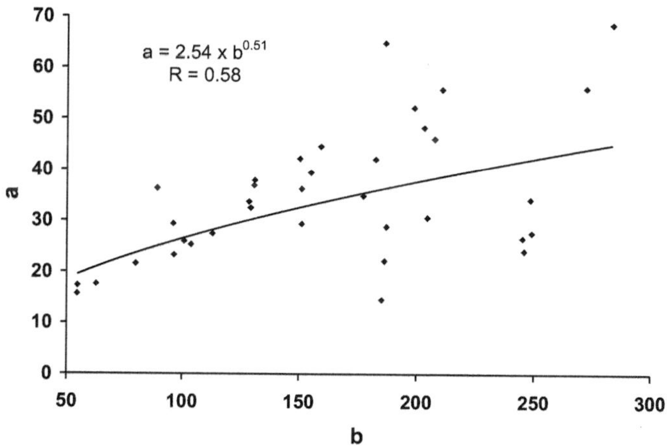

FIGURE 23.3.1.2 Relationships between the coefficient "a" and the maximum depth "b" (mm) within the infiltration basin during an individual storm.

Figure 23.3.1.2 give the design engineer the opportunity to estimate design depths related to different infiltration times and vice versa.

23.3.2 System Design Comparisons

The actual design for the infiltration basin (40 cm depth and 1.7 m^3 volume) was acceptable when compared to BRE (1991), CIRIA (Bettes, 1996), and ATV-DVWK (2002) guidelines. Signs of system failure (e.g., flooding of the nearby lawn and structural damage) have not yet been observed. However, strict application of all test guidelines (without adding a higher safety factor than recommended) would have led to system failures already during the first year of operation. This finding is in contrast to a previous investigation by Abbott and Comino-Mateos (2001), where British design guidelines did not fail. However, a perforated concrete ring soak-away to infiltrate stormwater runoff was used.

Figure 23.3.1.1 shows that all three official designs (Tables 23.3.1.1–23.3.1.3) failed at least once during the first 15 months of the 10-year design period. However, the period of study can be described as a particularly wet time (2160 mm precipitation between summer 2001 and spring 2002). Moreover, the study site is situated on a semi-exposed location approximately 240 m above sea level. In comparison, the annual average rainfall for Bradford (located in a valley) is approximately 830 mm (Meteorological Office, 2002).

Considering a maximum depth of 40 cm, the infiltration basin is associated with under-utilized storage capacity. However, observations have shown that a depth of 40 cm is justified considering that 20, 10, and 1 runoff events have led to the maximum design water depths of 21, 26, and 28 cm, respectively, being exceeded (Figure 23.3.1.1). It follows that there would have been no spare capacity for strict applications of the standard design guidelines.

23.3.3 Water Quality Management

The water quality of the attenuation pond (Tables 23.3.1.4 and 23.3.3.1) was acceptable for disposal (e.g., sustainable drainage) and recycling (e.g., irrigation, toilet flushing, and washing of cars), according to findings presented by Butler and Davies (2000) and Ellis et al. (2002).

TABLE 23.3.3.1 Summary Statistics: 24-h Sampling on March 15, 2002

		Mean			
Variable	**Unit**	**00:00–05:30**	**06:00–11:30**	**12:00–17:30**	**18:00–23:30**
Water temperature	°C	3.4	4.3	7.4	4.8
BOD_5[a]	mg/l	1.3	1.1	0.7	2.0
Nitrate-N[b]	mg/l	0.54	0.38	0.39	0.47
Ammonia-N	mg/l	0.06	0.03	0.04	0.07
Phosphate-P	mg/l	0.07	0.04	0.06	0.13
Total solids	mg/l	99.8	97.8	180.7	139.2
Conductivity	µS	36.0	42.0	43.1	41.3
Dissolved oxygen	mg/l	10.0	9.8	11.3	11.6
pH	–	7.54	7.57	7.89	8.01
Calcium[c,d]	mg/l	10.37	10.89	11.08	10.96
Magnesium[c,d]	mg/l	1.13	1.14	1.14	1.14
Manganese[c,d]	mg/l	0.006	0.006	0.005	0.010
Zinc[c,d]	mg/l	0.007	0.010	0.012	0.010
Zinc[c,e]	mg/l	0.009	0.006	0.006	0.006
Barium[c,d]	mg/l	1.57	1.57	1.66	1.72
		Standard Deviation			
Variable	**Total Number**	**00:00–05:30**	**06:00–11:30**	**12:00–17:30**	**18:00–23:30**
Water temperature	23	0.6	2.1	0.7	1.4
BOD_5[a]	36	0.8	0.6	0.5	1.1

TABLE 23.3.3.1 Summary Statistics: 24-h Sampling on March 15, 2002—cont'd

Variable	Total Number	Standard Deviation			
		00:00–05:30	06:00–11:30	12:00–17:30	18:00–23:30
Nitrate-N[b]	40	0.37	0.18	0.17	0.19
Ammonia-N	40	0.06	0.04	0.05	0.07
Phosphate-P	40	0.06	0.04	0.06	0.09
Total solids	43	18.8	14.6	59.1	31.3
Conductivity	24	1.2	1.5	1.7	2.0
Dissolved oxygen	24	0.7	1.2	0.8	1.3
pH	24	0.39	0.51	0.24	0.58
Calcium[c,d]	21	1.19	0.96	1.23	0.07
Magnesium[c,d]	21	0.08	0.07	0.04	0.06
Manganese[c,d]	21	0.002	0.002	0.002	0.009
Zinc[c,d]	21	0.003	0.005	0.009	0.001
Zinc[c,e]	39	0.004	0.004	0.003	0.002
Barium[c,d]	21	0.2	0.09	0.28	0.31

[a] *BOD_5 = 5-day biochemical oxygen demand at 20 °C.*
[b] *Concentrations include nitrite.*
[c] *Filtered sample (0.45 μm pore size).*
[d] *Analysis with the inductively coupled plasma optical emission spectrometer.*
[e] *Analysis with the electrothermal atomic absorption spectrometer.*

Figures 23.3.3.1 and 23.3.3.2 show results from the analysis of waters and digests by ICP-OES. Measured elemental concentrations were either low (boron, barium, calcium, magnesium, manganese, and zinc), close to the detection limit (aluminum, copper, and iron), or below the detection limit (most heavy elements were analyzed but not listed). Figure 23.3.3.2 indicates that the liquid phase (rainwater) within the treatment train can actually take up soluble contaminants from the sediment acting as a source, thereby polluting the liquid phase.

23.3.4 Twenty-Four-Hour Water Quality Monitoring

Manual sampling over 24 h was conducted on March 15, 2002 (Table 23.3.3.1). Most samples were taken either in 30-min intervals or hourly. Despite the cold climate, diurnal water quality variations (Wu and Mitsch, 1998) were

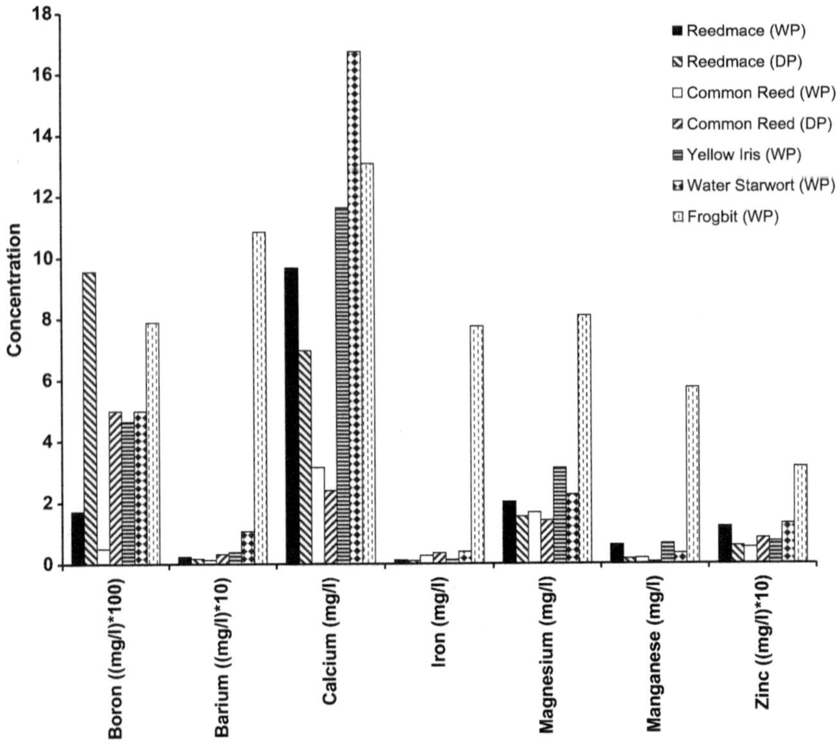

FIGURE 23.3.3.1 Summary of the ICP-OES results for different aquatic plants (dry weights) located in the attenuation pond and infiltration basin. Concentration readings for boron, barium, and zinc require division by a factor of 100, 10, and 10, respectively. WP = wet pond. DP = dry pond.

apparent for water temperature, 5-day biochemical oxygen demand (BOD_5), nitrate-N (including nitrite-N), ammonia-N, conductivity, pH, manganese, zinc, and barium.

Of the elements detected by ICP-OES, zinc was the only one that had a high overall standard deviation (approximately 0.0060 mg/l for triplicate analyses of each sample) as measurements were made close to the detection limit. However, analyzing the same samples using ETAAS reduced the overall standard deviation to 0.0036 mg/l (three replicates for each sample) as this technique is more sensitive for the determination of zinc. There were no significant fluctuations of zinc during the day.

23.3.5 Aquatic Plant Management

To prevent algal washout into the infiltration basin during storm events, it is necessary to control the rapid growth of algae with barley straw and by mechanical removal (P. Taylor, Director, Frogmat International Holdings,

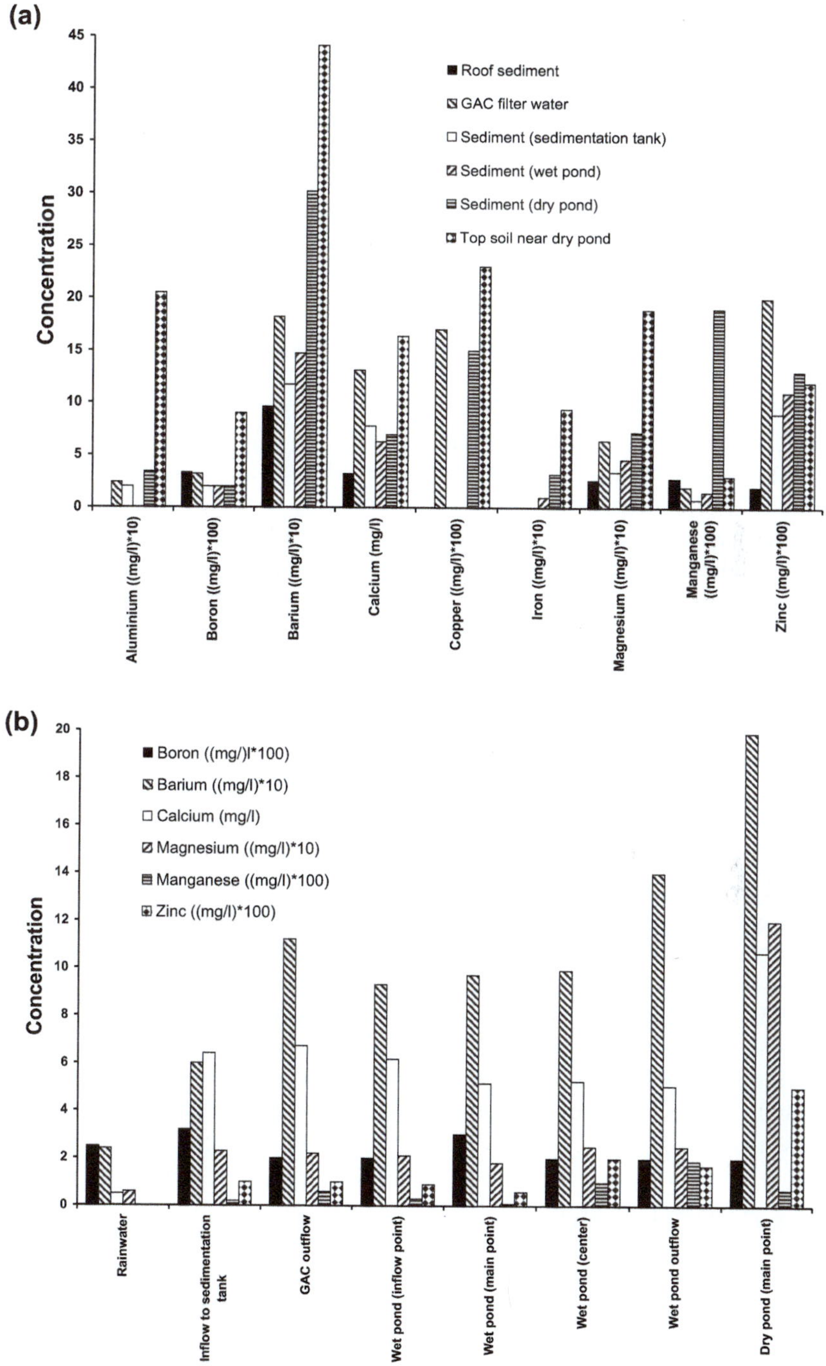

FIGURE 23.3.3.2 (a) Summary of ICP-OES results for the liquid phase of the pond system. Concentration readings for boron, barium, magnesium, manganese, and zinc require division by a factor of 100, 10, 10, 100, and 100, respectively. (b) Summary of ICP-OES results for the sediment phase of the pond system. Concentration readings for aluminum, boron, barium, copper, iron, magnesium, manganese, and zinc require division by a factor of 10, 100, 10, 100, 10, 10, 100, and 100, respectively. GAC = granular activated carbon.

personal communication). Algal washout allows decaying algae to block fine soil pores, thereby decreasing the infiltration capacity of the basin (see also Table 23.3.3.1).

In addition to biological (grazing zooplankton and tadpoles) and physical (removal by hand) algal control, biochemical control was also implemented (with barley straw extracts). The presence of small barley straw bales is likely to reduce algal growth by releasing a cocktail of phytotoxic chemicals (Everall and Lees, 1997).

The wet weights of litter removed during summer 2001, fall 2001, winter 2001/02, and spring 2002 were 1.1, 4.6, 1.0, and 6.2 kg, respectively. The number of occasions on which plant harvesting was carried out during these seasons were 6, 10, 3, and 14, respectively.

Litter production depends on the season (Scholz and Xu, 2002; Wu and Mitsch, 1998). Fresh biomass is usually produced in spring and summer and requires harvesting in late fall. Furthermore, algae (predominantly *Microspora* spp.) require frequent removal in spring. In contrast to summer and winter, the maintenance was more efficient during fall and spring because less labor (i.e., work input expressed as time) was required and more litter and algae were removed during these maintenance sessions.

Figure 23.3.3.1 indicates the elemental content of various aquatic plants. It can be seen that macrophytes located in the attenuation pond contain higher concentrations of calcium, magnesium, and manganese in comparison to macrophytes located in the infiltration basin. Furthermore, Frogbit was associated with high elemental concentrations in general. This was also the case for barium, iron, magnesium, manganese, and zinc.

All assessed aquatic plants were associated with detectable concentrations of magnesium, manganese, and zinc. Measured concentrations of these elements for aquatic plants were relatively higher than for the liquid phase (pond water) but comparable to associated sediment samples (Figures 23.3.3.1 and 23.3.3.2). However, a direct comparison between different loadings and concentrations (mg/kg and mg/l, respectively) is difficult. Nevertheless, plant harvesting reduced trace nutrients and elements (in addition to variables such as BOD_5 and SS) in the attenuation pond. This confirms findings by Wu and Mitsch (1998).

23.4 CONCLUSIONS

The case study described the successful design, operation, and maintenance of a novel stormwater pond system during the first 15 months of operation. However, the tested international design guidelines would only have been acceptable if local environmental conditions (precipitation and infiltration patterns) were fully considered or high safety factors (significantly higher than widely recommended) were applied.

Infiltration through the base of the infiltration basin was low (despite the presence of macrophytes) and should not have been considered during the design. The infiltration function for the infiltration basin was logarithmic and not linear, as indicated, mistakenly, by most international design guidelines.

The water quality of the attenuation pond was acceptable for recycling. Roof runoff did not require sophisticated treatment. Seasonal and diurnal variation in water quality for BOD_5, phosphate (only evidence for diurnal variation provided), DO, and pH were apparent.

Excess algal and litter biomass weighed approximately 2.3 kg wet biomass/m^2 of the attenuation pond surface area per annum. The biomass required mechanical removal to avoid eutrophication in the attenuation pond and low infiltration rates in the infiltration basin. Plant harvesting reduced BOD_5, nutrient, and trace element levels within the wetland system.

Chapter 24

Wetlands Treating Road Runoff

24.1 INTRODUCTION

Storm runoff from urban areas has been recognized as a major contributor to pollution of receiving urban watercourses (Park and Yoon, 2003). The principal pollutants in urban runoff are biochemical oxygen demand (BOD), suspended solids (SS), heavy metals, de-icing salts, hydrocarbons, and fecal coliforms (Scholz and Martin, 1998a; Scholz and Zettel, 2004; Scholz et al., 2005). Constructed wetlands integrated into a best management practice concept are a sustainable means of treating stormwater. Some systems are also more economical and energy efficient than traditional centralized treatment systems (Kadlec et al., 2000).

Metal bioavailability and reduction are controlled by chemical processes including acid volatile sulfide formation, organic carbon binding, and sorption in reduced sediments of constructed wetlands (Kadlec, 2002; Obarska-Pemkowiak and Klimkowska, 1999; Wood and Shelley, 1999). It follows that metals usually accumulate in the top layer (comprising fine aggregates, sediment, and litter) of vertical-flow and near the inlet of horizontal-flow constructed treatment wetlands (Cheng et al., 2002; Scholz and Xu, 2002; Vymazal and Krasa, 2003).

Physical and chemical properties of wetland soil and aggregates affecting metal mobilization include particle size distribution (texture), redox potential, pH, organic matter, salinity, and the presence of inorganic matter such as sulfides and carbonates (Backstrom et al., 2004).

Furthermore, sulfides and carbonates may combine with metals to form relatively insoluble compounds. Particularly the formation of metal sulfide compounds may provide long-term heavy metal removal because these sulfides will remain permanently in the wetland sediments as long as they are not re-oxidized (Cooper et al., 1996; Kadlec and Knight, 1996).

The major purpose of this experimental case study is to improve the design, operation, and management of vertical-flow constructed treatment wetlands to secure a high wastewater treatment performance during all seasons. The objectives are to assess:

1. The performance of vertical-flow constructed treatment wetland filters combined with a treatment pond (located above the litter zone) for gully pot liquor treatment;

Wetlands for Water Pollution Control. http://dx.doi.org/10.1016/B978-0-444-63607-2.00024-1

2. The compliance with water quality standards in terms of the reduction efficiencies of BOD, SS, nickel, copper, and nitrate;
3. The impact of environmental conditions such as variations of salt concentrations, pH, and the temperature on the treatment performance of constructed wetlands during all seasons;
4. The feasibility of conventional chemical pH adjustment to prevent heavy metal leaching during winter; and
5. The overall role of adsorption media and *Phragmites australis.*

24.2 SITE, MATERIALS, AND METHODOLOGY

24.2.1 Case Study Site

Twelve wetland filters (Figure 24.2.1.1) were located outdoors at The King's Buildings campus (The University of Edinburgh, Scotland) to assess the system performance in a cold climate (09/09/02–09/21/04). The first 12 days of operation were not analyzed because the water quality was not representative (start-up period). Inflow water, polluted by road runoff, was collected from randomly selected gully pots on the campus, the nearby predominantly housing estates, and two major urban roads. After mixing both the sediment and the water phase within the gully pot, water was collected by manual abstraction with a 2-l beaker.

FIGURE 24.2.1.1 Constructed treatment wetland rig (The King's Buildings campus; The University of Edinburgh) in May 2004.

24.2.2 Filter Design, Media Composition, and Limitations

Round drainage pipes were used to construct the above-ground filters. All 12 vertical-flow wetland filters (Figure 24.2.1.1) were designed with the following dimensions: height = 83 cm and diameter = 10 cm. The relatively small size of these wetlands makes them vulnerable to environmental changes. However, a literature search has revealed that the use of various small wetlands for experimental purposes is common to simulate industrial-scale systems, and that data variation is similar between small and large systems (Karathanasis et al., 2003; Scholz et al., 2002; Wießner et al., 2004; Zhao et al., 2004).

Above-ground filters experience higher temperatures and associated data variations than below-ground filters in summer. It follows that performance data might be overly optimistic for this season. However, above-ground filters are exposed to severe conditions such as relatively low temperatures in winter in comparison to below-ground systems that are insulated by the (potentially frozen) soil, debris, and litter and potentially snow (Picard et al., 2004; Scholz and Xu, 2002). Moreover, the macrophytes and most of the microbial biomass are usually inactive during winter. Therefore, freezing and thawing events have little impact on the filtration performance (Cooper et al., 1996; Kadlec and Knight, 1996).

Different packing order arrangements of filter media, plant rhizomes, and roots were used in the wetland filters (Tables 24.2.2.1 and 24.2.2.2). The outlet of each constructed wetland comprised a valve at the bottom of each filter. In September 2002, the calculated empty filter bed volumes were approximately 6.2, 6.4, 4.0, 4.1, 3.8, 4.1, 3.8, 4.0, 3.8, 4.0, 4.0, and 4.0 l for Filters 1–12, respectively. The filter volume capacities were measured by draining the filters entirely.

The inflow waters of Filter 2 as well as Filters 7–12 were dosed with hydrated copper nitrate ($Cu(NO_3)_2 \cdot 3H_2O$) and hydrated nickel nitrate ($Ni(NO_3)_2 \cdot 6H_2O$). Filters 1 and 2 (controls) are similar to wastewater stabilization ponds or gully pots (extended storage) without a significant amount of filter media (Table 24.2.2.2). In comparison, Filters 3, 5, 7, and 9 are similar to slow sand (Chapter 10) and gravel (Chapter 13) filters, and Filters 4, 6, 8, and 10 are typical reed bed filters (Chapter 20).

The reed bed filters contain gravel and sand substrate and native *P. australis* (Cav.) Trin. Ex Steud. (common reed), all of similar total biomass weight during planting and from the same local source. However, Filters 5, 6, 9, and 10 also contain adsorption media. Additional natural adsorption media (Filtralite and Frogmat) were used. Filtralite (containing 3% of calcium oxide (CaO)) with diameters between 1.5 and 2.5 mm is associated with enhanced metal and nutrient reduction (Brix et al., 2001; Scholz and Xu, 2002). Furthermore, Frogmat (natural product based on raw barley straw) has a high adsorption area and is therefore likely to be associated with a high heavy metal

TABLE 24.2.2.1 Systematic and Stratified Experimental Set-Up of Filter Content and Operation

Filter	Planted	Media[a]	Metals[b]	Aerated	High Load[c]
1	No	1	No	No	No
2	No	1	Yes	No	No
3	No	2	No	No	No
4	Yes	2	No	No	No
5	No	3	No	No	No
6	Yes	3	No	No	No
7	No	2	Yes	No	No
8	Yes	2	Yes	No	No
9	No	3	Yes	No	No
10	Yes	3	Yes	No	No
11	Yes	3	Yes	Yes	No
12	Yes	3	Yes	Yes	Yes

[a]1 = no media, 2 = standard, 3 = addition of Filtralite (light expanded clay) and Frogmat (barley straw).
[b]Addition of hydrated copper nitrate and hydrated nickel nitrate.
[c]Filter 12 received approximately 153% additional inflow in comparison to Filter 11.

reduction potential. The use of other filter media with high adsorption capacities such as activated carbon (Scholz and Martin, 1998a; Scholz et al., 2002) and oxide-coated sand (Sansalone, 1999) has been discussed elsewhere.

Filters 11 and 12 are more complex in their design and operation (Table 24.2.2.1). The top water layer of both filters is aerated (with air supplied by air pumps) to enhance oxidation (minimizing zones of reducing conditions) and nitrification (Cheng et al., 2002; Green et al., 1998; Obarska-Pempkowiak and Klimkowska, 1999). Filter 12 receives about 153% of Filter 11's mean annual inflow volume and load (Table 24.2.2.1). The hydraulic regime of Filter 12 differs from that of Filters 1–11 to identify the best filtration performance. A higher hydraulic load should result in greater stress on *P. australis* and biomass.

24.2.3 Environmental Conditions and Operation

The filtration system was designed to operate in batch flow mode to reduce pumping and computer control costs. All filters were periodically inundated with pretreated inflow gully pot liquor and drained to encourage air

TABLE 24.2.2.2 Packing Order of Vertical-Flow Wetland Filters

Height (cm)	F1	F2	F3	F4	F5	F6	F7	F8	F9	F10	F11	F12
61–83	W	W	W	W	W	W	W	W	W	W	AW	AW
56–60	W	W	6	6 + P	7	7 + P	6	6 + P	7	7 + P	7 + P	7 + P
51–55	W	W	6	6 + P	6	6 + P	6	6 + P	6	6 + P	6 + P	6 + P
36–50	W	W	4	4 + P	5	5 + P	4	4 + P	5	5 + P	5 + P	5 + P
31–35	W	W	3	3	5	5	3	3	5	5	5	5
26–30	W	W	3	3	4	4	3	3	4	4	4	4
21–25	W	W	2	2	3	3	2	2	3	3	3	3
16–20	W	W	2	2	2	2	2	2	2	2	2	2
11–15	2	2	2	2	2	2	2	2	2	2	2	2
0–10	1	1	1	1	1	1	1	1	1	1	1	1

Abbreviations: F = filter; W = water; P = *Phragmites australis* (Cav.) Trin. Ex Steud. (common reed); AW = aerated water; 1 = stones; 2 = large gravel; 3 = medium gravel; 4 = small gravel; 5 = Filtralite (light expanded clay product); 6 = sand (0.6–1.2 mm); 7 = Frogmat (barley straw).

penetration through the aggregates (Cooper et al., 1996; Gervin and Brix, 2001; Scholz and Xu, 2002).

Since September 22, 2003, the pH value of the inflow has been artificially raised by addition of sodium hydroxide (NaOH) to the sieved gully pot liquor to prevent a breakthrough of metals (see below and Tchobanoglous et al., 2003). It follows that the inflow pH was therefore increased from a mean pH value of 6.7 to 8.1 (Table 24.2.3.1).

24.2.4 Metal Nitrates

Copper and nickel were selected as additional heavy metals for investigation because they are commonly occurring contaminants from road vehicles and are not easily bioavailable (Cooper et al., 1996; Kadlec and Knight, 1996; Scholz and Xu, 2002). It follows that these metals are likely to accumulate within the sediment and debris of constructed wetlands. As the build up continues, metal toxicity increases, as does the risk of severe pollution due to leaching (Scholz et al., 2002).

Some heavy metals do accumulate easily in constructed wetlands but may be released if environmental conditions change; e.g., road gritting (containing salt) in winter. Such transformation processes are not well understood (Norrstrom and Jacks, 1998).

Copper nitrate and nickel nitrate were added to the inflow water of Filter 2 and Filters 7–12 to give total concentrations of dissolved copper and nickel of approximately 1 mg/l for each metal, which is comparable to the figures reported for urban water heavily contaminated with heavy metals and mine wastewater (Cooper et al., 1996; Kadlec and Knight, 1996; Mungur et al., 1997; Scholz and Xu, 2002).

Concerning the dosed inflow water, the background concentration for nitrate–nitrogen (including nitrite-nitrogen) was only approximately 0.497 mg/l. Therefore, introduced nitrate–nitrogen contributed to 65% (or approximately 0.917 mg/l) of the overall nitrate–nitrogen (including nitrite-nitrogen) load.

The filter volumes available for influent water differ among the filters owing to different filter media compositions (Table 24.2.2.2) and hydraulic regimes (Table 24.2.2.1). The mean annual total loading rates for the contaminated filters were therefore between 96 and 187 mg for each metal.

24.2.5 Metal Determinations

Metal concentrations were determined in the raw gully pot liquor, sieved (pore size of 0.25 mm) gully pot liquor (partially used as actual inflow water for some filters), contaminated (added metal nitrates) sieved gully pot liquor (partially used as actual inflow water for some filters), and the outflow waters from the experimental rig (Table 24.2.2.1). Raw gully pot liquor was sieved to simulate preliminary (Chapter 5) treated stormwater. This procedure is in line

TABLE 24.2.3.1 Primary Treated Gully Pot Effluent: Water Quality Variables after Contamination with Hydrated Copper Nitrate and Hydrated Nickel Nitrate

Variable	Unit	Number of Samples	Mean	SD[a]	Mean (Winter)	Mean (Summer)
09/22/02–09/21/03						
Nickel (dissolved)	mg/l	57	1.06	0.143	1.08	1.09
Copper (dissolved)	mg/l	58	1.03	0.036	1.04	1.02
BOD[b]	mg/l	58	61.1	49.29	43.8	86.9
Nitrate–nitrogen	mg/l	63	1.45	1.008	1.43	1.12
Ammonia-nitrogen	mg/l	63	1.65	2.058	1.72	2.11
Ortho-phosphate-phosphorus	mg/l	62	0.06	0.149	0.05	0.03
Suspended solids	mg/l	70	335.7	377.75	743.7	160.7
Total solids	mg/l	66	2995.5	6793.27	9403.9	376.5
Turbidity	NTU	71	311.7	479.65	690.6	162.1
Dissolved oxygen	mg/l	68	4.70	2.493	5.70	3.16
pH	–	71	6.69	0.411	6.89	6.72
Redox potential	MV	62	142.5	112.72	165.5	2.7
Conductivity	µS	71	5139.1	11,182.22	15,311.6	501.0
Temperature (air)	°C	70	12.3	5.96	7.2	17.9
Temperature (gully pot)	°C	69	10.7	5.78	4.8	18.0
09/22/03–09/21/04 (Artificial Increase of pH after 09/21/03)						
Nickel (dissolved)	mg/l	68	1.02	0.042	1.01	1.02
Copper (dissolved)	mg/l	66	1.02	0.018	1.01	1.02

Continued

TABLE 24.2.3.1 Primary Treated Gully Pot Effluent: Water Quality Variables after Contamination with Hydrated Copper Nitrate and Hydrated Nickel Nitrate—cont'd

Variable	Unit	Number of Samples	Mean	SD[a]	Mean (Winter)	Mean (Summer)
09/22/03–09/21/04 (Artificial Increase of pH after 09/21/03)						
BOD[b]	mg/l	73	89.2	55.5	89.7	66.6
Nitrate–nitrogen	mg/l	72	1.38	1.220	1.19	1.14
Ammonia-nitrogen	mg/l	72	1.45	1.934	1.89	1.35
Ortho-phosphate-phosphorus	mg/l	72	0.10	0.136	0.07	0.16
Suspended solids	mg/l	75	853.9	1420.85	1955.2	366.7
Total solids	mg/l	71	2141.8	3219.84	5296.4	794.7
Turbidity	NTU	78	274.5	358.57	546.2	143.6
Dissolved oxygen	mg/l	78	3.07	1.49	3.41	3.22
pH	–	78	8.07	1.082	8.25	8.73
Redox potential	MV	78	44.4	93.44	31.8	64.1
Conductivity	µS	78	2227.2	4620.82	6191.7	403.6
Temperature (air)	°C	155	13.7	6.46	7.1	20.6
Temperature (gully pot)	°C	75	12.0	5.65	6.0	18.9

[a]*Standard deviation.*
[b]*5-day at 20 °C N-allylthiourea biochemical oxygen demand.*

with common practice in the wastewater industry (Cooper et al., 1996; Tchobanoglous et al., 2003).

A Varian SpectrAA 400 atomic absorption spectrometer with a GTA-96 graphite furnace tube atomizer was used for the standard analysis of nickel and copper. Notched GTA partition tubes (coated) were applied. The carrier gas was argon.

24.2.6 Organic Strength, Nutrient, and Other Determinations

The BOD was determined in all water samples with the OxiTop IS 12-6 system, a manometric measurement device, supplied by the Wissenschaftlich-Technische Werkstätten (WTW), Weilheim, Germany. Nitrification was suppressed by adding 0.05 ml of 5 g/l N-allylthiourea (WTW Chemical Solution No. NTH 600) solution per 50 ml of sample water.

Nitrate was reduced to nitrite by cadmium and determined as an azo dye at 540 nm (using a Perstorp Analytical EnviroFlow 3000 flow injection analyzer) following diazotization with sulfanilamide and subsequent coupling with N-1-naphthyl ethylenediamine dihydrochloride (Allen, 1974).

Ammonia-nitrate and ortho-phosphate-phosphorus were determined by automated colorimetry in all water samples from reaction with hypochlorite and salicylate ions in solution in the presence of sodium nitrosopentacyanoferrate (nitroprusside) and reaction with acidic molybdate to form a phosphomolybdenum blue complex, respectively (Allen, 1974). The colored complexes formed were measured spectrometrically at 655 and 882 nm, respectively, using a Bran and Luebbe autoanalyzer (model AAIII). All other analytical procedures were performed according to American Standard Methods (Clesceri et al., 1998). Composite water samples were usually analyzed on Mondays, Wednesdays, and Fridays.

24.3 EXPERIMENTAL RESULTS AND DISCUSSION

24.3.1 Inflow Water Quality

Table 24.2.3.1 summarizes the water quality of the inflow and those filters artificially contaminated with heavy metals after the first and second year of operation. The pH of the inflow was artificially raised to assess its influence on the treatment performance and particularly on the potential breakthrough of heavy metals during the second winter.

Raw gully pot liquor was sieved (pore size of 2.5 mm) to simulate preliminary treatment (Tchobanoglous et al., 2003). Sieving resulted in a mean annual reduction of BOD and SS by approximately 12% and 22%, respectively.

The inflow data set was divided into two subsets (winter and summer) to assess the effect of seasonal variations (e.g., temperature) and road management (e.g., road gritting and salting) on the water quality. Most variables including BOD (except for the first year of operation), SS, total solids (TS), turbidity, and conductivity are high in winter compared to summer (Table 24.2.3.1).

24.3.2 Comparison of Annual Outflow Water Qualities

The overall filtration performance figures are summarized in Tables 24.3.2.1 and 24.3.2.2. Figure 24.3.2.1 shows selected inflow and selected outflow

TABLE 24.3.2.1 Mean and Standard Deviation of Outflow Water Quality Variables

Variable	Unit	Filter 1	Filter 2	Filter 3	Filter 4	Filter 5	Filter 6	Filter 7	Filter 8	Filter 9	Filter 10	Filter 11	Filter 12
Mean (09/22/02–09/21/03)													
BOD[a]	mg/l	37.2	43.4	16.4	30.7	23.3	33.5	12.7	22.0	21.8	37.6	20.1	19.8
Suspended solids	mg/l	174.6	189.0	120.8	130.4	132.4	127.3	96.7	92.5	85.8	79.1	73.2	163.5
Total solids	mg/l	2772.8	3602.5	2938.7	3266.2	2949.1	3990.4	2375.6	2118.4	1717.1	1808.8	1779.0	2941.7
Turbidity	NTU	79.8	89.0	15.0	39.0	35.4	25.6	11.2	20.6	27.3	34.3	17.5	43.5
Conductivity	µS	5148.5	6827.7	5920.8	5392.6	5809.9	5797.7	4912.2	4346.8	3808.8	3667.8	3528.5	5941.4
Mean (09/22/03–09/21/04; Artificial Increase of pH after 09/21/03)													
BOD[a]	mg/l	30.2	30.4	3.1	2.3	4.4	4.4	2.3	2.7	2.8	3.5	3.6	7.6
Suspended solids	mg/l	434.6	907.5	76.6	65.8	130.4	82.7	147.3	78.1	83.4	71.7	89.9	102.8
Total solids	mg/l	1690.3	1931.8	1379.6	1398.9	1372.9	1430.7	1390.3	1497.3	1509.4	1804.7	1578.5	1647.4
Turbidity	NTU	118.6	117.8	6.0	4.8	8.6	6.0	9.6	6.9	10.2	4.9	7.4	20.1
Conductivity	µS	2268.2	2356.0	2339.9	2260.3	2220.0	2507.6	2199.2	2484.2	2450.6	2459.6	2534.5	2495.3

Standard Deviation (09/22/02–09/21/03)													
BOD[a]	mg/l	39.89	41.95	13.09	31.24	22.74	30.33	13.88	25.03	20.08	35.76	14.04	17.63
Suspended solids	mg/l	365.53	440.59	265.83	252.75	249.34	273.44	225.06	197.81	168.08	166.03	144.40	307.73
Total solids	mg/l	7273.37	10,821.20	6886.58	6585.16	6216.36	8318.42	5948.07	4720.47	3580.56	3648.16	3659.61	6450.13
Turbidity	NTU	96.15	73.83	24.21	61.19	46.64	28.45	18.95	30.18	42.51	54.91	18.07	86.23
Conductivity	µS	13,473.42	19,920.19	13,442.84	11,789.78	12,286.69	12,860.16	11,672.86	9431.98	6319.37	6683.83	7077.11	11,479.77
Standard Deviation (9/22/03–9/21/04; Artificial Increase of pH after 9/21/03)													
BOD[a]	mg/l	21.24	37.06	7.23	3.11	8.11	6.66	3.86	3.83	4.00	4.49	4.24	7.83
Suspended solids	mg/l	449.80	3580.34	96.29	79.97	263.65	107.28	228.84	82.75	96.24	86.77	133.64	138.81
Total solids	mg/l	1507.80	2417.22	1722.92	1601.87	1445.83	1714.86	1623.29	1538.09	1716.03	2553.49	1905.64	2426.29
Turbidity	NTU	90.33	258.22	6.89	4.44	8.38	4.48	18.79	7.76	19.59	4.24	10.94	37.18
Conductivity	µS	2945.92	3000.52	2839.88	2565.04	2476.56	2857.82	2611.49	2572.43	2840.91	2741.12	3069.47	3638.43

[a] *5-day at 20 °C N-allylthiourea biochemical oxygen demand.*

TABLE 24.3.2.2 Removal (%) per Wetland Filter[a] of Outflow Variables

	Filter 1			Filter 2			Filter 3			Filter 4			Filter 5			Filter 6		
Variable	O[b]	W[c]	S[d]	O[b]	W[c]	S[d]	O[b]	W[c]	S[d]	O[b]	W[c]	S[d]	O[b]	W[c]	S[d]	O[b]	W[c]	S[d]
09/22/02–09/21/03																		
BOD[e]	*43*	*13*	82	*32*	*3*	80	76	*59*	85	*51*	*40*	84	*65*	*49*	88	*49*	*44*	86
SS[f]	*52*	*40*	81	*44*	*25*	*78*	*67*	*53*	92	*65*	*54*	86	*64*	*54*	92	*66*	*52*	88
TS[g]	9	N	N	N	N	21	4	N	6	N	1	N	3	N	N	N	N	N
Turb[h]	78	86	65	71	86	51	96	98	95	90	94	94	90	96	91	93	95	93
Cond[i]	2	N	N	N	N	2	N	N	N	N	N	N	N	N	N	N	N	N
09/22/03–09/21/04 (Artificial Increase of pH after 09/21/03)																		
BOD[e]	*66*	*69*	76	*63*	*44*	71	97	100	96	97	99	94	95	99	96	95	99	93
SS[f]	*49*	*74*	*52*	*15*	N	*76*	*91*	*91*	98	*92*	*92*	98	*85*	*91*	97	*90*	*91*	97
TS[g]	25	36	37	15	23	56	38	40	65	38	42	58	39	50	64	36	38	49
Turb[h]	57	*77*	48	61	68	62	98	99	95	98	99	98	97	97	96	98	99	98
Cond[i]	N	7	18	N	8	N	N	12	N	N	18	N	0	22	N	N	9	N

Variable	Filter 7			Filter 8			Filter 9			Filter 10			Filter 11			Filter 12		
	O[b]	W[c]	S[d]	O[b]	W[c]	S[d]	O[b]	W[c]	S[d]	O[b]	W[c]	S[d]	O[b]	W[c]	S[d]	O[b]	W[c]	S[d]
09/22/02–09/21/03																		
BOD[e]	79	53	94	*66*	*32*	92	*67*	*51*	93	*43*	*2*	89	69	68	89	69	65	84
SS[f]	*72*	*60*	98	*73*	*62*	91	*75*	*71*	95	*77*	*70*	92	*78*	*71*	88	*51*	*35*	86
TS[g]	20	15	25	29	32	N	42	48	19	39	45	2	40	43	N	2	N	28
Turb[h]	96	97	99	93	95	96	91	94	98	89	90	93	94	96	94	86	85	92
Cond[i]	4	N	N	15	11	N	26	36	N	28	32	N	31	29	N	N	N	N
09/22/03–09/21/04 (Artificial Increase of pH after 09/21/03)																		
BOD[e]	97	99	95	97	99	91	97	99	96	96	99	90	96	98	90	91	94	88
SS[f]	*81*	*89*	99	*90*	*91*	96	*91*	*90*	99	*91*	*91*	98	*89*	*87*	97	*87*	*87*	98
TS[g]	39	42	72	34	46	50	34	43	71	21	22	53	31	36	56	28	24	34
Turb[h]	97	98	99	98	98	98	97	99	98	98	99	99	98	97	99	93	93	97
Cond[i]	3	16	N	N	17	N	N	16	N	N	15	N	N	9	N	N	2	N

In italics: BOD > 20 mg/l and SS > 30 mg/l (outflow concentrations exceeding thresholds). N, negative removal (i.e., more output than input).
[a] *Change* (%) $= \frac{(\text{in}-\text{out})\times 100\ (\%)}{\text{in}}$, *where in = inflow and out = outflow.*
[b] *Overall mean.*
[c] *Mean of the winter.*
[d] *Mean of the summer.*
[e] *5-day at 20°C N-allylthiourea biochemical oxygen demand (mg/l).*
[f] *Suspended solids (mg/l).*
[g] *Total solids (mg/l).*
[h] *Turbidity (NTU).*
[i] *Conductivity (µS).*

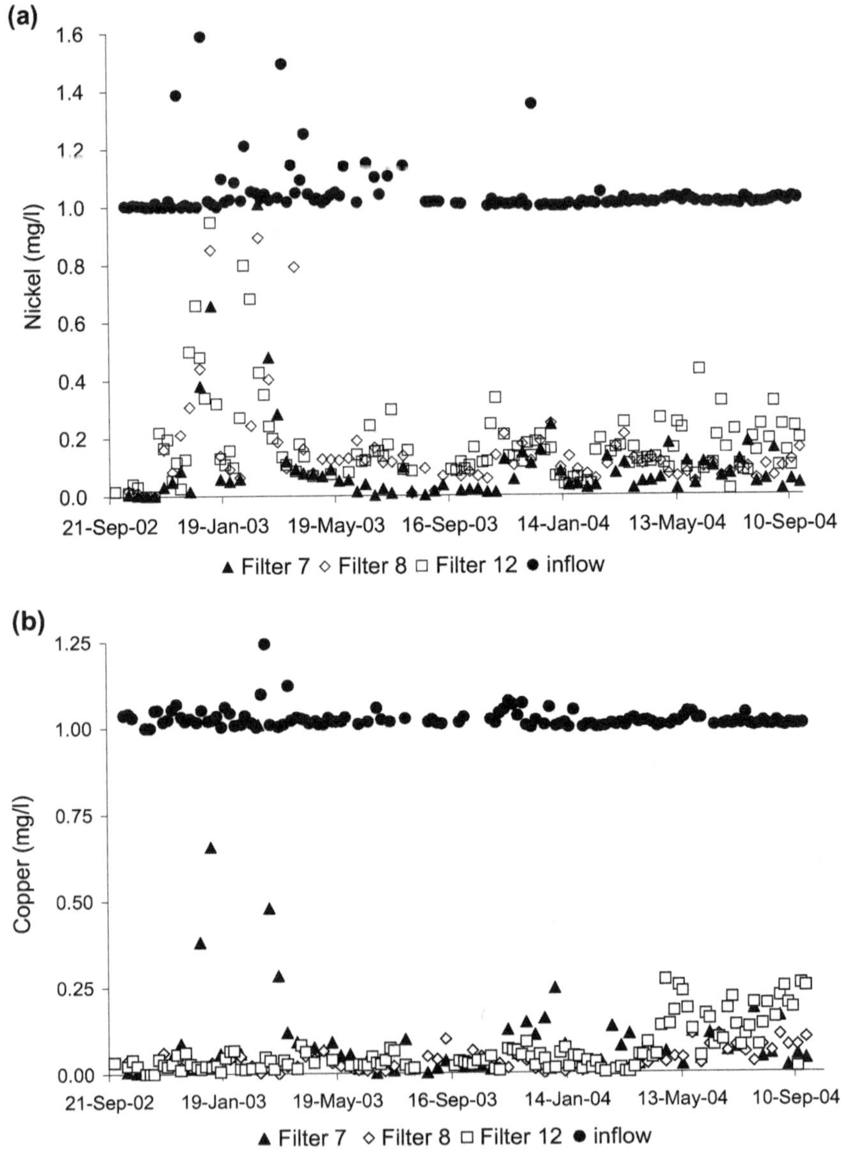

FIGURE 24.3.2.1 Daily (a) nickel and (b) copper concentrations (mg/l) in the inflow and outflows of Filters 7 (unplanted), 8 (planted), and 12 (planted and high loading).

concentrations for nickel and copper. Concerning BOD removal, the performances of all filters (except for Filters 1 and 2; extended storage) improved greatly over time. The reductions in BOD were also satisfactory for most filters, if compared to minimum American and European standards (<20 mg/l) for the secondary treatment of effluent (Tables 24.3.2.1 and 24.3.2.2).

Furthermore, the artificial increase of pH after the first year of operation had no apparent influence on the treatment performance of BOD. There is no obvious difference in performance between Filters 8 and 11, indicating that aeration did not contribute significantly to the removal of BOD (Tables 24.2.3.1 and 24.3.2.1).

Negative reduction rates for TS and conductivity were predominantly caused by road salting (and gritting) in late fall and winter. Any conventional filter system including constructed wetlands is unable to retain salts in high concentrations. Therefore, salts cannot be retained after a certain loading threshold, which is associated with a lag period, is exceeded. The lag period is predominantly a function of the buffering capacity of the biomass and the batch-flow operational mode (Section 24.3.3). It follows that after an initial positive removal period, the removal efficiencies are becoming negative (Norrstrom and Jacks, 1998). Furthermore, the dissolved solids fraction increases as microbial biomass mineralizes the organic contaminants. Conductivity correlates well with dissolved solids that contribute to a large proportion of the TS mass (Cooper et al., 1996; Scholz et al., 2002).

In contrast to previous researchers, who reported the worst seasonal performance for BOD removal during winter (Karathanasis et al., 2003), all filters with the exception of Filters 1 and 2 showed high BOD removal figures (>94%) even during the second winter. This suggests that soil microbes have the capacity to effectively decompose organic matter in winter despite relatively low temperatures.

Concerning other variables, reduction rates for SS also increased in the second year, although outflow concentrations frequently exceeded the threshold of 30 mg/l throughout the year except for summer. Turbidity values of the outflow decreased greatly over time. Despite the artificial increase of pH in the inflow, the pH of the outflow was approximately neutral and comparable to the first year of operation. Moreover, the pH of the outflow was relatively stable in the second year (standard deviation of approximately 0.18).

24.3.3 Heavy Metal Removal

Heavy metal removal efficiencies improved during the second year of operation (Figure 24.3.2.1). However, the reduction in metals was not sufficient to comply with American standards for secondary wastewater treatment (Clesceri et al., 1998). Dissolved nickel and dissolved copper concentrations should not exceed 0.0071 and 0.0049 mg/l, respectively (Tchobanoglous et al., 2003).

The decomposition of aquatic plants after fall, reducing soil conditions, road gritting and salting during periods of low temperatures, and acid rain contribute to increases of metal concentrations in the outflow (Norrstrom and Jacks, 1998; Sasaki et al., 2003). For example, high levels of conductivity were recorded in the filter inflow and outflows (Tables 24.2.3.1, 24.3.2.1, and 24.3.2.2), and the breakthrough of dissolved nickel was observed during the first winter (Figure 24.3.2.1(a)).

Concerning the effect of retention time on the treatment efficiency of metals, the heavy metal outflow concentrations of Filter 12 (higher loading rate) were slightly higher than the corresponding concentrations for the other filters. According to previous studies (Kadlec and Knight, 1996; Wood and Shelley, 1999), metal removal efficiencies for wetlands are highly correlated with influent concentrations and mass loading rates. Moreover, it was suggested that the formation of metal sulfides was favored in wetlands with long retention times. This may lead to more sustainable management of constructed treatment wetlands.

24.3.4 Link between pH and Treatment of Metals

After the increase of the inflow pH, mean reduction efficiencies for nickel increased during the second winter compared to the first winter; i.e., 90% and 65%, respectively, for Filter 7 (Figure 24.3.2.1(a)). Moreover, an obvious breakthrough of nickel was not observed during the second winter despite the presence of high salt concentrations in the inflow. This is likely to be due to the artificial increase of pH.

A high pH facilitates nickel precipitation. For example, nickel hydroxide ($Ni(OH)_2$) may precipitate at pH 9.1 if the corresponding metal concentration is 1 mg/l (Tchobanoglous et al., 2003), which is similar to the inflow concentrations of the spiked filters. Moreover, dissolved copper did not break through any constructed wetland filter throughout the study (Figure 24.3.2.1(b)).

All filters acted as pH buffers after pH increase, and pH levels were subsequently reduced. It can be assumed that this buffering capacity is greatly enhanced by the presence of active biomass rather than macrophytes (Kadlec and Knight, 1996; Sasaki et al., 2003). However, the outflow pH values for the planted filters recorded were slightly lower than those for the unplanted filters. For example, the overall mean pH value for Filter 7 (unplanted filter) is 7.31, and the corresponding value for Filter 8 (planted filter) is 6.98 during the second year of operation.

24.3.5 Analysis of Variance and Modeling

An analysis of variance has shown that all filters containing aggregates are relatively similar to each other with respect to most of their outflow variables. It follows that some filters could be considered as replicates (e.g., Filter 7 may be judged as a replicate of Filter 3 and vice versa) despite the differences in filter set-up (Tables 24.2.2.1 and 24.2.2.2). The P values of the pairs for Filters 3 and 7 are 0.20, 0.95, and 0.98 for BOD, SS, and turbidity, respectively. Pairs of data associated with $P \geq 0.05$ can be regarded as statistically similar.

The absence of filter replicates is not a serious issue, considering that the filter performances are similar (e.g., Table 24.3.2.2) and that the standard deviations (Table 26.3) of most outflow variables are comparable with large-scale systems (Cooper et al., 1996; Kadlec and Knight, 1996; Karathanasis et al., 2003; Vymazal, 2002). Moreover, a more detailed experimental study with three replicate filters, for example, would not be justified in terms of costs and potential scientific benefit gained.

24.4 CONCLUSIONS AND FURTHER WORK

Despite the highly variable water quality of road runoff, the filters showed great treatment performances, particularly with respect to the BOD reduction in a cold climate. Removal efficiencies for SS, in particular, improved over time, and dissolved copper was removed satisfactorily in comparison to values obtained from the literature. However, a breakthrough of dissolved nickel during the first winter of the first year of operation was observed. After creating an artificially high inflow pH of approximately 8 after one year of operation, nickel was successfully treated despite vulnerability to leaching when exposed to a high salt concentration during the second winter.

A high pH was apparently also linked to high removal efficiencies. The elevated pH had no apparent negative effect on the biomass, including macrophytes. Moreover, filters showed a great pH buffering capacity. Findings indicate that conventional pH adjustment can be successfully applied to constructed wetland systems for stormwater treatment.

The presence of Filtralite (adsorption filter media) and *P. australis* (macrophyte) did not result in an obvious reduction of metal concentrations in outflow waters. In comparison, operational conditions such as inflow pH and retention time were more important for the heavy metal treatment.

Furthermore, case-based reasoning of the experimental data set has been applied by Lee et al. (2005b). The purpose of this follow-up study was to show how case-based reasoning and other advanced modeling techniques (see also Chapters 26 and 30) can be applied for water quality control purposes.

Chapter 25

Combined Wetland and Below-ground Detention Systems

25.1 INTRODUCTION

The acronym for sustainable drainage system (a British phrase) is SuDS, which is also known as best management practice (an American phrase). A singular or series of management structures and associated processes designed to drain surface runoff in a sustainable approach to predominantly alleviate capacities in existing conventional drainage systems in an urban environment is defined as SuDS (Butler and Davies, 2000; CIRIA, 2000; SEPA, 1999).

New developments proposed for brownfield sites or the periphery of urban developments may be unable to obtain planning permission if existing local sewers have no spare capacity for stormwater drainage and if the stormwater discharge from the proposed site cannot be controlled. In the absence of suitable watercourses that can accommodate direct stormwater discharges, alternative technologies such as "at source" stormwater storage and detention systems are required (Butler and Davies, 2000).

Maintenance of all public SuDS structures above ground is usually the responsibility of the local authority (The Stationery Office, 1998). Above-ground SuDS structures are defined as swales, ponds, basins, and any other ground depression structures. In contrast, the maintenance of below-ground SuDS structures is commonly the responsibility of the regional water authority. Below-ground SuDS structures include culverts, infiltration trenches, filter strips, and below-ground detention systems (Butler and Davies, 2000; CIRIA, 2000; Nuttal et al., 1997).

Stormwater runoff is usually collected in gully pots that can be viewed as simple physical, chemical, and biological reactors. They are particularly effective in retaining suspended solids (SS) (Bulc and Slak, 2003). Currently, gully pot liquor is extracted once or twice per annum from road drains and transported (often over long distances) for disposal at sewage treatment works (Butler et al., 1995; Memon and Butler, 2002). A more sustainable solution would be to treat the entire road or car park runoff locally in potentially

Wetlands for Water Pollution Control. http://dx.doi.org/10.1016/B978-0-444-63607-2.00025-3

sustainable stormwater detention systems such as below-ground storage systems and stormwater ponds (Guo, 2001), reducing transport and treatment costs. Furthermore, runoff treated with stormwater detention systems can be recycled for irrigation purposes.

Below-ground stormwater storage and detention systems are defined as a subsurface structure designed to accumulate surface runoff, and from where water is released, as may be required to increase the flow hydrograph. The structure may contain aggregates with a high void ratio or empty plastic cells and may also act as a water recycler or infiltration device (Butler and Parkinson, 1997).

A below-ground stormwater detention system comprises a number of components forming a structure that is designed to reduce stormwater flow. The system captures surface water through infiltration and other methods. The filtered stormwater is usually stored below ground in a tank. The water is often cleaned and filtered before it is infiltrated or discharged to the sewer or watercourse via a discharge control valve. The system benefits include runoff reduction of minor storms, groundwater recharge, and pollution reduction. This detention system is predominantly applied in new developments.

The purpose of the case study is to advance knowledge and understanding by formulating design guidelines for vertical-flow stormwater detention systems treating road runoff predominantly by extended storage in a cold climate such as the southeast of Scotland. The objectives are to assess the following:

- The function of turf (absent vs present) and different aggregates such as Ecosoil as components of a primary treatment filtration stage located above the below-ground detention systems; and
- The overall passive treatment performance of vertical-flow stormwater detention systems.

25.2 MATERIALS AND METHODS

25.2.1 System Design and Operation

Five stormwater detention systems (Matrix Geo-Cell, provided by Atlantis Water Management Ltd) were located outdoors at The King's Buildings campus (The University of Edinburgh, Scotland) to assess the system performance during a relatively cold year (3/20/04–3/19/05). Inflow water, polluted by road runoff, was collected by manual abstraction with a 2-l beaker from randomly selected gully pots on the campus and the nearby main roads.

Five stormwater detention systems based on plastic cells (boxes with large void space) were used. Each system had the following dimensions: height = 85 cm, length = 68 cm, and width = 41 cm. Two plastic cells on top of each other made up one experimental detention system. The bottom cell (almost 50% full at any time) was utilized for water storage only. The top cell

TABLE 25.2.1.1 Systematic Experimental Set-Up of Gravel-Filled Stormwater Detention System Components and Operation

System	Planted	Additional Media Type	Natural Aeration Restricted
1	No	–	No
2	No	Sand	No
3	No	Sand and Ecosoil	No
4	No	Sand, Ecosoil, and block paving	Yes (due to block paving)
5	Yes	Sand, Ecosoil, and turf	No

contained the aggregates. Different packing order arrangements of aggregates and plant roots were used in the systems (Tables 25.2.1.1 and 25.2.1.2) to test the impacts of gravel, sand, Ecosoil, block paving, and turf on the water treatment performance.

The filtration system was designed to operate in vertical-flow batch mode. Manual flow control was practiced. Gully pot liquor compares well with concentrated road runoff (by a factor of at least 30, depending on spacing between gully pots) and was used in the experiment as a worst-case scenario liquid replacing road runoff. All detention systems (Tables 25.2.1.1 and 25.2.1.2) were, approximately twice per week, watered with 10 l of gully pot liquor as slowly as possible and were drained by gravity afterward to encourage air penetration through the soils (Cooper et al., 1996; Gervin and Brix, 2001). The relative quantity of gully pot liquor used per system was approximately 3.6 times the mean annual rainfall volume to again simulate a worst-case scenario. The hydraulic residence times were in the order of 1 h.

Biodegradation was enhanced by encouraging natural ventilation of the aggregates from the top via the natural air and from the bottom via the air pockets above the stored water and between the aggregates. Considering industrial-scale systems, vertical ventilation pipes should be installed to encourage passive ventilation as well.

25.2.2 Analytical Methods

The biochemical oxygen demand (BOD) was determined in the inflow and outflow water samples with the OxiTop IS 12-6 system (Wisenschaftlich-Technische Werkstätten (WTW), Weilheim, Germany), a manometric measurement device. The measurement principle is based on measuring pressure differences estimated by piezoresistive electronic pressure sensors. Nitrification was suppressed by adding 0.05 ml of 5 g/l N-allylthiourea (WTW Chemical Solution No. NTH 600) solution per 50 ml of sample water.

TABLE 25.2.1.2 Packing Order of the Stormwater Detention Systems Called Matrix Geo-Cells (provided by Atlantis Water Management Ltd)

Height (mm)	S1	S2	S3	S4	S5
861–930 (top)	Air	Air	Air	Paving and	Air
791–860	Air	Air	Air	6 mm gravel	Turf
751–790	Air	Air	Sand and Ecosoil	Sand and Ecosoil	Sand and Ecosoil
711–750	Air	Sand	Sand and Ecosoil	Sand and Ecosoil	Sand and Ecosoil
661–710	6 mm gravel	6 mm gravel	6 mm gravel	6 mm gravel	6 mm gravel
451–660	20 mm gravel	20 mm gravel	20 mm gravel	20 mm gravel	20 mm gravel
437–450	Sand	Sand	Sand	Sand	Sand
431–436	Geotextile	Geotextile	Geotextile	Geotextile	Geotextile
201–430	Air	Air	Air	Air	Air
0–200 (bottom)	Water	Water	Water	Water	Water

S = system.

Concerning the analysis of nutrients in the liquid phase, oxidized aqueous nitrogen was determined in all water samples as the sum of nitrate–nitrogen and nitrite-nitrogen. However, nitrite-nitrogen concentrations were significantly low. Nitrate was reduced to nitrite by cadmium and determined as an azo dye at 540 nm (using a Perstorp Analytical EnviroFlow 3000 flow injection analyzer) following diazotization with sulfanilamide and subsequent coupling with N-1-naphthyl ethylenediamine dihydrochloride (Allen, 1974).

Ammonia-nitrate and ortho-phosphate-phosphorus were determined by automated colorimetry in all water samples from reaction with hypochlorite and salicylate ions in solution in the presence of sodium nitrosopentacyanoferrate and reaction with acidic molybdate to form a phosphomolybdenum blue complex, respectively (Allen, 1974). The colored complexes formed were measured spectrometrically at 655 and 882 nm, respectively, using a Bran and Luebbe autoanalyzer (Model AAIII).

A Whatman PHA230 bench-top pH meter (for control only), a Hanna HI 9142 portable waterproof dissolved oxygen (DO) meter, an HACH 2100N turbidity meter, and a Mettler Toledo MPC 227 conductivity, total dissolved solids (TDS), and pH meter were used to determine pH, DO, conductivity, turbidity, and TDS, respectively. An ORP HI 98201 redox potential meter with a platinum-tip electrode HI 73201 was used to measure pH. Composite water samples were analyzed. All other analytical procedures were performed according to American standard methods (Clesceri et al., 1998).

Concerning the analysis of major nutrients in Ecosoil (aggregate supplied by Atlantis Water Management Ltd), 2 ml sulfuric acid (strength of 98%, v/v) and 1.5 ml hydrogen peroxide (strength of 30%, v/v) were used as an extraction medium (Allen, 1974). Approximately 0.1 g of each dried sample and the associated digestion media were placed in a tube and heated at 320 °C for 6 h. Aliquots were taken and digests were made up to 100 ml with distilled water.

For analysis of total nitrogen, the following procedure was adopted: Ammonium (present in the digest) reacts with hypochlorite ions generated by alkaline hydrolysis of sodium dichloroisocyanurate. The reaction forms monochloroamine, which responds with salicylate ions in the presence of sodium nitroprusside to form a blue indephenol complex. This complex is measured colorimetrically at 660 nm using a Bran and Luebbe autoanalyzer (Model AAIII).

For analysis of total phosphorus, the following procedure was used: ortho-phosphate (present in the digest) reacts with ammonium molybdate in the presence of sulfuric acid to form a phosphomolybdenum complex. Potassium antimonyl tartrate and ascorbic acid are utilized to reduce the complex, forming a blue color, which is proportional to the total phosphorus concentration. Absorption was measured at 660 nm using a Bran and Luebbe autoanalyzer (Model AAIII).

For the analysis of total potassium, the digest was analyzed by a flame atomic absorption spectrometer (Unicam 919, Cambridge, UK) at a wavelength of 766.5 nm and with a bandpass of 1.5 nm. Standards were prepared in 100-ml flasks using 2 ml concentrated sulfuric acid and 1.5 ml hydrogen peroxide (30% v/v) and made up to mark with de-ionized water. Cesium at a concentration of 100 mg/l was added to both standards and digests to overcome ionization.

Metal concentrations were determined in the raw gully pot liquor and the outflow waters from the experimental rig on June 16, 2004. Samples for metal determinations were stored at −19 °C until analysis.

Concerning the analysis of Ecosoil and grass cuttings, composite samples were collected and stored at −10 °C prior to analysis. After thawing, approximately 2.5 g of each sample was weighed into a 100-ml digestion flask to which 21 ml hydrochloric acid (strength of 37%, v/v) and 7 ml nitric acid (strength of 69%, v/v) were added. The mixtures were then heated on a Kjeldahl digestion apparatus (Fisons, UK) for at least 2 h. After cooling, all solutions were filtered through a Whatman Number 541 hardened ashless filter paper into 100-ml volumetric flasks. After rinsing the filter papers, solutions were made up to the corresponding mark with de-ionized water. The method was adapted from the section on nitric acid−hydrochloric acid digestion (Clesceri et al., 1998).

An inductively coupled plasma optical emission spectrophotometer (ICP-OES) called TJA IRIS, supplied by ThermoElemental (USA), was used to analyze selected wastewater, Ecosoil, and grass cutting samples. The purpose was to economically screen samples to determine various trace element concentrations and potential contaminants. The analytical precision (relative standard deviation) was typically between 5% and 10% for three individual aliquots.

25.3 RESULTS AND DISCUSSION

25.3.1 Comparison of Costs

The overall capital and maintenance costs were calculated for each detention system for the first year of operation. Maintenance included litter removal and grass cutting and was based on an area of 1000 m^2. Material prices were requested for a volume of 100 m^3 per aggregate to obtain realistic figures for a scaled-up detention system (industrial operation size).

The configuration of five systems had standardized cost ratios of approximately 1.0:1.1:1.2:1.3:1.6 based on Edinburgh prices in November 2005. However, the actual prices are subject to negotiation (e.g., quantities ordered) and fluctuations on the market.

25.3.2 Inflow Water Quality

Table 25.3.2.1 summarizes the inflow water quality. The standard deviations for all inflow parameters (except for DO, pH, and temperature) were high

TABLE 25.3.2.1 Gully Pot Liquor (Inflow to Systems): Water Quality Variables (3/20/04–3/19/05)

Variable	Unit	Total Number	Mean Spring[a]	Mean Summer[b]	Mean Fall[c]	Mean Winter[d]
BOD[e]	mg/1	78	33.5	29.0	31.8	27.6
Nitrate-nitrogen[f]	mg/l	71	0.5	1.8	0.7	0.3
Ammonia-nitrogen	mg/1	74	2.3	1.6	2.0	1.5
O-phosphate-phosphorus	mg/1	73	0.1	0.2	0.3	0.7
Suspended solids	mg/1	77	345.4	117.3	150.9	258.0
Total solids	mg/1	78	619.7	451.5	381.8	1604.7
Turbidity	NTU	84	89.2	57.4	67.2	126.2
Dissolved oxygen	mg/1	82	3.1	3.3	3.8	5.1
pH	–	84	6.79	7.16	7.02	7.38
Redox potential	mV	84	102.2	242.9	228.0	181.0
Conductivity	µS	83	285.8	142.0	149.0	239.3
Temperature (air)		84	16.2	19.3	11.1	8.1
Temperature (gully pot)	°C	84	14.6	19.1	9.6	4.9

[a] *3/20–6/20/04.*
[b] *6/21–9/21/04.*
[c] *9/22–12/20/04.*
[d] *12/21/04–3/19/05.*
[e] *5-day at 20 °C N-allylthiourea biochemical oxygen demand.*
[f] *Includes nitrite-nitrogen.*

owing to the random selection of gully pots and seasonal variations (Butler and Parkinson, 1997; Scholz, 2004a).

In general, the gully pot liquor was more polluted in winter in comparison to the other seasons. There are various reasons for this, including that the lower temperature in winter compared to other seasons results in a slower biodegradation rate within the gully pot (Table 25.3.2.1). Moreover, the retention time of the gully pot liquor in winter is likely to be shorter than in the remaining seasons owing to more frequent rainfall events. A shorter retention time correlated positively with a lower biodegradation rate (Butler and Davies, 2000; Clesceri et al., 1998; Scholz, 2004a).

25.3.3 Comparison of Outflow Water Quality

The overall filtration performance figures for spring and summer are summarized in Table 25.3.3.1, which should be compared with Table 25.3.2.1. Reduction efficiencies for BOD and SS (Table 25.3.3.1) are comparable to findings reported elsewhere (Bulc and Slak, 2003; Scholz, 2004a) for highway runoff treatment with constructed wetlands. The reductions of BOD (Table 25.3.3.1) were acceptable for most systems if compared to minimum American and European standards for the secondary treatment of (domestic) wastewater.

Biochemical oxygen demand, in contrast to SS (Table 25.3.3.1) outflow concentrations, did not exceed the American threshold of 30 mg/l (Tchobanoglous et al., 2003). However, some European standards or those of individual regional agencies (Cooper et al., 1996; Lim et al., 2003; Shutes et al., 2001) are more stringent (e.g., BOD < 20 mg/l). The BOD outflow concentration was also lower than the UK standard (Scholz, 2004a) of 20 mg/l for secondary treated wastewater (Table 25.3.3.1).

The mechanisms of improving the water quality are similar in all systems. The wetland filters on top of the detention cells retain solids and subsequently degrade organic matter. However, the greatest treatment performance is achieved owing to extended storage within the detention cells that function similar to a covered (no photosynthetic activity of algae) wastewater treatment pond.

A regression analysis has shown that BOD, ammonia-nitrogen, nitrate–nitrogen, and ortho-phosphate-phosphorus can be estimated with conductivity and total dissolved solids using a second-order polynomial equation. For example, BOD, nitrate–nitrogen, and ammonia-nitrogen can be determined with conductivity; the corresponding coefficients of determination (r^2) for Filter 4 are 0.60, 0.71, and 0.76, respectively. This would result in the reduction of costs and sampling effort. However, statistical relationships between other variables were not significant.

TABLE 25.3.3.1 Relative Change (%) of Outflow Variables in Spring and Summer

	System 1		System 2		System 3		System 4		System 5	
Variable	SP[a]	SU[b]	SP[a]	SU[b]	SP[a]	SU[b]	SP[a]	SU[b]	SP[a]	SU[b]
BOD[c]	93	90	94	89	88	90	91	91	94	91
NO_3[d]	−1479	−1001	−829	−1345	−481	−496	−562	−825	−850	−190
NH_4[e]	74	84	86	87	78	86	76	85	82	82
PO_4[f]	16	−133	12	−131	12	−101	8	−168	2	−208
SS[g]	28	74	31	80	32	79	31	79	25	79
Turbidity[h]	91	91	93	91	90	89	80	91	85	60
BOD[c]	94	92	89	94	94	87	92	93	93	88
NO_3[d]	−2374	−834	−2191	−1050	−731	−66	−1238	−472	−472	−363
NH_4[e]	63	77	93	84	65	71	97	38	93	32
PO_4[f]	−3	75	22	77	30	80	18	80	33	76
SS[g]	97	97	96	98	97	98	97	97	96	98
Turbidity[h]	96	96	96	95	93	93	93	95	94	94

SP, spring; SU, summer.
[a]*Mean of the spring (3/20–6/20/04).*
[b]*Mean of the summer (6/21–9/21/04).*
[c]*5-day at 20 °C N-allylthiourea biochemical oxygen demand (mg/l).*
[d]*Nitrate-nitrogen (mg/l).*
[e]*Ammonia–nitrogen (mg/l).*
[f]*Ortho-phosphate-phosphorus (mg/l).*
[g]*Suspended solids (mg/l).*
[h]*Turbidity (NTU).*

Furthermore, it has been suggested that mature and viable microbial biomass, in contrast to aggregates with high adsorption capacities (e.g., Ecosoil and turf), is responsible for the high overall filtration performances (Cooper et al., 1996; Scholz and Martin, 1998a). However, it is difficult to objectively classify a biological system as mature without having undertaken detailed microbiological work.

Finally, analysis by ICP-OES of selected inflow and outflow samples for a suite of cations showed that all waters generally contained low concentrations of heavy metals. Measured elemental concentrations were either low (barium, calcium, magnesium, and manganese) or close to the detection limit (iron), and for most heavy metals (including aluminum, copper, and cadmium) were below the detection limit. Dissolved zinc was the pollutant measured with the highest mean concentration. The mean inflow concentration for zinc was 0.14 mg/l, and the corresponding mean outflow concentration was 0.07 mg/l (standard deviation of 0.05 mg/l).

25.3.4 Ecosoil and Turf

Ecosoil did not contribute to elevated nutrient concentrations owing to very low total nitrogen, total phosphorus, and total potassium concentrations of 65, 46, and 1367 mg/kg, respectively. A previous soil quality analysis for areas in Glasgow, where SuDS were considered for implementation, indicated total nitrogen, total phosphorus, and total potassium concentrations of 1612, 605, and 4562 mg/kg, respectively (Scholz et al., 2005). It follows that Ecosoil does function only as a very weak fertilizer and that it is therefore unlikely to contribute to eutrophication after the release of the treated stormwater to the nearby watercourse.

Furthermore, Ecosoil contained only trace amounts of heavy metals (except for aluminum): 1036, 24, and 7 mg/kg dry weight of aluminum, zinc, and nickel, respectively. All other metal concentrations were below the detection limit of the instrument. However, even the aluminum concentrations are similar to values reported elsewhere for urban soil (Scholz et al., 2005).

The influence of turf (system 5) on the organic matter content of the outflow was studied. The BOD and SS concentrations within the outflow from the planted system, compared to the unplanted gravel and sand systems, were similar (Tables 25.3.2.1 and 25.3.3.1).

Moreover, grass on top of Filter 6 was cut when the length was >10 cm for optical reasons and to reduce the overall nutrient load. Total nitrogen, total phosphorus, and total potassium concentrations were 3001, 640, and 6909 mg/kg fresh weight, in that order. The presence and associated harvesting of grass seemed to have a positive effect on the overall nitrate−nitrogen outflow concentration that was lower for system 5, if compared to the remaining systems (Tables 25.3.2.1 and 25.3.3.1).

25.4 CONCLUSIONS AND FURTHER RESEARCH

The BOD outflow concentrations were mostly below the American threshold of 30 mg/l for secondary treated wastewater. The below-ground stormwater detention system did show signs of overloading, resulting in occasionally relatively high SS and nitrate–nitrogen concentrations. It follows that further treatment would be required. Moreover, denitrification was not completed, and longer retention times are therefore suggested. Nitrate–nitrogen was lower in the outflow of the planted system (turf on the top).

Hydraulic gully pot liquor (concentrated stormwater runoff) loads exceeded three times the mean annual rainfall for all systems. Therefore, it is likely that the SS concentration would be much lower in the field under "real" conditions.

An analysis of variance indicated that there was no significant difference between most systems in terms of their treatment performance (e.g., BOD and SS), despite their different set-ups. It follows that all systems function as covered wastewater stabilization ponds regardless of the corresponding soil filter type.

Sampling costs and effort can be reduced using relationships derived from regression analysis between expensive variables that can be substituted by low-cost ones. For example, BOD can be replaced by conductivity for internal control purposes.

Ecosoil did contain relatively low concentrations of nutrients and metals (except for aluminum). It follows that higher investment costs for more complex systems are not justified based on a water quality analysis alone.

However, further research related to the potential hydraulic and structural benefits of additional aggregates such as Ecosoil are required. It should be assessed if Ecosoil can help to evenly distribute the inflow water and if it is capable of supporting "moving" and variable heavy loads on paved areas. However, the latter would require a geotechnical assessment in a structural engineering laboratory.

Chapter 26

Modeling of Constructed Wetland Performance

26.1 INTRODUCTION

Constructed treatment wetlands are often seen as complex "black box" systems, and the processes within an experimental wetland are difficult to model owing to the complexity of the relationships between most water quality variables (Gernaey et al., 2004). However, it is necessary to monitor, control, and predict the treatment processes to meet environmental and sustainability policies, as well as regulatory requirements such as secondary wastewater treatment standards (Scholz, 2004a,b).

The measurement of biochemical oxygen demand (BOD) and suspended solids (SS) concentrations is widely applied for wastewater before and after treatment, as they give a general indication of the water quality status. However, taking BOD measurements can be both expensive (measurements are labor-intensive and capital costs of modern online equipment are relatively high: approximately £15,000) and only of historical value (results are not available until five days after the sample has been taken). Therefore, an indirect method of prediction of BOD and SS, if it could be made reliable enough, would be advantageous.

A variety of machine learning methods such as k-nearest neighbors (KNN) and artificial neural networks (ANN) have been widely used in a broad range of domains including wastewater treatment engineering. The KNN technique is based on a simple methodology and a memory-based model defined by a set of examples for which the outcomes are known. Moreover, the KNN model estimates the outcome by finding k examples that are closest in distance to the target point. Thus, the determination of the optimal value for k is essential in building the KNN model, because it should be the maximum number of neighbors with the minimum possible error (Ruiz-Jimenez et al., 2004).

The KNN model has been compared with advanced neural networks and tested for a wide range of areas such as medical diagnosis, chemical analysis, and remote sensing (Carpenter and Markuzon, 1998; Dong et al., 2005; Ruiz-Jimenez et al., 2004). In the case of the application of KNN models in the wastewater treatment industry, Belanche et al. (2000) employed a KNN model for reference purposes in predicting sludge bulking.

Wetlands for Water Pollution Control. http://dx.doi.org/10.1016/B978-0-444-63607-2.00026-5

Neural networks are relatively effective in simulating and predicting water treatment processes. The advantages of ANN are as follows: ease of use, rapid prototyping, high performance, minor assumptions, reduced expert knowledge required, nonlinearity, multidimensionality, and ease of interpretation (Werner and Obach, 2001).

Artificial neural networks such as feed forward neural networks were developed to predict the effluent concentrations including BOD, chemical oxygen demand (COD), and SS for wastewater treatment plants (Grieu et al., 2005; Hamed et al., 2004; Onkal-Engin et al., 2005) and to control water treatment processes automatically; e.g., by modeling the alum dose (Maier et al., 2004). These studies have shown that ANN could be applied to establish a mathematical relationship between variables describing a process state and different measured quantities.

Although ANN methods are cost-effective and highly reliable in analyzing processes, the traditional neural networks have suffered from their inherent drawbacks; i.e., overtraining, local minima, poor generalization, and difficulties in their practical application (Lu and Wang, 2005). The support vector machine (SVM), a supervised machine learning technique developed by Vapnik (1995), provides a novel approach to improve the generalization performance of neural networks.

Originally, SVM models have been applied for pattern recognition problems. However, along with the introduction of Vapnik's ε intensive loss function, SVM have also been extended to solve nonlinear regression estimation problems (Vapnik, 1995; Pai and Hong, 2005). SVM classifies the data based on the similarity between the examples measured by the similarity function or kernel function. This function can be chosen according to the problem at hand, thus making the algorithm flexible in handling a wide variety of problems (Dubey et al., 2005). Moreover, previous studies demonstrated that the SVM is superior to the conventional neural network in predicting chemical and biological variables (Liu et al., 2004; Lu and Wang, 2005). However, this novel method has not yet been applied in the field of wastewater treatment, including constructed treatment wetlands.

The self-organizing map (SOM), which is based on an unsupervised learning algorithm, uses powerful pattern recognition analysis and clustering methods, and at the same time provides excellent visualization capabilities (Garcia and Gonzalez, 2004). The SOM is able to map a structured, highly dimensional data set onto a much lower-dimensional network in an "orderly" fashion (Lu and Lo, 2004). It offers the distinctive ability to gather knowledge by detecting the patterns and relationships from a given data set, learning from relationships and adapting to change. The SOM potentially outperforms current methods of analysis because it can successfully deal with the nonlinearity of a system, handle "noisy" or irregular data, and be easily updated (Hong et al., 2003).

Interesting approaches of SOM have been reported in water quality assessment. The SOM models were developed to evaluate the state of water quality of a reservoir and to predict the trophic status of coastal waters, showing a strong ability to identify the diversity between data (Aguilera et al., 2001; Gevrey et al., 2004).

Moreover, Verdenius and Broeze (1999) used an SOM model as an indexing mechanism in case-based reasoning algorithms to control wastewater treatment processes, and it was employed to diagnose the diverse states of a wastewater treatment plant (Garcia and Gonzalez, 2004; Hong et al., 2003). These studies demonstrated that the SOM can assist a process engineer by analyzing multidimensional data and simplifying them into visual information, which can be easily applied to control plant performance. However, applications of SOM in water treatment process control are relatively new and were not implemented as much as traditional neural networks such as the free forward neural networks (Grieu et al., 2005; Hamed et al., 2004).

It follows that comparative studies of traditional KNN models with novel neural networks (e.g., SVM and SOM) applied to predict wastewater treatment performances are required to advance operation process control. Moreover, ANN should be used to find out if these models can be effectively applied to predict water quality variables such as BOD and SS effluent concentrations in wastewater treatment systems, using their potential for data classification and clustering.

The purpose of this study was to examine the goodness of applying KNN, SVM, and SOM to predict the outflow water quality of experimental constructed treatment wetlands by comparing the accuracy of these models. Additionally, this study describes how machine learning can be used for water treatment monitoring and assessment.

The objectives are to assess the following:

- The most appropriate method for assessing the performance of constructed treatment wetlands, considering both the accuracy of estimations and input costs;
- The potential of KNN, SVM, and SOM for analyzing biochemical performance data;
- The optimization of input variables associated with predictive models; and
- The potential use of KNN, SVM, and SOM as teaching tools to enhance the understanding of "black box" systems.

26.2 METHODOLOGY AND SOFTWARE

26.2.1 Experimental Data and Variables

Twelve wetland filters were operated to assess the system performance concerning the treatment of gully pot liquor in a cold climate. Gully pot liquor is concentrated surface runoff, which is detained in the wet gully pot until it overflows into the sewer owing to incoming surface runoff from new rainfall events. Design, operation, and monitoring methods of the system were explained previously (Scholz, 2004a,b). Different packing order arrangements of filter media and plant roots were used in the wetland filters, as described previously by Scholz (2004a). In comparison, Filters 3, 5, 7, and 9 are similar to gravel and slow sand filters, and Filters 4, 6, 8, and 10 are typical reed bed filters. Filters 11 and 12 are more complex in their design and operation.

Experimental data were collected by monitoring the effluent concentrations of the filters, including BOD and SS, for more than two years (9/9/02–9/21/04). The numbers of data were comparable to those used in other prediction models (Aguilera et al., 2001; Liu et al., 2004). These data were stored in the database together with up to six input variables: turbidity (NTU), conductivity (μS), redox potential (mV), outflow water temperature (°C), dissolved oxygen (DO) (mg/l), and pH (−). The corresponding output variable was either BOD (mg/l) or SS (mg/l). The input variables were selected according to their goodness of correlation with both BOD and SS (Scholz, 2003) because they are relatively cost-effective and easy to measure.

26.2.2 K-Nearest Neighbors

A KNN model used to predict the effluent BOD and SS concentrations of the wetland systems was created using MATLAB 7.0. Each KNN model is based on the mean of the outcomes of the k-nearest neighbors. The local similarity (i.e., the similarity of a past case and the problem case with respect to only one variable) is found via a mathematical function of the difference between each past case and a problem case.

The Gaussian function (bell-shaped curve) used to map the local difference onto the local similarity is defined in Eq. (26.2.2.1), which applies fuzzy theory (Dubois and Prade, 1998). This function has a tuning parameter, which is used to determine the flatness of the smoothing function:

$$f(x) = e^{\left(-0.5\left(\frac{x}{\alpha \times SDV_i}\right)^2\right)} \tag{26.2.2.1}$$

where:

$f =$ the function, which converts the local difference into the local similarity;
$x =$ the local difference between each past case and a problem case;
$\alpha =$ the tuning parameter; and
$SDV_i =$ the standard deviation of the local differences of variable i.

The global similarity, which is the similarity between the past case and the problem case, considering all variables of a past case, can be found from the local similarity of each variable. Each local similarity is first multiplied by a weighting factor that corresponds to the importance of that variable in predicting the output.

An algorithm proposed by Duch and Grundzinski (1999) was used to identify feature weightings. For the initial ranking of features, all weighting factors are set to 1, and evaluation with a single feature turned off (i.e., set to 0) is made for all features. Thus, ranking is done in the same way as the feature dropping selection method (Duch and Grudzinski, 1999). The important feature has a fixed weighting factor of 1, and the optimal weighting value for the second factor in the ranking is determined by the search procedure. The remaining factors are all fixed to 1. The search is implemented by means of the leave-one-out (LOO) cross-validation process (Kohavi, 2005).

When the global similarity of each past case with the problem case is found, the past cases can be selected by the first k closest cases. The tuning parameter α of the Gaussian function and the k value were determined by LOO cross-validation in the training phase (Duch and Grudzinski, 1999).

26.2.3 Support Vector Machine

Concerning SVM, the basic idea is to map original data into a feature space, which has a large number of dimensions via a nonlinear mapping function $\varphi_i(x)$, that is usually unknown and then carry out linear regression in the feature space (Vapnik, 1995). Hence, the regression addresses a problem of estimating a function based on a given data set (Eq. (26.2.3.1)). The SVM approximates the optimum decision function using Eq. (26.2.3.2). The regularized risk function R(C) is shown in Eq. (26.2.3.3). The ε-insensitive loss function is defined in Eq. (26.2.3.4). By introducing the kernel function, Eq. (26.2.3.2) can be transformed into the explicit Eq. (26.2.3.5).

$$G = \{(x_i, y_i)\}_i^l \qquad (26.2.3.1)$$

where:

G = data set;
x_i = input vector; and
y_i = desired values, which are produced from the nonlinear mapping function $\varphi_i(x)$.

$$f(x) = \sum_{i=1}^{l} w_i \phi_i(x) + b \qquad (26.2.3.2)$$

where:

$f(x)$ = decision function;
$\varphi_i(x)$ = nonlinear mapping function representing the features of inputs; and
w and b = coefficients, which are estimated by minimizing the regularized risk function (see below).

$$R(C) = C\frac{1}{l}\sum_{i=1}^{l} L_{\varepsilon}\left(y_i, f(x_i)\right) + \frac{1}{2}\|w\|^2 \tag{26.2.3.3}$$

where:

R(C) – regularized risk function;
C = regularized constant determining the trade-off between the training error and the model flatness;
$\frac{1}{l}\sum_{i=1}^{l} L_{\varepsilon}(y_i, f(x_i))$ = empirical error measured by the ε-insensitive loss function;
y_i = desired values;
$f(x_i)$ = decision function; and
$\frac{1}{2}\|w\|^2$ = measurement of function flatness.

$$L_{\varepsilon}\left(y_i, f(x_i)\right) = \left\{\left|y_i - f(x_i)\right| - \varepsilon, \left|y_i - f(x_i)\right| \geq \varepsilon \right. \tag{26.2.3.4}$$

where:

L_{ε} = ε-insensitive loss function;
y_i = desired values;
$f(x_i)$ $f(x_i)$ = decision function; and
ε = prescribed parameter.

$$f(x) = \sum_{i=1}^{l} (\alpha_i - \alpha_i^*) K\left(x_i, x_j\right) + b \tag{26.2.3.5}$$

where:

$f(x)$ = decision function;
α, α_i = Lagrange multipliers; and
$K(x_i, x_j) = \varphi(x_i) \times \varphi(x_j)$ = kernel function in the feature space.

For the kernel function, there are several design choices, such as the linear, polynomial, and radial basis functions (RBF) and the sigmoid kernel. However, most of the previous research selected the RBF kernel, which nonlinearly maps samples into a higher dimensional space, unlike the linear kernel (Dong et al., 2005). The RBF kernel is shown in Eq. (26.2.3.6):

$$K\left(x_i, x_j\right) = \exp\left\{-\gamma\left\|x_i - x_j\right\|^2\right\} \tag{26.2.3.6}$$

where:

$K(x_i, y_i)$ = kernel function; and
γ = kernel parameter.

More detailed theory on the SVM can be found in Vapnik (1995). In this study, the SVM^{light} was used owing to its fast optimization algorithms and good potential for regression (Joachims, 1999). Concerning the kernel function, the RBF kernel was selected to analyze the cases, which show a nonlinear relationship between input and output data set in this study. The RBF kernel contains the parameters γ, C, and ε (see above). There are no general rules determining these parameters (Lu and Wang, 2005). A five-fold cross-validation was conducted to find out appropriate parameters for training steps (Dong et al., 2005).

26.2.4 Self-Organizing Map

An SOM consists of neurons, which are connected to adjacent neurons by neighborhood relations. In the training step, one vector x from the input set is chosen, and all the weight vectors of the SOM are calculated using some distance measure such as the Euclidian distance (Kohonen, 2001). The neuron, whose weight vector is closest to the input x, is called the best-matching unit (BMU), subscripted here by c Eq. (26.2.4.1):

$$\|x - m_c\| = \min\{\|x - m_i\|\} \qquad (26.2.4.1)$$

where:

x = input vector;
m = weight vector; and
$\| \ \|$ = distance measure.

After finding the BMU, the weighting vectors of the SOM are updated so that the BMU is moved closer to the input vector. The SOM update rule for the weight vector of the unit is shown in Eq. (26.2.4.2). The detailed algorithm of the SOM can be found in Kohonen (2001) for theoretical considerations:

$$m_i(t+1) = m_i(t) + \alpha(t)h_{ci}(t)[x(t) - m_t(t)] \qquad (26.2.4.2)$$

where:

$m(t)$ = weight vector indicating the output unit's location in the data space at time t;
$\alpha(t)$ = the learning rate at time t;
$h_{ci}(t)$ = the neighborhood kernel around the "winner unit" c; and
$x(t)$ = an input vector drawn from the input data set at time t.

After the SOM has been trained, the map needs to be evaluated to find out if it has been optimally trained or if further training is required. The SOM quality is usually measured with two criteria: quantization error (QE) and topographic error (TE). The QE is the average distance between each data point and its BMU, and TE represents the proportion of all data for which the first and second BMU are not adjacent with respect to the measurement of topology preservation (Kohonen, 2001).

In this study, the SOM toolbox (version 2) for Matlab 5.0 developed by the Laboratory of Information and Computer Science at the Helsinki University of Technology was used (Vesanto et al., 1999). After training the map with different map sizes, the optimum map size was determined on the basis of the minimum QE and minimum TE. The prediction was implemented by finding the BMU in the map for each test data set.

26.3 RESULTS AND DISCUSSION

26.3.1 Performance Evaluation

When comparing the performances of different models, the scale-dependent measures based on the absolute error or squared error have been commonly used (Gevrey et al., 2004; Maier et al., 2004). In this study, the performances of each model were measured by the mean absolute scaled error (MASE) method proposed by Hyndman and Koehler (2005) because it is independent of the scale of the data, less sensitive to outliers, more easily interpreted, and less variable for small samples compared to most of the other methods (see above). The MASE is defined in Eq. (26.3.1.1).

$$\text{MASE} = \frac{\text{MAE}}{\frac{1}{n-1}\sum_{i=2}^{n}|m_i - m_{i-1}|} \tag{26.3.1.1}$$

where:

MASE = mean absolute scaled error;
MAE $= \frac{1}{n}\sum_{i=1}^{n}|m_i - p_i|$;
M_i = measured values;
p_i = predicted values; and
n = number of data sets.

26.3.2 Correlation Analysis

Table 26.3.2.1 summarizes the findings from a correlation analysis comprising input (turbidity, conductivity, redox potential, outflow water temperature, DO, and pH) and target (BOD or SS) variables. Correlations were all weak, except between BOD and turbidity, SS and turbidity, and SS and conductivity (all at 1% significance level). Therefore, turbidity and conductivity are likely to be the most important input variables.

26.3.3 Optimization of Input Variables

When analyzing the data set with the KNN model, the optimal k value and weighting factors for all variables were determined by LOO cross-validation. The selected k value was between 3 and 5 for most data sets, and weighting factors were different depending on the individual characteristics of the chosen case base.

TABLE 26.3.2.1 Correlation Coefficients from a Correlation Analysis Comprising Input (Column Headings) and Target (Row Headings) Variables

Variable	Turbidity (NTU)	Redox Potential (mV)	pH (−)	Conductivity (µS)	Temperature (°C)	Dissolved Oxygen (mg/l)
BOD[a] (mg/l)	0.413	−0.338	−0.271	0.254	−0.120	−0.074
Variable	Conductivity (µS)	Turbidity (NTU)	Temperature (°C)	Redox potential (mV)	Dissolved oxygen (mg/l)	pH (−)
SS[b] (mg/l)	0.930	0.509	−0.322	−0.308	−0.127	0.013

[a] *5-day at 20°C N-allylthiourea biochemical oxygen demand.*
[b] *Suspended solids.*

For example, when predicting the outflow BOD concentration for Filter 8, the k value was fixed at 3 and the weighting factors 1.00, 0.78, 1.00, 0.30, and 0.78 for water temperature (°C), redox potential (mV), conductivity (μS), pH (−), and turbidity (NTU), respectively, of the outflow were assigned.

When conducting the SVM analysis, the parameters C and ε were identified by the five-fold cross-validation approach (see above). Theoretically, a small value of C will underfit the training data because the weight placed on the training data is too small, thus resulting in a large error for the test data set. On the contrary, when C is too large, the SVM model will be overtrained (Dong et al., 2005). The ε is set to be 0.1 when varying C in the one-time search method (Cao et al., 2003). There exists an optimum point for C, as shown in Figure 26.3.3.1(a). Thus C was determined to be 10 at the point of the lowest mean absolute error (MAE) value.

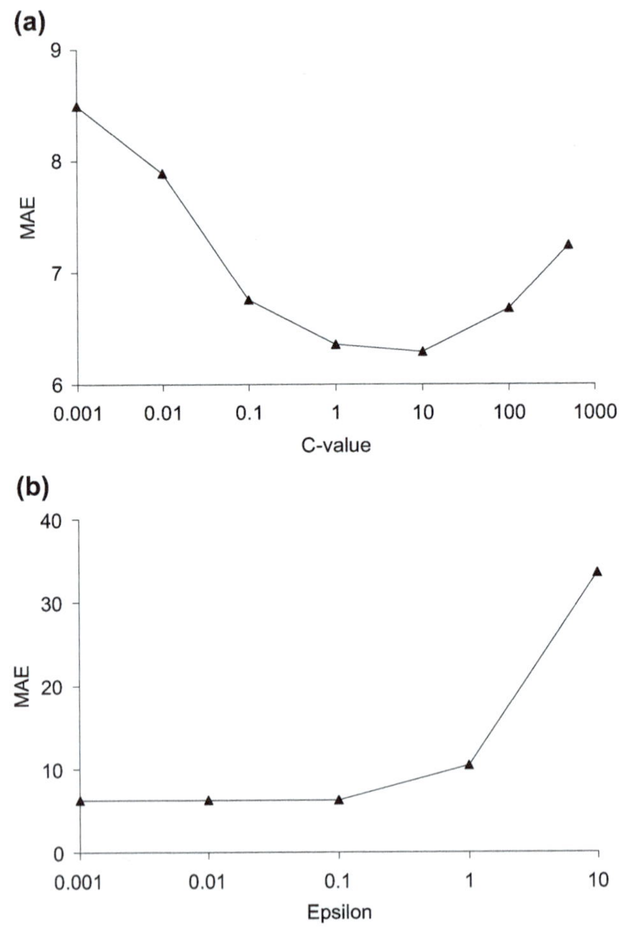

FIGURE 26.3.3.1 (a) C-value ($\varepsilon = 0.1$) and (b) epsilon (C = 10) versus the mean absolute error (MAE) of the training set based on five-fold cross-validation.

In general, the larger ε, the smaller the number of support vectors, and thus the sparser the representation of the solution. However, if ε is too large, it can deteriorate the accuracy of the training data set (Cao and Tay, 2003). According to Figure 26.3.3.1(b), parameter ε was fixed at 0.1. Furthermore, γ of the RBF kernel was set to 0.2 according to Eq. (26.3.3.1), as discussed elsewhere (Chang and Lin, 2005; Dong et al., 2005).

$$\gamma = \frac{1}{n} \tag{26.3.3.1}$$

where:

γ = parameter of the kernel function and
n = number of variables in the input data set.

In the SOM model, the map size is the most important factor in detecting the differences of data. If the map is too small, it might not explain some important differences. On the contrary, it is possible to overtrain the models (Leflaive et al., 2005). After creating maps with several different map sizes, the optimum map size, which has lower errors for both QE and TE, was chosen. For example, when predicting the BOD of Filter 8, the map size comprised 2304 units, as shown in Table 26.3.3.1.

Table 26.3.3.2 summarizes the findings of an input variable combination optimization exercise. The purpose was to identify the optimum combination of input variables by estimating both BOD and SS with as few input variables as possible to reduce the prediction error and the sampling costs and effort. Therefore, the data set contained the following input variables in order of their priority based on a correlation analysis (Table 26.3.2.1): turbidity, redox potential, pH, conductivity, and outflow water temperature in terms of their correlation with BOD; and conductivity, turbidity, outflow water temperature, redox potential, and DO in terms of their correlation with SS.

In the KNN model, best prediction results were obtained with five input variables for both BOD and SS. However, applying the SOM model, the BOD and SS predictions were most accurate with the single input variable turbidity and conductivity, respectively. The best combination of variables for each model is highlighted (underlined figures) in Table 26.3.3.2. Consequently,

TABLE 26.3.3.1 Quantization Error and Topographic Error for Different Self-Organizing Map Sizes

Map Size	608	1508	2006	2304	2613	2911
Quantization error	0.038	0.017	0.008	0.007	0.005	0.004
Topographic error	0.108	0.091	0.124	0.091	0.148	0.120

TABLE 26.3.3.2 Optimization of Input Attribute Combinations for Filter 8: k-Nearest Neighbors, Support Vector Machine, and Self-Organizing Map Applied to Predict the 5-Day at 20 °C N-Allylthiourea Biochemical Oxygen Demand (mg/l) and the Suspended Solids (mg/l) Concentrations

Variables	Number	KNN		SVM		SOM	
		MASE[a]	r^{2b}	MASE[a]	r^{2b}	MASE[a]	r^{2b}
BOD Prediction							
1	60	1.42	0.394	1.35	0.621	<u>0.29</u>	0.998
1 + 2	60	1.49	0.394	1.27	0.621	0.30	0.988
1 + 2 + 3	60	1.51	0.330	1.09	0.639	0.34	0.984
1 + 2 + 3 + 4	60	1.34	0.342	0.95	0.699	0.46	0.955
1 + 2 + 3 + 4 + 5	60	<u>1.01</u>	0.680	<u>0.94</u>	0.720	0.41	0.947
SS Prediction							
4	64	0.55	0.936	0.38	0.959	<u>0.29</u>	0.999
4 + 1	64	0.44	0.943	0.40	0.950	0.30	0.992
4 + 1 + 5	64	<u>0.37</u>	0.948	0.39	0.950	0.31	0.980
4 + 1 + 5 + 2	64	0.39	0.955	0.39	0.954	0.31	0.969
4 + 1 + 5 + 2 + 6	64	0.39	0.951	<u>0.36</u>	0.949	0.39	0.956

Note: The training and test data sets contained the following input variables: 1 = turbidity (NTU); 2 = redox potential (mV); 3 = pH (−); 4 = conductivity (μS); 5 = outflow water temperature (°C); 6 = dissolved oxygen (mg/l). The figures for the best combinations of variables are underlined.
[a] *Mean absolute scaled error.*
[b] *Prediction coefficient of determination.*

input variable combinations for each model were determined on the basis of this information.

26.3.4 Comparison of Applications

The prediction results of the outflow BOD and outflow SS using KNN, SVM, and SOM models are summarized in Table 26.3.4.1. Figures 26.3.4.1 and 26.3.4.2 visualize the BOD and SS prediction performances of the KNN, SVM, and SOM models for Filter 8 (typical UK reed bed according to Scholz (2004a,b)).

The results show that the BOD and SS concentrations are reasonably well predicted with given input variables. Despite the greater variability of SS in contrast to BOD (Scholz, 2004a,b), SS has a smaller MASE value between measured and predicted concentrations in comparison to BOD (Table 26.3.4.1).

The MASE from the BOD prediction with SOM, SVM, and KNN are 0.36, 0.84, and 0.79, respectively. While the performance of the SVM model was not superior to that of KNN, the SOM model gave excellent prediction performance values compared to the other models (Table 26.3.4.2; Figures 26.3.4.1 and 26.3.4.2).

To ensure the statistical significance of these findings, the prediction results were compared with each other and assessed by using an analysis of variance (ANOVA). The ANOVA threshold for statistically significant findings is $P < 0.05$. It follows that pairs of MASE associated with $P \geq 0.05$ can be regarded as similar. The analysis showed that the MASE of the SOM model was statistically different from those of SVM and KNN. However, the MASE of SVM and KNN were similar to each other, showing that the corresponding P value was >0.05. From the findings, it can be concluded that the SOM model outperformed the KNN and SVM models.

Figure 26.3.4.3 shows the regression analysis between measured and predicted BOD and measured and predicted SS for Filter 8 using the SOM model. The associated data set contained turbidity and conductivity for both BOD and SS predictions. The application of linear trend lines resulted in very good fits for both target variables; the prediction coefficients of determination are 0.998 for BOD and 0.999 for SS.

Concerning the supervised and unsupervised methods, a previous study has shown that the prediction accuracy of unsupervised neural networks is lower than the one for supervised networks, as generally expected (Lee et al., 2005a). However, this result indicated that the SOM model is superior to the SVM model. Moreover, the SOM model provides better prediction results with smaller input variables.

The outstanding performance of the SOM models is assumed to be attributed to the potential of clustering and classification of data. Particularly, in comparison with other neural networks, SOM was convenient for detecting

TABLE 26.3.4.1 **Constructed Treatment Wetlands: k-Nearest Neighbors, Support Vector Machine, and Self-Organizing Map Applied to Predict the 5-Day at 20°C N-Allylthiourea Biochemical Oxygen Demand (mg/l) and the Suspended Solids (mg/l) Concentrations**

		KNN		SVM		SOM	
Filter	Number	MASE[a]	r^{2b}	MASE[a]	r^{2b}	MASE[a]	r^{2b}
BOD Prediction							
3	55	1.02	0.461	0.98	0.384	0.77	0.997
4	58	1.01	0.681	1.07	0.713	0.17	0.994
5	59	0.58	0.696	0.67	0.666	0.35	0.994
6	61	0.76	0.798	0.83	0.643	0.44	0.997
7	58	0.78	0.609	0.77	0.503	0.85	0.999
8	60	1.01	0.680	0.94	0.720	0.29	0.998
9	57	0.76	0.544	0.90	0.431	0.20	0.998
10	60	0.71	0.744	0.87	0.513	0.41	0.999
11	59	0.79	0.645	0.96	0.451	0.37	0.985
12	115	0.88	0.342	0.87	0.213	0.24	0.997
3–12	642	0.79	0.550	0.84	0.463	0.36	0.888
SS Prediction							
3	60	0.49	0.973	0.33	0.966	0.17	0.999
4	61	0.38	0.957	0.28	0.954	0.08	0.998
5	60	0.66	0.938	0.28	0.978	0.20	0.999
6	64	0.57	0.941	0.32	0.940	0.17	0.998
7	57	0.32	0.892	0.38	0.883	0.08	0.995
8	64	0.37	0.948	0.36	0.949	0.29	0.999
9	62	0.53	0.882	0.51	0.883	0.54	0.996
10	65	0.48	0.922	0.51	0.949	0.62	0.998
11	64	0.42	0.839	0.36	0.895	0.48	0.999
12	113	0.83	0.803	0.68	0.770	0.47	0.953
3–12	670	0.54	0.844	0.43	0.883	0.31	0.932

[a]*Mean absolute scaled error.*
[b]*Prediction coefficient of determination.*

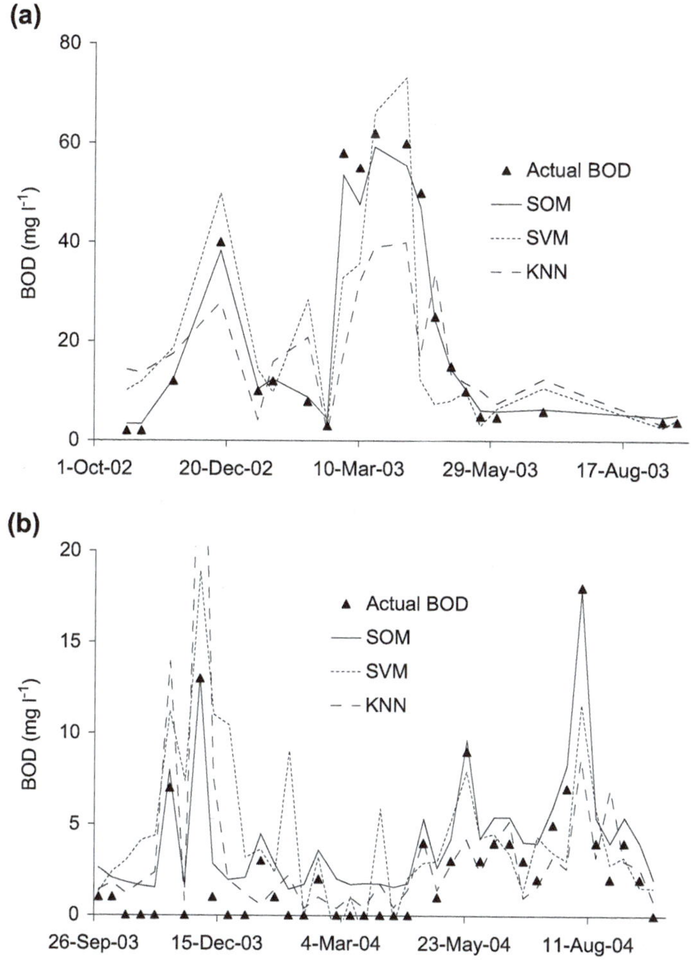

FIGURE 26.3.4.1 5-day at 20 °C N-allylthiourea biochemical oxygen demand (BOD) predicted by k-nearest neighbors (KNN), support vector machine (SVM), and self-organizing map (SOM) for (a) the first year (9/22/02–9/21/03) and (b) the second year (9/22/03–9/21/04) of wetland operation. The entry (32.6 mg/l on 12/3/03) for the KNN prediction is beyond the displayed range of BOD.

the outliers, which are displayed in particular parts of the map without affecting the remaining parts, because each outlier takes its place in one unit of the map, and only the weights of that unit and its neighbors are affected (Cereghino et al., 2001; Leflaive et al., 2005).

Additionally, the SOM model showed its high performance in visualization with respect to the relationship for nonlinear and complex biochemical data sets. Visualization gives better understanding of the relationships between

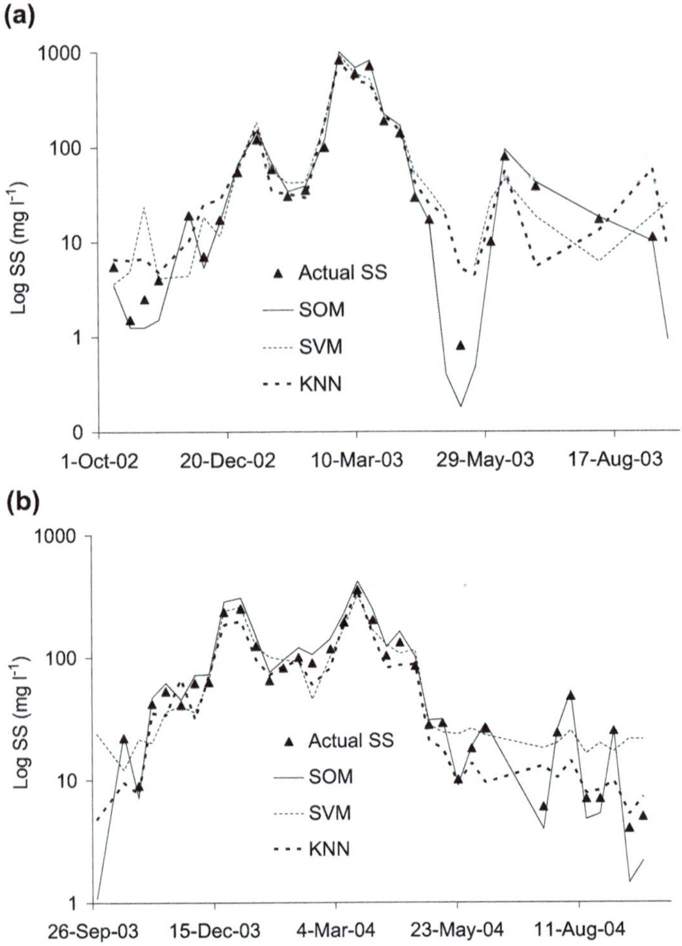

FIGURE 26.3.4.2 Suspended solids (SS) curve predicted by k-nearest neighbors (KNN), support vector machine (SVM), and self-organizing map (SOM) for (a) the first year (9/22/02–9/21/03) and (b) the second year (9/22/03–9/21/04) of wetland operation.

TABLE 26.3.4.2 Correct Prediction of Compliance (%) for the Estimation of the 5-Day at 20 °C N-Allylthiourea Biochemical Oxygen Demand (mg/l) and the Suspended Solids (mg/l) Concentrations

Variable	SOM[a]	SVM[b]	KNN[c]
Biochemical oxygen demand	97.0	87.9	87.2
Suspended solids	96.4	90.0	88.1

[a]*self-organizing map.*
[b]*support vector machine.*
[c]*k-nearest neighbors.*

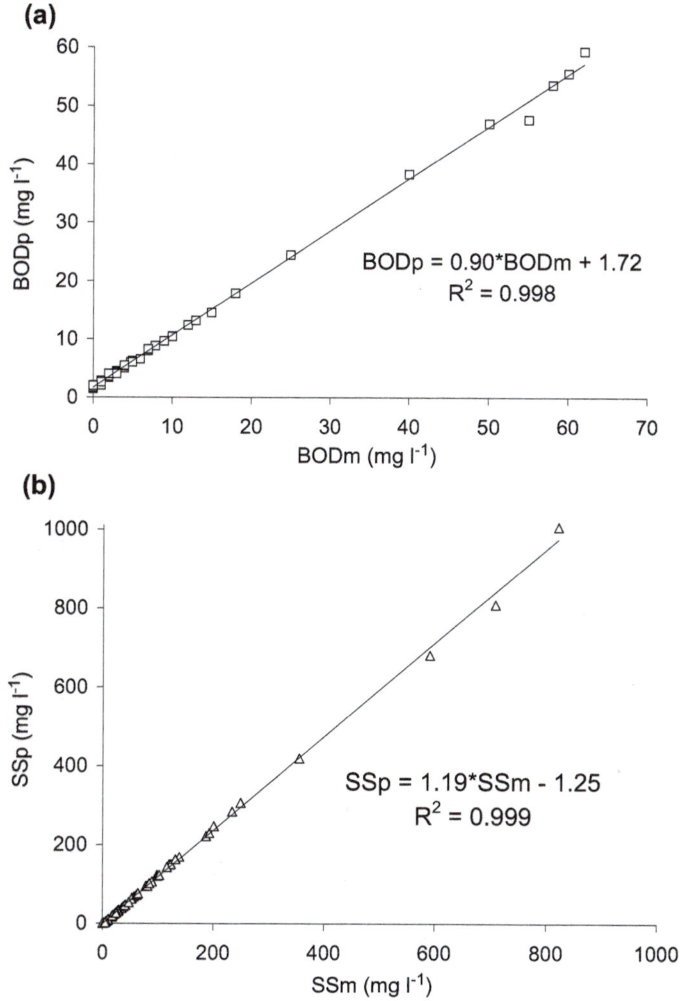

FIGURE 26.3.4.3 Regression analysis (a) between the measured 5-day at 20 °C N-allylthiourea biochemical oxygen demand (BODm) and the predicted BOD (BODp) applying self-organizing map (SOM) and (b) between the measured suspended solids (SSm) and the predicted SS (SSp) applying the SOM model for Filter 8.

most variables in biochemical processes. Figure 26.3.4.4 displays component planes on the trained map in gray scale.

The unified distance matrix (U-matrix) visualizes distances between neighboring map units and helps to identify the cluster structures of the map. Each component plane shows values for each variable with its corresponding unit. A cluster can be identified in the upper part of the U-matrix (Figure 26.3.4.4). In the equivalent area of each component plane, the temperature and redox potential are

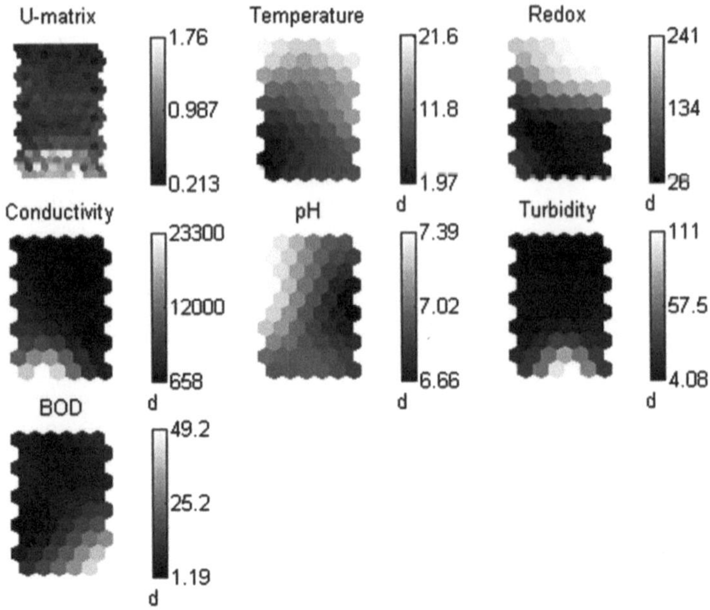

FIGURE 26.3.4.4 Visualization of variables (outflow water temperature (°C), redox potential (redox, mV), conductivity (μS), pH (−), turbidity (NTU), and 5-day at 20 °C N-allylthiourea biochemical oxygen demand (BOD) (mg/l)) on the self-organizing map trained to predict the BOD of Filter 8. The map size is 60 units. The U-matrix is the unified distance matrix.

high. On the other hand, the conductivity, turbidity, and BOD concentrations are low in the upper part of the planes. Figure 26.3.4.4 indicates that high BOD is associated with high temperature, high redox potential, low conductivity, and low turbidity.

The likelihoods of correct predictions, if the effluent concentrations are either below or above the thresholds for secondary wastewater treatment, are also shown in Table 28.5. The BOD and SS concentrations for compliance are 20 and 30 mg/l, respectively (Scholz, 2003). The correct predictions of compliance were all >87%. The probabilities are therefore all at least 0.37 higher in comparison to pure guessing (50%). Therefore, all models are well suited for relatively highly variable water quality data sets, such as those from constructed treatment wetlands (Scholz, 2004).

26.4 CONCLUSIONS

This chapter demonstrated the successful application of KNN, SVM, and SOM to typical "black box" systems such as constructed treatment wetlands governed by biochemical processes. The KNN, SVM, and SOM methodologies were successfully applied to predict water quality variables of constructed treatment wetlands. The BOD and SS, which are expensive to estimate, can be cost-effectively monitored by applying machine learning tools with input variables such as turbidity and conductivity. Their performances are encouraging, and the use of these models as management tools for day-to-day process control is therefore recommended. Moreover, little domain knowledge is required to build a model, and the input variables can be optimized by trial and error.

In terms of accuracy of prediction, the SOM model performed better using smaller input variables in comparison to both KNN and SVM models. Particularly, the SOM model demonstrated its potential to analyze the relationship between complex biochemical variables.

However, establishing the SOM model required more time, if compared to the KNN and SVM models, owing to the long searching process of the optimal map. Thus, in addition to the accuracy of model predictions, the resource that is required to build and test the model should be considered when selecting the optimal modeling tool.

Chapter 27

Infiltration Wetland Systems

27.1 INTRODUCTION

Conventional stormwater systems are designed to dispose of surface runoff as quickly as possible. This results in "end of pipe" solutions that often involve the provision of large interceptor and relief sewers, huge storage tanks at downstream locations, and centralized wastewater treatment facilities. These traditional civil engineering solutions often lead to flooding and environmental pollution during storms (Butler and Davies, 2000; Galuzzi and Pflaum, 1996; Scholz, 2004).

In contrast, sustainable drainage systems (SuDS) such as combined attenuation pond and infiltration pond systems (Ellis et al., 2002; EPA, 1999) can be applied as cost-effective local source control drainage solutions, delaying storm runoff and reducing peak flows. It is often possible to divert all storm runoff for infiltration or storage and subsequent water reuse. As runoff from roads is a major contributor to the quantity of surface water requiring disposal, this is particularly a beneficial approach where suitable ground conditions prevail (Butler and Davies, 2000). Furthermore, infiltration of storm runoff can reduce the concentration of diffuse pollutants such as leaves, feces, metals, and hydrocarbons, thereby improving the water quality of surface water runoff (Ellis et al., 2002; Scholz, 2003, 2004).

Despite the theoretical benefits of SuDS, the technical constraints associated with the design and operation of large features such as ponds have not been explored due to a lack of experimental data. The rainfall, runoff, and infiltration relationships for planted infiltration ponds treating road runoff have not been studied previously. There is also a lack of water quality management data and guidelines for unplanted and planted infiltration ponds that are operated in parallel. Considering the increase in popularity of SuDS, urban planners and developers need to understand the design and operation constraints of systems such as infiltration basins and ponds.

The aim is to assess an experimental infiltration pond system (case study) designed according to SuDS guidelines (CIRIA, 2000; EPA, 1999). The objectives are to:

- Identify technical constraints associated with the design and operation of novel planted infiltration ponds;
- Assess the rainfall, runoff, and infiltration relationships for unplanted infiltration ponds;

Wetlands for Water Pollution Control. http://dx.doi.org/10.1016/B978-0-444-63607-2.00027-7

- Assess the water quality, and its management for unplanted and planted infiltration ponds;
- Assess passive and active filamentous green algal control strategies, including the use of barley straw and *Carassius auratus*; and
- Promote the integration of SuDS into urban planning and development.

27.2 METHODS

27.2.1 Design of the Study Site

The pilot plant was designed considering the SuDS guidelines (Building Research Establishment, 1991; Bettes, 1996; CIRIA, 2000; ATV-DVWK-Arbeitsgruppe, 2002) of the British Research Establishment (BRE), Construction Industry Research and Information Association (CIRIA), and the German Association for Water, Wastewater, and Waste (ATV-DVWK). The return period for allowed flooding is 10 years. The total road area draining into the SuDS is approximately 446 m^2. The SuDS is based on a combined wetland and infiltration pond design (Figure 27.2.1.1).

Rainwater runoff from the University Road (catchment area) flows directly into a silt trap (Scholz and Zettel, 2004). The silt trap has a maximum capacity

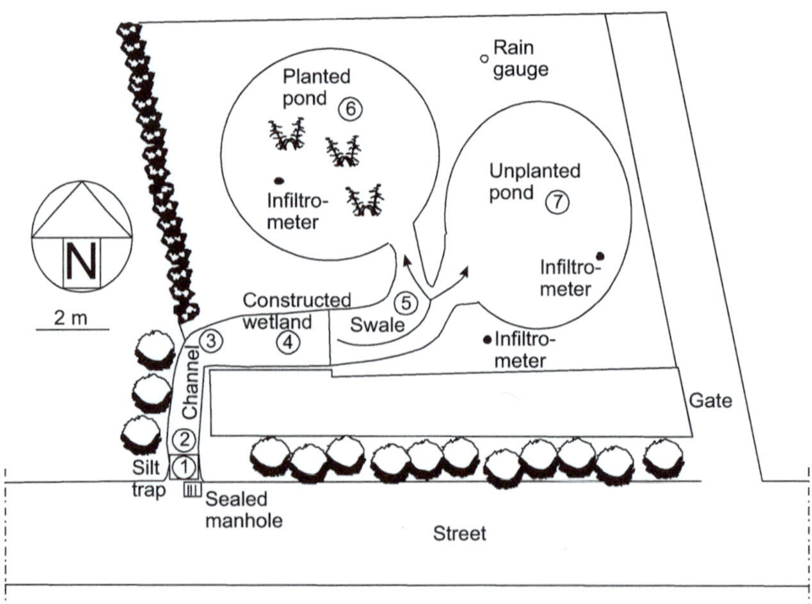

FIGURE 27.2.1.1 Case study: runoff flows from the road (only eastern part shown) into the silt trap (1), then via the gravel channel (2) into the constructed wetland (3 and 4), and finally via the swale (5) into the infiltration ponds (6 and 7).

of 0.2 m^3. Water from the silt trap overflows via a gravel ditch into the constructed wetland (volume of 2.7 m^3), which also serves the purpose of a below-ground storage tank. If the wetland attenuation system is full, storage water flows over a dry stonewall into a swale and finally into the infiltration ponds.

The ponds can accommodate maximum volumes of 9.7 m^3 each during heavy storm events before flooding of a nearby lawn would occur (Figure 27.2.1.1). The maximum depths of the constructed wetland and the unplanted and planted infiltration ponds are 0.85, 1.18, and 1.02 m, respectively.

In summer 2004, the dominant aquatic plants of the planted infiltration pond were *Typha latifolia* (Broadleaf Cattail), *Sparganium erectum* (Simple Stem Bur-reed), and *Iris pseudacorus* (Pale Yellow Iris). *T. latifolia* was planted in the infiltration pond to enhance infiltration. The constructed wetland was planted with *Phragmites australis* (Common Reed; Scholz et al., 2002), *I. pseudacorus*, and *T. latifolia*.

27.2.2 Hydrological Methods and Water Quality Analysis

The daily rainfall was monitored by a tipping bucket ARG 100 rain gauge. The infiltration rate of road runoff into the infiltration ponds was determined by using a single-ring infiltrometer. A PN 623–8001 (0.8–2.0 bar) Boart Longyear Interfels total pressure data logger was used to estimate water depth variations in the unplanted pond.

Grab samples at up to seven locations (Figure 27.2.1.1) were predominantly collected from the silt trap (1), constructed wetland (inflow (3) and outflow (4)), and both infiltration ponds (6 and 7). All analytical procedures to determine the water quality were performed according to American standard methods (Clesceri et al., 1998).

27.2.3 Fish Experiment Methodologies

Laboratory experiments with *C. auratus* and different aquatic plants (filamentous green algae, *Elodea canadensis* (Canadian Waterweed), and *Callitriche stagnalis* (Pond Water-starwort)) were carried out in eight aquarium tanks filled up with a mixture of filtered (sieve having a pore diameter of 250 μm) pond water (50%) and tap water (50%): Tank 1: filamentous green algae; Tank 2: *E. canadensis*; Tank 3: *C. stagnalis*; Tank 4: filamentous green algae and *E. canadensis*; Tank 5: *E. canadensis* and *C. stagnalis*; Tank 6: filamentous green algae and *C. stagnalis*; Tanks 7 and 8: filamentous green algae, *E. canadensis* and *C. stagnalis*.

All tanks (except for Tank 8; control) contained six *C. auratus* of similar weight. The overall biomass of plant food per tank was 600 g, and equal proportions of different plants were used. The experiment was stopped after the food sources dropped to $<$20 g of plant matter.

Twenty healthy *C. auratus* of approximately 180 g total weight were introduced into each infiltration pond on April 1, 2004. The ponds were covered with a plastic mesh to prevent animals such as *Ardea cinerea* (Grey Heron) and *Felis cattus* (Domestic Cat) from preying on *C. auratus*.

27.3 RESULTS AND DISCUSSION

27.3.1 Design and Operation of Infiltration Ponds

The critical storm durations for the BRE, CIRIA, and ATV-DVWK design calculations were 0.08, 0.25, and 0.08 h, respectively. The corresponding maximum infiltration pond height requirements were 0.38, 0.15, and 0.12 m, respectively. The estimated mean infiltration rate applied for all methods was 1.17 m/h. A storm duration of 15 min was associated with the maximum pond height and storage volume when CIRIA guidelines were applied (Bettes, 1996; CIRIA, 2000).

All calculated storage volumes (<1 m^3) are considerably lower than 9.7 m^3, so both ponds have more than sufficient storage volume each, compared to the theoretical design volumes. It seems that the infiltration rate of 1.17 m/h (standard deviation of 0.21 m/h) is an unrealistically high value for this SuDS system (case study only), considering the high spatial variability of the man-made soil at the study site. For example, the infiltration rate at water depths <0.8 m was frequently <0.05 m within 3 days (Figure 27.3.1.1).

The actual design depth for both infiltration ponds (1.10 m) was acceptable, when compared to the BRE (Building Research Establishment, 1991), CIRIA (Bettes, 1996), and ATV-DVWK (ATV-DVWK-Arbeitsgruppe, 2002) guidelines. Signs of system failure have not been observed. However, the strict

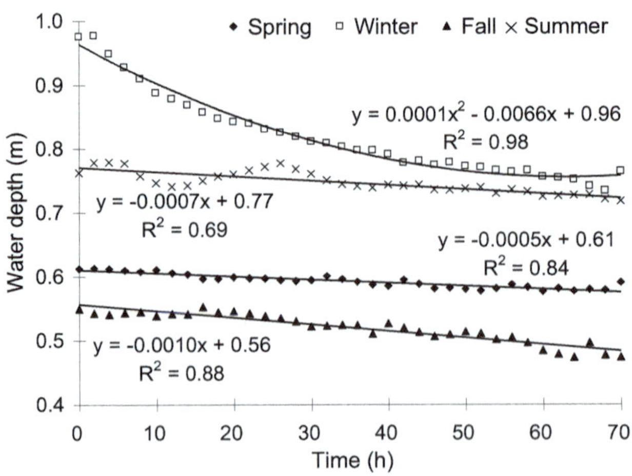

FIGURE 27.3.1.1 Water depth changes as a function of the infiltration rates: spring (April 1, 2004); summer (June 28, 2004); fall (September 30, 2004); and winter (January 28, 2004).

application of all test guidelines would have led to system failures, even during the first few months of operation. These findings are similar to a previous case study in England discussed by Scholz (2003).

27.3.2 Rainfall, Runoff, and Infiltration Relationships

Figure 27.3.1.1 gives an indication of representative water depth variations during periods of virtually no precipitation for all seasons. It can be seen that the infiltration is relatively low at water depths <0.8 m during all seasons except for winter. As a rule of thumb, Eq. (27.3.2.1) indicates the relationship between infiltration rate and water depth for water depth ranges between 0.3 and 1 m of the unplanted infiltration pond:

$$\text{water depth (m)} = 100\ \text{(h)} \times \text{infiltration rate (m/h)} + 0.5\ \text{(m)} \quad (27.3.2.1)$$

27.3.3 Water Quality Assessment and Management

The water qualities of the constructed wetland inflow, the unplanted pond, and the planted pond are shown in Tables 27.3.3.1–27.3.3.3, respectively. After 1 year of operation, the water quality of the unplanted infiltration pond (Table 27.3.3.2) was acceptable for disposal and recycling according to discussions, particularly on biochemical oxygen demand (BOD) and suspended solids (SS) threshold concentrations (Butler and Davies, 2000; Ellis et al., 2002; Scholz, 2003).

Variables indicating the organic strength of both ponds were frequently above international standards; i.e., thresholds of 20 mg/l for BOD and 30 mg/l for SS regarding secondary treatment of wastewater (Tchobanoglous et al., 2003). However, water quality monitoring is currently not required for closed systems with zero discharge (CIRIA, 2000).

The BOD and SS concentrations are high in the constructed wetland inflow and both infiltration ponds due to high loads of organic material such as leaves. Removal efficiencies for the SS were >93% and >89% for the unplanted and planted pond, respectively (Tables 27.3.3.2 and 27.3.3.3). The pH values of the wetland inflow and both ponds are neutral, similar, and stable due to algae control measures (see also Section 27.3.4).

Mats of algae are usually considered unpleasant in their appearance by the public. Therefore, it was necessary to use barley straw bales as a passive algae control method (Ball et al., 2001) for the second summer after system set-up. However, this method only provides a temporary solution and does not solve the problem of nutrient accumulation (nitrogen and phosphorus) within the pond sediment.

High temporal data variation indicates the need for at least weekly monitoring of SS to capture concentration peaks exceeding 30 mg/l. In comparison, the BOD concentrations were relatively stable (Scholz, 2003).

TABLE 27.3.3.1 Summary Statistics: Water Quality of the Inflow to the Constructed Wetland (April 1, 2003–September 30, 2004)

		Mean			
Variable	**Unit**	**Time[a]**	**Time[b]**	**Time[c]**	**Time[d]**
Temperature	°C	11.9	12.3	8.9	14.0
BOD_5[e]	mg/l	17.0	12.3	26.4	16.0
SS[f]	mg/l	385.2	104.4	208.8	601.0
Ammonia-N	mg/l	0.6	0.5	0.5	0.8
Nitrate-N	mg/l	2.5	1.6	1.9	3.2
Phosphate-P	mg/l	0.2	0.0	0.1	0.3
Conductivity	μS	242.2	204.9	276.9	228.9
Turbidity	NTU[h]	224.0	133.5	90.5	368.6
DO[g]	mg/l	4.2	3.7	6.0	2.9
pH	—	7.0	7.0	7.0	7.1
Variable	**No.**	**Standard Deviation**			
Temperature	104	5.5	5.5	5.5	5.5
BOD_5[e]	86	18.2	18.2	18.2	18.2
SS[f]	95	600.3	600.3	600.3	600.3
Ammonia-N	67	0.8	0.8	0.8	0.8
Nitrate-N	60	8.3	8.3	8.3	8.3
Phosphate-P	67	0.2	0.2	0.2	0.2
Conductivity	105	193.2	193.2	193.2	193.2
Turbidity	106	442.3	442.3	442.3	442.3
DO[g]	100	6.4	6.4	6.4	6.4
pH	106	0.6	0.6	0.6	0.6

[a] *4/1/2003–9/30/2004.*
[b] *4/1/2003–9/30/2003.*
[c] *10/1/2003–3/31/2004.*
[d] *4/1/2004–9/30/2004.*
[e] *5-day at 20 °C N-allylthiourea biochemical oxygen demand.*
[f] *Suspended solids.*
[g] *Dissolved oxygen.*
[h] *Nephelometric turbidity unit.*

TABLE 27.3.3.2 Summary Statistics: Water Quality of the Unplanted Pond (April 1, 2003–September 30, 2004)

Variable	Unit	Mean			
		Time[a]	Time[b]	Time[c]	Time[d]
Temperature	°C	11.8	13.4	6.4	15.6
BOD_5[e]	mg/l	19.0	35.9	17.6	14.2
SS[f]	mg/l	26.4	29.2	24.1	27.1
Ammonia-N	mg/l	0.4	1.4	0.2	0.3
Nitrate-N	mg/l	1.6	0.3	1.3	2.2
Phosphate-P	mg/l	0.3	0.1	0.3	0.3
Conductivity	μS	215.6	246.8	207.1	210.2
Turbidity	NTU[h]	17.9	17.3	9.0	25.7
DO[g]	mg/l	5.1	3.8	7.4	3.5
pH	–	7.3	7.0	7.3	7.4
Variable	**No.**	**Standard Deviation**			
Temperature	106	5.5	2.3	4.1	3.8
BOD_5[e]	82	20.4	22.1	25	10.3
SS[f]	92	50.3	24.3	60.1	49.7
Ammonia-N	65	1.2	2.6	0.3	0.7
Nitrate-N	63	6.4	0.1	2.9	9.0
Phosphate-P	70	0.4	0.1	0.5	0.4
Conductivity	107	121	87.9	156.4	95.2
Turbidity	108	36.9	14.1	19.8	50.6
DO[g]	102	7.9	3.3	12.1	1.2
pH	108	0.4	0.3	0.5	0.2

[a]*4/1/2003–9/30/2004.*
[b]*4/1/2003–9/30/2003.*
[c]*10/1/2003–3/31/2004.*
[d]*4/1/2004–9/30/2004.*
[e]*5-day at 20 °C N-allylthiourea biochemical oxygen demand.*
[f]*Suspended solids.*
[g]*Dissolved oxygen.*
[h]*Nephelometric turbidity unit.*

TABLE 27.3.3.3 Summary Statistics: Water Quality of the Planted Pond (April 1, 2003–September 30, 2004)

		Mean			
Variable	**Unit**	**Time[a]**	**Time[b]**	**Time[c]**	**Time[d]**
Temperature	°C	11.6	13.2	6.6	15.1
BOD_5[e]	mg/l	26.5	26.3	41.8	13.9
SS[f]	mg/l	40.9	38.2	64.7	23.5
Ammonia-N	mg/l	0.2	0.6	0.3	0.1
Nitrate-N	mg/l	0.5	0.5	0.8	0.3
Phosphate-P	mg/l	0.2	0.1	0.2	0.3
Conductivity	μS	288.9	324.7	311.6	254.3
Turbidity	NTU[h]	17.6	22.6	16.4	16.7
DO[g]	mg/l	4.8	5.8	6.3	3.2
pH	–	7.2	7.1	7.2	7.2
Variable	**No.**	**Standard Deviation**			
Temperature	106	2.3	4.1	3.8	2.3
BOD_5[e]	82	19.4	43.4	17.5	19.4
SS[f]	92	64.9	129.7	34.4	64.9
Ammonia-N	65	1.0	0.3	0.1	1.0
Nitrate-N	63	0.6	2.6	0.8	0.6
Phosphate-P	70	0.1	0.1	0.3	0.1
Conductivity	107	101.4	124.0	47.7	101.4
Turbidity	108	24.9	16.9	40.5	24.9
DO[g]	102	2.7	7.9	1.3	2.7
pH	108	0.3	0.4	0.1	0.3

[a]4/1/2003–9/30/2004.
[b]4/1/2003–9/30/2003.
[c]10/1/2003–3/31/2004.
[d]4/1/2004–9/30/2004.
[e]5-day at 20 °C N-allylthiourea biochemical oxygen demand.
[f]Suspended solids.
[g]Dissolved oxygen.
[h]Nephelometric turbidity unit.

Nutrient harvesting was undertaken. The silt trap was regularly emptied and the total wet weight of silt was 55 kg per annum. Aquatic plants of 20 kg were harvested in November 2003. The purpose was to reduce the input of additional nutrients that would otherwise be released from decaying leaves into the planted pond. Furthermore, plant harvesting leads to a prolonged lifetime of the infiltration system by indirectly increasing the storage volume available for sediment during storm events (Scholz, 2003).

27.3.4 Active Control of Algae with Goldfish

Carassius auratus is classified as an herbivore, with wild specimens predominantly feeding on plants. This particularly applies to closed pond systems (Seaman, 1979). Therefore, *C. auratus* could be used to control aquatic weeds (Caquet et al., 1996) and potentially algae in ponds (Richardson et al., 1992). This hypothesis was subsequently tested in the laboratory.

Figure 27.3.4.1 indicates that the mean weights and associated standard deviations of *C. auratus* for each tank appear to be similar at the beginning and end of the controlled laboratory experiment. Moreover, an analysis of variance showed that all tanks were significantly similar ($P < 0.05$) concerning their associated fish mean weights, which were measured on eight occasions each.

Carassius auratus seem to feed on all three aquatic plants, including filamentous green algae such as *Odeogonium* spp. and *Ulothrix* spp., which were also naturally present in the pond system. However, the greatest drop in fish weights was observed for Tank 3, containing *C. stagnalis* only. Considering that the associated standard deviation was also the lowest, and that *C. stagnalis* belongs to the group of floating aquatic plants that are only

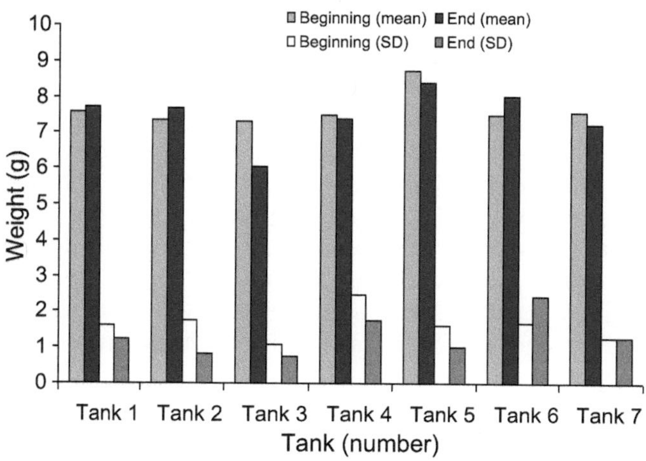

FIGURE 27.3.4.1 Comparison of means and standard deviations (SD) for weights of *C. auratus* (Common Goldfish) at the beginning (February 9, 2004) and end (March 31, 2004) of the laboratory experiment.

accessible by fish below the water level, it can be assumed that *C. auratus* would not have chosen this plant as a preferred food source, if alternative submerged aquatic plants would have been abundant.

Concerning the field experiment, relatively high numbers of filamentous green algae (Chlorophyta) were counted in pond samples taken on March 29, 2004. The dominant alga present was *Odeogonium capillare*, which is cosmopolitan in freshwater. *Odeogonium capillare* can form mats in small ponds and is often mistaken for the more common *Cladophora glomerata* (Blanket Weed).

Carassius auratus was introduced into both ponds on April 1, 2004, to control filamentous green algae and to increase public acceptance of SuDS. Concerning algae samples taken on October 4, 2004, both ponds were less dominated by *O. capillare* in comparison to March 2004.

Moreover, the unplanted pond developed a greater diversity of filamentous green algae if compared to the planted pond. This may be due to the absence of macrophytes that would compete with algae for nutrients. Moreover, large macrophytes provide shade, leading to a reduction of sunlight penetrating the water and subsequently reducing the growth of algae.

27.3.5 Integration of Wetlands into Urban Planning and Development

The environmental impacts of urban constructions can be reduced with innovative approaches such as smart growth as well as sustainable urban and green planning (Galuzzi and Pflaum, 1996). Flood protection management and recreational value can be improved by integrating infiltration pond design and operation (in contrast to conventional drainage) into the urban planning and development processes (Campbell and Ogden, 1999; Scholz, 2003). Recreational activities may include watching birds and ornamental fish such as *C. auratus*, walking, fishing, boating (sufficiently large ponds only), holding picnics, and teaching children about aquatic ecology (Galuzzi and Pflaum, 1996).

The confidence of town planners toward SuDS and public acceptance of infiltration ponds and/or wetlands can both be increased by correct dimensioning of systems to avoid flooding (Zheng and Baetz, 1999), enhance water pollution control by using a robust pretreatment train (e.g., silt trap, constructed wetland, and swale), and control algae by biological (e.g., *C. auratus*) and not chemical (e.g., copper sulfate) means (Ellis et al., 2002). Moreover, stormwater can be reused for watering gardens and flushing toilets as part of an urban water resources protection strategy.

27.4 CONCLUSIONS

This case study described the design and operation of a novel stormwater pond system during the first 18 months of operation (April 2003–October 2004).

During this period, the SuDS would only have complied with design guidelines if local environmental conditions such as spatial infiltration patterns had been fully considered. Infiltration through the base of both ponds was virtually absent (despite the presence of macrophytes in the planted pond) and should therefore be neglected during the design.

The water qualities of the infiltration ponds were generally unacceptable for water reuse immediately after the set-up period of the SuDS. Biochemical oxygen demand and SS concentrations, in particular, frequently exceeded recognized international secondary wastewater treatment standards, which are, however, not applicable for closed systems with zero discharge. At least weekly water quality monitoring was required to capture temporal data variability.

A bloom of filamentous green algae dominated by *O. capillare* during spring 2004 was observed. Barley straw reduced the growth of algae but requires removal before it starts to degrade, releasing nutrients. Moreover, experiments with *C. auratus* have shown that they eat filamentous green algae equally as well as *C. stagnalis* and *E. canadensis*. The presence of tall macrophytes reduced the growth and biodiversity of filamentous green algae.

A successful integration of SuDS into urban regeneration and development can be achieved, if potential design, operation, and water quality management problems are addressed during the planning phase. Moreover, current international guidelines for the design and management of infiltration ponds require alterations to avoid system failure, as shown in this case study.

Chapter 28

Retrofitting of Sustainable Drainage Systems in the Presence of Vegetation

28.1 INTRODUCTION

28.1.1 Ecosystem Services and Urbanization

The benefits human beings may obtain from the semi-natural environment (e.g., urban green space) can be referred to as ecosystem services (Millennium Ecosystem Assessment, 2005). Defra (2011) defines ecosystem services as the benefits individuals gain from the goods and services produced by nature and its natural systems. The natural resources such as timber and water, and functioning natural systems such as healthy fertile soils, clean water and air, and a regulated climate, are essential for human well-being, security, and economic prosperity (Millennium Ecosystem Assessment, 2005).

A number of official documents, such as the Natural Environment White Paper (Defra, 2011), the UK National Ecosystem Assessment (2011), and The Economics of Ecosystems and Biodiversity (TEEB) (2011), have identified four broad categories of ecosystem services generally referred to as supporting, regulating, provisioning, and cultural. The impacts of supporting services on nature take place over a long period of time and are indirectly beneficial to human life. They refer to all ecosystem services that provide a basic infrastructure of life (UK National Ecosystem Assessment, 2011), and it is on supporting services that all other ecosystem services that do not fall within this category depend for their own continuation.

The ecosystem services classed within the category of regulating services are very diverse and include all those ecosystems that provide benefits through the regulation of ecosystem services. The goods that are obtained from ecosystem services are referred to as provisioning services (UK National Ecosystem Assessment, 2011).

Ecosystem services present due to environmental settings that provide recreational areas where individuals can interact with nature and with each other (UK National Ecosystem Assessment, 2011) and find spiritual fulfillment and

Wetlands for Water Pollution Control. http://dx.doi.org/10.1016/B978-0-444-63607-2.00028-9

mental development are known as cultural services. Sander and Haight (2012) estimated the economic value of cultural ecosystem services in an urban area using hedonic pricing. They discovered that many aspects of the aesthetic environment significantly impacted on home sale prices.

Urban areas are increasingly being faced with issues such as elevated surface runoff and more heat islands, which result in alterations of the local energy exchange and hydrology, thus having a negative impact on regulating services for climate, soil and water quality, and noise. In the UK, roughly 30% of ecosystem services have been identified as declining, while many others are considered to be in a reduced or degraded condition (UK National Ecosystem Assessment, 2011).

A more resilient semi-natural environment needs to be created to protect ecosystems to counteract the negative impacts of urbanization. This can be achieved by bettering habitat management to improve the quality and size of existing wildlife sites, creating physical corridors to improve connections between sustainable drainage system (SuDS) sites, creating new habitats by planting more urban trees, and lessening the strain on wildlife by reviving the wider environment (Lawton et al., 2010).

28.1.2 Urban Trees and Linked Ecosystem Services

The Construction Industry Research and Information Association (CIRIA) (2012) promotes the planting of large urban trees. Researchers claim that on average the annual net benefit of planting large tree species is 44% greater than for a medium tree species and 92% greater than for a small tree species (McPherson et al., 1999). They also state that it takes less than five years from planting for the net benefits of large trees to outweigh net costs. However, some trees at certain locations may have obvious associated negative aspects. These include health and safety issues such as fall of fruits and pine cones on humans; damage to the built environment due to vigorous root growth; blockage of gutters, downpipes, gullies, and permeable pavements by fruits, pine cones, and leaf litter; and deterioration of driving conditions due to leaf cover on roads in fall or the occasional visual obstruction of bends due to trees' large size. Moreover, this leads to maintenance activities such as road sweeping to remove the leaves or branch pruning.

A list of ecosystem service variables and their respective categories used in this chapter is provided in Table 28.1.2.1. The listed ecosystem services have been reinterpreted to make them relevant to combined permeable pavement and tree systems retrofitted in urban areas and are categorized in broad agreement with TEEB (2011) and other guidance documents, such as Moore and Hunt (2012) as well as Scholz and Uzomah (2013). TEEB (2011) proposed a comprehensive list of ecosystem service variables of a generic nature, while Moore and Hunt (2012) chose a smaller set of variables particularly adapted for SuDS, such as constructed wetlands and ponds.

TABLE 28.1.2.1 Universal Ecosystem Service Categories and Variables for Combined Permeable Pavement and Tree Systems (after Uzomah and Scholz (2013))

Category	Variable	Generic Ecosystem Service Variable Description
Supporting services	1. Habitats for species (HS)	Urban habitats should provide everything that an animal needs to survive: food, water, and shelter. Each ecosystem provides habitats that can be essential for a species' life cycle. Migratory species including birds and insects all depend upon different ecosystems during their movements.
	2. Maintenance of genetic diversity (MGD)	Genetic diversity distinguishes different breeds or races from each other, providing the basis for locally well-adapted cultivators. Some urban habitats have an exceptionally high number of species (biodiversity hotspots), which makes them more genetically diverse than others.
Regulating services	3. Local climate and air quality regulation (LCAR)	Trees lower the temperature by providing shade and influence water availability (e.g., evapotranspiration). Trees and other plants also play an important role in regulating air quality by removing pollutants from the atmosphere (e.g., filtration and absorption of particulates and mono-nitrogen oxides).
	4. Carbon sequestration and storage (CSS)	Ecosystems regulate the climate by storing greenhouse gases such as carbon dioxide through burial and sediment accretion. As trees grow, they remove carbon dioxide from the atmosphere.
	5. Moderation of extreme events (MEE)	Ecosystems and living organisms create buffers against natural disasters, thereby preventing or reducing damage from extreme weather events or natural hazards including floods, storms, and landslides. Trees stabilize slopes. Flooding may be reduced through regulating runoff.
	6. Storm runoff treatment (SRT)	Physical, chemical, and biological treatment takes place within permeable pavement systems. Trees filter effluents such as stormwater runoff. Through the biological activity of microorganisms in the soil and sediment, most waste is broken down; thereby pathogens are eliminated, and the level of nutrients and pollution is reduced.

Continued

TABLE 28.1.2.1 Universal Ecosystem Service Categories and Variables for Combined Permeable Pavement and Tree Systems (after Uzomah and Scholz (2013))—cont'd

Category	Variable	Generic Ecosystem Service Variable Description
	7. Erosion prevention and maintenance of soil fertility (EPMSF)	Tree cover provides a vital regulating service by preventing soil erosion. Soil fertility is essential for plant growth and agriculture, and well-functioning ecosystems supply soil with nutrients required to support plant growth.
	8. Pollination (P)	Insects and wind pollinate plants including trees, which is essential for the development of fruits and seeds. Animal pollination is an ecosystem service mainly provided by insects but also by some birds.
	9. Biological control (BC)	Ecosystems are important for regulating pests and vector-borne diseases that attack plants, animals, and people. Ecosystems regulate pests and diseases through the activities of predators and parasites.
Provisioning services	10. Food (F)	Food comes principally from managed urban horticulture (e.g., fruit trees).
	11. Raw materials (RM)	Some trees deliver a great diversity of materials for construction and fuel, including wood, biofuels, and plant oils that are directly derived from wild and cultivated plant species.
	12. Freshwater (FW)	Ecosystems play a vital role in providing cities with drinking water, as they ensure the flow, storage, and purification of water. Trees influence the quantity of water available locally.
	13. Medicinal resources (MR)	Some tree products may be used as traditional medicines or provide raw materials for the pharmaceutical industry.
Cultural services	14. Recreation, and mental and physical health (RMPH)	Wildlife viewing, walking, jogging, and playing sports in green spaces is a good form of physical exercise and helps people to relax. The role that green space plays in maintaining mental and physical health is increasingly becoming recognized, despite difficulties of measurement.

TABLE 28.1.2.1 Universal Ecosystem Service Categories and Variables for Combined Permeable Pavement and Tree Systems (after Uzomah and Scholz (2013))—cont'd

Category	Variable	Generic Ecosystem Service Variable Description
	15. Tourism and area value (T)	Ecosystems and biodiversity play an important role for local tourism, which in turn provides considerable economic benefits. Cultural tourism and ecotourism can also educate people about the importance of biological diversity. The value of properties in the area may be positively affected by the presence of an attractive permeable pavement site with trees and street furniture.
	16. Aesthetic and educational appreciation, and inspiration for culture, art, and design (AEAICAD)	Language, knowledge, and the natural environment have been intimately related throughout human history. Biodiversity, ecosystems, and (predominantly natural) landscapes have been the source of inspiration for art, culture, and, increasingly, science. Trees within urban green spaces may provide soothing and educational benefits and a sense of beauty for some observers. Some attractive urban areas may also promote health and well-being.
	17. Spiritual experience and sense of place (SESP)	Some urban forms may be considered to have a religious meaning. Trees are a common element of some major religious and traditional knowledge. They can become important for creating a sense of belonging.

28.1.3 Example Case Study

An example case study (Scholz and Uzomah, 2013) to test the generic tools discussed in this chapter has been chosen. Greater Manchester, a sub-region in the northwest of England, is made up of 10 local authorities. The six authorities of relevance for this study (in order of decreasing importance: Manchester, Salford, Trafford, Bury, Oldham, and Tameside) are highlighted in Figure 28.1.3.1. The whole of Greater Manchester covers a total surface area of 1300 km^2 and is home to approximately 2.5 million people.

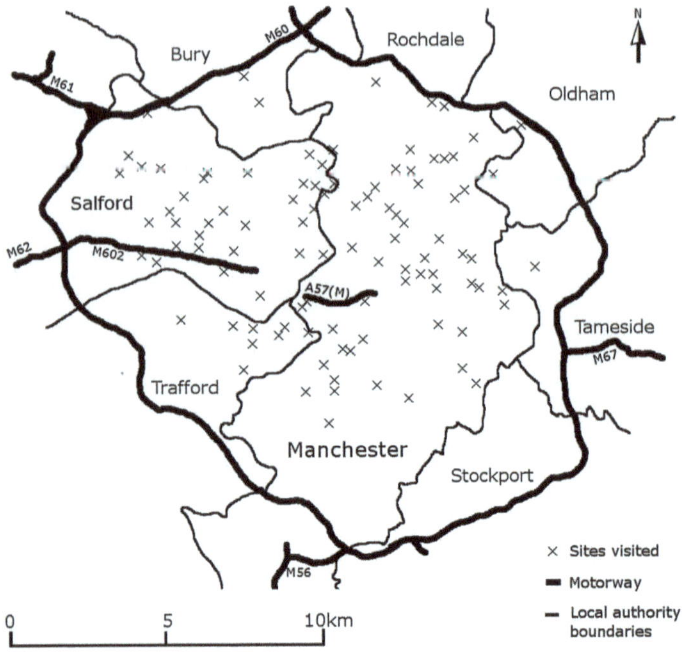

FIGURE 28.1.3.1 Map indicating all potential combined permeable pavement and tree sites assessed in Greater Manchester.

28.1.4 Rationale, Aim, and Objectives

Particularly in times of recession and spending cuts in the public sector, rapid and inexpensive expert assessment systems supporting planners currently undergo a revival in the context of ecosystem services. The application of ecosystem service values to a new area such as sustainable drainage is a novel contribution to knowledge and understanding. The timely and applied nature of such expert systems has a strong appeal, particularly for urban and landscape planners interested in the total environment.

Established tools evaluating a range of SuDS techniques for retrofitting of drainage systems already exist (e.g., CIRIA, 2004). However, sophisticated tools focusing on the retrofitting of permeable pavements on sites with existing trees and taking into account a wide range of ecosystem service variables (including functions associated with vegetation such as trees) do not exist.

With an increasing public awareness of the importance of ecosystems and the services they provide for humans, Chapter 28 aims to develop a unique and rapid decision support tool based on novel ecosystem service variables for retrofitting of combined permeable pavements and tree systems in densely populated areas. The key objectives to achieve this aim are:

- to compile a comprehensive data set of sites within an example case study area (Greater Manchester) where retrofitting of permeable pavements in the presence of trees would be possible;
- to identify all generic ecosystem service variables relevant for permeable pavement retrofitting;
- to broadly categorize these ecosystem service variables under the four established categories of supporting, regulating, provisioning, and cultural;
- to assess the suitability of potential combined permeable pavement and tree system sites within Greater Manchester based on traditional "community and environment" variables;
- to assess the suitability of potential combined permeable pavement and tree system sites within Greater Manchester based on the new ecosystem service variables;
- to compare the assessment outcomes based on the traditional and new variables with each other;
- to assess a combination of the traditional and new approach for sites within Greater Manchester; and
- to outline a simple methodology to determine trees that may impact on the retrofitting of permeable pavement systems.

28.2 METHODOLOGY

28.2.1 Site Assessment

A total of 100 sites and corresponding catchment areas that were large enough for the retrofitting of combined permeable pavement and tree systems to have a positive urban drainage impact were identified by studying Ordnance Survey and Google maps of Greater Manchester (representative case study area). Moreover, discussions with local authorities, United Utilities (water authority), and major private landowners (e.g., local authorities) regarding suitable combined permeable pavement and tree sites were held. A map of Greater Manchester highlighting all sites visited was created using the computer software GNU Image Manipulation Program (Figure 28.1.3.1). The main areas targeted within Greater Manchester were Salford and Manchester, the two most built-up local authorities in the conurbation (Section 28.1.3). A number of sites within the innermost parts of Trafford were also visited.

The site assessment template was based on a combination of the frameworks developed by Scholz (2006) and Scholz et al. (2006) for retrofitting of SuDS techniques in Glasgow, Edinburgh, and elsewhere as well as the CIRIA guidelines (CIRIA, 2004, 2007). Each potential SuDS site was assessed during a site visit by a group of experts (2−5 team members) to reduce subjectivity (Munoz-Pedreros, 2004), supported by a desk study. A relative measure of certainty expressed in percentage points was given to each variable to indirectly measure the reliability of the assessment. Only values higher than 50% were considered to be acceptable.

General site information such as site number and name, postcode, grid reference numbers, location name, names of the inspection team members, site acceptability for combined permeable pavement and tree systems, and presence of existing permeable pavements and trees was recorded. Photos of the key site features were taken for each potential SuDS site and its catchment. Land ownership information such as number of owners, ownership type (private or public), and estimated site value (£) was noted. Proportions (%) of site classification categories including development, regeneration, retrofitting, and recreation were estimated. Surrounding area characteristics such as descriptions of the neighborhood to the north, south, east, and west; current and future site use; total area of the catchment (m^2); and catchment shape were recorded. The location and distance (m) of the sites to the nearest sewer, storm pipe, stream, river, canal, pond, lake, and sea were described by the team, if located within a reasonable distance from the catchment. Estimated current and future surface permeability (%) for the land categories grass, trees, shrubs, and impermeable surfaces of the proposed SuDS site and its catchment were noted. The researchers estimated the proportions (%) of current and future roof runoff for the categories institutional, commercial, industrial, high-density housing, medium-density housing, low-density housing, and other. Furthermore, estimated proportions (%) of current and future road runoff for the categories car park, motorway, primary road (or dual carriageway), A road, B road, tertiary road, and other were also recorded. For each subcatchment, the area (m^2) and the gradient in the two main directions having an angle of 90° to each other in the horizontal plain were recorded as well. Hydrogeological information such as contaminated land (present or absent), soil infiltration (low, medium, or high), and groundwater level (below or above 2 m depth) was noted. Additional remarks regarding current drainage techniques and potential problems regarding the implementation of future SuDS techniques may be added. Finally, the SuDS technology feasibility proportion (%) for combined permeable pavement and tree systems was estimated.

28.2.2 Ecosystem Services

The potentials of new quantitative and qualitative approaches to assessing ecosystem services have been explored by Busch et al. (2012). Each ecosystem service variable was described qualitatively and quantitatively with the help of five bins (category groupings). A team of experts has to assign percentage points per variable based on comparable experience. Bin 1 always describes the characteristics of sites that are considered to be of lowest ecosystem service value (0–20%), while bin 5 constantly identifies characteristics of sites that are considered to be of greatest ecosystem service potential (80–100%). Moreover, a qualified measure of certainty expressed in percentage points was given by a team of experts based on their comparable experience to each variable to indicate the reliability of the assessment; the higher the value given, the more certain were the group of assessors. Only values greater than 50% were considered to be acceptable to progress to the next estimation without conducting further studies.

A weighting system specific to the needs of a particular region or stakeholder group could be introduced by providing weights for individual variables after consultation with a team of experts. For example, variables of low relevance, such as medicinal resources in Greater Manchester, could be assigned a low weight of, for example, (1), while variables with a medium (e.g., recreation, and mental and physical health) or high (e.g., moderation of extreme events) relevance could be assigned a medium (2) or high (3) weight, respectively. However, such a weighting system has not been introduced for this case study to keep the example simple.

28.2.3 Traditional "Community and Environment" Variables

The site assessment was inspired by the SuDS decision support key and the SuDS decision support matrix developed for the Glasgow and Edinburgh SuDS retrofitting studies (Scholz, 2006; Scholz et al., 2006). All corresponding variables characterizing the site were determined using these tools. However, the method used to determine which SuDS technique, including permeable pavements, is likely to be most suitable for a particular site was based on the expert tool published by CIRIA (2004).

The guideline C609 (CIRIA, 2004) proposes for the selection of SuDS techniques a number of criteria related to hydrology, land use, physical site features, community and environment, and economics and maintenance. The criteria have been adapted from a technical report previously authored by Ellis et al. (2003). Each SuDS technique is scored according to each criterion from 1 to 5 where 1 is the lowest and 5 is the highest score. The scores are then tallied up and the SuDS technique, such as permeable pavements, obtaining the highest score is considered to be the practice best suited to the case study site.

Use is only made of the criteria falling under the category "community and environment," which comprises the traditional variables safety, pond premium, aesthetics, wildlife habitat, and acceptance. Eliminating the other categories that do not directly relate to ecosystem services introduces bias in the overall assessment, but this was seen as acceptable for the purpose of this chapter, which focused only on categories of interest to ecosystem services.

The option "combined permeable pavement and tree system" was assessed for each suitable SuDS site according to the following criteria definitions: safety (probable danger to people and pets), water premium (potential rise of the cost of property within the area), aesthetics (likely improvement of the visual aesthetics of that area), wildlife habitat (likely ecological impact of wildlife within the wider SuDS area), and acceptance (expected adequacy of proposal to the local community). Scores for each criterion range from 1 to 5, where 1 refers to a SuDS technique being very unsuitable, and 5 signifies a SuDS technique being very suitable for that particular criterion. The minimum and maximum overall scores for all criteria were 0 and 25, respectively. For the purpose of this chapter, the focus was only a combination of permeable pavements with existing mature trees.

28.2.4 Comparison between the Traditional and Revised Methods

A comparison between a site assessment based on the community and environment variables with the new ecosystem service variables has been performed. Considering that the latter method comprises more variables (17 instead of 5) and higher maximum points (100 instead of 5) per variable can be awarded in comparison to the former, not absolute total scores but adjusted scores (values for established approach times 68) were compared with each other to avoid bias toward the new approach.

28.2.5 Combination of the Traditional and New Approaches

This section describes the last assessment method, which is a combination of the traditional and new approach by replacing the traditional criteria "aesthetic" and "wildlife habitat" with the four ecosystem service categories (not the 17 variables). These two traditional criteria are included in the new ecosystem service assessment. Thus the SuDS techniques are scored according to the following criteria: safety, water premium, and acceptance as well as provisioning, regulating, supporting, and cultural services. The same scoring system applied above was adopted during the reassessment of evaluating which SuDS technique is likely to be most appropriate for a particular site. Only those techniques that were most suitable (i.e., first preferences) for particular sites were subsequently recommended to the local authorities, United Utilities, and private landowners for implementation. For the purpose of this chapter, however, the focus was on the suitability of the SuDS technique "combined permeable pavement and tree system."

28.2.6 Vegetation Determination

Areas at sites suitable for retrofitting of SuDS techniques including permeable pavement systems were determined. This chapter focuses on sites where mature trees were already present in close proximity (within 10 m) to the proposed permeable pavement system. No restrictions were made regarding tree species and its current or projected growth characteristics, and no assumptions were made regarding environmental interactions between any prospective pavement system and any corresponding existing trees.

Trees with a stem diameter of more than 10 cm measured at breast height (1.5 m above ground level) were determined by the survey team for all case study sites in Greater Manchester in fall 2012. Trees were identified by assessing the overall shape, leaves, bark, buds, and flowers (if available). Only healthy trees with a good survival likelihood that were located within a strip of 10 m surrounding the site where permeable pavements could be retrofitted were assessed using a very wide range of standard tree determination guides (e.g., Woodland Trust, 2012) combined with expert judgment.

28.3 RESULTS AND DISCUSSION

28.3.1 Site Assessment Outcomes

Only 16% of the example case study sites were existing SuDS sites. Most of these were of poor design with low ecosystem service value. There was a clear need for introducing SuDS to improve drainage in all cases, which is realistic, because most sites are owned by the public hand (52% public, 23% private and public, and 25% private) and not by multiple private owners. The estimated site values were relatively low (26% low, 32% low to medium, 29% medium, 11% medium to high, and 2% high), which supports SuDS implementation due to low competition with private investors.

Most sites were not just potential retrofitting sites but a combination of different potential types: combinations of at least two categories (81%), retrofitting only (15%), development only (2%), regeneration only (2%), and recreation only (0%). This was reflected in the current site use, which was dominated by public or at least publicly accessible areas: park (41%), car park (33%), institutional building (12%), disused land (8%), and field (6%). This scenario is unlikely to change in the near future, according to consultations with site owners and local authorities (park, 56%; car park, 32%; institutional building, 12%; disused land, 0%; field, 0%). The recent trend toward converting "empty spaces" into parks is a positive development.

The current site assessment indicated that 21 sites were fully paved (usually with tarmac). The remaining 79 sites were unpaved. Permeable pavements would be suitable to replace 20 and 46 currently paved and unpaved sites, respectively. The sites that are currently unpaved and where permeable pavement would be suitable would drain the precipitation covering the paved area. The retrofitted permeable pavements would function as future car parks and pavements.

The catchment size (in 1000 m^2) of most sites was relatively small (<25, 29%; $25 \leq 50$, 32%; $50 \leq 100$, 20%; $100 \leq 150$, 6%; and ≥ 150, 13%), indicating that SuDS may only make a minor contribution toward resolving the urban drainage problem. The vast majority of potential SuDS sites (63%) had only sewers, while the remaining sites had both sewers and storm pipes (37%).

The current site permeability (%) was high (<20, 24%; $20 \leq 40$, 2%; $40 \leq 60$, 1%; $60 \leq 80$, 3%; and ≥ 80, 70%). In contrast, the current catchment permeability (%) was low (<20, 21%; $20 \leq 40$, 13%; $40 \leq 60$, 17%; $60 \leq 80$, 30%; and ≥ 80, 19%), indicating a good potential for infiltration.

The present runoff area proportions were as follows: roads (48%), mainly roofs and roads (11%), and mainly roofs (41%). Despite the high proportion of roofs, the slopes (%) of current sites were rather flat (<0.5, 20%; $0.5 \leq 1.5$, 53%; $1.5 \leq 2.5$, 10%; $2.5 \leq 3.5$, 7%; and ≥ 3.5, 10%). This is likely to lead to reduced leveling costs when retrofitting permeable pavement systems. Nevertheless, leveling of at least some parts of the site is required in 57% of all cases to allow for adequate flow.

The majority of all SuDS case study sites (83%) comprised only one catchment. Only 13% and 5% of all sites had two and more than two subcatchments, respectively. This finding indirectly supports the statement that many sites were flat (see above), which is beneficial for permeable pavement systems.

Only one site (1% of all sites) was seriously contaminated. It follows that permeable pavement systems used as infiltration devices are suitable for Greater Manchester. The soil infiltration rates were predominantly of medium magnitude (96%). Only 3% of the sites had low rates. Even fewer sites (1%) had high infiltration rates. The groundwater level was usually relatively high, which is not a concern for permeable pavement systems.

28.3.2 Strengths and Weaknesses of the Site Assessment

The strengths of this site screening tool include quick, simple, inexpensive, user-friendly, and easy-to-understand assessments of site data and physical site features; a comprehensive data set of key variables hosted within a widely available nonexpert database such as Microsoft Excel; and the acceptability of the site, which does not discriminate against current site use. However, the following weaknesses of the methodology may apply: its too-simplified approach leaves wide room for interpretation and personal bias, and the assessment is not dynamic by nature, ignoring various possible future scenarios. These two negative points also apply, however, to the traditional approach proposed by CIRIA (2004).

28.3.3 Assessment of Ecosystem Service Variables

This study, combining ecosystem service assessments for sustainable drainage techniques with an expert system applied to a large database of real case studies, is unique. Danso-Amoako et al. (2012) assessed large sustainable flood retention basins located within the wider Greater Manchester area.

However, they only focused on a subset of the ecosystem service variables (Scholz and Yang, 2010) proposed in Table 28.1.2.1. Moreover, Gill et al. (2007) and White and Alarcon (2009) discussed green infrastructure in the context of climate change and planning policies associated with sustainable drainage in Greater Manchester, respectively, but were not concerned with the influence of ecosystem service variables on decision making.

While Table 28.1.2.1 outlines the universal application of the ecosystem service variables, this section discusses the specific application of the more relevant variables for an example case study. A brief explanation regarding the key new ecosystem service variable suitable for permeable pavement retrofitting, with particular reference to selected sites in Greater Manchester, is given below (see also Table 28.1.2.1 for generic descriptions).

Habitats for species (HS): A public park with a highly permeable surface coverage of diverse vegetation and structure-rich semi-natural watercourses will provide a great area for wildlife benefits, thus scoring highly. This habitat is, unfortunately, rare in Greater Manchester.

Local climate and air quality regulation (LCAR): Factors likely to reduce temperature and improve air quality in urban areas, such as the density of tree coverage and the presence of surface water, were assessed. A site that is entirely covered by dense trees and contains surface water such as a pond will receive a relatively high ecosystem service potential value for this variable. The density of trees, as well as the presence of water bodies, is rather low in the example case study area.

Carbon sequestration and storage (CSS): This assessment was predominantly based on the density of tree coverage. Mature woodlands are rare within the study area.

Moderation of extreme events (MEE): The ability of a potential SuDS site to manage extreme events such as flooding, drought, and fire was assessed. Sites that can mitigate runoff and store water that can subsequently be used as a resource will score highly. Thus, sites where permeable pavements can be implemented will be associated with high bin numbers. There is a high potential for retrofitted SuDS in Greater Manchester to score high on MEE.

Storm runoff treatment (SRT): The evaluation of this variable is based on the ability of a potential SuDS site to break down pollutants from surface runoff by physical, chemical, and/or biodegradation processes. Some trees have a high potential to degrade pollutants. Thus the corresponding ecosystem service potential will be relatively high. There is an extraordinary potential for retrofitted SuDS in the study area to score relatively high on SRT.

Erosion prevention and maintenance of soil fertility (EPMSF): The potential of a SuDS site to protect its underlying soil from the harmful pollutants associated with surface runoff was evaluated. A site that is covered by dense and mature trees will have a high ecosystem service potential associated with this variable. There is a good opportunity for erosion prevention by a combination of permeable pavements and trees in Greater Manchester.

Pollination (P): The assessment was based on the likely presence of animals capable of encouraging and/or conducting pollination, such as bees and butterflies. Sites that score high for this variable are associated with semi-natural green spaces such as parks, small woods, and fields, which act as a habitat to such animals. P usually scores low in Greater Manchester and most other cities.

Biological control (BC): The assessment was based on the potential presence of predatory animals capable of regulating pests and diseases in the surrounding area. Sites that score high for this variable are associated with large parks and fields. The picture is rather mixed within the study area.

Freshwater (FW): The quantity and quality of surface runoff that a site is expected to receive was assessed. There is a great potential for FW to score high in terms of quantity across the study area. However, the water quality will be a function of the permeable pavement design and the presence of mature and dense vegetation such as trees.

Recreation, and mental and physical health (RMPH): The probable ability of a site to provide an area where people can interact with others through a wide range of activities including sport was considered. Large areas with cafes and space to play sports will score highest for this variable. There is an underutilized potential for RMPH in Greater Manchester, mainly due to cultural reasons.

Tourism and area value (TAV): This assessment was based on whether or not the attributes of a site are substantially attractive enough for people to come and visit the area from nearby neighborhoods. A park that is considered spectacular to Greater Manchester may attract a relatively high number of local visitors, whereas a site such as the Manchester City Stadium will attract visitors from around the world. An attractive site is likely to result in higher house and land prices. There is a considerably underutilized potential for TAV to score high in Greater Manchester, mainly due to the presence of a few large parks suffering from underinvestment.

Aesthetic and educational appreciation, and inspiration for culture, art, and design (AEAICAD): This evaluation is founded on the magnitude of a potential SuDS site in terms of its appeal to a high number of people of diverse backgrounds by creating areas where individuals can come, reflect and find inspiration for a range of things. There is a considerably underutilized potential for AEAICAD to score high in Greater Manchester, mainly due to public underinvestment.

28.3.4 Strengths of the New Ecosystem Services Assessment Approach

The new ecosystem service variables adapted for combined permeable pavement and tree systems are partly based on the categories previously published by TEEB (2011). The strengths of the new approach, particularly in comparison to the community and environment methodology adopted by Ellis et al. (2003) and CIRIA (2004), include a novel, innovative, and generic approach truly based on universal ecosystem service variables and not on ecological engineering understanding. Furthermore, each variable has been clearly defined by the assessment team, thus leading to a quick and numerical (rather than a qualitative) assessment. An inexpensive, user-friendly, and easy-to-understand evaluation is enabled. A comprehensive data set of the key variables characterizing a site hosted within a widely available nonexpert database is available. Finally, the overall ecosystem service potential of a site is expressed through an individual value.

Moore and Hunt (2012) proposed an alternative system based on the following ecosystem service variables for existing (not proposed, as with the new approach) constructed wetlands and ponds used as SuDS: hydraulic, water quality, greenhouse gas regulation, air quality, climate, food, raw material, recreation, education, aesthetics, and biodiversity. In comparison, the ecosystem service variables selected in this chapter go beyond those proposed by Moore and Hunt (2012), who did not specifically consider HC (although referred to under biodiversity), MEE (although partly addressed under hydraulic), EPMSF, P, BC, FW, MR, TAV, AEAICAD (however, education is an independent variable), and SESP.

28.3.5 Limitations of the New Ecosystem Services Assessment Approach

Spatial assessments of landscape functions including urban ecosystem services by experts are an attractive challenge (Willemen et al., 2008). There are some potential weaknesses of the ecosystem services assessment approach. Subjectivity and aggregation are generic limitations of an expert-based system, which can be addressed by involving expert groups and determination of uncertainty values for all estimations (Munoz-Pedreros, 2004; Scholz and Yang, 2010; Danso-Amoako et al., 2012). Furthermore, some ecosystem service variables are not always applicable in the UK, because the proposed system has been designed to be universal and generic. There is also a strong perceived (falsely; see below) bias toward natural sites and "soft" SuDS, in contrast to urban sites and "hard" SuDS such as traditional permeable pavements. Finally, there is a possibility of multicollinearity among variables (McMinn et al., 2010).

Some of the above limitations, such as subjectivity, are also inherent in the traditional assessment approach (CIRIA, 2004, 2007). However, multicollinearity might be a more relevant challenge with the proposed ecosystem services approach due to the use of more variables. Multicollinearity is a statistical phenomenon in which at least two predictor variables in a multiple regression model correlate highly with each other. Multicollinearity does not reduce the predictive power of a regression model. However, the model may not give correct results regarding any individual predictor or indicate which predictors are redundant. Considering that any tests for multicollinearity are case study-dependent, the inevitable bias associated with a case study does not allow for objective testing unless the number of case studies is very high and there is an adequate geographical spread to reduce bias. Nevertheless, a principal component analysis was carried out to identify redundant variables in order to reduce the risk of multicollinearity (McMinn et al., 2010). However, all ecosystem service variables (Table 28.1.2.1) were considered to be necessary for the proposed expert system.

28.3.6 Comparison of Assessment Methods

Gill et al. (2007), McMinn et al. (2010), and Lundy and Wade (2011) recognize the need for a holistic assessment of green infrastructure, including urban watercourses, in the UK. Moreover, Lundy and Wade (2011) propose the integration of all relevant sciences to sustain ecosystem services. Assessment methods have evolved from previously representing only the views of a few stakeholder groups such as planners and civil engineers to representing as many views as are voiced nowadays. The transition from the traditional (particularly CIRIA, 2004) to the proposed novel ecosystem services assessment approach for SuDS is a good example.

This section compares the assessment approaches discussed above with each other. The data obtained for the suitability of sites for the retrofitting of SuDS differ greatly, depending on the approach used to carry out the assessment. All visited sites were considered suitable for the retrofitting of SuDS when the traditional assessment based on "community and environment" (CIRIA, 2004) variables was carried out. This differs greatly from the assessment performed using the new ecosystem services variables, where nearly half the sites visited are valued as having a relatively low ecosystem services potential, making them of limited use for retrofitting of permeable pavement systems.

This can be explained by the fact that most ecosystem service variables do relate well to the natural environment, such as biologically diverse parks (41%), and not to the built environment, like impermeable car parks (33%). This relationship reduces the number of sites suitable for retrofitting of SuDS, as car parks only perform well with respect to three ecosystem service variables. The presence of public parks did not pull up the overall suitability of retrofitting sites because the parks were usually small in size (30% of sites were $<25{,}000\ m^2$) and low in tree coverage (7%) and the presence of surface water (no streams; river, 11%; canal, 21%; and standing water, 8%) of the associated catchment was limited.

The permeable pavement option competed against the following alternative SuDS techniques in terms of preference for all case study sites: filter strip, swale, green roof, pond, constructed wetland, infiltration trench, soak-away, infiltration basin, below-ground storage, and water playground. Considering that this chapter focuses on permeable pavement systems in combination with existing trees, the results for the other SuDS techniques are not shown to maintain focus.

The SuDS preferences were determined in a two-step process. Initially, the total scores for the SuDS technique assessed for every site were calculated for the three approaches. Considering that the maximum possible scores for each approach are different, a direct comparison of the SuDS technique popularity between the approaches is not meaningful. Therefore, the relative proportions for the SuDS technique have been expressed in percentage points to allow for a direct comparison between approaches. This procedure allows for the easy comparison between the three assessment methods applied to all selected sites in Greater Manchester (Figure 28.1.3.1). Furthermore, a full rank-order correlation analysis was not considered to carry great weight, because uncertainties associated with less preferred SuDS options were considerably higher than those associated with the top three choices.

Permeable pavements were favored using the assessment based on the traditional and combined approaches. The new ecosystem services and the traditional assessment approaches have the lowest and highest intersite variability, respectively. The relatively high variability associated with the traditional approach cannot be explained by factors relating to specific planning policies for Greater Manchester (White and Alarcon, 2009).

Table 28.3.6.1 shows a comparison of the three assessment approaches in terms of relative scores for all sites, indicating the evaluation differences between methods. Some sites were assessed as not (or less) suitable according to one (or two) method(s) (e.g., site 24 received 0 points for both the traditional and new approaches) and (highly) suitable according to another method (e.g., site 24 received 16 points for the combined approach). A site associated with a first preference could be replaced by another one with a different preference without variation of the total score. It follows that two methods could appear to be giving the same results while actually they assess the sites differently.

28.3.7 Vegetation Suitable for Urban Areas

Table 28.3.7.1 shows an overview of key identified trees and their suitability for urban permeable pavement sites in Greater Manchester and other cities with similar temperate and oceanic climates. The suitability has been determined based on expert judgment and a literature review. However, opinions on individual characteristics were sometimes diverse and controversial. The most generically suitable trees were *Acer platanoides* (Norway Maple), *Acer pseudoplatanus* (Sycamore), *Alnus glutinosa* (Common Alder), *Betula pendula* (Silver Birch), *Cupressus* × *leylandii* (Leyland Cypress or Leylandii), *Robinia pseudoacacia* (Black Locust or False Acacia), *Platanus* × *acerifolia* (London Plane), *Quercus palustris* (Pin Oak), and *Tilia* × *europaea* (Common Lime).

TABLE 28.3.6.1 Comparison of Assessment Approaches in Terms of Relative Scores for All Sites (after Uzomah and Scholz (2013))

Site	CE	ES	C	Site	CE	ES	C
1	14	5	14	30	8	3	14
2	17	5	16	31	12	5	0
3	15	5	15	32	0	0	15
4	17	5	16	33	16	5	0
5	0	0	0	34	0	0	15
6	0	0	0	35	15	5	16
7	0	0	0	36	4	2	5
8	17	5	16	37	15	5	16
9	0	0	0	38	17	5	15
10	0	0	0	39	16	5	16
11	0	0	0	40	17	5	13
12	0	0	0	41	14	5	16
13	0	0	8	42	17	5	15
14	9	3	7	43	15	5	15
15	14	5	16	44	10	3	10
16	17	5	14	45	15	5	17
17	14	5	7	46	17	5	15
18	0	0	8	47	16	5	16
19	16	5	11	48	17	5	13
20	0	0	7	49	14	5	17
21	15	5	0	50	17	5	0
22	0	0	0	51	0	0	9
23	0	0	0	52	17	5	0
24	0	0	16	53	0	0	0
25	17	5	0	54	0	0	15
26	0	0	7	55	16	5	16
27	14	5	16	56	17	5	0
28	17	5	0	57	0	0	0
29	0	0	0	58	0	0	16

TABLE 28.3.6.1 Comparison of Assessment Approaches in Terms of Relative Scores for All Sites (after Uzomah and Scholz (2013))—cont'd

Site	CE	ES	C	Site	CE	ES	C
59	17	5	16	80	17	5	16
60	17	5	13	81	0	0	0
61	14	5	16	82	0	0	0
62	17	5	14	83	17	5	16
63	15	5	7	84	17	5	16
64	0	0	0	85	0	0	0
65	0	0	0	86	0	0	0
66	0	0	0	87	0	0	0
67	16	5	15	88	17	5	16
68	0	0	0	89	0	0	0
69	16	5	15	90	0	0	0
70	0	0	0	91	16	5	15
71	16	5	15	92	17	5	16
72	0	0	0	93	16	5	15
73	0	0	0	94	0	0	0
74	0	0	0	95	0	0	0
75	0	0	0	96	0	0	0
76	0	0	0	97	0	0	0
77	17	5	16	98	0	0	0
78	17	5	16	99	17	5	16
79	14	5	13	100	17	5	16

CE, community and environment approach; ES, ecosystem service approach; C, combined approach.

In contrast, the most generically unsuitable trees for streets and permeable pavements were *Aesculus hippocastanum* (Horse Chestnut), *Populus* spp. (Poplar), *Salix babylonica* (Weeping Willow), *Tilia platyphyllos* (Large-leafed Lime), and *Ulmus procera* (English Elm).

At least 66% of the trees determined are suitable for permeable pavements. The trees with the highest proportions are *A. pseudoplatanus* (34%), *T.* × *europaea* (26%), *B. pendula* (4%), *A. glutinosa* (1%), and *C.* × *leylandii*

TABLE 28.3.7.1 Overview of the Potential Suitability of Identified Trees with a Proportion of More Than 1% for Permeable Pavement Sites in Greater Manchester (after Uzomah and Scholz (2013))

Species	Common English Name	Suitability for Urban Environment and Particularly for Permeable Pavements	Soil Type	Site Characteristics	Features	Planted under Wires?	Minimum Strip Width (m)	Mature Crown Spread (m)	Tree Height (m)	Reference[a]
Acer platanoides	Norway maple	Suitable for suburban developments, recreation areas, small parks, and streets, roads, and pavements	Recommended for wet soils, clay soils, chalk soils, and dry sandy soils; unsuitable for industrial soils	Strongly recommended for smoke and fume sites; recommended for seaside and exposed sites	Strongly recommended for flowers and leaves	No	2.0	12	12–25	2–4
Betula pendula	Silver birch	Suitable for parks, roadsides, streets, and pavements	Strongly recommended for dry sandy soils and industrial soils; recommended for clay soils; unsuitable for wet soils and chalk soils	Strongly recommended for exposed sites; recommended for seaside, smoke, and fume sites, small spaces	Not recommended for bark	No	1.5	10	18–25	2–4
Crataegus monogyna	Common hawthorn	Suitable for urban fringe woodlands and large parks,	Strongly recommended for industrial soils;	Strongly recommended for seaside and exposed sites;	Not recommended for flowers and fruits	No	1.5	8	12–15	1, 2, and 4

		and only sometimes for roadsides, streets, and pavements	recommended for wet soils, clay soils, and chalk soils; unsuitable for dry sandy soils	recommended for smoke and fumes sites						
Fagus sylvatica	Common beech	Suitable for urban fringe woodlands, large parks, and roadsides	Strongly recommended for chalk soils and dry sandy soils; unsuitable for industrial soils	Recommended for seaside; strongly recommended for exposed sites; unsuitable for smoke and fume sites	Not recommended for leaves	No	1.8	12	15–49	2–4
Fraxinus excelsior	Common ash	Suitable for urban fringe woodlands, large parks, and roadsides	Strongly recommended for chalk soils; recommended for clay soils; not recommended for wet soils	Strongly recommended for exposed sites and smoke and fume sites; recommended for seaside	Nothing particular	No	1.8	9–15	15–46	2 and 3
Ilex spp.	Holly	Some streets and roads	Strongly recommended for clay, chalk, and dry sandy soils; recommended for wet soils	Strongly recommended for seaside, exposed sites, and smoke and fume sites	Not recommended for fruits	No	1.5	2–9	8–15	2 and 4
Populus spp.	Poplar	Not to be planted near buildings	Strongly recommended for wet and clay soils;	Recommended for seaside, exposed sites,	Nothing particular	No	2.0	3–7	15–20	3–5

Continued

TABLE 28.3.7.1 Overview of the Potential Suitability of Identified Trees with a Proportion of More Than 1% for Permeable Pavement Sites in Greater Manchester (after Uzomah and Scholz (2013))—cont'd

Species	Common English Name	Suitability for Urban Environment and Particularly for Permeable Pavements	Soil Type	Site Characteristics	Features	Planted under Wires?	Minimum Strip Width (m)	Mature Crown Spread (m)	Tree Height (m)	Reference[a]
			recommended for chalk soils; unsuitable for dry sandy soils	and smoke and fume sites						
Prunus spp.	Cherry	Suitable for suburban developments, recreation areas, small parks, roads, streets, and pavements	Strongly recommended for clay soils; recommended for chalk soils and dry sandy soils	Recommended for seaside, smoke and fume sites; unsuitable for exposed sites	Recommended for flowers; not recommended for fruits	Often yes	1.5–2.5	2–7	6–8	2–4
Tilia × europaea	Common lime	Parks, avenues, streets, and pavements	Recommended for wet and clay soils; not recommended for industrial soils	Recommended for seaside, exposed sites, and smoke and fume sites	Not recommended for leaves	No	2.0	7–17	15–50	1, 3, and 4

[a]*Reference examples: 1, Essex County Council (2012); 2, Hibberd (1989); 3, Woodland Trust (2012); 4, Seattle Department of Transportation (2012).*

(1%). In comparison, only about 4% of the identified trees are unsuitable for retrofitting of permeable pavements. The highest proportions were associated with *Populus* spp. (2%), *A. hippocastanum* (1%), and *S. babylonica* (1%). Approximately 30% of trees identified are likely to have a neutral impact on permeable pavement retrofitting.

The minimum strip width for planting trees to protect tree roots and nearby building structures is specified in Table 28.3.7.1 (column 8). If practitioners follow the guidelines in Table 28.3.7.1, permeable pavements and trees are less likely to have a negative impact on each other.

Some readers may get the impression that the case study is somehow unrepresentative due to the strong presence of a few species on a small number of sites. However, *A. pseudoplatanus* was only dominant (i.e., highest individual tree counts) on five sites. Furthermore, *T.* × *europaea* dominated just two sites.

Tree diversity was lacking in Greater Manchester, because only two species accounted for 60% of all trees specified. However, it is not unusual to have large numbers of popular species planted in cities. For example, in Hong Kong, the top 10 dominant species are 56% of the population. *Aleurites moluccana* constitutes 13% of the tree population (Jim, 1996). In Chicago, the 10 most common species account for 46% of the urban trees (Nowak et al., 2010). In Nordic cities, 30–90% of all trees planted belong to a single species (Sæbø et al., 2003). In other European cities, only three to five genera usually accounted for 50–70% of all street trees planted (Pauleit et al., 2002).

28.4 CONCLUSIONS AND RECOMMENDATIONS

A unique and rapid decision support methodology based on novel ecosystem service variables for retrofitting of combined permeable pavements and tree systems in densely populated areas was developed. Generic ecosystem service variables suitable for combined permeable pavement and tree systems were determined for a representative example case study area and were broadly categorized under the four established categories of supporting, regulating, provisioning, and cultural services. Due to the high density of built-up areas, the overall ecosystem service scores were relatively low. However, the right choice of trees is likely to enhance the uptake of permeable pavements in the future.

The suitability of sites for combined permeable pavement and tree retrofitting was assessed based on traditional "community and environment" variables and the new ecosystem service variables. The proposed new technique received relatively high scores for the traditional but not for the new approach. A combination of the traditional and new approach provides findings that are similar to the traditional approach.

A simple methodology to determine trees located close to potential permeable pavement sites was developed. Retrofitting of combined permeable pavement and tree systems is generally possible for a high number of sites within a densely built-up area such as Greater Manchester. However, the case study area was dominated by only a few tree species (Sycamore and Common Lime). This finding is similar to previous results for other conurbations. Unfortunately, a few tree species are unsuitable for the successful operation of permeable pavement systems due to factors such as unsuitable fruits (Horse Chestnut) and destructive root systems (some Poplar species and Weeping Willow).

Spatial modeling of ecosystem services is an attractive challenge. More research is recommended to further develop the ecosystem service assessment approach into a numerical model. Additional urban but also rural case studies with a large number of sites could be assessed to test the robustness of the new approach and to subsequently refine it. The fact that newly planted trees are ignored by the proposed method, because they have only a minor positive effect over many years, has been recognized as a challenge for the correct application of ecosystem service variables.

A weighting system for the variables as a function of any SuDS technique (permeable pavements in this case study) could easily be introduced to reduce the impact of what may be perceived as less relevant ecosystem service variables. However, this would introduce extra bias, considering that, for example, an engineer would have a different weighting system (possibly biased toward structural integrity of the pavement) than an ecologist (perhaps greater emphasis on habitat and biodiversity), sociologist (more oriented to assessing social and cultural services), or geographer (greater awareness of spatial physical variability). It follows that a truly trans-disciplinary collaboration is required to build a balanced assessment method for supporting multifunctional SuDS retrofitting and design aimed at improving the whole life quality of urban systems.

Chapter 29

Expert Tool Based on Ecosystem Variables for Retrofitting of Wetland Systems

29.1 INTRODUCTION

In times of recession and spending cuts in the public sector, rapid and inexpensive computer-based expert assessment systems supporting urban and landscape planning usually undergo a revival in the context of ecosystem services (Dobbs et al., 2011; Scholz et al., 2012). Estimating rather than measuring complex ecosystem service variables reduces the overall cost and length of a project considerably. Euliss et al. (2011) show the successful integration of estimated ecosystem service variables within models used for decision support processes.

The application of ecosystem service variables to a new area such as sustainable drainage is a novel contribution to knowledge and understanding. The applied nature of such expert systems should have a strong appeal, particularly to environmental managers and urban and landscape planners.

With an increasing public awareness of the importance of ecosystems and the services they provide for humans (Butler and Davies, 2004; Scholz, 2010), the aim of Chapter 29 is to develop a rapid decision support tool for a range of professions based on estimated ecosystem service variables for retrofitting of sustainable drainage systems (SuDS) in densely populated areas. The key objectives to achieve this aim are:

- To broadly categorize identified generic ecosystem service variables relevant for SuDS retrofitting under the four established categories of supporting, regulating, provisioning, and cultural;
- To compare the suitability of potential SuDS sites within Greater Manchester based on the traditional "community and environment" variables, new ecosystem service variables, and a combination of the traditional and new approach for sites within Greater Manchester;

Wetlands for Water Pollution Control. http://dx.doi.org/10.1016/B978-0-444-63607-2.00029-0

- To evaluate the variability of estimated example variables and the learning process of estimation by nonexperts; and
- To develop a decision support tool for SuDS retrofitting taking into account the perspectives of drainage engineers, developers, ecologists, planners, and social scientists.

Traditional drainage frequently leads to flooding and pollution challenges in the lower catchment (Scholz, 2006, 2010). The implementation of SuDS, also known as best management practices (more recently called stormwater control measures in the United States), and water-sensitive urban design features (Australia) can help to alleviate these problems. The main objective of SuDS is to mimic the natural drainage into the ground, as closely as possible, prior to its development (CIRIA, 2007). Most SuDS techniques are able to do this in a number of different ways, such as attenuation of runoff before entering the watercourse, storage of water in natural contours, infiltration of partially treated runoff into the ground, and evapotranspiration of water by vegetation (CIRIA, 2010; Scholz, 2006, 2010).

Sustainable drainage techniques should reduce the impact of urbanization on the quantity and quality of surface runoff, while increasing amenity and biodiversity opportunities at the same time. Some of the techniques control surface runoff through infiltration, detention, attenuation, conveyance, and water harvesting (CIRIA, 2007; Scholz et al., 2006, 2012). In general, they make use of physical, chemical, and/or biodegradation processes to improve the quality of surface runoff by minimizing the amount of stormwater-based pollutants washed into nearby watercourses (Eriksson et al., 2007; Scholz, 2010). However, improvement opportunities with respect to ecosystem services, including amenity and biodiversity, by introducing SuDS are frequently ignored by engineers (Scholz, 2010).

Costanza et al. (1997) introduced the concept of ecosystem services, associated values, and corresponding categories. The ecosystem approach is a strategy for the integrated management of land, water, and living resources that promotes conservation and sustainable use in an equitable way. Increasing human population size, economic growth, and global consumption patterns place pressure on environmental systems. It follows that the provisioning of ecosystem goods and services is affected (Seppelt et al., 2011).

The concept of ecosystem services stayed much the same until de Groot et al. (2002) published a framework diagram and a table in an attempt to distinguish between ecosystem functions, processes, goods, and services. Brown et al. (2007) subsequently defined ecosystem services as the results of ecosystem processes that either directly sustain or enhance human life or maintain the quality of ecosystem goods.

The benefits that the public may obtain from the urban environment can also be considered as ecosystem services (Millennium Ecosystem Assessment, 2005; Busch et al., 2012; Moore and Hunt, 2012). Defra (2011) defines ecosystem services as the benefits individuals gain from the goods and services produced by nature and its natural systems. The natural resources and functioning natural systems such as fertile soils, clean water (Walsh et al., 2012) and air, and a regulated climate are essential for human well-being, security, and economic development (Millennium Ecosystem Assessment, 2005).

A number of official documents such as the Natural Environment White Paper (Defra, 2011), the UK National Ecosystem Assessment (2011), and The Economics of Ecosystems and Biodiversity (TEEB) (2011) have identified four broad categories of ecosystem services generally referred to as supporting, regulating, provisioning, and cultural. All existing ecosystem services are strongly linked to one another and to other types of ecosystem services.

Tzoulas et al. (2007) provided a detailed literature review on ecosystem services in the urban environment. A list of ecosystem service variables is provided in Table 28.1.2.1. The listed ecosystem services have been reinterpreted to make them relevant to SuDS retrofitted in urban areas and are categorized in broad agreement with TEEB (2011) and other guidance documents, such as Moore and Hunt (2012). TEEB (2011) proposed a comprehensive list of ecosystem service variables of generic nature, while Moore and Hunt (2012) chose a smaller set of variables particularly adapted for constructed wetlands and ponds.

Ecosystem service assessment is dynamic considering that the built environment constantly changes (Eigenbrod et al., 2011) and the scientific knowledge of associated processes develops further. For example, surface permeability and green roof runoff estimates may be different in the future. It follows that SuDS recommendations will change over time.

An example case study to test the generic tools discussed in Chapter 29 has been chosen. Greater Manchester, a subregion in the northwest of England, is made up of 10 local authorities. The six authorities of relevance for this study (in order of decreasing importance: Manchester, Salford, Trafford, Bury, Oldham, and Tameside) have been highlighted in Figure 28.1.3.1. The whole of Greater Manchester covers a total surface area of 1300 km^2 and is home to approximately 2.5 million people. Salford and Manchester form the core of the conurbation and are the most densely built-up areas in Greater Manchester. The remaining eight local authorities form an urban fringe around Salford and Manchester and are considerably less urbanized (White and Alarcon, 2009).

29.2 METHODOLOGY

29.2.1 Site Assessment

A total of 100 sites that were large enough for the retrofitting of SuDS to have a positive urban drainage impact (i.e., improved drainage, water quality, and enhanced biodiversity and recreation) were identified by assessing Ordnance Survey and Google maps of Greater Manchester. Local authorities, United Utilities (water authority), and major private landowners were consulted regarding suitable SuDS sites. A map of Greater Manchester highlighting all sites visited was created using the computer software GNU Image Manipulation Program (Figure 28.1.3.1). The purpose on focusing the study on Greater Manchester was to demonstrate that the implementation of SuDS even within densely built-up cities is possible.

The site assessment template was based on the frameworks developed by Scholz (2006) and Scholz et al. (2006) for retrofitting of SuDS techniques in Glasgow, Edinburgh, and elsewhere as well as the Construction Industry Research and Information Association (CIRIA) guidelines (CIRIA, 2004, 2007). Each potential SuDS site was assessed during a site visit by a group of experts (2–5 team members) to reduce subjectivity (Munoz-Pedreros, 2004), supported by a desk study.

General site information such as site number and name, postcode, grid reference numbers, location name, names of the inspection team members, site acceptability for SuDS, and presence of existing SuDS was recorded. Photos of the key site features were taken for each potential SuDS site and its catchment. Land ownership information such as number of owners, ownership type (private or public), and estimated site value (£) was noted.

Proportions (%) of site classification categories including those designated for development, regeneration (e.g., former industrial sites and run-down neighborhoods), and recreation were estimated. The team also recorded surrounding area characteristics such as descriptions of the neighborhood to the north, south, east, and west, current and future site use, measured total area of the catchment (m^2), and catchment shape. Location description and measured distance (m) to the nearest sewer, storm pipe, stream, river, canal, pond, lake, and sea were recorded, if located within a reasonable distance within or at the border of the catchment.

Current and future surface permeability (%) for the land categories grass, trees, shrubs, and impermeable of the proposed SuDS site and its catchment were estimated. Proportions (%) of current and future roof runoff for the categories institutional, commercial, industrial, high-density housing, medium-density housing, low-density housing, and other were determined. The team also estimated proportions (%) of current and future road runoff for the categories car park, motorway, primary road (or dual carriageway), A road, B road, tertiary road, and other.

For each subcatchment, area (m^2) and gradient in the two main directions having an angle of 90° to each other in the horizontal plain were determined. Hydrogeological information such as contaminated land (present or absent), soil infiltration (low, medium, or high), and groundwater level (below or above 2 m depth) was assembled by the team. Additional remarks regarding current drainage techniques and potential problems regarding the implementation of future SuDS techniques were made if and when appropriate. The information collected with the standard site assessment template supports the assessment team in determining the variables required for the traditional and ecosystem services approaches.

29.2.2 Assessment of Ecosystem Service Variables

New quantitative and qualitative approaches of assessing ecosystem services have been assessed by Busch et al. (2012). In addition to the standard variables, Table 28.1.2.1 shows an overview of the proposed 17 new ecosystem service variables that were also determined for the 100 potential SuDS sites (Figure 29.2.2.1). These variables belong to the four established ecosystem service categories (Costanza et al., 1997) of supporting, regulating, provisioning, and cultural.

Each ecosystem service variable is described qualitatively and quantitatively in Table 28.1.2.1 with the help of five bins (category groupings). Bin 1 always describes the characteristics of sites that are considered to be of lowest ecosystem service value, while bin 5 constantly identifies characteristics of sites that are considered to be of greatest ecosystem service potential. Table 28.1.2.1 can be adapted by the reader for his or her own case studies of concern. More quantitative guidance can be introduced to cater for any specific situation.

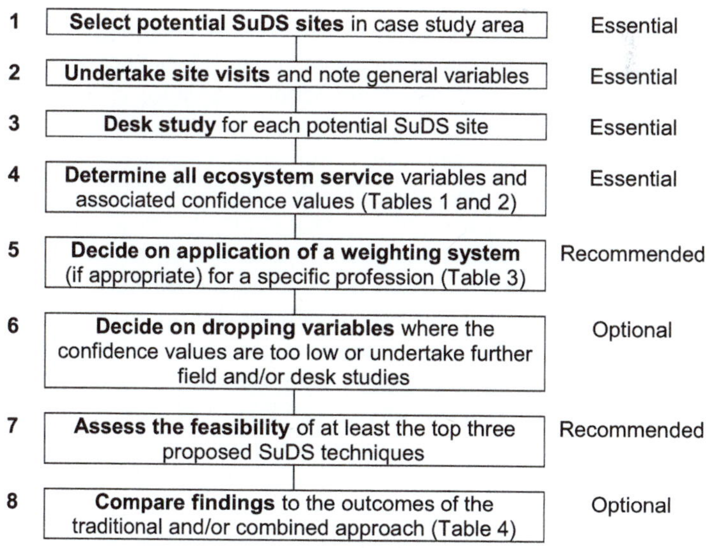

FIGURE 29.2.2.1 Overview of the essential, recommended, and optional steps of the new ecosystem services assessment approach for retrofitting of sustainable drainage systems (SuDS) in urban areas (after Uzomah et al., 2014).

29.2.3 SuDS Variable Determination Approaches

The traditional approach follows guideline C609 (CIRIA, 2004), which bases the selection of a SuDS technique on a number of criteria related to hydrology, land use, physical site features, community and environment, and economics and maintenance. The criteria have been adapted from a technical report previously authored by Ellis et al. (2003). Scores for each criterion range from 1 to 5, where 1 refers to a SuDS technique being very unsuitable, and 5 signifies a SuDS technique being very suitable for that particular criterion. The scores are then tallied up, and the SuDS technique obtaining the highest score is considered to be the practice best suited to the case study site. The minimum and maximum overall score for all criteria were 0 and 25, respectively.

For the purpose of Chapter 29, use is only made of the criteria falling under the category "community and environment," which comprises the conventional variables of safety, pond premium, aesthetics, wildlife habitat, and acceptance (CIRIA, 2004). Eliminating the other categories that do not directly relate to ecosystem services introduces bias in the overall assessment, but this was seen as acceptable for the purpose of this chapter, which focused only on categories of interest to ecosystem services. Each SuDS technique was assessed for each suitable SuDS site according to the following criteria definitions: safety (probable danger to people and pets), water premium (potential rise of the cost of property within the area), aesthetics (likely improvement of the visual aesthetics of that area), wildlife habitat (likely ecological impact of wildlife within the wider SuDS area), and acceptance (expected adequacy of proposal to the local community).

In comparison, the new ecosystem services methodology described in Table 28.1.2.1 was used to determine numerical values (i.e., scores) using expert judgment for the new ecosystem service variables (Figure 29.2.2.1). A comparison between a site assessment based on the community and environment variables with the new ecosystem service variables has been performed. Considering that the latter method comprises more variables (17 instead of 5) and higher corresponding maximum values per variable (100 instead of 5) than the former, not absolute total scores but adjusted scores (i.e., values for established approach times an adjustment factor of 68) were compared with each other to avoid bias toward the new approach, which would otherwise score too highly. Furthermore, weighting factors (Table 29.2.3.1) that correspond to user preference based on professional background may be applied to adjust each ecosystem service variable.

The third and last assessment method is a combination of the traditional and new approach by replacing the traditional criteria "aesthetic" and "wildlife habitat" with the four ecosystem service categories (not the 17 ecosystem service variables). These two traditional criteria are included in the new ecosystem service assessment. Thus the SuDS techniques are scored according

TABLE 29.2.3.1 Proposed Weights as a Function of User Preference Based on Professional Background

Category of Service	Variable	Weights				
		Drainage Engineer	Developer	Ecologist	Planner	Social Scientist
Traditional Assessment Approach (CIRIA, 2004)						
	Safety	3	3	1	3	3
	Pond premium	1	3	3	2	2
	Aesthetics	1	3	1	3	2
	Wildlife habitat	1	1	3	2	2
	Acceptance	3	3	1	3	3
New Ecosystem Services Approach						
Supporting	1. Habitats for species (HS)	1	1	3	2	2
	2. Maintenance of genetic diversity (MGD)	1	1	3	1	1
Regulating	3. Local climate and air quality regulation (LCAR)	1	1	3	2	3
	4. Carbon sequestration and storage (CSS)	1	1	3	1	1
	5. Moderation of extreme events (MEE)	3	3	2	3	2
	6. Storm runoff treatment (SRT)	3	2	2	2	2
	7. Erosion prevention and soil fertility (EPSF)	2	2	2	2	2
	8. Pollination (P)	1	1	3	1	1
	9. Biological control (BC)	1	1	3	2	2

Continued

TABLE 29.2.3.1 Proposed Weights as a Function of User Preference Based on Professional Background—cont'd

Category of Service	Variable	Weights				
		Drainage Engineer	Developer	Ecologist	Planner	Social Scientist
Provisioning	10. Food (F)	1	1	1	1	2
	11. Raw materials (RM)	1	1	1	1	2
	12. Freshwater (FW)	3	1	2	2	2
	13. Medicinal resources (MR)	1	1	1	1	2
Cultural	14. Recreation, and mental and physical health (RMPH)	2	2	1	2	3
	15. Tourism and area value (TAV)	1	3	1	2	3
	16. Aesthetics, education, culture, and art (AECA)	1	2	1	2	3
	17. Spiritual experience and sense of place (SESP)	1	2	1	2	3
Combined Approach						
	Safety	3	3	1	3	3
	Pond premium	1	3	3	2	2
	Acceptance	3	3	1	3	3
	Supporting services	1	1	3	2	2
	Regulating services	3	2	3	2	2
	Provisioning services	1	1	1	1	2
	Cultural services	1	2	1	2	3

to the following criteria: safety, water premium, and acceptance as well as provisioning, regulating, supporting, and cultural services. The same scoring system applied above was adopted during the reassessment of evaluating which SuDS technique is likely to be most appropriate for a particular site. Only those techniques that were most suitable (i.e., first preferences) for particular sites were subsequently recommended to the local authorities, United Utilities, and private landowners for implementation.

29.2.4 Uncertainties of the Rapidly Estimated Variables

A relative measure of certainty (also known as a confidence value), expressed in percentage, was given by the assessors to each variable determined with the help of Table 28.1.2.1, indicating the reliability of their subjective assessment; the higher the value given, the more certain was the group of assessors (Figure 29.2.2.1; step 4). Only values greater than 50% were acceptable and inconsistencies were removed.

29.2.5 Variability of Estimated Variables and the Learning Process of Estimation

The learning process of estimation, undertaken by a relevant student (nonexpert) cohort example, was assessed with the help of a three-stage questionnaire survey based on a PowerPoint presentation. The variability associated with their assessments, as learning progressed, was also monitored.

Six pictures for each variable, representing virtually the whole numerical spectrum—i.e., very low (close to 0) to very high (close to 100) values—of possible answers, were selected for the questionnaire. The pictures were taken from actual case study sites and did not contain any misleading or irrelevant information, such as distracting objects of random occurrence in the foreground.

Fifty-one typical full-time undergraduate and taught postgraduate students, who were broadly familiar with the overall case study area and were studying for the module water resources technology in their third year at The University of Salford, were asked to assign values to each picture associated with a particular variable. About 70% of the students were white males born around 1990.

The questionnaire was divided into three different parts to test progressive learning. The same pictures had to be assessed at each stage. However, the order of pictures shown was changed at random. About 15 seconds were allocated for each picture. At stage 1, students had to assign values that they had to benchmark against their personal perception. Subjects had to make reasonable assumptions about what is a low or high value for a particular variable. At stage 2, students were aware of the range of possible scenarios for each variable and had the opportunity to refine their first choices purely based on their memory. In the final stage, all pictures associated with a particular variable were shown at the same time, allowing for direct picture comparisons and value readjustments. Each mean score per picture was compared to a target score (guidance value), which was determined by the authors based on professional drainage engineering perception.

29.2.6 Decision Support for Various Professions

The proposed methodology can be adjusted to take into account perspectives of drainage engineers, developers, ecologists, planners, and social scientists, as defined by Blockley (2005). A weighting system (Table 29.2.3.1) specific to the needs of a particular stakeholder group (Figure 29.2.2.1) was introduced by providing weights for individual variables after consultation with different teams of academics representing different professions (drainage engineers, developers, ecologists, planners, and social scientists) within the authors' university. Note that academics rather than consultants and contractors were consulted, considering that they most likely have a greater and better overview of their subject discipline.

Table 29.2.3.1 indicates weights from the viewpoint of drainage engineers, developers, ecologists, planners, and social scientists. A simple weighting system (1, low; 2, normal; 3, high) has been proposed.

29.3 RESULTS AND DISCUSSION

29.3.1 Site Assessment Outcomes and Associated Strengths and Weaknesses

Only 16% of the example case study sites were existing SuDS sites. Most of these were of poor traditional design with low corresponding ecosystem service value. There was a clear need for introducing SuDS to improve drainage in all cases, which is realistic, because most sites are owned by the public hand (52% public, 23% private and public, and 25% private) and not multiple private owners. The estimated site value was relatively low (26% low, 32% low to medium, 29% medium, 11% medium to high, and 2% high), which supports SuDS implementation due to low competition with private investors.

Most sites were not just potential retrofitting sites but a combination of different potential types: combinations of at least two categories (81%), retrofitting only (15%), development only (2%), regeneration only (2%), and recreation only (0%). This was reflected in the current site use, which was dominated by public or at least publicly accessible areas: park (41%), car park (33%), institutional building (12%), disused land (8%), and field (6%). After discussions with landowners and planners, the authors assume that this scenario is unlikely to change considerably in the foreseeable future, say the year 2030: park (56%), car park (32%), institutional building (12%), disused land (0%), and field (0%). The trend toward converting "empty spaces" into parks is a positive development. The catchment size (in 1000 m^2) of most sites was relatively small (<25, 29%; 25 ≤ 50, 32%; 50 ≤ 100, 20%; 100 ≤ 150, 6%; and ≥150, 13%), indicating that SuDS may only make a minor contribution toward resolving the urban drainage problem.

The vast majority of potential SuDS sites (63%) had only sewers, while the remaining sites had both sewers and storm pipes (37%). Additional catchment drainage options were dominated by sewers: sewer (100%), storm pipe (2%), stream (0%), river (11%), canal (21%), pond (6%), and lake (2%). The low proportion of natural and flowing receiving watercourses makes the introduction of large-scale infiltration techniques attractive. However, infiltration trenches and soak-aways did not score high on ecosystem service variables compared with the traditional community and environment variables.

The current SuDS site permeability (%) was high (<20, 24%; 20 ≤ 40, 2%; 40 ≤ 60, 1%; 60 ≤ 80, 3%; and ≥80, 70%). In contrast, the current catchment permeability (%) was low (<20, 21%; 20 ≤ 40, 13%; 40 ≤ 60, 17%; 60 ≤ 80, 30%; and ≥80, 19%), indicating a good potential for infiltration devices, which, however, scored low on ecosystem service variables (see above).

The present runoff area proportions were as follows: roads, 48%; mainly roofs and roads, 11%; and mainly roofs, 41%. Despite the high proportion of roofs, the slopes (%) of current sites were rather flat (<0.5, 20%; 0.5 ≤ 1.5, 53%; 1.5 ≤ 2.5, 10%; 2.5 ≤ 3.5, 7%; and ≥3.5, 10%). Nevertheless, leveling of at least some parts of the site is required in 57% of all cases to allow for adequate flow by gravity. This finding indicates that linear SuDS structures and SuDS treatment trains (combinations of at least two SuDS techniques with each other) are not suitable for most sites. Furthermore, the majority of all SuDS case study sites (83%) were composed only of one catchment. Only 13% and 5% of all sites had two and more than two subcatchments, respectively. This finding indirectly supports the statement that too many sites were rather too flat (see above).

Only one site was seriously contaminated, and the suitability for some unlined SuDS techniques was therefore limited. The soil infiltration rates were predominantly of medium magnitude (96%). Only 3% of the sites had low rates, which limited the use of some infiltration-based SuDS techniques. Even fewer sites (1%) had high infiltration rates. The groundwater level was almost always relatively high.

The traditional and new approach of estimating variables used for assessing SuDS retrofitting was based on a further development of a rapid site assessment adapted from Scholz (2006) and Scholz et al. (2006). The strengths of this site screening tool are as follows:

- Quick, simple, inexpensive, user-friendly, and easy-to-understand assessment of site data and physical site features;

- Comprehensive data set of the key variables hosted within a widely available nonexpert database such as Microsoft Excel; and
- Acceptability of site does not discriminate against current site use.

However, the following weaknesses of the methodology may apply:

- Too-simplified approach leaves wide room for interpretation and personal bias; and
- The assessment is not dynamic by nature, ignoring various possible future scenarios.

These two negative points also apply, however, to the site assessment required by the traditional approach proposed by CIRIA (2004).

29.3.2 Strengths and Limitations of the New Ecosystem Services Assessment Approach

The new ecosystem service variables adapted for SuDS are partly based on the categories previously published by TEEB (2011). The strengths of the new approach, particularly in comparison to the community and environment methodology adopted by Ellis et al. (2003) and CIRIA (2004), are as follows:

- Novel, innovative, and generic retrofitting approach truly based on universal ecosystem service variables and not on conventional engineering understanding;
- The tool recognizes that various professions have different priority variables and therefore suggests different weights;
- Each bin in Table 28.1.2.1 is clearly defined, thus leading to a quick and numerical (rather than a qualitative) assessment for each site within a case study area by the same team of experts;
- The assessment is inexpensive, user-friendly, and easy to understand;
- Comprehensive data set of the key variables characterizing a site hosted within a widely available nonexpert database; and
- The ecosystem service potential of a site is expressed through a single numerical value.

Moore and Hunt (2012) proposed an alternative system based on the following ecosystem service variables for existing (not proposed as part of the new approach) constructed wetlands and ponds used as SuDS: hydraulic, water quality, greenhouse gas regulation, air quality, climate, food, raw material, recreation, education, aesthetics, and biodiversity. In comparison, the ecosystem service variables selected in this chapter go beyond those proposed by Moore and Hunt (2012), who did not specifically consider HS (although referred to under biodiversity), MEE (although partly addressed under hydraulic), EPSF, P, BC, FW, MR, TAV, AECA (however, education is an independent variable), and SESP.

The potential weaknesses of the ecosystem service assessment approach are:

- Subjectivity and aggregation are the generic limitations of any expert system, which has been addressed by involving expert groups and introducing uncertainty values for all estimations (Munoz-Pedreros, 2004; Scholz and Yang, 2010; Danso-Amoako et al., 2012). An uncertainty value (often expressed in %) is the number assigned by an assessor describing his or her confidence in making an assessment;
- Some ecosystem service variables may not always be applicable in the case study area because the proposed system has been designed to be universal and generic;
- Strong falsely perceived bias toward natural sites and "soft" (e.g., ponds and wetlands) SuDS in contrast to urban sites and "hard" (e.g., permeable pavements and below-ground storage systems) SuDS; and
- Multicollinearity among some variables (McMinn et al., 2010).

Some of the above limitations are also inherent in the traditional assessment approach (CIRIA, 2004, 2007). However, multicollinearity might be a more relevant problem with the proposed multivariable-based ecosystem services approach. Considering that any tests for multicollinearity are case study-dependent, the inevitable bias associated with a case study does not allow for objective testing. Nevertheless, a principal component analysis was carried out to identify redundant variables to reduce the risk of multicollinearity (McMinn et al., 2010). Results indicate that all ecosystem service variables (Table 28.1.2.1) were justified.

29.3.3 Comparison of Assessment Methods

Gill et al. (2007), McMinn et al. (2010), and Lundy and Wade (2011) recognize the need for a holistic assessment of green infrastructure, including urban watercourses, in the UK. Moreover, Lundy and Wade (2011) propose the integration of all relevant sciences to sustain ecosystem services. Assessment methods have evolved from previously representing only the views of a few stakeholder groups, such as planners and civil engineers, to representing as many views as are voiced nowadays. The transition from the traditional particularly CIRIA (2004) to the proposed novel ecosystem services assessment approach for SuDS is a good example.

The data obtained for the suitability of sites for the retrofitting of SuDS differ greatly, depending on the approach used to carry out the assessment. All visited sites were considered suitable for the retrofitting of SuDS when the traditional assessment based on "community and environment" (CIRIA 2004) variables was carried out. This differs greatly from the assessment performed using the new ecosystem service variables, where nearly half of the sites visited are valued as having a relatively low ecosystem service potential, making them of limited use for retrofitting of most SuDS techniques. This finding can be used to prioritize sites for SuDS retrofitting.

This difference can be explained by the fact that most ecosystem service variables do relate well to the natural environment, such as biologically diverse parks (41%), and not to the built environment, like impermeable car parks (33%). This relationship reduces the number of sites suitable for retrofitting of most SuDS, as car parks usually only perform well with respect to three ecosystem service variables (MEE, SRT, and FW; Table 28.1.2.1). The presence of public parks did not pull up the overall suitability of retrofitting sites because the parks were usually small in size (30% of sites were <25,000 m^2) and low in tree coverage (7%) and the presence of surface water (no streams; river, 11%; canal, 21%; and standing water, 8%) of the associated catchment was limited. However, the introduction of a weighting system (Table 29.2.3.1) that puts bias toward what could be seen by a drainage engineer as more important variables for SuDS (e.g., flood control as part of MEE and water quality control considered by SRT) could increase the suitability of sites for retrofitting.

A comparison of all assessment approaches in terms of proposed SuDS techniques for Greater Manchester was made. The preferences were determined in a two-step process. Initially, the total scores for each SuDS technique assessed for every site were calculated for the three approaches. Considering that the maximum possible scores for each approach are different, a direct comparison of individual SuDS technique popularity between the approaches is not meaningful. Therefore, the relative proportions for each SuDS technique have been expressed in percentage points to allow for a direct comparison between approaches and preferences. Note that there were many occasions where more than one SuDS technique had the same order of preference.

The proposed procedure allows for easy comparison between the three methods applied to all selected sites in Greater Manchester (Figure 28.1.3.1). Furthermore, a full rank-order correlation analysis was not considered to carry great weight because uncertainties associated with less preferred SuDS options were considerably higher than those associated with the top three choices.

The new ecosystem services and the traditional assessment approaches usually have the lowest and highest intersite variability, respectively. The relatively high variability for most variables such as ponds and constructed wetlands cannot be explained by factors relating to specific planning policies for Greater Manchester (White and Alarcon, 2009). Ponds are associated with the greatest intersite variability for all three approaches because of their potentially relatively small size and great popularity (Scholz, 2004c, 2010; Scholz et al., 2006), particularly with the traditional approach. However, it would not be right to assume that the Greater Manchester case study findings necessarily apply to other areas as well.

A comparison with the traditional approach of determining community and environment variables indicates that infiltration trenches, soak-aways, and below-ground storage systems are usually less preferred than permeable pavement systems (Tota-Maharaj and Scholz, 2012), regardless of the professional perspective for the first preferences of the ecosystem services approach. The first preferences of the combined approach for the SuDS techniques permeable pavement, filter strip, and constructed wetland are more similar numerically to the traditional than to the new ecosystem services approach. The opposite is usually the case for ponds, infiltration trenches, soak-aways, and below-ground storage solutions.

Permeable pavements scored relatively high on ecosystem service variables, which contradicts the common belief that there has to be a strong bias toward natural and soft techniques when using ecosystem service assessment techniques (Butler and Davies, 2004; Scholz, 2010; Tota-Maharaj and Scholz, 2012). Nevertheless, permeable pavements usually attract high values for variables such as SRT and MEE. Nevertheless, these specific findings relate to the Greater Manchester case study area and therefore might not apply to other areas.

29.3.4 Certainty Estimations Based on Drainage Engineering Expert Opinion

The estimation of certainties has to be performed consistently. The confidence-related data reported in this chapter were based on expert drainage engineering judgment only. Human judgment may vary considerably (O'Connor, 1989). In comparison, a group's level of judgment usually outperforms that of an average individual (Schultze et al., 2012).

Knowledge used by engineers is not entirely of scientific nature. Understanding is built up over time as a result of continuously improving judgment (Ferguson, 1992; Holt, 1997). Therefore, the drainage engineering judgment and associated confidence values expressed in this study were regularly reviewed to increase consistency and to teach team members.

29.3.5 Variability of Estimated Variables and the Associated Learning Process

The variables aesthetics and land costs were determined relatively well. Absolute differences between target scores and mean scores for the third estimation stage were between 2% and 20% and between 0% and 26% per picture, respectively. In comparison, habitats for species (differences between 9% and 78%) and safety (differences between 7% and 30%) were associated with higher but still acceptable estimated errors. However, the cohort had serious difficulties in estimating land sizes (difference up to 447%). Nevertheless, this is not considered to be a serious problem, because land size can be measured accurately using maps in practice.

Learning only improved clearly for land size estimation between the three stages. The authors expected to identify a clear reduction in variability (indicated by the standard deviation) as learning progressed, which was, however, not the case. Both unexpected observations were promising for the proposed decision support tool relying partly on estimations because it shows that even complex variables such as aesthetics and land cost can be estimated well simply by intuition and very limited experience. This should result in similar site assessments across the region.

29.3.6 Decision Support Tool Addressing Different Professional Perspectives

Various professions may have preferences for different variables. Therefore, the new tool gives any user the opportunity to select a weighting system of greatest relevance to his or her profession. The introduction of corresponding bias can be avoided by not selecting any weighting system (Figure 29.2.2.1).

Strong variations in preference were recorded. For the ecosystem services and combined approach, the bias toward permeable pavements and ponds is apparent, regardless of subject discipline. In comparison, the traditional approach strongly favors soak-aways, with the exception of the ecological perspective. However, below-ground storage systems also score high for most subject disciplines.

29.4 CONCLUSIONS AND RECOMMENDATIONS

A rapid computer-based assessment methodology for retrofitting of SuDS was successfully introduced to reduce the currently high level of subjectivity in practice. Retrofitting of SuDS is possible for a high number of sites within a densely built-up area such as Greater Manchester. Generic ecosystem service variables suitable for SuDS were determined and were broadly categorized under the four established categories of supporting, regulating, provisioning, and cultural. Due to the high density of built-up areas, the overall ecosystem service scores were relatively low.

The suitability of sites for SuDS retrofitting was assessed based on traditional "community and environment" variables and the new ecosystem service variables. The old approach favors infiltration trenches, soak-aways, and below-ground storage systems. In comparison, the new approach promotes permeable pavement systems regardless of the professional perspective. All sites were suitable for the retrofitting of SuDS when the traditional assessment based on "community and environment" variables was carried out. In comparison, the ecosystem services approach shows that nearly half of the sites visited are valued as having a relatively low ecosystem service potential, making them of limited use for retrofitting of most SuDS techniques. This finding can be used to prioritize sites for SuDS retrofitting, which is particularly important during difficult financial times.

Findings show that estimations based on expert judgment were rather similar for a virtually untrained example cohort of engineering students. The introduction of a weighting system as a function of different professional bias leads to the preferred selection of some SuDS techniques by several professions.

More research is recommended to develop the ecosystem services assessment approach further. Additional urban but also rural case studies with a larger number of sites could be assessed to test the robustness of the new approach and to subsequently refine it. The tool should also be tested at a relatively small scale, such as in a neighborhood or housing development in different cities, towns, and villages. This would allow for the assessment of variability as a function of scale and local peculiarities for individual SuDS techniques.

Larger cohorts of experts and nonexperts could be used to refine the estimation methodology and weighting systems. Specific weighting systems for the ecosystem service variables as a function of different climatic regions and cultures could be introduced.

Chapter 30

Sustainable Drainage System Model

30.1 INTRODUCTION

30.1.1 Sustainable Drainage Systems

Prevention of flooding in urban areas caused by inadequate drainage systems has become a significant problem. With an increase in development of greenfields and urban regeneration, potential damage due to prolonged flooding could easily lead to substantial financial losses (Schmitt et al., 2004). The concept of source control for the treatment of stormwater runoff from impermeable surfaces has become widely accepted among drainage engineers in both North America and Europe (Ellis et al., 2004; Villarreal et al., 2004).

Over the past 20 years, the use of best management practices (BMP) in the United States and sustainable drainage systems (SuDS) in the UK has been instrumental in reducing both the detrimental impact of polluted runoff to the water quality of receiving watercourses and flooding caused by increased urbanization and traditional stormwater drainage systems. Both BMP and SuDS attempt to mimic the drainage patterns of the natural watershed and can also provide substantial treatment needed to improve the quality of the water discharged to an acceptable standard (Jefferies et al., 1999).

Wetlands generally comprise a basin with shallow water and aquatic vegetation, which may provide biofiltration. Constructed wetlands remove pollutants through sedimentation, plant uptake, microbial decomposition, and filtration. Stormwater wetlands are specifically constructed to treat surface water. They may feature a forebay for suspended solids (SS) capture and can be designed to allow extended detention during low flows. There is a high proportion of emergent vegetation in relation to open water despite possible benching or stepping of the sides to promote varying water depths. They are not normally designed to provide significant attenuation, but temporary storage may be provided above the permanent water level.

Ponds increase the duration of the flow hydrograph with a consequent reduction in peak flow. Also known as stormwater ponds, retention ponds, and wet extended detention ponds, they can be used for attenuation, detention,

Wetlands for Water Pollution Control. http://dx.doi.org/10.1016/B978-0-444-63607-2.00030-7

retention, storage, infiltration, and recreational purposes. Wet ponds have a capacity greater than the permanent pond volume, which permits storage of the influent (i.e., stormwater runoff). An overflow outlet to an existing watercourse or sewer is provided for excess flow. These are basins that have a permanent pool of water in the base for treating incoming stormwater runoff. The primary pollutant removal mechanisms are the settling out of solids and biological degradation activity in the pond. Typically, a residence time of one day gives sufficient treatment, as a retention period of one day is believed to improve the water quality.

Whenever possible, wet ponds should be designed to provide enhanced amenity and wildlife benefits. Performance and maintenance can also be improved by installing a fore bay to trap sediments and a wetland around the fringe of the pond perimeter. A catchment of 2 ha is sufficient to maintain the wet pool, which suggests that ponds can be used on most sites. The only restriction is the space required, which limits the use of ponds in constricted city-center or other urban sites. A liner is required to maintain the wet pool, if the soils below the pond are highly permeable. Alternatively, fully lined ponds can be used if infiltration is not desired.

An infiltration pond takes runoff from a development and allows it to infiltrate into the ground. The pond has sufficient storage volume so that the runoff may be allowed to empty from the device into the ground over a specific period of time (usually a period of one day to half empty) to provide storage for runoff from any following storms. Infiltration systems reduce the volume of water that has to be disposed of through sewers and provide recharge of groundwater that may maintain water levels in local watercourses. In this manner, they are similar to dry unlined ponds.

An infiltration basin is a dry depression structure designed to promote the infiltration of surface water runoff through the surface of the ground, with an emergency spillway (no standard outlet structure exists) to an existing watercourse or sewer. It can be landscaped to give added aesthetic and amenity value. A drawback of infiltration basins is the space required for infiltration, which limits their application in dense urban developments. Infiltration basins should be restricted to a maximum catchment area of 4 ha to reduce the risk and consequences of premature clogging. The removal of pollutants occurs on the infiltration basin's base and also in the underlying soils, where pollutants are filtered and absorbed onto soil particles in the unsaturated zone. In comparison, an infiltration pond is smaller, may permanently contain surface water, and is planted.

Standard swales are generally grassed stormwater conveyance channels that use biofiltration and limited infiltration to remove pollutants. When compared to a conventional ditch, a swale channel is broad and shallow and can provide temporary storage for stormwater and reduce peak flows. Swales can form a network within a development scheme, linking SuDS features and/or conveying water (i.e., predominantly runoff) to a watercourse or sewer. Typically, they are located next to highways, but they can also be constructed in landscaped areas within car parks and elsewhere. It is possible to replace conventional piped drainage with swales and subsequently remove the need for kerbs and gullies.

Shallow swales can be used if the groundwater table is high and/or there are potential contamination issues with respect to deeper soil layers. However, groundwater should be at least 1 m below the base, and swales should be lined, if contamination is detected and leaching of contaminants is possible. Lined swales should also be used if infiltration is thought to cause ground instability, foundation problems, or even flooding of below-ground building structures.

Swales can be designed for infiltration and pollutant removal but predominantly transport water from a site. Pollutants can be removed for frequent small storm events, but for less frequent storms (return periods of 1 in 2 or 1 in 10 years), swales act as a storage and conveyance mechanism. It may be difficult to incorporate swales in steeply sloping sites, where it is difficult to retard flow rates. Swales require a relatively slow flow of water over them to remove pollutants and prevent erosion. The speed of flow is dependent on the soil type but should not exceed 2 m/s for any situation.

Green roofs can reduce the volume and rate of rainfall runoff by using vegetation placed on top of a building to retain water. Below this vegetative surface, there may be soil or substrate, drainage protection, waterproofing, and insulation layers.

The extensive green roof covers the entire roof area with low-growing and low-maintenance plants. They are accessible only for maintenance. Extensive green roofs typically comprise a 25- to 125-mm-thick soil layer in which a variety of hardy, drought-tolerant, and low plants are grown.

Intensive green roofs are landscaped environments for recreation that include plants or small trees and are usually publicly accessible. They may also include water features and storage of rainwater for irrigation. Intensive roofs generally impose far greater loads on the roof structure and require significant on-going maintenance. They may be used over podium decks to underground car parks.

Infiltration trenches are shallow excavated trenches that have been backfilled with stone to create a below-ground reservoir. Stormwater runoff that is diverted into the trench gradually infiltrates into the surrounding soil and possibly to the water table. A perforated pipe can be incorporated in the base or a number of perforated pipes at designated depths to promote infiltration.

It is important that infiltration trenches are protected from sediment loads during and after construction until the surrounding runoff area has developed mature ground cover to minimize erosion and sediment transport. The performance of infiltration trenches can be improved and clogging of the voids between the permeable materials reduced by providing pretreatment in the form of grass filter strips. This will filter particles out of runoff before reaching the infiltration trench. However, such an arrangement may not be practical if infiltration trenches are located at the sides of highways.

Soak-aways are subsurface structures into which surface water is conveyed. Factors for soak-away success are the size of the area to be drained, the infiltration rate of the soil, and the depth to the water table. If the water table is high (i.e., near the surface), a soak-away is not practical. The site of a soak-away must also be located at least 5 m away from any building, and it must be designed so that it will not saturate the foundations of any structure.

Soak-aways take the form of prefabricated sections and are often laid on a simple ring foundation or a strip footing of concrete that supports the weight of the chamber section. A cover is essential to allow access for cleaning and inspection. A soak-away should not require any maintenance, but it is good practice to check the structure every few months for silting or contamination. If significant silt build-up occurs on a regular basis, it may be worthwhile to construct a "catch pit" to intercept silt from the inlet. Any build-up of silt at the base of the soak-away can be removed manually during dry conditions when the soak-away is empty.

Filter strips are vegetated strips of land that act as buffers by accepting storm runoff from adjacent impermeable areas. The runoff is designed to sheet flow across the filter strip sufficiently slowly to filter out sediment and associated pollutants. The main restriction to the use of filter strips is the large space that they require. They are most suited to small car parks, roads, and similar areas but are not generally suitable for use in dense urban developments. If sheet flow changes to concentrated flow, as is possible over impervious surfaces, then the velocity is too great for filtration to be effective; thus filter strips should be used to treat only very small drainage areas. Filter strips have treatment mechanisms similar to those of swales, but runoff should be evenly distributed across the surface of these structures.

Grassed filter strips behave in a similar way to grassed channels, and many of the design methods and criteria for swales can also be applied to filter strips. They can be used in most ground conditions, but must be constructed to provide an even and consistent slope with no depressions that will cause localized ponding or promote flow in channels.

Pervious surfaces are pavement constructions that allow rainwater to infiltrate through the surface and into the underlying construction layers. Water can be stored prior to infiltration to the ground, reused, or released to a surface watercourse or other drainage system. Pervious surfaces can be either porous or permeable.

Permeable pavements can be used to reduce runoff rates and volumes from urban areas. They attenuate the runoff while preserving the value of the area for urban development. The surface infiltration rate should be greater than the design rainfall intensity and include any runoff from adjacent impermeable areas. The storage volume can be calculated based on the volume and porosity of the underlying storage layer. The storage volume should take between one and two days to empty.

A below-ground storage structure should be designed to accumulate surface water runoff and to release it after a lag period to prolong the runoff hydrograph and to decrease the peak flow. The structure may contain aggregates (or plastic detention cells) and also act as a water recycler or infiltration device. Subsurface storage means that predevelopment runoff levels can be sustained without constructing ponds that use valuable space and create potential health and safety hazards.

Below-ground stormwater storage means that above-ground space can be put to alternative use. Depending on the site, the space saved could be used for parking lots, more area for the construction of buildings, or enhanced landscaping. This SuDS option is particularly useful for dense city centers or retail areas, where the space for other SuDS options is limited. Furthermore, below-ground storage systems can be successfully combined with pervious or permeable surfaces to provide a SuDS treatment train. Various permeable pavement solutions are on the market.

A water playground may be designed as a small raised pond-like feature integrating play facilities for children and acting as an amenity feature to increase public acceptance with little impact to surface water runoff attenuation. Water playgrounds are environments in which people interact with water and each other. Children can participate by turning valves, pulling levers, and controlling the direction and flow of water in a park or other recreational setting.

Water playgrounds are ideal additions to city and residential parks and are also well suited to commercial and retail installations. Properly designed and maintained water playgrounds are as safe as commercial swimming pools and should only be installed when the water quality and cleanliness is of a suitable standard to be handled by humans.

30.1.2 Sustainable Drainage Impact on Water Quantity and Quality

Rapid urbanization and its consequent increase in impermeable surface areas and changes in land use generally result in problems of flooding and heavy pollution of urban streams and other receiving watercourses. This is frequently coupled with groundwater depletion and a threat to natural water resources (Andoh and Declerck, 1997).

The traditional drainage system is no longer adequate in many areas, especially in high-density housing estates. The subsequent effects can be life threatening and may lead to damage to buildings and other infrastructure assets and disrupt the functioning of business and society (Mark et al., 2004; Schmitt et al., 2004). Through SuDS techniques, water is either infiltrated or conveyed more slowly to watercourses or sewer systems via ponds, swales, filter strips, or other sustainable installations, which could reduce the peak flow during heavy storms (Pettersson, 1997).

Stormwater in open systems forms the basis for recreation and the development of ecosystems with a diverse animal and plant life. In many ways, the principles of alternative stormwater management adopt natural processes taking place in the environment and adapt them to urban conditions and requirements. At the same time, the effects of pollution are taken into account (Astebøla et al., 2004).

Butler and Parkinson (1997) mentioned that SuDS should maintain a good public health barrier and avoid local or distant pollution of the environment. Nonpoint sources of pollution are difficult to identify and control and are one of the main reasons that urban rivers fail to reach the water quality objectives set for them, while SuDS techniques are available to help combat diffuse pollution (Mitchell, 2005). For instance, the runoff from roofs and streets contributes between 50% and 80% of heavy metals to the total mass flow in domestic sewage (Butler and Parkinson, 1997).

30.1.3 Development and Regeneration in Glasgow

Glasgow is associated with many areas for which the need for SuDS is becoming apparent. In 2003, a new City Plan was adopted, outlining a large number of urban areas, known as New Neighborhoods, in which redevelopment should occur. The main urban drainage concern for Scottish Water (water utility) and Glasgow City Council was the lack of sewer system capacities for additional surface runoff.

With future development and regeneration activities, this situation is likely to become worse, and the need for SuDS implementation becomes vital to Glasgow's continuing expansion. New guidelines incorporating a decision support tool that identifies SuDS techniques that are feasible for a particular site over a range of boundary and environmental conditions are required.

Furthermore, a previous consultancy project for the East End of Glasgow did not demonstrate how SuDS implementation could lead to a significant reduction of potential combined sewer overflow spills. However, the desk study did not incorporate actual site conditions such as the percentage of built-up areas or the local topography of the landscape, including watercourses and hills.

30.1.4 SuDS in Edinburgh

In 2004, construction projects in Edinburgh's outskirts were dominated by greenfield site development activities. City regeneration projects were predominantly restricted to brownfield sites. Current and proposed building projects within the city boundary were the City Centre, the Granton Waterfront, Edinburgh Park, South Gyle, and Sighthill. These areas were designated to become commercial or high-density residential developments.

Within the city, there are designated recreational areas that can be used to drain runoff after SuDS implementation. This process of utilizing recreational space is a less controversial form of SuDS retrofitting. As development and regeneration activities increase, the amount of impermeable surfaces also increases. It follows that there is a pressing need to use SuDS techniques to control runoff by ground infiltration or storage, which has beneficial impacts on downstream catchments (City of Edinburgh Council, 1999).

Flooding within the city of Edinburgh is mostly caused by rainfall in the upper river catchment areas, which lie outside the boundaries of the city. Recent flooding has occurred predominantly in the following catchments and localities: River Almond and Gogar Burn; Ferry Burn; South Queensferry; Linn Mill Burn; Braid Burn and Figgate Burn; Burdiehouse, Niddrie, and Brunstane Burn (City of Edinburgh Council, 1999).

30.1.5 Aims and Objectives

The main aims are to establish new SuDS decision support tools for planners and to compare the outcomes for Edinburgh and Glasgow with each other. The detailed objectives include the following:

- To identify suitable SuDS sites within Glasgow and Edinburgh;
- To classify qualitatively and quantitatively sites suitable for different SuDS technologies;
- To outline a SuDS decision support model for development, regeneration, and retrofitting sites;
- To publish the model and to evaluate feedback from users;
- To compare "linear" with case-based reasoning (CBR) models to assess the usefulness of the SuDS database for decision making; and
- To identify SuDS technologies for representative construction sites that could be used for demonstration purposes.

30.2 SITES, METHODOLOGY, AND MODELING

30.2.1 Overview of Sites in Glasgow and Edinburgh

Figure 30.2.1.1(a) shows a map of Glasgow highlighting the spatial distribution of 79 sites that were identified as potentially suitable for the implementation of SuDS. In comparison, Figure 30.2.1.1(b) shows a map of Edinburgh highlighting 103 sites that were identified for the same reason. All sites were classified into four groups (development, regeneration, retrofitting, and recreation) used for planning purposes (Table 30.2.1.1).

30.2.2 Decision Support Matrix and Weighting System

The SuDS decision support matrix comprises dominant and supplementary criteria, which aim to identify the level of SuDS implementation feasibility.

(a)

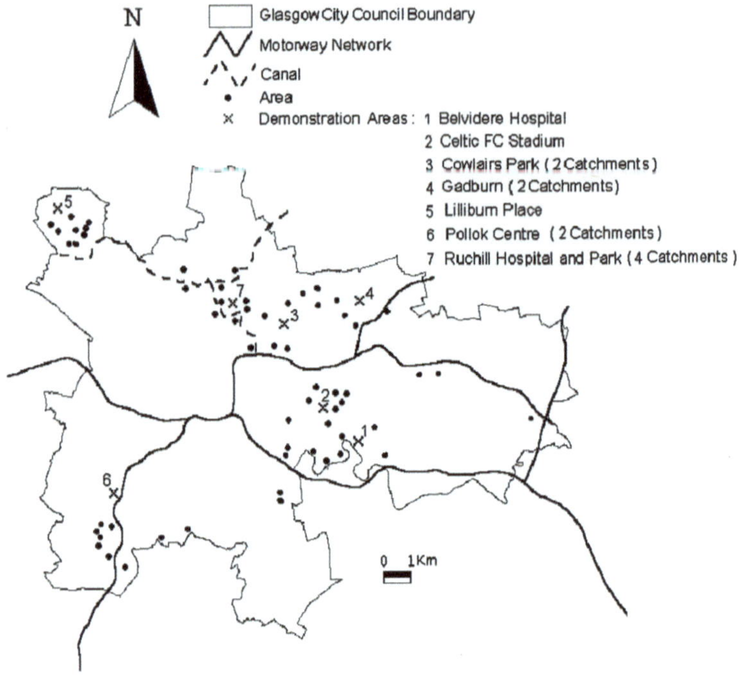

(b)

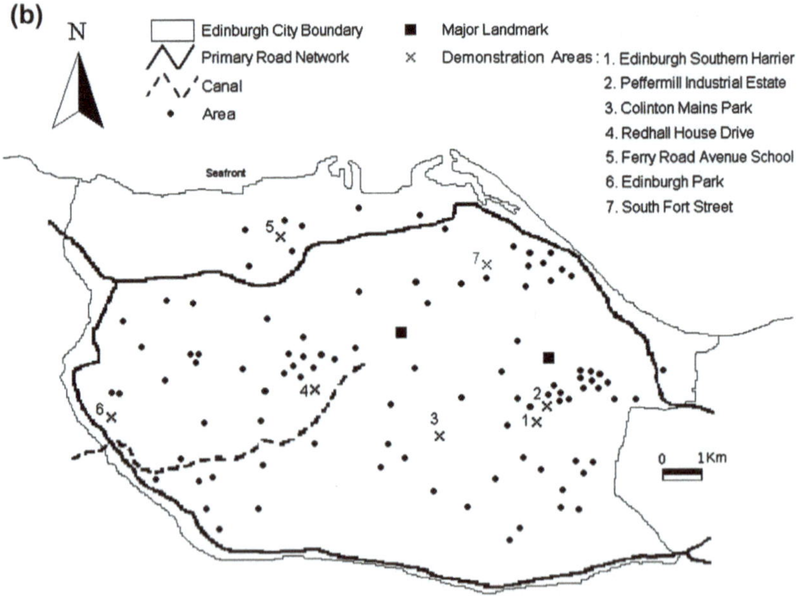

FIGURE 30.2.1.1 Proposed sustainable drainage systems for (a) Glasgow and (b) Edinburgh.

TABLE 30.2.1.1 Database Summary for Glasgow (79 Sites) and Edinburgh (103 Sites) Representing the Current Situation but Taking into Account Already Planned Construction Work

Contamination	Yes: 9/0	No: 91/100			
Possible SuDS area (%)	≤20: 1/8	20 < x ≤ 60: 23/4	60 < x ≤ 80: 23/1	>80: 53/87	
Land values	Low: 19/33	Low-medium: 30/19	Medium: 39/35	Medium-high: 9/9	High: 3/4
Runoff quantity	Low: 27/46	Low-medium: 24/22	Medium: 42/25	Medium-high: 5/6	High: 3/1
Runoff quality	Poor: 5/0	Average: 52/39	Good: 43/61		
Roof runoff (%)	≤50: 63/6	>50: 37/94			
Car park runoff (%)	≤50: 85/88	>50: 15/12			
Road runoff (%)	≤50: 65/88	>50: 35/12			
Roads[a]	Motorways: 9/0	Primary road: 15/12	A road: 27/18	B road: 14/3	Others: 35/67
Drainage type	Sewer only: 60/58	Watercourse: 39/40	Not currently detected: 1/2		
Groundwater	High: 22/15	Low: 62/68	Currently not determined: 16/17		
Soil infiltration	High: 62/68	Low: 22/15	Currently not determined: 16/17		
Impermeable surface (%)	≤20: 3/53	20 < x ≤ 40: 52/14	40 < x ≤ 60: 16/7	60 < x ≤ 80: 19/8	>80: 10/18

Continued

TABLE 30.2.1.1 Database Summary for Glasgow (79 Sites) and Edinburgh (103 Sites) Representing the Current Situation but Taking into Account Already Planned Construction Work—cont'd

Catchment size (m^2)	≤50,000: 30/46	50,000 < x ≤ 100,000: 28/30	100,000 < x ≤ 200,000: 32/13	x > 200,000: 10/11	
Slope (x in 100 m)	≤1: 19/21	1 < x ≤ 5: 49/53	5< x ≤10: 16/19	x > 10: 16/7	
Ownership	Council: 62/39	Private: 19/35	Council and private: 19/26		
Ecological impact	Yes: 24/26	No: 76/74			
Acceptance warning	Green flag: 82/81	Orange flag: 18/17	Red flag: 0/2		
Site classification	Development: 35/21	Regeneration: 35/38	Retrofitting: 11/17	Retrofit with parks: 19/24	

Note: Proportions are expressed in %.
[a] *Higher classified roads take precedence over any lower classified roads, and score the associated bin entry.*

Dominant criteria (Table 30.2.2.1) specify the technical conditions that must be satisfied for corresponding SuDS implementation.

Supplementary criteria (Table 30.2.2.1) should be satisfied but play a less crucial role in identifying the most suitable SuDS. Nevertheless, they could finally decide on the most appropriate SuDS technique, because their relative importance is a function of the weighting system outlined in Table 30.2.2.2. Supplementary criteria were weighted according to their relative importance for each SuDS technique. The corresponding variables for each SuDS technique were weighted between 0 (unimportant or not applicable) and 3 (very important) according to their relative importance to each other and with regard to the constraints identified for each SuDS technique.

30.2.3 Decision Support Model

The optimization model for SuDS implementation in Glasgow and Edinburgh utilized the site classification data to suggest a suitable SuDS solution either as an individual SuDS technique or as a combination of two techniques in the form of a SuDS treatment train. The calculations are performed by a Microsoft Excel steady-state spreadsheet model. The relative feasibility of the 16 selected SuDS techniques (including key combinations) is specified for each selected SuDS demonstration site in terms of a matrix and associated weighting system.

All site classification data for the 182 sites were inputted into the SuDS decision support tool database, and the following operations were carried out in Microsoft Excel:

- All raw site information data are transferred into the input data spreadsheet, which corresponds to the SuDS decision support matrix.
- If all the dominant criteria for a particular SuDS technique are satisfied by the site data, the final SuDS technique is determined by a weighting system, which is based on supplementary criteria.
- Otherwise, the model would directly indicate that the SuDS technique (or a combination of two SuDS techniques) is nonapplicable, which means that it should not be used for the corresponding site because of unfavorable site conditions.

All supplementary SuDS variables relevant for the corresponding site are checked to obtain cumulative sums of weightings for each SuDS technique. If a site meets all requirements for a particular SuDS technique, the highest possible cumulative sum will be obtained. The actual cumulative sum can be compared with the highest possible cumulative sum to indicate the level of suitability of a particular SuDS technique for the corresponding site. The higher the sum, the more suitable the SuDS technique is likely to be for the corresponding site.

TABLE 30.2.2.1 Dominant (d) and Supplementary (s) Sustainable Drainage System Decision Support Criteria

SuDS Technique	Catchment Size (m^2)	Area for SuDS (m^2)	Future Runoff (1–5)	Runoff Quality	Land Contamination
Wetland	>50,000 (s)	>5000 (d)	>2 (d)	Average (d)	No (d)
Pond	>15,000 (s)	>50 (d)	–	Average (d)	No (d)
Lined pond	>15,000 (s)	>50 (d)	–	–	–
Infiltration basin	>15,000 (s)	>50 (d)	<4 (s)	Average (d)	No (d)
Standard swale	–	>200 (s)	<4 (s)	Average (s)	No (d)
Shallow swale	–	>200 (s)	<3 (s)	Average (s)	No (s)
Filter strip	>15,000 (s)	>500 (d)	<4 (s)	Average (d)	No (d)
Soak-away	>3000 (s)	>200 (s)	<3 (s)	Average (s)	No (d)
Infiltration trench	>3000 (s)	>50 (s)	<4 (d)	Average (s)	No (d)
Permeable pavement	–	–	<4 (s)	Average (s)	No (d)
Below-ground storage	>5000 (s)	>50 (d)	<5 (s)	–	–
Water playground	>3000 (s)	>10 (s)	<3 (s)	Good (d)	No (d)
Green roof	–	>20 (s)	<4 (s)	Good (s)	–
Swale + pond	>20,000 (s)	>300 (d)	–	Average (d)	No (d)
Shallow swale + pond	>20,000 (s)	>250 (d)	<4 (s)	Average (d)	No (d)
Infiltration trench + below-ground storage	>8000 (s)	>150 (d)	<5 (s)	Average (s)	No (d)

SuDS Technique	Land Value	Ownership Fragmented	Groundwater Level	Slope (× m in 100 m)	Maximum Slope
Wetland	<4 (d)	No (s)	–	–	15 (s)
Pond	<4 (s)	–	–	–	20 (s)
Lined pond	<4 (s)	–	–	–	20 (s)
Infiltration basin	<4 (s)	–	Low (d)	–	30 (s)
Standard swale	<4 (s)	No (s)	Low (s)	>1 (s)	10 (s)
Shallow swale	<4 (s)	No (s)	–	>1 (s)	15 (s)
Filter strip	<4 (d)	No (d)	Low (d)	>2 (s)	40 (s)
Soak-away	<5 (s)	–	Low (d)	>1 (s)	25 (s)
Infiltration trench	–	No (s)	Low (d)	>1 (s)	15 (s)
Permeable pavement	–	–	–	–	20 (s)
Below-ground storage	–	–	Low (d)	–	15 (s)
Water playground	–	–	–	–	20 (s)
Green roof	–	No (s)	–	–	20 (s)
Swale + pond	<4 (s)	No (s)	Low (s)	>1 (s)	10 (s)
Shallow swale + pond	<4 (s)	No (s)	–	>1 (s)	15 (s)
Infiltration trench + below-ground storage	–	No (s)	Low (s)	>1 (s)	15 (s)

Continued

TABLE 30.2.2.1 Dominant (d) and Supplementary (s) Sustainable Drainage System Decision Support Criteria—cont'd

SuDS Technique	Soil Infiltration	Ecological Impact	Impermeable Area < (%)	Impermeable Area > (%)	Drainage to Watercourse or Sewer
Wetland	–	Yes (s)	40 (s)	–	Yes (d)
Pond	–	–	65 (s)	–	Yes (d)
Lined pond	–	–	65 (s)	–	Yes (d)
Infiltration basin	High (d)	–	60 (s)	–	–
Standard swale	–	–	85 (s)	–	Yes (s)
Shallow swale	–	–	90 (s)	–	–
Filter strip	High (s)	–	50 (s)	–	–
Soak-away	High (d)	–	90 (s)	–	–
Infiltration trench	High (d)	–	–	30 (s)	–
Permeable pavement	–	–	–	30 (s)	–
Below-ground storage	–	–	–	40 (s)	Yes (d)
Water playground	–	–	–	–	–
Green roof	–	–	–	–	Yes (d)
Swale + pond	–	–	55 (s)	–	Yes (d)
Shallow swale + pond	–	–	60 (s)	–	Yes (d)
Infiltration trench + below-ground storage	–	–	–	40 (s)	Yes (d)

TABLE 30.2.2.2 Weightings for the Corresponding Supplementary SuDS Decision Support Criteria

SuDS Technique	Catchment Size (m^2)	Area for SuDS (m^2)	Future Runoff (1–5)	Runoff Quality	Land Contamination
Wetland	2	0	0	0	0
Pond	1	0	0	0	0
Lined pond	1	0	0	0	0
Infiltration basin	2	0	2	0	0
Standard swale	0	1	3	2	0
Shallow swale	0	1	2	2	1
Filter strip	2	0	3	0	0
Soak-away	2	2	3	2	0
Infiltration trench	1	1	0	2	0
Permeable pavement	0	0	2	2	0
Below-ground storage	1	0	2	0	0
Water playground	1	1	2	0	0
Green roof	0	1	3	2	0
Swale + pond	2	0	0	0	0
Shallow swale + pond	2	0	3	0	0
Infiltration trench + below-ground storage	1	0	3	3	0

Continued

TABLE 30.2.2.2 Weightings for the Corresponding Supplementary SuDS Decision Support Criteria—cont'd

SuDS Technique	Land Value	Ownership Fragmented	Groundwater Level	Slope (× m in 100 m)	Maximum Slope
Wetland	0	2	0	0	1
Pond	2	0	0	0	1
Lined pond	2	0	0	0	1
Infiltration basin	2	0	0	0	1
Standard swale	1	3	2	2	2
Shallow swale	1	2	0	2	2
Filter strip	0	0	0	2	1
Soak-away	2	0	0	1	1
Infiltration trench	0	2	0	2	2
Permeable pavement	0	0	0	0	1
Below-ground storage	0	0	0	0	2
Water playground	0	0	0	0	1
Green roof	0	2	0	0	1
Swale + pond	2	2	2	1	1
Shallow swale + pond	2	2	0	1	1
Infiltration trench + below-ground storage	0	2	2	1	2

SuDS Technique	**Soil Infiltration**	**Ecological Impact**	**Impermeable Area < (%)**	**Impermeable Area > (%)**	**Drainage to Watercourse or Sewer**
Pond	0	0	1	0	0
Lined pond	0	0	1	0	0
Infiltration basin	0	0	1	0	0
Standard swale	0	0	1	0	2
Shallow swale	0	0	2	0	0
Filter strip	2	0	1	0	0
Soak-away	0	0	2	0	0
Infiltration trench	0	0	0	1	0
Permeable pavement	0	0	0	1	0
Below-ground storage	0	0	0	1	0
Water playground	0	0	0	0	0
Green roof	0	0	0	0	0
Swale + pond	0	0	1	0	0
Shallow swale + pond	0	0	1	0	0
Infiltration trench + below-ground storage	0	0	0	2	0

The actual sum was divided by the maximum possible sum to get a SuDS suitability index between 0 and 1. The suitability indices for each SuDS technique were classified into four classes suggesting different actions to the decision maker. The class boundaries were determined empirically and are based on the SuDS expert knowledge and understanding of the author, his research team, and senior planners in Glasgow and Edinburgh.

An index below 0.3 would indicate that the SuDS technique should not be used for the corresponding site because of unfavorable conditions (i.e., nonapplicable). An index between 0.3 and 0.75 would indicate satisfactory conditions for the implementation of a particular SuDS technique (i.e., applicable). Conditions are good for the range between 0.75 and 0.95 (i.e., recommended option). If the index is >0.95, SuDS implementation is strongly encouraged because the conditions for implementation are very good (i.e., best option), for instance, no harm to the environment, elegant engineering solution, and very cost effective.

30.2.4 Prevalence Rating Approach for Sustainable Drainage Techniques

The majority of potential construction sites will be suitable for the implementation of at least two different SuDS techniques. For these sites, more than one individual SuDS technique might be possible, or even more than one combination of different SuDS techniques might be suitable. Each site should be associated with a recommended SuDS option, which is achieved by using a scaling method named Prevalence Rating Approach for SuDS Techniques (PRAST).

The PRAST method rates all SuDS options on a scale by their attribute of alleviating runoff volume while being environmentally sympathetic. Table 30.2.4.1 shows that the PRAST scale promotes the most favorable SuDS technique. The PRAST scale comprises civil engineering and sustainability ratings. The civil engineering perception rating is influenced by issues such as water management, flood attenuation, and health and safety, but not by ecological issues, for example. In contrast, environmentalists are also concerned with flood control, but the main emphasis is on using sustainable materials and methods and increasing the beneficial impact of SuDS on the urban ecology. Environmentalists wish to increase or create natural habitats by using SuDS, which are also visually sympathetic to the mature landscape. However, Table 30.2.4.1 is only an example showing the methodology.

There is a simple weighting method for these two rating systems to allow the user to compromise different points of view. For example, if the users are civil engineers, they might prefer to increase the weighting given to the civil engineering rating. Alternatively, more weighting might be given to the sustainability rating by environmentalists or ecologists.

TABLE 30.2.4.1 Example Showing the Methodology for the PRAST Scale

Original PRAST Number	SuDS Techniques	Civil Engineering Rating (Weighing Value = 4/10)	Sustainability Rating (Weighing Value = 6/10)	Final Weighted Score	Final PRAST Position
14	Standard swale + pond	90	97	942	1
15	Shallow swale + pond	90	90	900	2
2	Attenuation/detention pond	65	85	770	3
1	Wetland	40	100	760	4
3	Lined attenuation/ detention pond	65	80	740	5
4	Infiltration basin/pond	70	75	730	6
5	Standard swale	55	80	700	7
6	Shallow swale	60	70	660	8
16	Infiltration trench + below-ground storage	70	60	640	9
13	Green roof	60	50	540	10
10	Permeable pavement	60	35	450	11
7	Filter strip	20	60	440	12
8	Soak-away	70	25	430	13
12	Water playground	15	50	360	14
11	Below-ground storage	65	10	320	15
9	Infiltration trench	50	15	290	16

30.2.5 Case-Based Reasoning Model

Case-based reasoning is a method of problem solving that has arisen out of the field of artificial intelligence (Chapter 26) and aims to recreate the robust problem-solving technique often used by humans within the constraints of a computer program (Aamodt and Plaza, 1994; Arditi and Tokdemir, 1999). When a human encounters a problem, he or she tends to remember similar situations encountered in the past and the methodology in which solutions were found. By recalling these events, it becomes possible to reuse the previous solution(s) to solve the current problem, perhaps adjusting the methodology and outcome slightly to meet the specific requirements of the new task (Aamodt and Plaza, 1994).

Case-based reasoning works very similarly to the human logic of data handling. A data set of past experiences (e.g., allocation of SuDS techniques for sites) that may be useful to solve a particular type of query (e.g., SuDS planning for a new site) is kept in a database. The difficulty in CBR is the design of a system that is capable of recalling past experiences that would provide useful information when a new problem is introduced to the system.

In CBR terminology, the process of finding a solution to a former problem is referred to as a case, which is stored in the system's case base. Each case should be stored within the case base systematically and consistently. The chosen structure is referred to as the case representation (Arditi and Tokdemir, 1999).

When a new problem is introduced to the CBR system, it should be represented in the same format as the stored cases, and then the process of deciding which of the past cases may be of use in finding a solution to this problem can begin. The main assumption underlying a CBR methodology is that similar problems will have similar solutions. It follows that the most useful cases in the case base will be those that are most similar to the problem case (Kaster et al., 2005).

The concept of similarity is fundamental in CBR theory, making inexact matching possible, which is required when previously unseen problems arise. A mechanism is implemented within the system that is capable of recalling past cases that are most closely matched to the problem presented in terms of the variable(s) used to describe the cases. Therefore, the variables used should be carefully chosen so that the solutions recalled will also be relevant to the problem case. Once the most similar cases have been selected, the predicted solution is found using an adaptation or learning process (Aamodt and Plaza, 1994).

Concerning SuDS decision-making issues, CBR could provide a systematic and effective approach to arrive at a more realistic output (if compared to human judgment alone) through establishing a large database, which could include all SuDS installation implementations throughout Scotland or even the whole of Great Britain and the United States in the future. For the Glasgow and Edinburgh SuDS Management Project, 182 sites were available to build up a database of cases, which has a high variety of different sites. A suitable SuDS technique for a new site is suggested based on a comparison between the new site information and previous cases. The recommended SuDS option and the PRAST output can be obtained through both the "linear" and the CBR models.

30.3 RESULTS AND DISCUSSION

30.3.1 Decision Support Model Output

The SuDS decision support model output is based on the raw data of 182 sites (Table 30.2.1.1 and Figure 30.2.1.1). Sites in Edinburgh have more roof runoff in comparison to sites in Glasgow because of the relatively higher housing density. The proportion of properties owned by the Council is higher in Glasgow, if compared to Edinburgh, reflecting the high proportion of the population close to the poverty line. Glasgow has considerably more regeneration sites compared to Edinburgh. Therefore, more SuDS retrofitting is required in Edinburgh due to a lack of affordable open space.

The SuDS variables land value and runoff quantity are influenced greatly by property purchasing costs and available storage volumes, respectively. They are the most influential variables for both cities. Land value is estimated as a relative variable. The high runoff volumes in Glasgow reflect higher rainfall depths and larger catchment areas in comparison to Edinburgh (Table 30.2.1.1).

Table 30.3.1.1 shows the proportions of four SuDS decision support model categories (i.e., nonapplicable, applicable, recommended option, and best option) for different SuDS techniques applied for sites in Glasgow (G) and Edinburgh (E). Ponds (G: 59%; E: 60%), lined ponds (G: 65%; E: 61%), and permeable pavements (G: 57%; E: 71%) obtained high proportions for best SuDS options, considering individual SuDS techniques for both Glasgow and Edinburgh. Moreover, ponds combined with swales (G: 19%; E: 29%) or shallow standard swales (G: 35%; E: 38%) are the most likely choices for SuDS combinations. This output is reasonable considering general technical judgment.

Ponds (or lined ponds) increase the duration of the flow hydrograph with the consequence of a reduction in the peak flow, which is considered to be the most effective SuDS technique to control stormwater quantity and quality (Pettersson, 1997). Permeable pavements require little space and could be applicable in most situations. Standard and shallow swales are generally grassed stormwater conveyance channels that use biofiltration and limited ground infiltration to remove pollutants. Swales can also form a network within a SuDS development scheme, linking SuDS techniques and conveying runoff to a watercourse or sewer.

Comparing the output from Glasgow and Edinburgh with each other, Edinburgh sites are more suitable for below-ground SuDS techniques such as infiltration trenches, permeable pavements, and below-ground storage facilities. Below-ground storage is more suitable for retrofitting of sites in Edinburgh due to a lack of affordable open space. For both cities, the wetlands

TABLE 30.3.1.1 Proportions of Weightings for the Linear SuDS Decision Support Model and CBR Model Applied for Glasgow and Edinburgh (%)

	SuDS Decision Support Model				CBR Model			
Sites (%)	–	x	xx	xxx	–	x	xx	xxx
Glasgow								
Wetland	80	8	3	10	82	8	4	6
Pond	13	9	19	59	10	16	30	43
Lined pond	1	11	23	65	3	16	34	47
Infiltration basin	34	19	11	35	33	15	23	29
Standard swale	8	10	61	22	6	11	48	34
Shallow swale	0	25	53	22	0	25	51	24
Filter strip	54	10	10	25	49	16	19	15
Soak-away	29	10	33	28	28	35	6	30
Infiltration trench	37	10	30	23	35	11	39	14
Permeable pavement	8	13	23	57	5	15	42	38
Below-ground storage	23	14	30	33	22	20	30	28
Water playground	58	10	4	28	61	11	5	23
Green roof	4	14	47	35	6	48	15	30
Swale + pond	13	24	44	19	13	29	34	24
Shallow swale + pond	13	16	35	35	11	19	34	35
Infiltration trench + below-ground storage	9	29	54	8	5	29	57	9
Edinburgh								
Wetland	85	9	4	2	88	7	3	2
Pond	5	11	24	60	2	16	39	44
Lined pond	4	11	24	61	2	15	39	45
Infiltration basin	16	28	9	48	15	19	30	36
Standard swale	1	13	38	49	0	10	37	53
Shallow swale	0	15	45	41	0	12	47	42
Filter strip	32	17	24	27	29	20	36	15
Soak-away	16	8	35	42	15	23	17	46
Infiltration trench	21	5	39	35	19	4	51	25

TABLE 30.3.1.1 Proportions of Weightings for the Linear SuDS Decision Support Model and CBR Model Applied for Glasgow and Edinburgh (%)—cont'd

	SuDS Decision Support Model				CBR Model			
Sites (%)	–	x	xx	xxx	–	x	xx	xxx
Edinburgh								
Permeable pavement	1	8	20	71	0	6	39	55
Below-ground storage	17	7	48	29	15	14	53	18
Water playground	40	14	7	40	44	12	15	30
Green roof	4	13	34	50	8	32	16	45
Swale + pond	7	30	34	29	4	30	44	22
Shallow swale + pond	7	31	24	38	2	25	45	28
Infiltration trench + below-ground storage	6	15	62	17	2	16	64	18

– = not applicable; x = applicable; xx = recommended option; xxx = best option.

option obtained a high proportion of "nonapplicable" entries, although wetlands are considered to be the most ecological SuDS technique by environmentalists. However, traditional constructed wetlands usually require relatively large construction areas and high ecological impact potentials.

The 14 demonstration sites in Glasgow and Edinburgh were selected based on the rationale that demonstration sites should represent different geographical areas, types of land use, site classification types, and SuDS techniques to be implemented. Moreover, the detailed design of demonstration sites would help engineers and planners to have a better understanding of SuDS.

30.3.2 Prevalence Rating Approach Analysis

Both individual and combinations of SuDS techniques were rated on the PRAST scale (Table 30.2.4.1). Rating methods based on civil engineering and sustainability perceptions were explored. With respect to Glasgow and Edinburgh, civil engineers are predominantly concerned with reducing the peak flow to alleviate the risk of flooding further downstream and minimizing the space for SuDS implementation to utilize the majority of available land for construction. Correspondingly, environmentalists are predominantly concerned with increasing the ecological impact of SuDS to enhance biodiversity.

For the purpose of this modeling example, civil engineering rating obtained 4 out of 10 and sustainability rating 6 out of 10 weighting points. The final SuDS technique positioning after PRAST implementation (example only) is shown in Table 30.3.2.1. Findings show that combinations of ponds and swales

TABLE 30.3.2.1 Results of the Sustainable Drainage System Decision Support Model and Case-Based Reasoning Model for the Implementation of SuDS Techniques Using the PRAST Scale (Table 30.2.4.1) for Both Glasgow and Edinburgh Sites

		Glasgow Sites/Edinburgh Sites			
		SuDS Decision Support Model		CBR Model	
PRAST Position	SuDS Techniques	Number of Sites	% of Sites (%)	Number of Sites	% of Sites (%)
1	Standard swales + ponds	15/30	19/29	19/23	24/22
2	Shallow swales + ponds	14/9	18/9	11/6	14/6
3	Attenuation/detention ponds	18/23	23/22	5/17	6/17
4	Wetlands	0/0	0/0	0/0	0/0
5	Lined attenuation/ detention ponds	4/1	5/1	3/1	4/1
6	Infiltration basins/ponds	1/1	1/1	0/2	0/2
7	Standard swales	3/12	4/12	9/22	11/21
8	Shallow swales	1/0	1/0	2/2	3/2
9	Infiltration trenches + below-ground storage	1/3	1/3	4/0	5/0
10	Green roof	5/9	6/9	2/9	3/9
11	Permeable pavement	10/13	13/13	11/16	14/16
12	Filter strips	0/0	0/0	0/0	0/0
13	Soak-aways	0/0	0/0	0/1	0/1
14	Supplementary water playgrounds	0/0	0/0	0/0	0/0
15	Below-ground storage	6/0	8/0	7/1	9/1
16	Infiltration trenches	0/0	0/0	0/0	0/0
17	No SuDS possible	1/2	1/2	6/3	8/3

obtained the highest score. In comparison, the approach rates infiltration trenches as the least favorable SuDS technique because of the frequently insufficient engineering performance and low environmental benefit (Jefferies et al., 1999).

The PRAST scale was applied to all selected SuDS possibilities and recommended the highest-ranked SuDS techniques as the solution. The numbers and corresponding proportions of individual and dual SuDS techniques derived from the PRAST analysis are shown in Table 30.3.2.1. Noticeably, the combination of standard swales (or shallow swales) with ponds and ponds alone have the greatest proportion of recommended SuDS solutions for sites in both Glasgow and Edinburgh. Wetlands, filter strips, and some other SuDS features did not score at all. However, these techniques were potentially suitable for some sites, but a different SuDS technique, which featured higher on the PRAST scale, was more suitable. Moreover, there is one site in Glasgow and there are two sites in Edinburgh that are not associated with any recommended SuDS technique, mainly because of steep slopes on these sites (Table 30.3.2.1).

30.3.3 Case-Based Reasoning Model Output

The output of the SuDS decision support model incorporating PRAST is shown in Tables 30.3.1.1 and 30.3.2.1. The output from the CBR model is similar to the SuDS decision support model output. The novel idea is to use CBR in the decision-making process for SuDS installations. However, a larger case base would have been an advantage in order to increase the relevance of this modeling methodology.

30.4 CONCLUSIONS

A survey of 182 sites in Glasgow and Edinburgh indicated that it is feasible to implement different SuDS techniques and short SuDS treatment trains within both cities, which are fundamentally different from each other. A general SuDS implementation matrix (including dominant and supplementary criteria), which can be adapted to other cities and countries, has been outlined.

A practical SuDS decision support model based on a support matrix (SuDS variable specifications for SuDS techniques) and an associated weighting system have been developed to give practitioners a tool to assess the suitability of different SuDS techniques for a particular site with and without applying his or her own judgment.

The modeling outcome indicates that ponds (or lined ponds) and permeable pavements are the most frequently proposed SuDS techniques for Glasgow and Edinburgh, and ponds combined with standard swales or shallow swales are the most recommended SuDS combinations. Fourteen SuDS demonstration areas, which are representative of both different sustainable drainage techniques and geographical areas available for development, regeneration, or retrofitting within Glasgow and Edinburgh, have been identified.

The PRAST, which allows for the application of different proportions of civil engineering and sustainability judgment used in SuDS decision making to be made, was developed. Ponds linked to swales (or shallow swales) were placed at the top (most suitable) after PRAST ranking was undertaken. They were also associated with the highest proportion of recommended SuDS solutions for both Glasgow and Edinburgh. A separate CBR model was compared to the SuDS decision support model. The outputs of both models were similar, indicating that the linear model works similarly to the human mind.

Chapter 31

Natural Wetlands Treating Diffuse Pollution

31.1 INTRODUCTION

31.1.1 Background of the Case Study

Phosphorus concentrations in most freshwater bodies of European countries such as Germany showed a decreasing trend after the implementation of advanced technical solutions in sewage treatment plants (e.g., phosphate stripping). These measures only slightly reduced total nitrogen concentrations. As a consequence, their concentrations have altered little since 1980 (European Environment Agency, 1999; Umweltbundesamt, 2001).

Presently, agriculture is frequently the main source of diffuse nitrogen and phosphorus pollution inputs to water bodies in Western Europe (European Environment Agency, 1999). A further reduction of nitrogen concentrations in freshwater bodies requires interventions during all stages of nitrogen flow through the landscape. Nitrogen loss from unsaturated soil to the saturated soil layers can be reduced by the adoption of best land use practice (Frede and Dabbert, 1998; Reiche, 1994). A reduction of nitrate occurs during transport through groundwater, peat lands, and wetlands (Arheimer and Brandt, 1998). Burt (2003) has reviewed the monitoring of nitrate for selected hydrological systems.

Quantifying nutrient retention in natural and constructed wetlands, and in constructed treatment wetlands, is an important scientific and environmental task when enhancing water quality (Mitsch and Gosselink, 2000; Scholz and Xu, 2002; Vymazal et al., 1998). In general, the abiotic and biotic conditions in wetlands support several biogeochemical and physical processes that result in reduced nutrient concentrations in the outflowing compared to the inflowing water. Therefore, wetland and river restoration are frequently suggested as an effective measure to combat eutrophication of aquatic ecosystems and to reduce nutrient loads to the sea.

Some environmental authorities have demonstrated sustainable innovations in water resources protection; e.g., the Environment Ministry of Schleswig–Holstein (Northern Germany) has implemented a novel peat land rehabilitation

Wetlands for Water Pollution Control. http://dx.doi.org/10.1016/B978-0-444-63607-2.00031-9

program to utilize the high denitrification potential of degenerated and minerotrophic peat lands for the reduction of nitrate input into aquatic ecosystems and to simultaneously improve the habitat conditions. The realization of both goals required changes to the water management strategy to adapt it to the site-specific geohydrological conditions. The effect of raised groundwater levels and extensive land use on the water quality of heavily vegetated and groundwater-fed ditches was assessed in a riparian peat land located in the Eider River Valley (an internationally important wetland case study area in Schleswig–Holstein).

In groundwater discharge areas such as the River Eider Valley, groundwater buffer zones are an effective means for reducing the total nitrogen concentration (predominantly nitrate–nitrogen) before the polluted load reaches the River Eider, predominantly via small ditches (Trepel and Kluge, 2002). Rivers themselves support nitrogen removal processes, depending, for example, on macrophyte composition and stream morphology (Eriksson and Weisner, 1999).

31.1.2 Nutrient Transformations and Removal Processes

A further reduction of nutrient input into aquatic water bodies can be achieved by two strategies: through a reduction of nutrient losses by the adoption of a best land use practice and by using natural nutrient removal and transformation processes. A successful adoption of the latter strategy requires a basic understanding of the involved processes.

In the nitrogen cycle, ammonification, nitrification, and denitrification are the three dominant transformation processes. Ammonification is the conversion from organic nitrogen to ammonia-nitrogen. Ammonia-nitrogen is used by the aquatic plants and microorganisms for new biomass development. Ammonification is slower in anaerobic than in aerobic soils because of the reduced efficiency of heterotrophic decomposition in anaerobic environments. Ammonification also depends on the pH being within an optimum range of approximately 6.5–8.5.

Furthermore, nitrification transforms ammonia-nitrogen to nitrate–nitrogen. This transformation has two steps. Ammonia-nitrogen originates from decomposed plants and animals. Nitrification (i.e., the conversion of ammonia-nitrogen to nitrate–nitrogen) is important because plants often take up nitrate–nitrogen preferentially to ammonia-nitrogen. However, this transformation requires oxygen.

Denitrification is the process in which nitrate–nitrogen is reduced to gaseous nitrogen. This transformation is supported by facultative anaerobes. These organisms are capable of breaking down oxygen-containing compounds such as nitrate–nitrogen to obtain oxygen in an anoxic environment. A more detailed discussion on the nitrogen cycle, including nitrogen transformations, has been published by Nuttal et al. (1997) and Davidson et al. (2002).

31.1.3 Aim and Objectives

The aim of this chapter is to assess the temporal and spatial hydraulic and biochemical functioning of groundwater-fed ditch ecosystems and their contribution to the overall catchment and, particularly, nutrient dynamics of a riparian peat land located in the River Eider Valley. The objectives are as follows:

- To characterize all watercourses and classify ditches at a particular stretch of the River Eider;
- To determine ditch discharges by carrying out experiments with temporary weirs at different hydraulic loading rates and different flow obstruction covers in ditches;
- To assess the water quality of the selected ditches and the River Eider between May and September;
- To describe the effect of summer flooding on the overall water quality of the buffer zone;
- To characterize the vegetation at the case study site, particularly at those areas close to the ditches;
- To assess the effect of macrophytes and channel morphologies (river and ditches) on nutrient retention; and
- To assess if reduced ditch maintenance is an option to achieve nutrient reduction.

The data gathered should help society to understand the challenges of holistic wetland system management, diffuse pollution, and the linking scales in catchment management.

31.2 MATERIALS AND METHODS

31.2.1 Case Study, Sampling, and Data Analysis

The River Eider Valley is a riparian freshwater wetland of 257 ha in northern Germany (10°3′ east and 54°11′–54°15′ north; Figure 31.2.1.1). The River Eider Valley is the name of a specific site located between the villages Schmalstede (upstream) and Techelsdorf near Flintbek (downstream) and not the name for the valley of the River Eider as a whole. This area is located 10 km south of Kiel (capital of the federal state of Schleswig-Holstein). The climate is usually humid and cool temperate, with an annual mean temperature of approximately 8 °C and a mean annual precipitation of approximately 800 mm/year (Trepel et al., 2003).

The River Eider Valley peat land is hydrologically characterized by groundwater inflow from the surrounding hills and river water inflow from the upstream catchment basin with an area of 120 km (van der Aa et al., 2001; Jensen et al., 2001). Mire genesis was dominated first by

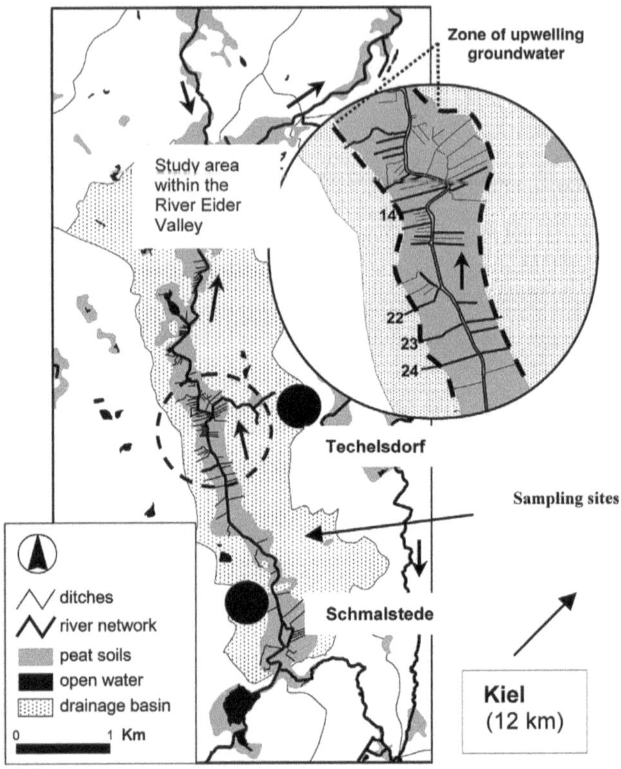

FIGURE 31.2.1.1 Overview showing the study area within the River Eider Valley, the River Eider, the main ditches with measurable discharge since June 2003, and the selected representative ditches (Ditches 14, 22, 23, and 24). The zones of upwelling groundwater on each side of the river are indicated by broken lines.

river water inflow from the upstream catchment area and in a later successional stage by groundwater inflow at the mire margins (Trepel and Kluge, 2002).

Active drainage with ditches and tile drains, land use intensification, cutting of turf by hand (Klove, 2001), and river regulation have resulted in severe degeneration of the peat land. Altering of the water flow patterns led to the mobilization of nutrients from decaying peat vegetation and resulted in an increased nutrient load for most ditches and some river stretches (Bockholt et al., 1992; Trepel et al., 2003).

The data analysis was performed using Microsoft Excel and Minitab. Statistical methods such as summary statistics, correlation analysis, significance tests, and analysis of variance (ANOVA), outlined by Fowler and Cohen (1998), were applied.

31.2.2 Ditches and the Channel of the Receiving River

The study area is a specific site within the River Eider Valley. Figure 31.2.1.1 gives an overview of the study area showing the River Eider, the main ditches, and the locations of the studied selected representative ditches (Ditches 14, 22, 23, and 24). The River Eider has a discharge range between 0.3 and 4 m^3/s (Trepel et al., 2003).

Since the eighteenth century, reed cutters have mowed the vegetation in the River Eider and the adjacent ditches to keep the riverbed open and to reduce the risk of late summer flooding (Hanssen, 1842). Before the decline in agriculture, summer flooding was a serious problem for the farming community making use of the fertile drained meadows in the River Eider Valley for cattle grazing.

The water management plan of the Federal State of Schleswig-Holstein, discussed by Trepel et al. (2003), aimed to cease the regular mowing of instream (predominantly *Sparganium emersum* L.) and in-ditch macrophyte vegetation to increase the hydraulic residence time. Consequently, the river water level increased in the riparian wetland valley. This led to deliberate flooding of the lower parts of the ditches and lowland meadows. The overall purpose of the water management plan is the reduction of peat mineralization by raising water levels, which may lead to the establishment of new peat-forming vegetation types in the future (Trepel et al., 2003). Researchers assessed alternative land and water quality control management methods as part of this plan.

For different ditch cross-sections, physical parameters such as ditch water depth, width, wetted perimeter, and flow obstruction cover were noted at 5-m intervals along the ditch from the outlet at the River Eider toward the upper ditch stretch (part of the ditch where water is first observed). Flow obstruction cover is predominantly due to vegetation obstructing the flow in a vertical plane located in a 90° angle to the flow direction. The obstruction cover is estimated in percentage by eye. A chartered drainage engineer (i.e., the author of this book and his research team) carried out the assessment because data creation is subjective. The horizontal flow obstruction cover area variable (measured at least twice for all selected ditches) depends on various environmental and management conditions such as season, hydrology, and ditch maintenance. Standard hydrological methods are described by Chaudhry (1996).

For the purpose of this case study, a ditch has been defined as an open channel with the following criteria:

- A small individual watercourse (i.e., not part of a wider network of different types of watercourses);
- A watercourse that discharges into the River Eider throughout the year;
- Channel length >50 m;

- Channel width ≥10 cm;
- Water depth ≥3 cm throughout the ditch; and
- Not-overgrown watercourse; i.e., watercourse without a flow obstruction cover ≥95%.

A total of 160 small watercourses including ditches (24%) in the River Eider Valley were hierarchically classified (six levels) according to their geometry (width and depth), flow obstruction cover, hydrogeology, hydrochemistry, and hydraulic flow properties (Figure 31.2.2.1).

The following 4 out of 39 individual ditches have been selected for further studies: Ditches 14, 22, 23, and 24 (Figure 31.2.1.1). The four selected ditches show common characteristics of groundwater-fed ditches (Trepel et al., 2003). Springs along the upper stretches of the selected ditches were clearly visible. The groundwater table at lower ditch stretches was approximately 0.5 m below the ground. However, the groundwater table rose gradually during summer flooding. A detailed investigation of the groundwater table was beyond the scope of this study but has been discussed elsewhere (Jensen et al., 2001; Trepel, 2001; Trepel and Kluge, 2002; van der Aa et al., 2001). Water samples were collected before summer flooding during low flow conditions to reduce the risk of groundwater and river water influencing the water quality of the ditch water due to mixing.

However, the selected ditches are representative for different groups of classes of ditches (Figure 31.2.2.1). Ditch 22 represented main ditches with a width ≥150 cm and a depth ≥100 cm (regardless of the flow conditions). Ditches 14, 23, and 24 were examples of small ditches. Ditches 14 and 23

<table>
<tr><td colspan="8">1a. Ditch width < 150 cm (145)</td></tr>
<tr><td colspan="8">2a. Ditch depth < 100 cm (140)</td></tr>
<tr><td colspan="3">3a. Flow obstruction cover < 50% (17)</td><td colspan="5">3b. Flow obstruction cover ≥ 50% (123)</td></tr>
<tr><td colspan="3">4a. Groundwater-fed (16)</td><td colspan="2">4a. Groundwater-fed (54)</td><td colspan="3">4b. Not groundwater-fed (69)</td></tr>
<tr><td>5a. Conductivity < 500 μS (5)</td><td colspan="2">5b. Conductivity ≥ 500 μS (11)</td><td colspan="2">5b. Conductivity ≥ 500 μS (46)</td><td colspan="2">5a. Conductivity < 500 μS (25)</td><td>5b. Conductivity ≥ 500 μS (44)</td></tr>
<tr><td>6a. Ditch (1) = Ditch 14</td><td>6a. Ditch (1) = Ditch 23</td><td>6b. Not a ditch (10)</td><td>6a. Ditch (9) includes Ditch 24</td><td>6b. Not a ditch (37)</td><td>6a. Ditch (6)</td><td>6b. Not a ditch (19)</td><td>6b. Not a ditch (37)</td></tr>
</table>

<table>
<tr><td colspan="2">1b Ditch width ≥ 150 cm (15)</td></tr>
<tr><td>2a. Ditch depth < 100 cm (10)</td><td>2b. Ditch depth ≥ 100 cm (5)</td></tr>
<tr><td></td><td>3a. Flow obstruction cover < 50% (3)</td></tr>
<tr><td></td><td>4a. Groundwater-fed (3)</td></tr>
<tr><td></td><td>5a. Conductivity < 500 μS (2)</td></tr>
<tr><td></td><td>6a. Ditch (1) = Ditch 22</td></tr>
</table>

FIGURE 31.2.2.1 Hierarchical classification of 160 small watercourses in the River Eider Valley on six classification levels with two subgroups each. Only individual groups with ≥10 watercourses and/or groups containing subgroups that include a selected representative ditch (i.e., Ditches 14, 22, 23, and 24) are shown. Highlighted values (bold) in brackets indicate the within-group number of watercourses.

represented examples of small ditches with fairly low flow obstruction covers (<50%). In contrast, Ditch 24 had a high flow obstruction cover (>50%). The vegetation distribution was estimated in May 2003 because plants could be determined easily and the ditches were not influenced by the activity of the research team (see below for experiments undertaken). In comparison with the other selected ditches, Ditch 24 had a high flow obstruction cover. The conductivity of Ditch 14 in comparison to Ditch 23 was relatively low (<500 μS).

The threshold of 500 μS for conductivity was selected on practical grounds and was based on the empirical observation that relatively unpolluted ditches, in contrast to relatively polluted ditches, in Northern Germany had corresponding conductivity values much lower and much higher, respectively, than this threshold. However, in contrast to other waters such as rainwater runoff, unpolluted groundwater, and freshwater from formerly glaciated uplands, all measured conductivity values were relatively high and all ditch water could be regarded as eutrophic (Mitsch and Gosselink, 2000).

Furthermore, the land use adjacent to the four selected ditches was different. The land surrounding Ditch 14 was not utilized by the farming community for at least 10 years. Less extensive cattle grazing was practiced near Ditch 22, and more extensive cattle grazing was undertaken near Ditches 23 and 24.

31.2.3 Discharge Determination for Open Channels

Temporary triangular-shaped galvanized metal weirs were used to estimate the discharge of the selected ditches at different cross-sections. The time taken for a small bucket of 12 l or a large bucket of 20 l, with selection depending on ditch geometry, to fill up was timed. A minimum of three runs each were noted and the average discharge during steady-state conditions was calculated. Furthermore, the total roughness (or total resistance) coefficient k_M ($m^{1/3} \times s$) of watercourses (with and without vegetation) was determined empirically according to the International Institute for Land Reclamation and Improvement (1964) and Chaudhry (1996).

31.2.4 Water Quality Analyses

The water quality of water samples collected in the late morning hours was determined only between May and September (warm period in Schleswig-Holstein). The water analysis for all nutrients was undertaken using a four-channel continuous flow auto-analyzer with auto-sampler and automatic predilution (Skalar, San plus System, Model 4000, Type 1121). Nitrate–nitrogen and nitrite–nitrogen, ortho-phosphate, and ammonia-nitrogen in water samples were determined directly after sampling according to the German standard methods DIN EN ISO 13395 (Deutsches Institut für

Normung, 1996a), DIN EN 1189 (Deutsches Institut für Normung, 1996b), and DIN EN ISO 11732 (Deutsches Institut für Normung, 2003), respectively. These analytical procedures were in agreement with American standard methods (Clesceri et al., 1998).

The following additional water quality parameters (instruments used are indicated in brackets) were measured in agreement with standard methods (Clesceri et al., 1998) in the field at 5-m intervals along the selected ditches: conductivity (WTW TetraCon 325), pH (WTW SenTix 41), redox potential (WTW SenTix ORP), dissolved oxygen (DO) (WTW OxiCal SL), and temperature (WTW TetraCon 325). No replicates were taken during the main study period (June–September 2003), unless stated otherwise.

31.2.5 Vegetation Characterization

Vegetation relevees (Braun-Blanquet, 1932) were carried out near the ditch outlets into the channel of the River Eider and at intervals of 50 m along each representative ditch. Aquatic plants located <1 m away from each wetted ditch embankment were included in the survey. The nomenclature of the plant species coincides with Wisskirchen and Häupler (1998).

31.3 RESULTS

31.3.1 Characteristics of Watercourses in the River Eider Valley

Figure 31.2.2.1 gives a further overview of the characterization of small watercourses in the River Eider Valley. The majority of the watercourses in this valley were small and fully covered with macrophyte vegetation. This fact reflects the reduced interest of the local farmers in maintaining the watercourses, because they are not using the adjacent peat lands for agricultural purposes anymore.

Pictures of Ditch 23 (Figure 31.3.1.1) before and after the removal of vegetation give an indication of total roughness. Within-ditch vegetation has been reduced, but algal growth has increased owing to the release of nutrients during mechanical vegetation removal. The estimated k_M values of both the vegetated and excavated Ditch 23 are about 1 and 10 $m^{1/3} \times s$, respectively (Chaudhry, 1996). Furthermore, Figure 31.3.1.2 shows a comparison of the "multiplication factors" of Ditches 14, 22, 23, and 24 for the water depth and ditch width, respectively, preceding and following flooding. The multiplication factors indicate the increases of the corresponding ditch dimensions due to flooding.

31.3.2 Water Quality during Spring and Summer

The summer of 2003 was one of the hottest in northern Germany in 40 years. Daytime temperatures were frequently around 30 °C in the

FIGURE 31.3.1.1 Pictures of Ditch 23 at cross-section 50 m: (a) preceding (July 22, 2003) and (b) following (August 11, 2003) the removal of aquatic vegetation and associated silt.

shade, and there were only a few days with significant rainfall during this study. This hot climate had an obvious influence on the hydraulic properties and water quality characteristics of the ditches; e.g., low water levels, high plant productivity, and elevated water temperatures. However, the four selected ditches were groundwater-fed, and the water temperatures at upstream reaches were therefore relatively constant between 9 °C and 10 °C, while the temperatures downstream were influenced by the air and river temperature and were therefore between 15 °C and 21 °C during daylight.

Table 31.3.2.1 summarizes the water quality of the small watercourses and the river in the River Eider Valley. Water samples were taken regularly (twice a week, on Monday and Thursday) during the late morning hours in May and June between 1999 and 2003. The water quality of all watercourses was characterized by a slightly alkaline pH. Total nitrogen (particularly nitrate–nitrogen) concentrations in these waters were high (Chaudhry, 1996). According to the German river water quality classification scheme (Länderarbeitsgemeinschaft Wasser, 1998), the River Eider is classified as considerably contaminated (between 2.5 and 5.0 mg/l nitrate–nitrogen). In comparison, most small ditches were classified as slightly contaminated (between 1.0 and 1.5 mg/l nitrate–nitrogen).

Grohmann et al. (2002) have reviewed international nitrate–nitrogen threshold values. In the European Union, the nitrate–nitrogen threshold value for potable water is 11.3 mg N/l. This value has been exceeded for some upper stretches of the ditches (see also Figure 31.3.2.1 for Ditch 24).

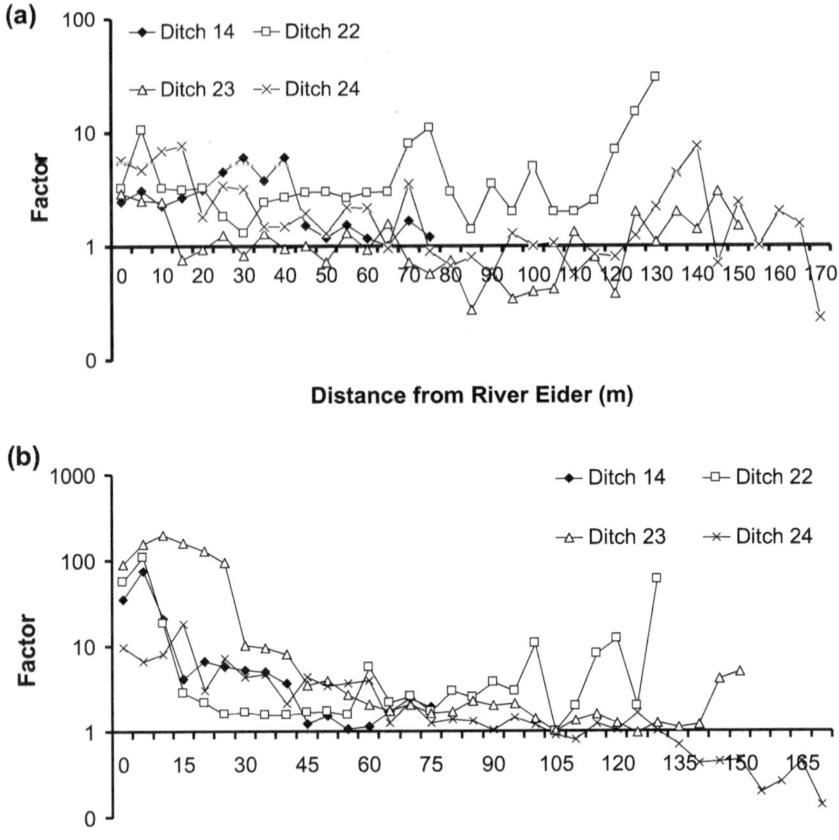

FIGURE 31.3.1.2 Comparison of the "multiplication factors" for (a) the water depth and (b) the ditch width before and after summer flooding for Ditches 14, 22, 23, and 24. The multiplication factors indicate the increase of the physical dimensions due to flooding.

A broad suite of parameters was analyzed for 24 ditches. Mean biochemical oxygen demand (BOD), ammonia-nitrate, nitrite-nitrogen, nitrate–nitrogen, and ortho-phosphate-phosphorus concentrations were 6.0, 0.2, 0.0, 1.6, and 0.0 mg/l, respectively. Table 31.3.2.2 shows the summary statistics for the BOD and nutrient concentrations of different stretches of the selected Ditches 22, 23, and 24, which are statistically significantly different from each other (see below). Elevated BOD and nitrate concentrations for upstream groundwater-fed ditch cross-sections were apparent (Table 31.3.2.2), regardless of the observation that all ditches belonged to different classes (Figure 31.2.2.1). A spatial pattern is apparent where the nitrate–nitrogen concentration declined by about 60% within the ditch. Ditches that were not groundwater-fed (Figure 31.2.2.1) always had nitrate–nitrogen values <2.5 mg/l.

TABLE 31.3.2.1 Water Quality Comparison between Small Watercourses and the River Eider

Variable	Statistics	Schmalstede (Upstream)	Watercourses (in the Valley)	Flintbek (Downstream)
Water temperature (°C)	Mean	13.8	12.0	13.6
	Standard deviation	1.68	2.39	1.86
	Count	17	72	17
pH (−)	Mean	7.9	7.5	7.7
	Standard deviation	0.10	0.26	0.12
	Count	17	56	17
Nitrate–nitrogen[a] (mg/l)	Mean	3.1	1.5	2.2
	Standard deviation	0.53	2.39	0.28
	Count	17	101	17

Note: The small watercourses include ditches located within the River Eider Valley. Watercourse data were collected in May and June between 1999 and 2003. River Eider data were collected in May and June between 1999 and 2002. Ammonia-nitrogen, nitrite-nitrogen, and ortho-phosphate-phosphorus concentrations have also been measured, but their mean concentrations were <0.1 mg/l.
[a]*Considering screening data for groundwater-fed open ditches collected in summer 2003 only, the nitrate–nitrogen concentration for these open ditches is higher than the corresponding one for the River Eider.*

Figures 31.3.2.2–31.3.2.4 show comparisons of the conductivity, redox potential, and DO values for selected ditches with distances from their confluence with the River Eider. Obvious statistical data variations within and between most ditches were noted. Considering conductivity, Ditch 22 was associated with relatively low values (Figure 31.3.2.2).

Figures 31.3.2.3 and 31.3.2.4 show relatively low redox potentials and DO concentrations for Ditches 22 and 23. The mechanisms responsible for the slight decrease of the redox potential for Ditch 23 have not been investigated. However, it has to be noted that Ditch 23 had the highest discharge despite a small channel profile (Figure 31.2.2.1).

An ANOVA for Ditches 14, 22, 23, and 24 with respect to the variables conductivity (Figure 31.3.2.2), redox potential (Figure 31.3.2.3), DO (Figure 31.3.2.4), and pH was performed. The findings confirmed that the ditches were statistically significantly different in terms of their standard water quality characteristics ($P < 0.01$). This explains the high standard deviations

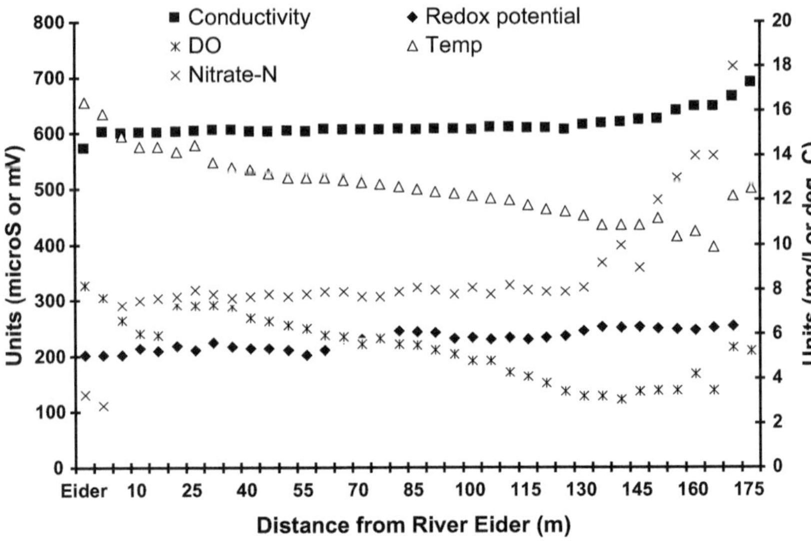

FIGURE 31.3.2.1 Comparison of standard water quality variables (easy and cost-efficient to measure) and nitrate–nitrogen (difficult and cost-intensive to measure) for Ditch 24 on July 22, 2003. Single data values have been presented.

in Table 31.3.2.2. However, the analysis was only performed for the lower 90 m of each ditch (19 cross-sections in total) because the lower stretches always provided sufficient sample water.

Figure 31.3.2.1 shows a comparison between standard water quality variables and nitrate–nitrogen for Ditch 24. According to the German river classification scheme (Ländesarbeitsgemeinschaft Wasser, 1998), the upper stretches of the ditches are classified as "highly contaminated" (between 10 and 20 mg/l nitrate–nitrogen) owing to the high nitrate contamination. The correlation coefficients between nitrate–nitrogen and the redox potential and temperature were 0.62 and −0.55, respectively, and are statistically significant ($P < 0.05$).

31.3.3 Ditch Vegetation

The overall ditch vegetation was dominated by tall and productive plant species such as *Phalaris arundinacea* L. and *Carex acutiformis* Ehrh. These species indicate nutrient-rich conditions. The species composition of the ditches reflected the land use intensity of the adjacent fields. All four selected ditches were dominated by *P. arundinacea*. However, the growth of *C. acutiformis* Ehrh. was also dominant for the most extensively used Ditch 14 (Figure 31.3.3.1).

TABLE 31.3.2.2 Summary Statistics for the Unfiltered Biochemical Oxygen Demand and Filtered Nutrient Concentrations of Different Stretches of Ditches 22, 23, and 24 (Combined Data for July 9, 2003)

Statistics	Mean	SD[b]	Mean	SD[b]	Mean	SD[b]
Stretch	0 m (near Eider)		30 m		60 m	
BOD[a] (mg/l)	6.0	2.74	1.8	0.40	5.2	2.39
Ammonia-nitrogen (mg/l)	0.13	0.081	0.07	0.037	0.09	0.039
Nitrate–nitrogen (mg/l)	2.8	0.33	5.0	2.99	4.2	1.92
Stretch	90 m		120 m		150 m (near spring)	
BOD[a] (mg/l)	2.0	1.75	14.6	1.76	14.8	Ned[c]
Ammonia-nitrogen (mg/l)	0.09	0.023	0.13	0.030	0.13	0.050
Nitrate–nitrogen (mg/l)	4.8	1.86	5.3	1.32	13.1	2.01

Note: The water level for all corresponding sections was ≥3 cm. Nitrite-nitrogen and ortho-phosphate-phosphorus concentrations have also been measured, but their mean concentrations were below 0.1 mg/l.
[a]*7-day at 20 °C N-allylthiourea (nitrification inhibitor) biochemical oxygen demand.*
[b]*Standard deviation.*
[c]*Not enough data (only one value for Ditch 24).*

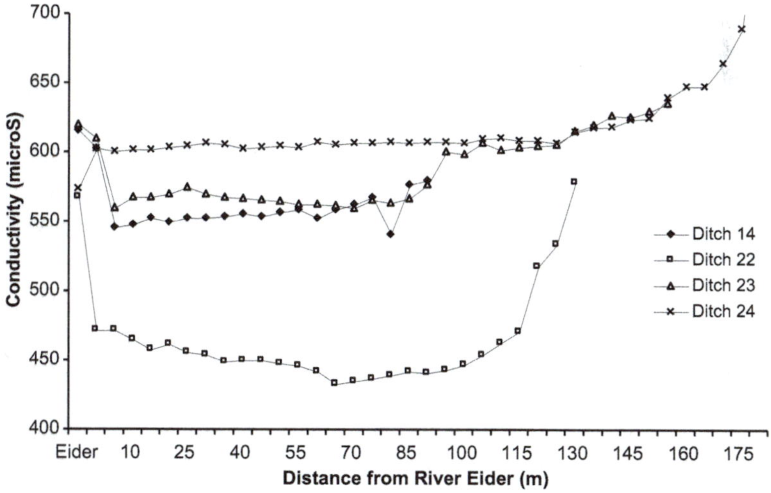

FIGURE 31.3.2.2 Comparison of the conductivity for the selected Ditches 14, 22, 23, and 24. The low conductivity of Ditch 22 has been influenced by dilution via a wetland (i.e., cut-off former river stretch) in the north that is associated with low conductivity values. The conductivity axis starts at 400 μS to highlight data variation.

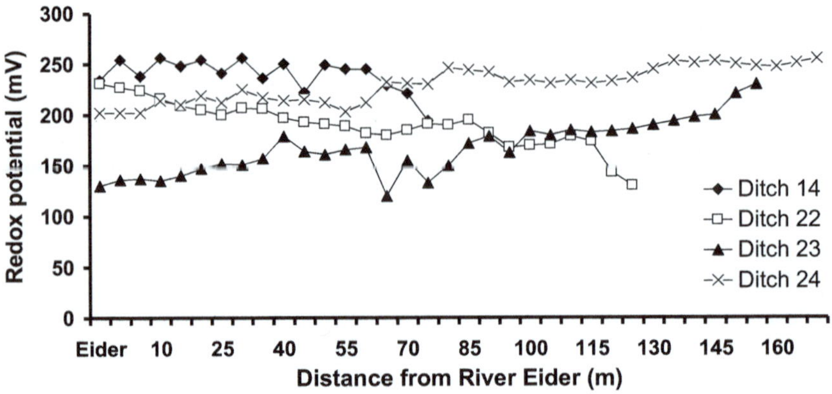

FIGURE 31.3.2.3 Comparison of the redox potential for the selected Ditches 14, 22, 23, and 24. Redox potentials were only recorded for water levels ≥4 cm.

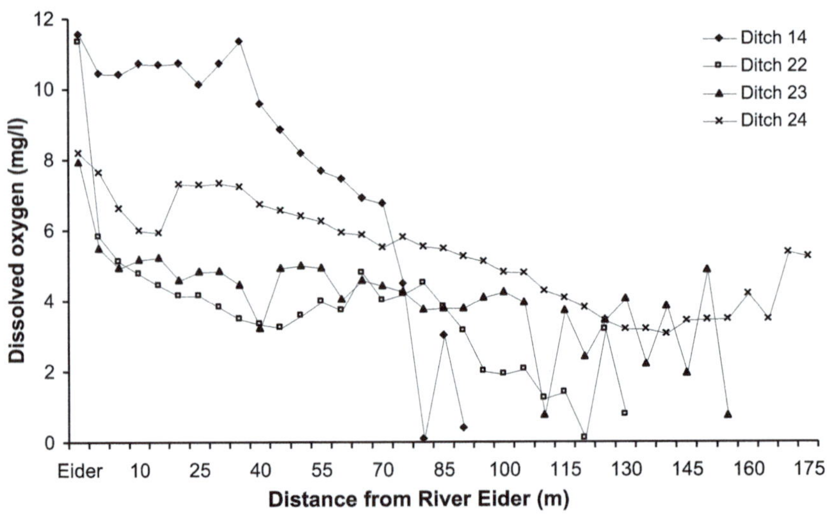

FIGURE 31.3.2.4 Comparison of the dissolved oxygen for the selected Ditches 14, 22, 23, and 24.

Furthermore, the natural flow obstruction cover was at least 70% upstream of cross-section 60 m for all selected ditches. Elevated flow obstruction cover percentages after ditch clearance (manual vegetation removal) were merely due to the development of meander within the silt bed upstream of cross-section 100 m.

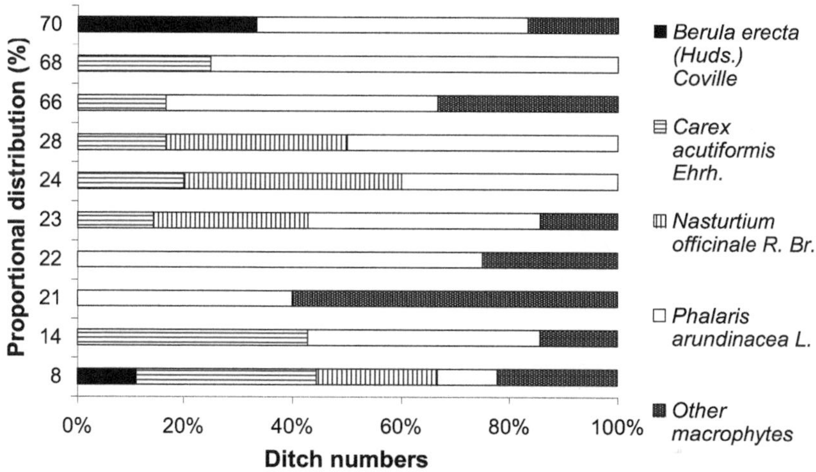

FIGURE 31.3.3.1 Comparison of the dominant macrophytes within or near (1 m of each side of the embankments) 10 randomly selected ditches (160 watercourses including 39 ditches in total) of the study area.

31.4 DISCUSSION

31.4.1 Disappearance of Ditches Due to Vegetation Growth

The River Eider Valley is characterized by a mosaic of different fields; all of them have been drained in the past, predominantly using ditches. In recent years, the authorities have promoted ceasing of both mowing of ditch vegetation and dredging of ditches to encourage the rise of the water table (see also Section 31.1.1).

Owing to the deliberate absence of an active overall ditch maintenance plan, ≥85% of all watercourses in the River Eider Valley were small in terms of their width (<150 cm) and depth (<100 cm). Greater than 75% of all watercourses were heavily vegetated and therefore had flow obstruction covers of ≥50% in May 2003 (Figure 31.2.2.1). Nevertheless, approximately 25% of all watercourses were classified as individual ditches with a major drainage purpose (average discharge >0.5 l/s). However, ditches with a low discharge still have a minor drainage function and influence the nearby diversity and composition of vegetation.

All watercourses were small (average depth below 0.8 m) and densely vegetated because they have not been managed (e.g., ceasing of mowing and dredging) for up to 10 years. The agricultural land was only used extensively for cattle and horse grazing. It follows that predominantly species from the *Phragmitetalia australis* group (macrophytes like reeds) and rushes grow along the sides of the ditches, and a very dense overgrowth of aquatic plants in the center of the cross-section of the ditches can be noted (Figure 31.3.3.1). The corresponding estimated k_M values were therefore generally $<5\ m^{1/3} \times s$ (International Institute for Land Reclamation and Improvement, 1964).

The water quality analysis shows a great variation of nutrient concentrations within the different ditches (standard deviation shown in Table 31.3.2.1). However, high carbon and nutrient losses (particularly nitrate–nitrogen) to the drainage water is apparent (Table 31.3.2.1). A discussion on the effective monitoring of nitrate–nitrogen change is beyond the scope of this chapter but can be found elsewhere (Bockholt et al., 1992; Burt, 2003).

As the ditches continue to overgrow, it is likely that there will be virtually no more ditches with a significant drainage function in the River Eider Valley in about a decade. The increase of the water table will create a shallow freshwater lake near the river and a groundwater-fed buffer zone between the hill slope and the river. Both developments will reduce further peat degradation at higher locations.

31.4.2 Water Quality Variations within the Study Area

The hydrological conditions in the study area between the villages Techelsdorf and Schmalstede are characterized by groundwater flow from the upwelling spring area down to the River Eider (Jensen et al., 2001; Trepel and Kluge, 2002) through predominantly vegetated ditches and silted-up tile drains. The presence of the ditches reduces nutrient retention and biochemical transformation (e.g., denitrification as described by Mohamed et al. (2003)) within the peat land. The rewetting of the ditch embankments and peat soils increased the potential of sedge reeds to accumulate nutrients in the peat. The water quality analysis showed great variation of nutrients within the different ditches (Figures 31.3.2.1 and 31.3.2.2).

A comparison between the overall water quality of the River Eider and the water quality of all small watercourses indicated that the total discharge from the small watercourses usually does not pollute the River Eider in terms of nutrient input (Table 31.3.2.1). This is partially the case due to the good water quality (except for high nitrate contamination) of the groundwater and the presence of heavy aquatic vegetation that takes up nutrients. As the ditches continue to overgrow and the water table continues to increase, it is likely that nitrate–nitrogen will be reduced further. The inflow from the ditches located in the River Eider Valley into the River Eider contributes to approximately 10% of the overall River Eider discharge (Trepel and Kluge, 2004).

Nitrate–nitrogen concentrations above 1 mg/l (see also Table 31.3.2.1) are usually an indication for man-made pollution from fertilized agricultural fields located on nearby hills. Peat land use for cattle grazing purposes requires drainage and fertilization. Nitrate concentrations were higher for grazed (Ditches 23 and 24: 3.5–6.5 mg/l) than ungrazed ditches (Ditches 14 and 22: 0.1–0.7 mg/l). Peat lands have a high denitrification potential (Davidson et al., 2002) and are therefore suitable for natural pollution load reduction.

Elevated conductivity concentrations for ditches in the River Eider Valley are usually associated with diffuse environmental pollution such as urban and agricultural runoff. The conductivity within all ditches was fairly constant (except for elevated levels near the springs) but on different concentration levels for each ditch. For example, the moderately vegetated and wide Ditch 22 was associated with relatively low values (450 μS), in contrast to Ditches 14, 23, and 24 (550–600 μS). This observation can be explained by the presence of a high productive wetland (i.e., cut-off former stretch of the River Eider) to the north of Ditch 22. However, further studies on this part of the wetland were beyond the scope of this study.

The redox potentials of Ditches 14 and 23 increased from the springs toward the river. The opposite trend was apparent for Ditches 22 and 24 (Figure 31.3.2.3). A transformation of SO_4^{2-} to H_2S (note that the smell of "rotten eggs" was not detected) may take place well below 180 mV.

Most representative ditches show typical characteristics of groundwater-fed ditches with dominant macrophyte communities (see below). Groundwater-fed ditches are associated with a constant temperature, relatively high nitrate–nitrogen concentration, relatively low oxygen concentration, low siltation, and a relatively moderate flow obstruction cover. Usually, these groundwater-fed ditches are deep, have a constant water level, and are associated with moderate temperature fluctuations. Furthermore, these ditches are usually mesotrophic to eutrophic in lowlands such as the River Eider Valley.

31.4.3 Vegetation Characterization

Table 31.4.3.1 gives an overview of the vegetation characteristics of the selected Ditches 14, 22, 23, and 24. The ditch vegetation is dominated by tall and productive plant species such as *P. arundinacea* L., *C. acutiformis* Ehrh., *Glyceria maxima* (Hartm.) Holmb., or *Urtica dioica* L. s. str. These species indicate nutrient-rich conditions (Wisskirchen and Häupler, 1998). Species typical for less nutrient-rich conditions (*Carex paniculata* L. or *Carex cespitosa* L.) occur only rarely at the top and middle sections of ditches.

The ditches referred to in Figure 31.3.3.1 have been selected at random to give an unbiased overview of macrophyte communities associated with the ditches of the River Eider Valley. Figure 31.3.3.1 indicates the cover and dominance of the *Phragmitetea australis* group (macrophytes) close to the drainage ditches. This is particularly the case for Ditches 14, 23, 24, and 66. *P. arundinacea* (Figure 31.3.3.1) stands are typical for drained peat soil. These stands predominantly occur near the riverbed and the lower end of the ditches (Ditches 21, 29, 33, and 69), where flooding results in higher water levels.

The growth of submerged *S. emersum* L. in the River Eider leads to localized flooding during storm events. However, a cluster analysis revealed no

TABLE 31.4.3.1 Vegetation Characteristics of Selected Representative Ditches (Ditches 14, 22, 23, and 24)

Ditch Number/Distance to River Eider (m)		14	14	14	22	22	22	23	23	23	24	24	24
Species	**Group**	**0**	**50**	**100**	**0**	**50**	**100**	**0**	**50**	**100**	**0**	**50**	**100**
Poa trivialis L. s. l.	MA[a]					2	3				1		
Ranunculus repens L.	MA[a]	1	1										
Caltha palustris L.	MO[b]							1	1				
Carex cespitosa L.	MO[b]	2	2					1					
Epilobium hirsutum L.	MO[b]			2									
Filipendula ulmaria (L.) Maxim.	MO[b]		2	1		2	2		2	1			
Juncus effusus L.	MO[b]		1				2						
Lysimachia nummularia L.	MO[b]									1		1	
Myosotis scorpioides agg.	MO[b]				2			1	1		2		
Scirpus sylvaticus L.	MO[b]						2						
Berula erecta (Huds.) Coville	PH[c]		1		1								
Calamagrostis canescens agg.	PH[c]		2										
Carex acutiformis Ehrh.	PH[c]	2	3	2					2			2	
Carex paniculata L.	PH[c]			2		3			2				
Equisetum fluviatile L.	PH[c]					1	2						

Galium palustre L. s. l.	PH^c		1						1			1	
Glyceria maxima (Hartm.) Holmb.	PH^c			3	1					1			
Nasturtium officinale R. Br.	PH^c			1	1			2	3		2		3
Persicaria hydropiper (L.) Delarbre	PH^c	1			2			2	1		2	1	
Phalaris arundinacea L.	PH^c	3	3	2	3	2	2	3	3	3	3	3	1
Rorippa palustris (L.) Besser	PH^c				1						1		
Stellaria alsine Grimm	PH^c	1			1						1		
Valeriana officinalis L. s. str.	PH^c		1						1				
Galeopsis tetrahit L.	AR^d				1	1					1		
Galium aparine L.	AR^d		1	1	1	1	1			2			
Urtica dioica L. s. str.	AR^d	1		2		2	2			3			
Alopecurus geniculatus L.	LP^e				1			1					
Glyceria fluitans (L.) R. Br.	LP^e		1		1								
Lemna minor L.	LE^f				2			2					

Note: Classes were indicated as: rare = 1 (<1% cover), present = 2 (1%–30% cover), and dominant = 3 (>30% cover).
[a]*Molinio-Arrhenatheretea.*
[b]*Molinietalia.*
[c]*Phragmitetea australis.*
[d]*Artemisietea vulgaris.*
[e]*Lolio-Potentillion.*
[f]*Lemnetea.*
Nomenclature after Wisskirchen and Häupler (1998).

significant link between different ditch groups and stretches influenced by different hydrological conditions and aquatic plant species. This might be explained with the observation that most plant species were cosmopolitan. However, a detailed discussion of the influence of flooding and other hydrological changes on the vegetation community dynamics is beyond the scope of this chapter.

31.4.4 Hydraulic Changes Due to Summer Flooding

Groundwater-fed ditches were less overgrown than ditches supplied by surface runoff only (Figure 31.2.2.1). In fact, most ditches that were not fed by groundwater and first dredged in the nineteenth century have disappeared (Trepel and Kluge, 2002). A two-way flow between ditch water and groundwater was not significant for the upper ditch stretches, considering discharge values of <2 l/s and that multiple springs were clearly visible in each selected ditch. However, interactions between ditch, river, and groundwater were likely during flooding.

Studies in the River Eider Valley have shown that the water quality within groundwater-fed ditches is relatively constant over time for the season sampled, with the exception of the stretch near the River Eider where mixing between ditch and Eider water takes place. Water quality variables such as redox potential and conductivity did not change significantly during flooding.

Mixing at the lower ditch stretch was due to the fluctuations of the River Eider water level (approximately 80 cm per annum). In general, the reasons for high water levels during summer and winter are vegetation growth (predominantly *S. emersum* L.) and precipitation (increased discharge from upstream urban regions), respectively.

For the headwater zone of the upper River Eider, the slope usually drains directly into the river channel. A small floodplain separating slope and river channel is located further downstream in the River Eider Valley. The hydrological function of the floodplain in the valley is predominantly to provide a conduit for slope drainage. Moreover, land drainage via ditches and tile drains connects the slope to the channel. It follows that the existence of the floodplain is of little hydraulic importance, because slope drainage short-circuits the floodplain. However, heavily vegetated and silted-up ditches contribute, together with the vegetated River Eider, to the flooding of the adjacent floodplains.

Ceasing macrophyte mowing resulted in localized flooding of the embankments of the River Eider. For example, Figure 31.3.1.2(a) shows a comparison of the water level near the outlet arrangement of selected ditches before and during summer flooding in the absence of significant precipitation. This finding confirms observations made earlier by Jensen et al. (2001) and Trepel and Kluge (2002).

31.5 CONCLUSIONS AND RECOMMENDATIONS

The findings of the watercourse classification exercise indicated the dominance of small and heavily vegetated ditches. A minority of predominantly large ditches were supplied with groundwater at their upper stretches. The ceasing of aquatic plant mowing resulted in heavy vegetation growth within both the selected ditches and the River Eider, leading to a rise in the water level of approximately 0.8 m.

It is likely that ceasing ditch and river maintenance will lead to the disappearance of the ditches and subsequently to a rise of the water table in the neighboring agricultural fields, which is intended by the federal state government. This strategy should result in the improvement of the water quality of the River Eider. The likely formation of a shallow freshwater lake in the River Eider Valley will subsequently lead to the flooding of the riparian wetland and to a decline of peat degradation.

The natural horizontal flow obstruction area cover was at least 70% upstream of cross-section 60 m for the four selected representative ditches. Elevated flow obstruction cover values after ditch clearance were merely due to the development of meanders within the silt bed upstream of cross-section 100 m. The total roughness values of both the vegetated and excavated ditch stretches were relatively low in comparison to values for well-maintained open channels reported in the hydraulic literature. However, it would be difficult for an inexperienced scientist or engineer to correctly identify values by eye for the different ditch classes. Moreover, the total roughness is also a function of season influencing vegetation growth.

Significant hydrochemical differences within the selected ditches were noted. For example, the DO and water temperature within the ditch waters increased along the ditches toward the River Eider. Elevated BOD and nitrate–nitrogen concentrations for upstream ditch cross-sections were also apparent. Peat degradation and groundwater infiltration were likely sources of diffuse pollution.

Statistically significant water quality data variations between most ditches were also noted. For example, elevated conductivity concentrations indicated usually diffuse environmental pollution. The conductivity within all ditches was fairly constant (except for elevated levels near the springs), but on different concentration levels. Ditches that were not dominated by groundwater infiltration always had nitrate–nitrogen values <2.5 mg/l. Nitrate concentrations were higher for grazed than for ungrazed ditches.

A comparison between the water qualities of the River Eider and a representative selection of small watercourses indicated that the overall discharge from the small watercourses usually does not pollute the River Eider in terms of their average nutrient concentrations during May and June. This finding shows that ceasing ditch maintenance has no negative impact on the overall water quality of the River Eider.

The River Eider flooded large parts of the riparian wetland and subsequently discharged into the ditches during late summer. This was due to the growth of *S. emersum* L. in the River Eider, leading to an increase of the water level. However, the growth of vegetation within the ditches also resulted in elevated water levels. It follows that the river floodplain was flooded. The creation of a temporary shallow freshwater lake is anticipated, if ceasing of watercourse maintenance continues for another decade.

In particular, the *Phragmitetea australis* group (tall and productive macrophytes) dominated the ditch ecosystem. All four selected ditches were heavily vegetated by *P. arundinacea* L. Moreover, the growth of *C. acutiformis* Ehrh. was also dominant for the most extensively used ditch. Nevertheless, the wetland vegetation communities near the ditches could not be used meaningfully to further classify the identified groups of ditches within a water engineering and drainage context. No statistical data link between the cosmopolitan vegetation and the hydrology of the selected ditches was apparent even 10 years after ceasing ditch maintenance.

Further work on plant species, which could be used as water quality and quantity indicators, is recommended. These plants would need to be rather cosmopolitan and easy to identify by laymen to make a practical difference. Moreover, the long-term ecosystem service benefits and potential disservices associated with a shallow lake should be assessed. However, gaining the buy-in of the local population by promoting active local stakeholder engagement measures is crucial to success.

Chapter 32

Integrated Constructed Wetlands for Pollution Control

32.1 INTRODUCTION

The first research work related to the free-water surface integrated constructed wetlands (ICW) database has been published elsewhere (Carroll et al., 2005; Dunne et al., 2005a,b), providing clear evidence that phosphorus removal correlates positively with an increase in wetland area. Current mass pollutant removal by surface flow constructed wetland treatment of agricultural dirty water usually varies between 30% and virtually 100% of the phosphorus input, depending on its type (Braskerud, 2002; Newman et al., 2000; Reddy et al., 2001).

However, there are also some other references indicating insufficient phosphorus removal for constructed treatment wetlands. For example, constructed wetlands were not efficient in removing total phosphorus in Nordic countries, especially during periods of high loading rates (Smith et al., 2006). Good initial removal efficiency of total phosphorus, which, however, became less effectively removed after five years of operation, was reported by Tanner et al. (1999). The variation in wetland sizes and configurations makes it difficult to compare phosphorus removal performances.

The litter from decaying macrophytes provides surface area for attachment of biofilms and is therefore important for microbial processes such as the transformation of nutrients in wetlands. For most aquatic systems, the bulk microbial conversions are undertaken by microorganisms immobilized in sediments, microbial mats, and biofilms attached to solid surfaces. Therefore, sediment and litter samples should be collected to gain an insight into the microbial transformations taking place in removing nutrients from ICW.

The aim of Chapter 32 is to assess the ICW performance, putting special emphasis on the specific water treatment potential of selected ICW examples. The objectives are:

- to analyze and assess ICW performance;
- to assess the corresponding phosphorus and nitrogen removal capacity;
- to identify the presence or absence of ammonia-oxidizing and denitrifying groups of organisms; and
- to recommend ICW design details for subsequent guidelines.

Wetlands for Water Pollution Control. http://dx.doi.org/10.1016/B978-0-444-63607-2.00032-0

32.2 CASE STUDY SITES, MATERIALS, AND METHODOLOGIES

32.2.1 Sites within the Case Study Catchment

The entire Annestown-Dunhill case study catchment (near Waterford, Ireland) is divided into subcatchments serving each farmyard curtilage and comprises 19 working farms. In 1999 and 2000, 12 ICW (e.g., Figure 32.2.1.1) were designed, and construction was carried out to intercept and treat farmyard dirty water. Additionally, a sewage treatment ICW (site 7) was constructed to intercept the discharge from a septic tank (Table 32.2.1.1).

The ICW comprised several wetland cells operated in series, and their total sizes ranged between approximately 3000 and 23,000 m^2. Components of farmyard dirty water discharged to the wetlands were variable. The runoff typically consisted of yard and dairy washings and rainfall on open yard and farmyard roofed areas, along with silage and manure effluents. All ICW were in operation for at least seven years. ICW sites 1, 2, 5, 6, 7, 8, and 13 were subject to surface runoff and/or groundwater infiltration (Table 32.2.1.1). This led to reductions of toxic ammonia concentrations due to dilution, which supports healthy plant growth.

32.2.2 Sampling and Analytical Methods

Grab samples for each wetland cell inlet and outlet were taken on an approximately fortnightly basis. Water analysis was conducted at the Waterford County Council water laboratory using standard methods (APHA, 1998).

To better understand nutrient removal processes in different parts and components of selected ICW examples, sediment and litter samples were

FIGURE 32.2.1.1 Irish integrated constructed wetland example comprising four wetland cells and one polishing pond. *Photo taken by Dr Rory Harrington.*

TABLE 32.2.1.1 Site Characteristics of Farms and Corresponding Integrated Constructed Wetland Systems in the Anne Valley (near Waterford, Ireland), after Scholz et al. (2010)

ICW No.	Farm Enterprise	Farmyard Area (m^2)	Effective ICW Area (m^2)	ICW-to-Yard-Area Ratio
1	Dairy	4500	3906	0.9
2	Dairy	14,750	22,966	1.6
3	Dairy; beef	5400	10,288	1.9
4	Dairy	9200[c]	10,327	1.1
5	Dairy; tillage	4000	3940	1.0
6	Dairy	9800	12,691	1.3
7	Sewage	n/a	3075	n/a
8	Beef	2300	3940	2.0
9	Mixed	4800	7964	1.7
10	Mixed	2100	4375	2.1
11	Dairy	5000	7676	1.5
12	Mixed	13,600[d]	10,748	0.8
13	Sheep; tillage	5000	5610	1.1

ICW No.	Aspect Ratio of ICW Cells[a]	Dairy Washings[b] (Cow Number)	Roof Water	Spring Water Input
1	2.0	Yes (60)	Yes	Yes
2	3.1	Yes (60)	Yes	Yes
3	9.0	Yes (50)	Yes	No
4	6.6	Yes (100)	Yes	No
5	4.3	Yes (35)	Yes	Yes
6	4.7	Yes (80)	No	Yes
7	4.0	n/a	No	Yes
8	2.5	No	Yes	Yes
9	3.0	Yes (55)	Yes	No
10	3.0	Yes (50)	Yes	No
11	5.6	Yes (77)	Yes	No
12	3.6	Yes (85)	Yes	No
13	2.3	No	Yes	Yes

n/a = not applicable.

[a]*Nonvegetated first and last cells were excluded. Total cell numbers for ICW systems 1–13 were 9, 4, 6, 6, 4, 6, 4, 5, 5, 5, 4, 7, and 6, respectively.*

[b]*A "Yes" entry indicates discharges to the wetland and a "No" entry indicates that there were none.*

[c]*500 m^2 added in 2003.*

[d]*2000 m^2 and 1100 m^2 added in 2004 and 2005, respectively.*

collected in April and May 2007 from three different wetlands (ICW 3, 9, and 11), frozen, and sent off to Linköping University (Sweden) for subsequent molecular microbiological analysis. Samples from ICW 3 and 9 were taken on April 24, 2007, while samples for ICW 11 were taken on May 8, 2007.

Mostly triplicate field sediment and litter samples were collected 1 m to the left, 1 m in front, and 1 m to the right of each sampling point located near the inlet of each ICW cell and the outlet of each ICW system. Sediment samples were collected with a sediment sampler (diameter 4 cm). Green plant material was removed from litter samples before collection. Samples were stored at −20 °C before analysis.

32.2.3 Deoxyribonucleic Acid Extraction

Sediment and litter samples were subjected to deoxyribonucleic acid (DNA) extraction using a FastDNA® SPIN kit for Soil (Bio 101, Inc., La Jolla, CA, USA). Samples (0.25 g) were suspended in a sodium phosphate buffer supplied with the FastDNA® SPIN kit as stipulated by the manufacturer and homogenized for 180 s with a hand-held blender (DIAX 900 Homogeniser Tool G6, Heidolph, Kelheim, Germany).

DNA was extracted from soil samples by bead beating, a procedure in which soil aggregates are disrupted and bacterial cells are lysed mechanically. Bead beating was extended to 3×30 s to achieve good homogenization of the samples. The subsequent centrifugation was prolonged to 2×5 min and the centrifugation after washing with SEWS-M, a salt and ethanol wash solution (Qbiogene, Inc., USA), was extended to 5 min. The extracted DNA was stored at −20 °C.

32.2.4 Polymerase Chain Reaction

The ammonia-oxidizing bacterial community was investigated using group-specific polymerase chain reaction (PCR) primers targeting the 16S ribosomal ribonucleic acid (rRNA) gene, while the denitrifying bacterial community was assessed using the functional gene nitrous oxide reductase (nosZ), which is the gene for the terminal enzyme in denitrification (Hallin et al., 1999; Sundberg et al., 2007). A primer is a nucleic acid strand that serves as a starting point for DNA replication and is required because most DNA polymerases (i.e., enzymes that catalyze the replication of DNA) cannot synthesize a new DNA strand from scratch; rRNA is one of the three major types of ribonucleic acid (RNA) and is part of ribosomes and composed of RNA of different sizes, such as 5S, 16S, and 23S in prokaryotes.

The extracted DNA from all samples was diluted 10-fold to avoid inhibition of the PCR by humic substances. This was determined by testing for different dilution ratios. PCR amplification was undertaken using forward and reverse primers for ammonia-oxidizing bacteria. The PCR was performed on a

PTC-100TM thermal cycler (MJ Research Inc., San Francisco, CA, USA) in a 50 μl mixture (Sundberg et al., 2007b).

For the nitrous oxide reductase, the forward and reverse primers targeting the nosZ gene were used in the subsequent PCR. The PCR was performed on a PTC-100TM thermal cycler in a 50 μl mixture including 1.33 U of Taq polymerase and 5 μl of the supplied buffer (including 1.5 mM $MgCl_2$; Roche Diagnostic GmbH, Mannheim, Germany), each nucleotide at a concentration of 200 μM, the primers at 0.125 μM each, 600 ng/μl bovine serum albumin, and 2 μl of the DNA template (Sundberg et al., 2007a).

32.2.5 Agarose Gel Electrophoresis

The PCR products of DNA extraction and PCR reactions were examined by agarose gel electrophoresis. The agarose was melted by heating the mixture (agarose and buffer), and subsequently pouring it in the agarose gel casting tray. The solidified gel was covered with an electrophoresis buffer before running electrophoresis. The electrophoresis buffer was the same as the one used to prepare the agarose.

The PCR products and dye supplied with the DNA extraction kit (2 μl of dye and 4 μl of PCR products) were placed into the loading wells formed by the gel comb. The first well of each row was loaded with 2 μl of Gene Ruler (1 kb DNA ladder; 1000 base pairs for ammonia-oxidizing bacteria and nosZ) and 4 μl of distilled water. The electrophoresis was run for 40 min at 120 V (Owl Scientific, Inc., Woburn, MA, USA). The gel was then placed in ethidium bromide solution (immersed for 15 min) located in the fume cupboard and washed subsequently with tap water. The ethidium bromide-stained gel was then visualized by ultraviolet illumination.

32.3 RESULTS AND DISCUSSION

32.3.1 Wetland Water Quality

Water quality data for all wetland influents and effluents for the monitoring period between August 2001 and August 2007 were analyzed. Molybdate reactive phosphate (MRP) concentration reductions were generally greater than 90% (Table 32.3.1.1), with the most notable exception of ICW site 7, which is likely to be due to the high hydraulic loading rates and its similarity with a sewage treatment works. Owing to the high loading rates and treatment of domestic wastewater from a relatively high population of approximately 200, it exhibited only a 30% MRP reduction (i.e., an indication for an undersized plant).

No obvious signs of phosphate saturation along with wetland maturation were observed. There were also no significant differences between cold (winter) and warm (summer) months (seasons). Phosphorus is likely to accumulate owing to soil sorption and other processes within the sediment and detritus of

TABLE 32.3.1.1 Summary Data for Molybdate Reactive Phosphate (mg/l), after Scholz et al. (2010)

Flow	Statistic	ICW Site Number						
		1	2	3	4	5	6	7
Inflow	Max.	918.00	28.30	31.70	60.00	99.30	40.00	16.40
	Min.	0.20	0.20	0.05	1.45	0.01	0.20	0.65
	Mean	75.69	15.46	15.22	22.32	14.33	10.76	7.51
	SD	198.91	6.734	9.730	11.515	20.613	9.202	3.211
	N	25	28	72	82	25	27	33
Outflow	Max.	1.30	0.73	11.47	4.70	3.01	0.96	10.35
	Min.	0.01	0.01	0.02	0.02	0.01	0.01	0.16
	Mean	0.22	0.27	3.43	1.66	0.24	0.13	5.25
	SD	0.384	0.231	2.334	0.940	0.639	0.218	2.977
	N	28	28	76	64	28	28	64
% Phosphate reduction		99.71	98.22	77.46	92.56	98.34	98.83	30.08

		ICW Site Number					
Flow	Statistic	8	9	10	11	12	13
Inflow	Max.	5.00	124.71	42.80	50.96	250.00	4.49
	Min.	0.05	0.01	0.14	0.35	0.91	0.02
	Mean	1.46	8.57	7.16	11.55	43.53	0.94
	SD	1.250	13.754	9.357	10.05	58.061	1.413
	N	26	98	69	131	37	21
Outflow	Max.	0.29	1.59	0.36	4.41	5.22	0.32
	Min.	0.00	0.03	0.01	0.00	0.01	0.00
	Mean	0.04	0.47	0.06	0.94	0.74	0.06
	SD	0.063	0.307	0.094	0.628	0.916	0.089
	N	27	74	24	131	65	24
% Phosphate reduction		97.21	94.52	99.16	91.86	98.30	93.34

min. = minimum; max. = maximum; SD = standard deviation; N = number of entries.

the wetland cells (i.e., sinks for pollutants), as discussed by Carroll et al. (2005) and Dunne et al. (2005a). Plants were not harvested to remove phosphorus.

Concerning the inflow mean water quality values, chemical oxygen demand (COD) was 1813 ± 3763 mg/l, standard five-day biochemical oxygen demand (BOD) was 539 ± 1041 mg/l, suspended solids (SS) was 679 ± 2856 mg/l, ammonia-nitrogen was 61 ± 121 mg/l, MRP was 15 ± 21 mg/l, and there were 321,600 ± 384,800 *Escherichia coli* colony-forming units (CFU) per 100 ml. This indicates very high water quality variability concerning the farmyard dirty water discharged to the ICW systems.

The mean effluent water quality was 78 ± 73 mg/l COD, 18 ± 21 mg/l BOD, 24 ± 46 mg/l SS, 1.1 ± 2.6 mg/l ammonia-nitrogen, 1.3 ± 1.6 mg/l MRP, and 228 ± 306 *Escherichia coli* CFU per 100 ml during the monitoring period. This suggests that most ICW systems had a high capacity to remove pollutants. If the two farm sites (ICW sites 8 and 13) without dairy washings had been excluded, the BOD concentration of water entering the other ICW systems would have been approximately 1800 mg/l. This value is consistent with soiled water such as dairy washings diluted with cleaner water, including yard and roof runoff (European Community, 2006).

32.3.2 Flows

Hydraulic fluxes from extraneous sources were minimal at ICW 11. Therefore, flow meters were installed at the inlets to cells 1 and 2 and the outlet at this example site. Mean daily inflows and outflows from the ICW system during the monitoring period (April 2003–July 2007) were 5.97 ± 7.35 m^3/day and 1.59 ± 4.47 m^3/day, respectively. Although there was usually inflow to the ICW system between August and December, there was no outflow from it during most of these periods. It follows that there was only a short discharge period to receiving surface waters.

The patterns of flow continuously recorded at ICW 11 were similar to those estimated for other ICW systems, where instantaneous flows were measured manually. Decreases in flow through the wetlands were recorded during the summer, and none of the 12 ICW systems were discharging to receiving watercourses.

During dry periods, the loss of water from ICW due to evapotranspiration and slow infiltration created an increased storage capacity (freeboard) within most ICW cells. This provided a buffering capacity for wet weather periods such that waters are stored within the ICW system prior to discharge during storm events. Furthermore, groundwater contamination due to the infiltration of partially treated runoff was not observed in nearby wells.

32.3.3 Design Considerations

The variations in ICW design and the observed variations in effluent quality allowed examination of the influence of some key design factors on water treatment performance. Mean effluent MRP concentrations (Table 32.3.1.1) were considered to be an appropriate indicator of the treatment performance. The ICW were built with a mean size of 1.3 times the farmyard area and with a minimum of four cells. Most (10 out of 12) of the ICW built to this design produced a good-quality effluent with MRP concentrations below the nominal target concentration of 1 mg/l (Table 32.3.1.1). Thus the sizing of at least 1.3 times interception area and a minimum of four (approximately equally sized) vegetated cells are taken to be the primary design criteria. ICW systems 3 and 4, which did not meet the desired effluent quality, had long, narrow ponds (i.e., high individual aspect ratio). The mean aspect ratio is defined as the mean length of the wetland system divided by the mean width. In this study, individual aspect ratios ranged from 2 to 9. A linear regression analysis ($MRP = 0.45 \times$ aspect ratio) between the aspect ratio and the effluent MRP concentrations had a coefficient of determination of 0.55 ($P < 0.05$). An aspect ratio of less than 4 resulted in good MRP removal. These are considered to be the minimum criteria for good runoff treatment performance (i.e., effective nutrient removal). Ideally, for system robustness, farm ICW should be built to a larger relative size (twice the interception area) and with more than four cells.

The observed impact of the ICW total cell number, cell sizing, and cell aspect ratio on effluent quality reflects the fact that the effluent should stay as long as possible in the ICW (and also have a flow velocity as low as possible) to allow the ICW ecosystems to develop, thrive, and be able to remove pollutants (Keffala and Ghrabi, 2005; Scholz, 2006). This is achieved by round wetland cells (which are, however, not economical if land costs are high) and low gradients, thus allowing the pollutant plume to spread as slowly as possible throughout each ICW cell.

32.3.4 Nitrogen Transformation Processes

In comparison to ammonia oxidizers, denitrifiers were more abundant in most of the collected litter and sediment samples (Table 32.3.4.1). Since the nitrate concentrations within the ICW systems were low, it is likely that oxygen and nitrate have served as electron acceptors in the lower layer of the wetland cells, and this might have promoted the growth of denitrifying bacteria. Each ICW system contained denitrifying bacteria, but they were present in varying quantities. For example, ICW 11 had lower denitrifying bacteria numbers than both ICW 3 and ICW 9. Samples analyzed for ICW 3 and ICW 9 did not indicate the presence of ammonia-oxidizing bacteria.

Most denitrifiers are heterotrophs. The supply of organic carbon by macrophytes raised the overall heterotrophic activity, leading to the consumption of oxygen. Thus, oxygen availability in the sediment was reduced, and subsequently denitrification was supported (Bastviken et al., 2005).

TABLE 32.3.4.1 Bacterial Community Abundance in Integrated Constructed Wetland Litter and Sediment Samples Compared to Corresponding Mean Ammonia-Nitrogen and Nitrate–Nitrogen Concentrations within the Top Water Phase, after Scholz et al. (2010)

ICW No.	Ammonia-Oxidizing Bacteria[a]	Denitrifying Bacteria[a]	Ammonia-Nitrogen (mg/l)	Nitrate–Nitrogen (mg/l)
3	0	73	13.95	0.27
9	0	80	5.20	1.69
11	27	53	9.73	1.73

[a]*Relative presence (%) of bacterial community (0 = absent; 100 = present) for multiple samples.*

Ammonia-oxidizing bacteria were present at ammonia-nitrogen concentrations between approximately 5 and 20 mg/l. Denitrifying bacteria were observed at nitrate–nitrogen concentrations between 0.1 and 4.5 mg/l. In comparison to ammonia-oxidizing bacteria, more denitrifiers were present in most ICW systems, while ammonia-oxidizing bacteria were found in samples collected from ICW 11 only.

32.3.5 Comparison of Ammonia Oxidizers and Denitrifiers

Interesting differences between three selected ICW example sites are discussed in this section (see also Table 32.3.4.1). For example, concerning ammonia-oxidizing bacteria, samples analyzed from ICW 3 and ICW 9 did not indicate the presence of ammonia-oxidizing bacteria. Concerning denitrifying bacteria, ICW 3 had lower denitrifying bacteria numbers than ICW 9. Furthermore, ICW 9 had a higher aquatic plant cover density than ICW 3. It follows that the decaying macrophytes within ICW 9 contributed to organic matter, which became a source of carbon and therefore energy for denitrifying bacteria.

Concerning ICW 9 versus ICW 11, samples obtained from ICW 9 did not indicate the presence of ammonia-oxidizing bacteria, while samples taken from ICW 11 indicated the presence of these bacteria. There was a reduced availability of organic matter at the bottom of ICW 11, which led to decreased numbers of heterotrophic bacteria and consequently created conditions in which ammonia-oxidizing bacteria proliferated. Regarding denitrifying bacteria, ICW 9 had higher denitrifying bacteria numbers than ICW 11. Furthermore, ICW 9 had a higher plant density than ICW 11. The high number of denitrifying bacteria was linked to high concentrations of nitrate. The decaying plants also contributed to organic matter that became a source of energy for denitrifying bacteria.

32.4 CONCLUSIONS AND RECOMMENDATIONS

Findings suggest that integrated constructed wetlands were capable of treating farmyard dirty water and that they provided a sustainable management option to effectively reduce nutrient and contaminant loss from farmyards to watercourses. Significant concentration reductions in suspended organic matter, nutrients, and fecal bacteria between ICW influents and effluents were observed. Surface discharges from the ICW sites had seasonal patterns. None of the farm ICW had surface discharges during most summer months.

In terms of ICW design for the southeast of Ireland, the effective ICW area should be sized from the farmyard area by applying a multiplying factor of at least 1.3 to the latter. The optimal theoretical ICW cell aspect ratio (to achieve an effluent MRP concentration below 1 mg/l) was approximately 1:1; i.e., the ICW wetland cells become relatively round. The small aspect ratio is specific for ICW comprising planted cells, which are characterized by very low flow velocities. The recommended minimum number of cells within each ICW was four. However, constructing a greater number of cells with greater surface areas was advised, as both the treatment performance and the aesthetic appeal of the ICW system increases with increasing total wetland area.

The proposed design results in relatively large treatment systems, which are unsuitable if land prices are high (e.g., inner-city areas). However, the absence of both a liner as well as standard engineering control and operational structures makes these systems particularly affordable for farmers.

The number of denitrifying bacteria detected in different ICW systems was higher than the number of ammonia-oxidizing bacteria. The high vegetation density impeded the water flow at the surface. Anoxic conditions at the wetland bottom (coupled with a high organic carbon matter content available from the decaying macrophytes) created conditions that were favorable for denitrifying bacteria. It follows that sediment and litter within the ICW supported denitrification.

Further research is required on decreasing the costs for molecular microbiological analysis to make the proposed techniques more widely available. Too little is also known about the ecology and engineering application of many new microbes that have been discovered since 2005. Engineers are usually only interested in organisms that pose a risk to human health or that can be used to treat waters.

Chapter 33

Infiltration and Contaminant Migration beneath Integrated Constructed Wetlands

33.1 INTRODUCTION

Across the world, sustainable and effective wastewater treatment methods that are also inexpensive are being sought to address problems of pollution to both surface and groundwater, as well as to protect quality of life. All liquid wastes should be treated to a specified standard before being discharged into a water body. This has further given impetus to the development of alternative wastewater treatment techniques that meet regulatory criteria, while also providing a more environmentally sustainable option.

Among the different treatment systems now available worldwide, the use of constructed treatment wetlands has attracted more attention; these have evolved as a viable alternative to traditional wastewater treatment systems largely because they are easy to operate, require low maintenance, and have low investment costs (Hammer et al., 1993; Davis, 1995; Scholz, 2010). Vertical-flow, horizontal-flow, and surface-flow wetlands based on soil, sand, and/or gravel (Sun et al., 1999; Scholz, 2010) have been used worldwide for many years to treat wastewater from point and diffuse sources.

In Ireland, a specific design concept of constructed treatment wetlands, namely ICW, has been developed over the last two decades by the ICW Initiative of the Irish Department of Environment, Heritage and Local Government. The ICW concept endeavored to achieve water treatment, landscape fit, and biodiversity enhancement targets through an innovative wetland design methodology. As described by Scholz et al. (2007), the ICW concept employs the free water surface flow (FWS) constructed wetlands model and incorporates the concept of restoration ecology, specifically mimicking the structure and processes of natural wetlands. Most ICW are characterized by a multicelled configuration with sequential through-flow and are based on the holistic and interdisciplinary use of land to improve water quality. In general, ICW have shallow water depths (10–30 cm) and contain many plant species. This facilitates microbial and animal diversity (Nygaard and Ejrnæs, 2009;

Wetlands for Water Pollution Control. http://dx.doi.org/10.1016/B978-0-444-63607-2.00033-2

Jurado et al., 2010) and is generally aesthetically appealing, which enhances recreation and amenity values.

Classically, ICW systems are built by utilizing resources from the site so that they will not require the installation of expensive artificial liners such as plastic or concrete to impede wastewater infiltration and the consequent contaminants loading into the nearby groundwater system. The local on-site subsoil materials are reworked and compacted to form a low-permeability liner for the ICW cells. Typically, ICW are underlain by at least 1 m of subsoil, with the upper 0.5 m enhanced, where highly permeable subsoil is encountered, to a hydraulic conductivity of 1×10^{-8} m/s (Dunne et al., 2005a,b; Harrington et al., 2007; Scholz et al., 2007; Carty et al., 2008). However, because the soil is a dynamic system that has been known to transmit water contaminants through its pore space, the question of wastewater infiltration from ICW cells as a potential source of groundwater pollution has often been raised.

Particularly in Ireland, concerns are raised regarding the movement of nitrogen compounds into local drinking water supplies, where approximately 26% of all the drinking water supply is provided by groundwater (EPA, 2008). The limit for maximum nitrogen concentrations in groundwater is founded on health-based grounds at 10 mg NO_3−N/l, as defined by World Health Organization (WHO) guideline values (WHO, 2008). Ultimately, groundwater flow can transport this nutrient-rich infiltration water to surrounding and nearby surface waters, where water quality degradation may occur (Hathaway et al., 2010). For ICW, contaminant loading represents the rate at which ammonia-nitrogen, nitrate−nitrogen, and other soluble compounds flow from the wetland into the underlying soils. Actual measurements of infiltration from ICW cells under field conditions are therefore needed to determine the risk of groundwater contamination. Furthermore, an understanding of the factors governing infiltration under field conditions is required to improve the design, construction, and management of ICW systems.

Nevertheless, wastewaters received by ICW systems often have high suspended solids content as well as organic matter and related compounds. These can develop a seal at the soil−water interface and may potentially further impede infiltration beneath the ICW cells. Mechanisms for this phenomenon include physical clogging of soil pore spaces by the entrapment of sludge and particulates of an organic nature, the growth of polysaccharides and other microbial by-products, or chemical sealing from salt-induced dispersion of clay (Chang et al., 1974; Hills, 1976; Barrington et al., 1987; Barrington and Madramootoo, 1989; Davis et al., 1973; Kadlec and Knight, 1996; Maulé et al., 2000; Scholz, 2006). Several studies have concluded that this sealing effect can sufficiently limit wastewater infiltration to rates that would not significantly contaminate the shallow groundwater (Hills, 1976; Ritter et al., 1984; Miller et al., 1985; Rowsell et al., 1985).

Moreover, factors such as the toxicity of the wastewater constituents and their input loading occurring beneath the ICW cells under field conditions, as well as the intrinsic vulnerability of the underlying aquifers, could additionally influence the potential impact of ICW on groundwater quality. With the main aquifers in Ireland occurring in fissured or fractured bedrock formations overlain by subsoil of variable thickness and permeability (Misstear and Daly, 2000), the subsoil properties beneath ICW cells could be important in determining the fate of the infiltrating contaminants.

The importance of subsoil has been demonstrated by examining chemical profiles beneath old animal waste lagoons, where it has been shown that concentrations of ammonia-nitrogen (NH_4—N) and other cations were higher near the lagoon liner and then decreased with depth (Miller et al., 1976; Culley and Phillips, 1989; Maulé and Fonstad, 1996; Perschke and Wright, 1998; Ham, 2002). On the other hand, anions like chloride (Cl^-), which are not adsorbed by soil clays, can reach much greater depths. Therefore, the risk of groundwater contamination is not only dependent on the infiltration rate but also governed by the chemical concentrations in the wastewater, the depth to the water table, and soil properties beneath the ICW (Ham, 2002).

While several studies, such as Dunne et al. (2005a,b), Mustafa et al. (2009), Kayranli et al. (2010), Dong et al. (2011), and Dzakpasu et al. (2011), have indicated that ICW systems can provide an effective method of wastewater treatment from both point and diffuse sources, the question still remains as to whether the use of local soil materials to line ICW cells is capable of providing effective protection to the underlying and associated groundwater. Furthermore, limited studies have been carried out concerning the potential infiltration of water contaminants from full-scale constructed wetlands into the subsoil.

The aims addressed in Chapter 33 are to evaluate infiltration and the rate of contaminant loading occurring underneath full-scale earthen-lined ICW cells and to assess the implications in terms of potential groundwater pollution, as well as the overall importance of infiltration in the water contaminant reduction within ICW. The specific objectives are (1) to measure the rate of infiltration and estimate the daily volume of water discharged beneath ICW cells; (2) to estimate the apparent permeability of compacted earthen liners; and (3) to evaluate water contaminant loading rates occurring beneath ICW cells.

33.2 CASE STUDY, MATERIALS, AND METHODS

33.2.1 Study Site

The ICW treatment system at the center of the study is located within the walls of Castle Leslie Estate at Glaslough in County Monaghan, Ireland (06°53′37.94″ W, 54°19′6.01″ N). The site is surrounded by woodland and required sensitive development in terms of landscape fit, as well as biodiversity, amenity and habitat enhancement.

The ICW (Figure 33.2.1.1) comprises a small pumping station, two sludge cells, and a sequence of five shallow vegetated wetland cells. Hydraulic

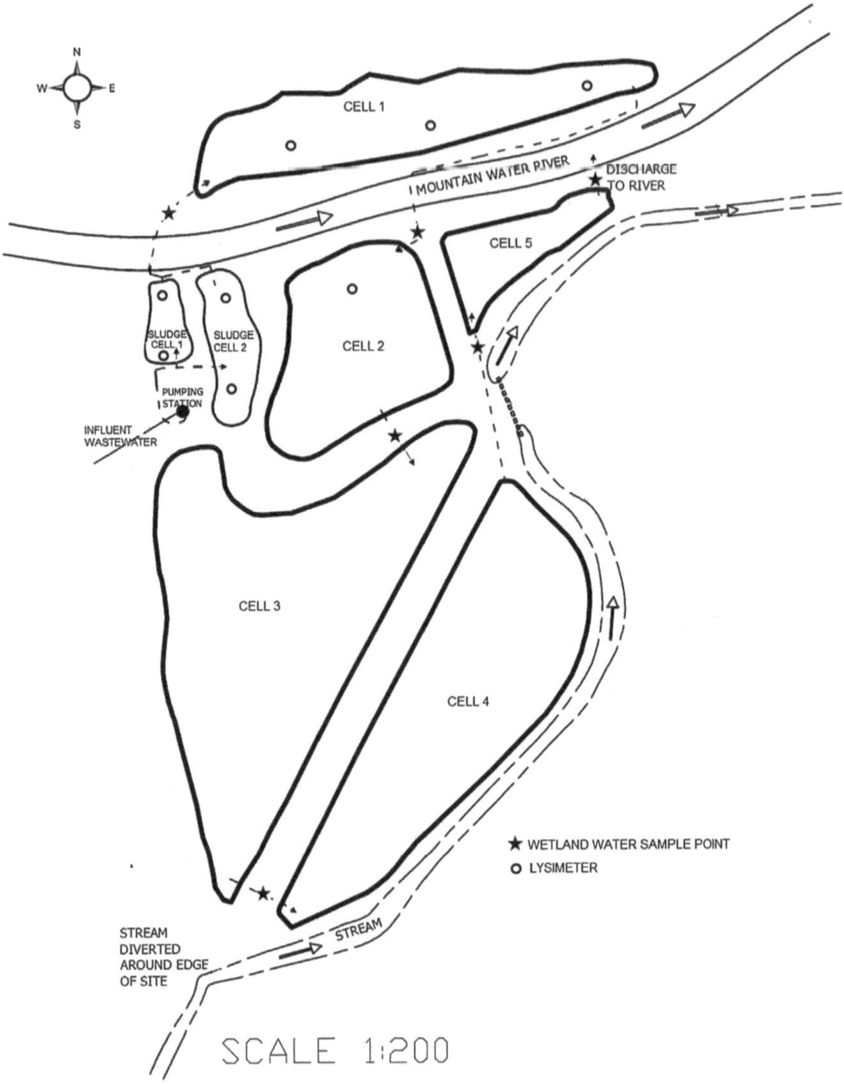

FIGURE 33.2.1.1 Schemata of ICW located at Glaslough in Ireland showing all sampling points (after Dzakpasu et al., 2012).

characteristics of the wetland cells are presented in Table 33.2.1.1. The wetland system was commissioned in October 2007 to treat combined sewage from Glaslough village and to improve the water quality of the Mountain Water River, which flows through the site. The design capacity of the ICW system is 1750 population equivalent. The functional water area of the ICW cells is 3.25 ha, within a curtilage area of 6.74 ha. The wetland cells have no artificial lining. Excavated local soil material was used to construct the base of

TABLE 33.2.1.1 Hydraulic Characteristics of the First Three ICW Cells

ICW Section	Area (m^2)	Depth (m)	Volume (m^3)	Wastewater Inflow (m^3/day)	Effective Flow Rate (m^3/day)	nHRT (days)
Sludge cell	285	0.45	128.3	112.9	113.2	1
Cell 1	4664	0.42	1958.9	96.9	103.0	19
Cell 2	4500	0.38	1710.0	106.3	109.7	15

nHRT = nominal hydraulic retention time.

the wetland cells and compacted to a thickness of 500 mm to form a low-permeability liner.

A site investigation conducted in September 2005 indicated a soil coefficient of permeability of 9×10^{-11} m/s (IGSL Ltd, Business Park, Naas, Co. Kildare, Ireland). The soils at the study area comprise a mixture of coarse- and fine-grained materials, namely alluvium, organic soils, tills, and gravel. Subsoil samples collected from the first 2 m near the sludge cell were classed as sandy gravelly clay, whereas those collected near cell 1 and cell 2 were classed as sandy silt and silty clay, respectively (IGSL Ltd, Business Park, Naas, Co. Kildare, Ireland). Typically, the mineral content of the soils is in the following range: Al^{3+}, 4.1%–5.0%; Fe^{2+}, 2.01%–2.5%; Ca^{2+}, 0.301%–0.45%; Na^+, 0.501%–0.65%; K^+, 0.751%–1.00%; Mg^{2+}, 0.41%–0.5%; organic C, 6.1%–10%, and pH, 5.01–5.5 (Fay et al., 2007). The main ICW system is flanked by the Mountain Water River and the Glaslough Stream.

Influent primary domestic wastewater from the village is pumped directly into a receiving sludge cell. The system contains two sludge cells that can be used alternately so that one can be de-sludged without interrupting the whole treatment process. The purpose of the sludge cell is to retain the suspended solids contained in the influent wastewater. In this way, the buildup of sludge in the wetland cells, which could otherwise decrease the capacity of the cells, is prevented. From the sludge cell, the wastewater subsequently flows by gravity sequentially through the five earthen-lined cells, and the effluent of the last cell discharges directly into the adjacent Mountain Water River.

The wetland cells were planted in a club pattern and the main plantings were *Carex riparia* Curtis, *Phragmites australis* (Cav.) Trin. ex Steud., *Typha latifolia* L., *Iris pseudacorus* L., and *Glyceria maxima* (Hartm.) Holmb. This currently includes a complex mixture of *Glyceria fluitans* (L.) R.Br., *Juncus effusus* L., *Sparganium erectum* L. emend Rchb, *Elisma natans* (L.) Raf., and *Scirpus pendulus* Muhl.

33.2.2 Instrumentation

During construction of the ICW, eight large zero-tension pan lysimeters were constructed and placed in the ground underneath the first wetland cells to provide a sample of the infiltrating water. In zero-tension pan lysimeters, there are no wicks exerting suction on the sampled soil and, therefore, they collect soil water only when the subsoil above the sampler is saturated, allowing free drainage. In this way, the extraction of bound moisture that would otherwise be unavailable to groundwater recharge is prevented. There are two lysimeters in each of the sludge cells, three in cell 1, and one in cell 2. This is because these four cells contain the most polluted water within the ICW, and so these cells are the most likely to contaminate the groundwater. The study presented in this chapter covered one sludge cell, cell 1, and cell 2. These were the first three ICW cells in sequence.

The lysimeters used in this study are 900 mm in diameter. They are large circular high-density polyethylene cylinders with the tops open. Installation was similar to that described by Martin et al. (1999). At the site where the lysimeters were to be placed, the soil was removed, layer by layer, and separated into individual piles. Once all of the soil was taken away, the lysimeters were set in place, approximately 500 mm below the soil surface beneath the wetland cells, and the cells were filled with soil, again layer by layer, so as to best represent the soil in its natural condition. Careful installation of these lysimeters into the undisturbed subsoil directly beneath the wetland cells was essential to avoid creating artificial preferential flow paths and to ensure that they were located within the potential effluent plume. The lysimeters were lined with a glass microfiber filter, a layer of washed pea gravel, and a plastic mesh. Drainage tubing was then connected to each lysimeter to allow draining when the collection bucket became full.

33.2.3 Water Sampling and Analysis

Infiltration water was collected weekly between November 2009 and May 2011 by attaching 2 l polyethylene bottles to the open end of each drainage tubing, which extended beyond the ICW cells into a dug-out area in the ground near the banks of the adjacent river, allowing water to flow under gravity to the bottles. Typically, the sample bottles were left out on day 1 and collected a day later. Collected water samples were transported directly to the laboratory for immediate analysis. The volumes of water collected were recorded for the time period, typically 24 h, or converted to determine an average daily flow, as shown in Eq. (33.2.3.1), as discussed by United States Bureau of Reclamation (USBR, 1997). The rate of infiltration was calculated as the linear velocity of flow (v) to the lysimeters using Eq. (33.2.3.2).

$$Q = \frac{V}{t} \tag{33.2.3.1}$$

where Q = volumetric flow rate (m^3/day), V = volume of water that was collected from each lysimeter (m^3), and t = amount of time elapsed between sampling periods (days).

$$v = \frac{Q}{A} \tag{33.2.3.2}$$

where v = average linear velocity of flow (m/day), and A = cross-sectional area of lysimeter perpendicular to water flow (m^2). Area of the lysimeters used is 0.64 m^2.

33.2.4 Hydraulic Conductivity of Soil Liner

When the volumetric data were transformed into daily flows, the apparent hydraulic conductivity of the soil liners was estimated using Darcy's law. According to this law, the flow of water through a porous material is proportional to the hydraulic gradient and the hydraulic conductivity, as shown in Eq. (33.2.4.1), which has been discussed elsewhere (Freeze and Cherry, 1979; Cedergren, 1997; Fetter, 2001; Jury and Horton, 2004).

$$Q = \mathrm{k}A\left(\frac{dh}{dl}\right) = \mathrm{k}iA \tag{33.2.4.1}$$

where Q = volume of water per unit time (m^3/day), i = hydraulic gradient (dimensionless), A = cross-section of flow area (m^2), and k = hydraulic conductivity (m/day).

The hydraulic gradient (i) was determined by the Green-Ampt (1911) equation (Chin, 2000). This approach assumes the following: a wetting front extends to some depth, l, under ponded water conditions; water infiltrates into the soil as "piston flow," resulting in a sharply defined wetting front between the infiltration zone and soil at the initial water content; the wetting front moves downward as a sharp interface; the soil is saturated above the wetting front and the water content is equal to the porosity; and the water content below the wetting front is equal to some lower initial value. The hydraulic gradient is approximated by Eq. (33.2.4.2):

$$i = \frac{H_o + L + h_{wf}}{L} \tag{33.2.4.2}$$

where H_o = head of water in wetland (m), and L = depth of wetting front beneath ICW cell (m). In this study, L is a constant that represents the depth to the lysimeter. h_{wf} = average capillary head at the wetting front (m).

33.2.5 Wetland System Water and Hydrological Monitoring

A suite of automated sampling and monitoring instrumentation such as the ISCO 4700 refrigerated automatic wastewater sampler (Teledyne Isco, Inc., Lincoln, Nebraska, USA) was used for weekly wetland water sampling.

Additionally, all water flows into and out of each ICW cell were measured and recorded with Siemens electromagnetic flow meters FM MAGFLO and MAG5000 (Siemens Flow Instruments A/S, Nordborgrej, Nordborg, Demark) and their allied computer-linked data loggers. Mean flows were recorded at 1 min interval frequency. A weather station is located beside the inlet pump sump to measure local temperature, precipitation, and evapotranspiration. Overall, precipitation and wastewater outflow from the preceding cells were considered as input sources to each ICW cell. Evapotranspiration and water infiltration were assumed to be water losses. Precipitation and evapotranspiration were measured as the amount of water falling on, or evaporating from, the wetland cell surface, respectively. The groundwater table in this area is approximately 2 m below the ground level. Therefore, groundwater discharge to the ICW cells was assumed to be negligible.

33.2.6 Water Quality Analysis

The water samples were analyzed weekly for several parameters, including chemical oxygen demand (COD), ammonia-nitrogen (NH_3−N), nitrate−nitrogen (NO_3−N), and molybdate reactive phosphate (MRP), at the Monaghan County Council wastewater laboratory in Ireland, by following Standard Methods for the Examination of Water and Wastewater (APHA, 1998), using kits supplied by HACH Lange (HACH Company, Loveland, Colorado, USA). The reactor digestion method, followed by colorimetric analysis (Method 8000), was used to measure COD, whereas NH_3−N and NO_3−N were determined by the Nessler method (Method 8038) and the cadmium reduction method (Method 8171), respectively. MRP was determined by the ascorbic acid method (Method 8048). All analyses were done using the HACH DR/2010 portable data logging spectrophotometer (HACH Company).

33.2.7 Contaminant Loading Rates

The contaminant mass loading in the ICW cells was calculated using the dynamic water budget approach shown in Eq. (33.2.7.1), which has been discussed by Kadlec and Knight (1996), as well as by Kadlec and Reddy (2001). The hydraulic loading rate was calculated using Eq. (33.2.7.2). The effective flow rate was determined by applying Eq. (33.2.7.3). The mass removal efficiencies (%) for contaminants were calculated using Eq. (33.2.7.4). Contaminant fluxes from the wetland cells in infiltration water (g/m^2/day) were determined by multiplying the average contaminant concentration for a given lysimeter with daily infiltration rates, as shown in Eq. (33.2.7.5).

$$L_o = q \times C_o \qquad (33.2.7.1)$$

where L_o = contaminant inflow loading rate (g/m/day), q = hydraulic loading rate (m/day), and C_o = inflow concentration of contaminant in ICW cell (g/m).

$$q = \frac{Q}{A} \tag{33.2.7.2}$$

where Q = effective flow rate in ICW cell (m^3/day) and A = area of ICW cell (m^2).

$$Q = Q_o + (P - ET - I)A \tag{33.2.7.3}$$

where Q_o = inlet wastewater volume flow rate in ICW cell (m^3/day), P = precipitation rate (m/day), ET = evapotranspiration rate (m/day), and I = infiltration rate in ICW cell (m/day).

$$\text{Removal efficiency} = \frac{(Q_oC_o) - (Q_eC_e)}{(Q_oC_o)} \times 100 \tag{33.2.7.4}$$

where Q_e = outlet wastewater volume flow rate in ICW cell (m^3/day), and C_e = outflow concentration of contaminant in ICW cell (g/m).

$$\text{Loading rate} = \frac{Q_f \times C_f}{A_l} \tag{33.2.7.5}$$

where Q_f = daily water volume infiltrating beneath wetland cell (m^3/day), C_f = contaminant concentration in infiltration water (g/m^3), and A_l = cross-sectional area of lysimeter (m^2).

33.2.8 Statistical Analysis

Data distributions were tested for normality using the Shapiro–Wilk normality test (Shapiro and Wilk, 1965). Statistically significant differences were determined at $\alpha = 0.05$, unless otherwise stated. Comparisons of means were by paired-samples t-tests and one-way analysis of variance (ANOVA). Tukey HSD (Honestly Significant Difference) tests were applied for post-ANOVA pairwise comparisons to identify significant differences among means. All statistical analyses were performed using Minitab 16 statistical software (Minitab Inc, State College, Pennsylvania, USA) and IBM SPSS Statistics 20 (IBM Corporation, Armonk, New York, USA).

33.3 RESULTS AND DISCUSSION

33.3.1 Infiltration Rates

The average rates of infiltration recorded throughout the study period (November 2009 to May 2011) were variable in all three ICW cells. Cell 2 recorded significantly higher ($P < 0.05$) average rates of infiltration than cell 1 and the sludge cell, and this ranged between 1.75×10^{-10} and 2.27×10^{-8} m/s. Cell 1 and the sludge cell recorded lower rates of infiltration (Table 33.3.1.1)

TABLE 33.3.1.1 Average Rates of Infiltration beneath ICW Cells and In situ Apparent Hydraulic Conductivity of ICW Cell Liners Recorded between November 2009 and May 2011 (after Dzakpasu et al., 2012)

		Infiltration Rate ($\times 10^{-9}$ m/s)		Apparent Hydraulic Conductivity ($\times 10^{-9}$ m/s)	
ICW Section	n	Mean	SD	Mean	SD
Sludge cell	45	4.34	3.66	3.80	3.10
Cell 1	51	3.67	1.42	3.53	1.35
Cell 2	50	10.2	4.54	9.78	4.25

n = number of samples; SD = standard deviation.

and were not significantly different ($P > 0.05$) from each other. The significantly lower infiltration rates recorded in cell 1 and the sludge cell could be partly attributable to the characteristics of wastewater received by these two cells, which contained higher concentrations of suspended solids and of organic matter and related compounds than cell 2, and these could develop a seal at the soil–water interface to potentially impede the infiltration rates (Chang et al., 1974; Hills, 1976; Barrington et al., 1987; Barrington and Madramootoo, 1989; Davis et al., 1973; Kadlec and Knight, 1996; Maulé et al., 2000; Scholz, 2006).

As part of a laboratory study involving seven different Saskatchewan soils with clay contents ranging from 9% to 33% collected from subsoil materials at a depth of 3 m below natural ground level, Maulé et al. (2000) showed that swine waste reduced seepage through compacted animal lagoon liners by two or three orders of magnitude and that this effect was independent of soil liner texture. Fresh swine manure from local pen gutters was used with a total N of 6100 mg/l and electrical conductivity of 39 mS/cm. According to Maulé et al. (2000), flow was restricted by a thin, 3- to 8-mm-thick "black layer" that developed at the liner–sludge interface and reduced the effective hydraulic conductivity of the whole liner to approximately 1×10^{-9} m/s. Similarly, Hills (1976) showed that dairy waste with total solids content between 1690 and 16,073 mg/l and chemical oxygen demand of 38,000 mg/l, which were diluted 5:1 with tap water, caused large reductions in flow-through liners constructed from loam, silt-loam, sand-loam, and clay-loam soils containing additives such as bentonite, sodium carbonate, and sodium tripolyphosphate, during the initial 16 weeks of waste addition. DeTar (1979) indicated that the infiltration rate of liquid dairy manure (which did not contain the heavy fibers that tend to settle out immediately after mixing) into four different soils, each having different steady-state infiltration rates with water, decreased as the total solids content of the manure increased. This sealing effect can be sufficient to limit wastewater infiltration from ICW cells (Hills, 1976; Ritter et al., 1984; Miller et al., 1985; Rowsell et al., 1985).

The rates of infiltration increased slightly within the flow path from the influent point to the effluent in the sludge cell and cell 1. In the sludge cell, $2.44 \pm 1.17 \times 10^{-9}$ m/s and $4.69 \pm 3.75 \times 10^{-9}$ m/s (mean $\pm$ SD) were recorded, respectively, in the lysimeters near the influent and effluent points. Similarly, $3.75 \pm 5.44 \times 10^{-9}$ m/s infiltration was recorded in the lysimeter near the influent point in cell 1, and $4.05 \pm 1.29 \times 10^{-9}$ m/s in the one near the midpoint of the cell. Infiltration in the lysimeter near the effluent point of cell 1 was slightly higher ($6.72 \pm 1.39 \times 10^{-7}$ m/s). However, the quality of infiltration water collected from this lysimeter was similar to that of the groundwater quality near the area and therefore suggested a possible discharge of groundwater into the lysimeter.

Overall, the average infiltration rates recorded in all three ICW cells were lower than those in other earthen-lined wastewater holding ponds. Several researchers, such as Demmy et al. (1993), found the seepage rate of an unlined dairy holding pond in a clay soil to be in the order of 2×10^{-7} m/s. Miller et al. (1985) also reported an infiltration rate of 1×10^{-8} m/s beneath a large earthen storage pit containing dilute liquid beef manure. Similarly, Rowsell et al. (1985) measured infiltration of liquid beef manure in various soils and found that the infiltration rate decreased rapidly with time and reached approximately 1×10^{-8} m/s in all soils.

Culley and Phillips (1982) performed percolation tests using liquid dairy manure on sand, loam, and clay soils and found that the hydraulic conductivity of all soils reduced to about 3.4×10^{-8} m/s. In a field study, Ham and DeSutter (1999) found that the average seepage rate from swine waste lagoons in southwestern Kansas was 0.8–1.1 mm/day (9×10^{-9} to 1×10^{-8} m/s) when the compacted soil liners, which were constructed from silt-loam and silty-clay soils, were 0.30–0.46 m thick, with the lowest rate occurring at the site with a 0.46-m liner augmented with bentonite.

There was no significant seasonal variation in average wastewater infiltration beneath the ICW sludge cell and cell 1 ($P > 0.05$). However, seasonally, the average wastewater infiltration beneath cell 2 was significantly variable ($P < 0.05$). Slightly higher infiltration was recorded in winter than in the other seasons in cell 1 and cell 2. The relatively higher hydraulic loading in this season might have led to a slightly higher hydraulic head in the ICW cells and consequently an overall slightly higher infiltration rate. On the other hand, in the sludge cell, slightly higher infiltration was recorded in summer and spring than in fall and winter. Similar seasonal variations have been reported by Asare et al. (1993), where lower field saturated hydraulic conductivity and matrix potential were measured on a silt-loam soil in spring as compared with summer. Higher values in summer were attributed to drier conditions in the subsurface, which, therefore, created more macropores. The slightly higher infiltration recorded in summer during this study could be attributed to flow paths created by macropores formed by plant roots and burrowing invertebrates, as well as a lower viscosity of water (Larson et al., 2000). The lowest infiltration rate of 3.7×10^{-10} m/s and 1.8×10^{-10} m/s was recorded, respectively, for the sludge cell and cell 2 in May 2011.

33.3.2 Hydraulic Conductivities

Apparent hydraulic conductivity (K) of ICW cell liners was estimated for the first three cells using Darcy's law and the total volumes of infiltration water, which were collected from the lysimeters. Mean apparent K of $3.8 \times 10^{-9} \pm 3.1 \times 10^{-9}$ m/s and $3.5 \times 10^{-9} \pm 1.4 \times 10^{-9}$ m/s (Table 33.3.1.1) were recorded, respectively, for the sludge cell and cell 1, which were significantly lower than the approximate maximum soil coefficient of permeability of 1×10^{-8} m/s required to both establish the ICW soil and provide adequate protection to groundwater resources (Dunne et al., 2005a,b; Carty et al., 2008; DOEHLG, 2010). Cell 2 recorded an apparent K of $9.8 \times 10^{-9} \pm 4.3 \times 10^{-9}$ m/s, which was slightly lower than the maximum soil coefficient of permeability required for ICW. The lower apparent K in all three cells suggests that the ICW cells do not constitute an immediate threat to the nearby groundwater system.

33.3.3 Wetland Water Fluxes

During the study period, the ICW received 1490.6 mm of precipitation in total. The highest precipitation rate (211.7 mm) was recorded in November 2009 and the lowest (28.8 mm) in May 2010. Wastewater inflow rates to each of the ICW sections considered in this study were 112.9 m^3/day for the sludge cell, 96.9 m^3/day in cell 1, and 103.0 m^3/day in cell 2. The highest daily wastewater inlet flows observed for the ICW sludge cell and cell 1 were 930.3 and 613.1 m^3, respectively, on November 19, 2009, and that for cell 2 was 421.0 m^3 on September 7, 2010. During the study period, daily outlet flow volumes for each of the ICW cells mostly equaled or exceeded inflow volumes during high flow events, as a result of precipitation inputs.

33.3.4 Contaminant Concentrations and Input Loading in Wetland Cells

The ICW operates as a sequential multicellular structure, with the wastewater passing sequentially from the sludge cell to cell 1 and cell 2. The effluent water quality characteristics from the sludge cell represent the influent to cell 1. Similarly, the effluent quality for cell 1 represents the influent to cell 2.

Overall, NH_3–N was recorded as the dominant form of nitrogen received by the ICW. The input NH_3–N load was increased slightly in the sludge cell, possibly due to mineralization of organic compounds. Cell 1 and cell 2 recorded 26% and 43% removal, respectively, indicating a significant reduction in the first two cells of the ICW. It has been confirmed by Kadlec et al. (2000) and Carty et al. (2008) that nutrient reductions occur predominantly within the initial wetland cells. The NO_3−N loading rates were lower in comparison to NH_3−N. Retention rates of approximately 49% and 34%, respectively, were recorded in cell 1 and cell 2. Generally, NO_3−N concentrations within the ICW system were low. It is likely that nitrate and oxygen provided electron acceptors in the lower layer of the ICW cells (Eriksson and Weisner, 1996).

The MRP loading rate in all three ICW cells varied slightly during the study period. Approximately 10%, 18%, and 29% of the input loadings to the sludge cell, cell 1, and cell 2, respectively, were achieved. The retention rate for MRP was relatively low in each of the three cells, probably because phosphorus is generally considered to be the most wetland area-dependent parameter (Reddy et al., 1999; Pant et al., 2001). The input COD was effectively reduced significantly in the sludge cell (40%) and cell 1 (62%), leaving a loading rate of only 2.7 $g/m^2/day$ in cell 2. The high COD removal rate may be attributable to high plant diversity in the ICW cells and the relatively lower water level, both of which enhance root zone oxygen and carbon (Zhu and Sikora 1995; Stottmeister et al., 2003; Sun et al., 2009).

33.3.5 Contaminant Concentrations and Input Loading beneath Wetland Cells

When the average contaminant loads received by each section of the ICW over the study period were compared to the respective loading rates after infiltrating through 0.5-m depth in the subsoil, significant differences ($P < 0.05$) were found for all parameters (Table 33.3.5.1). The nitrogen in the infiltration water discharged mainly in ammonic form. It is possible that nitrogen was nitrified in the upper layer of the subsoil and then denitrified within anoxic microsites (Parkin, 1987) in the sporadically saturated conditions below, where biological oxygen demand exceeds supply, leaving significantly lower total nitrogen loads to the groundwater.

The average weekly NH_3−N concentrations beneath cell 2 were usually higher than those in cell 1 and the sludge cell. During the study period, this ranged from 4.3 mg N/l to 27.8 mg N/l in cell 2 and represented almost all the soluble nitrogen. The highest concentrations were recorded in summer and were significantly higher ($P < 0.05$) than in the other seasons. It is possible that macropores in the soil liners of cell 2, created by plant roots and burrowing invertebrates, may have created preferential flow paths, causing rapid flow of ammonia-rich wastewater into the subsurface. Nevertheless, there was a steady decline in NH_3−N concentration after the summer period.

Other researchers have reported that nitrogen concentrations decrease in the warmer summer months and increase in the fall and winter (Dickey and Vanderholm, 1977). In cell 1 and the sludge cell, slightly lower NH_3−N concentrations were recorded in summer and fall than in winter and spring.

TABLE 33.3.5.1 Average Contaminant Loading to Subsoil underneath ICW Cells Recorded between November 2009 and May 2011 (after Dzakpasu et al., 2012)

ICW Section	n	NH_3–N (10^{-3} g m^2/day)		NO_3–N (10^{-4} g m^2/day)		MRP (10^{-5} g m^2/day)		COD (10^{-2} g m^2/day)	
		Mean	SD	Mean	SD	Mean	SD	Mean	SD
Sludge cell	37	3.8	3.76	6.0	5.43	3.3	3.60	7.0	6.08
Cell 1	50	1.8	1.86	7.8	7.86	2.7	2.10	2.5	1.10
Cell 2	48	10.8	5.34	7.2	11.45	6.8	8.10	6.3	3.57

n = number of samples; SD = standard deviation.

Cell 1 recorded the lowest concentrations at all times of the year. Furthermore, slightly higher NO_3–N concentrations were recorded in the infiltration water beneath the sludge cell and cell 2 during fall and winter than in spring and summer, whereas slightly lower concentrations were recorded beneath cell 1 during winter than during the other seasons.

Overall, higher average NO_3–N concentrations were recorded in infiltration water beneath cell 1 than beneath the other cells. In an ICW treating farmyard runoff, Mustafa et al. (2009) found a reduced availability of organic matter at the bottom of ICW cell 1, where the greatest biofilm formation took place and the maximum biodegradation of organic matter occurred. This led to decreased numbers of heterotrophic bacteria and consequently created conditions that favored the growth of ammonia-oxidizing bacteria. This, therefore, may have led to high nitrification rates, consequently leading to slightly higher NO_3–N concentrations leaching through the infiltration water in cell 1 of this ICW than in the other cells. In addition, macropores such as root channels may facilitate bypass flow around the root zone area of most active denitrification (Buss et al., 2005).

The average total NH_3–N load after 0.5-m depth of unsaturated subsoil represented approximately 0.02%, 0.20%, and 1.53% of the total input loading to the sludge cell, cell 1, and cell 2, respectively. This demonstrated that only a small volume of soil is required to attenuate the NH_3–N exported in the infiltration water. Several researchers have shown that NH_4^+–N concentrations are elevated directly beneath the compacted liner and then abruptly decrease to trace levels at lower depths (Miller et al., 1976; Culley and Phillips, 1989).

Ham (2002) reported that ammonia and organic nitrogen distributions under lagoons in Kansas showed concentrations in excess of 1000 mg/kg within the first meter and then decreased with depth. Similarly, average NO_3–N loads recorded after 0.5-m depth of unsaturated subsoil were approximately 0.03%, 0.97%, and 2.47% of the total $NO_3{-}N$ loading for the sludge cell, cell 1, and cell 2, respectively. There is a slight difference with respect to nitrogen loading to the subsoil when comparing the discharge from the first three ICW cells into the subsoil.

The results from this study, therefore, provide additional support to findings by other researchers that the potential total nitrogen load to groundwater underneath ICW cells receiving partially treated wastewater is slightly higher than that for cells receiving primary wastewater. This assertion is further strengthened by the slightly lower $NH_3{-}N$ concentrations in infiltration water within the flow path as the wastewater flowed from the inlet to the outlet of cell 1. Approximately 7.3 ± 6.89 mg/l, 4.5 ± 2.02 mg/l, and 1.6 ± 0.71 mg/l, respectively, were recorded underneath the inlet, midpoint, and outlet of cell 1. Nevertheless, in the sludge cell 6.8 ± 1.92 mg/l and 8.4 ± 3.34 mg/l were measured underneath the inlet and outlet points, respectively, possibly due to mineralization of organic compounds that occurred in that cell.

When compared with the influent wastewater in each ICW section, the MRP concentration was significantly ($P < 0.05$) reduced after 0.5-m depth of unsaturated subsoil. There was no significant variation ($P > 0.05$) in the MRP concentrations recorded beneath the three ICW cells. However, slightly higher concentrations were recorded during spring than during the other seasons in cell 1 and cell 2.

The total MRP loading to the subsoil represented approximately 0.002%, 0.03%, and 0.08% of the total MRP load received in the sludge cell, cell 1, and cell 2, respectively. This indicates that phosphate is largely removed by the subsoil, possibly as a result of the mineralogy of the respective soils (Robertson, 2003, 2008). The environmental quality standard for phosphorus in rivers in Ireland is set at 0.035 mg/l. With such low MRP loading recorded at the first 0.5 m beneath ICW cells and possible further MRP removal down the subsoil column, as well as potential dilution of MRP concentrations by groundwater, which is a function of aquifer characteristics and recharge in the area, ICW may not pose a considerable threat to the nearby groundwater system.

In this study, the biodegradable organic loading to subsoil from ICW cells was measured as COD. The main process for removal of soluble biodegradable organics in the subsoil is adsorption, followed by microbial degradation (Wilhelm et al., 1994; Reed et al., 1995; Siegrist et al., 2000). Overall, the COD concentrations beneath the ICW cells were significantly reduced ($P < 0.05$) after 0.5-m depth of unsaturated subsoil. The average weekly COD concentration recorded in subsoil beneath the sludge cell was significantly higher ($P < 0.05$) than beneath cell 1 and cell 2, which reflected the higher organic matter concentration received in that cell. The highest concentrations recorded in summer, however, were not significantly higher ($P > 0.05$) than in the other seasons. Moreover, there was no significant variation ($P > 0.05$) in the COD concentrations recorded beneath cell 1 and cell 2. The total COD loading to the subsoil beneath the sludge cell, cell 1, and cell 2 represented approximately 0.03%, 0.32%, and 2.30% of the total COD load received in the respective ICW sections.

33.4 CONCLUSIONS AND RECOMMENDATIONS

The average infiltration rates from the first three ICW cells were 4.3×10^{-9}, 3.7×10^{-9}, and 1.0×10^{-8} m/s. Variation among sites was marginal. Infiltration rates from ICW sludge cell and cell 1, which received wastewater with slightly higher organic content, were similar and were significantly lower than that in cell 2. This provided additional support to findings by other researchers that organic sludge on the bottom of the ICW cells reduced flow through the soil liners that largely limited infiltration, especially considering that soils with similar characteristics were used to line all the ICW cells.

The apparent hydraulic conductivity of the soil liners of the ICW cells was 3.8×10^{-9}, 3.5×10^{-9}, and 9.8×10^{-9} m/s for the first three cells, respectively. Conductivity was slightly lower in the first two cells than in the third. For most of the contaminants considered in this study, less than 0.5% of the influent loading to the first two ICW cells was lost through infiltration to subsoil beneath the respective cells, as compared to approximately 2% in the third. Overall, the amount of infiltration occurring beneath the ICW cells increased from the proximal cells to the distal one. Higher resultant contaminant loading was found in the distal cell, which received partially treated wastewater. This indicates that each ICW cell has the potential to impact differently on the underlying groundwater. This effect is slightly less with the cells receiving the most contaminated wastewater than with the most distant wetland cells, which receive the least contaminated water.

The results from this study represent a small number of samples from domestic wastewater treatment with ICW in Ireland. Additional measurements may be needed in different geographical areas with different climatic conditions, soil characteristics, and wastewater types. The nearby groundwater system should be routinely monitored to fully understand the potential impacts from ICW systems. Moreover, there is scope for further studies on the assessment of the processes and time required for possible self-sealing to occur on different soil materials that may be used as ICW cell liners.

Chapter 34

Seasonal Assessment of Vertical-Flow Wetlands Treating Domestic Wastewater

34.1 INTRODUCTION

Constructed treatment wetlands are engineered wastewater purification systems that encompass biological, chemical, and physical processes, which are all similar to processes occurring in natural treatment wetlands. They are implemented for environmental pollution control to treat a variety of wastewaters, including industrial effluents, urban and agricultural runoff, animal wastewaters, sludge, and mine drainage (Scholz and Xu, 2002; Picard et al., 2005; Scholz and Lee, 2005). Recently, some large-scale wetland systems have also been successfully applied to treat domestic wastewater (Scholz, 2010; Dong et al., 2011). However, there are few long-term and controlled studies involving domestic wastewater, owing to health and safety concerns.

Vertical-flow wetland systems are often associated with poor nitrogen removal (Scholz, 2010). However, Fan et al. (2013) discovered that intermittent aeration combined with high influent chemical oxygen demand (COD) to nitrogen ratios could achieve high nitrogen removal in vertical-flow systems. Moreover, the importance of wetland plants in nitrogen update within these systems has been reported by Korboulewsky et al. (2012).

Vertical-flow constructed wetlands are widely regarded as the state of the art in wetland technology worldwide (Scholz and Xu, 2002; Tsihrintzis et al., 2007; Scholz, 2010). However, an inevitable drawback limiting the competitiveness and efficiency of vertical-flow wetland systems is the concomitant biological clogging and physical clogging, induced as a result of excessive formation of biomass from degradation of pollutants and macrophyte litter and retention of inert suspended fine particles, respectively. Clogging is often a seasonal phenomenon and usually leads to the deterioration of water quality. Temporal variations in wetland performance often depend on the corresponding macrophyte growth rates (Picard et al., 2005; Kouki et al., 2009). However, Merlin et al. (2002) reported that the influence of temperature seems very weak, because there were no significant seasonal variations of the process efficiency with respect to their wetland system.

Wetlands for Water Pollution Control. http://dx.doi.org/10.1016/B978-0-444-63607-2.00034-4

The process design of vertical-flow constructed wetlands has involved a casual and empirical approach to mitigating clogging by selection or recommendation of values for many parameters whose relationship to substrate clogging are assumed or poorly understood. The resultant lack of understanding hinders the ability of operators to maintain good hydraulic and treatment performances. While Sani et al. (2013b) were concerned about the short-term hydraulic processes of clogging with respect to this experimental set-up, Chapter 34 focuses on the long-term and, particularly, the seasonal treatment performance of the same system. This chapter is particularly important considering that there is conflicting information regarding the influence of seasons on wetland removal efficiencies (Merlin et al., 2002; Song et al., 2006; Tunçsiper et al., 2009).

Numerical modeling of constructed wetlands has gained popularity since the early years of the 21st century. Wetlands are complex systems in which varied seasonal physical, chemical, and biological processes such as sedimentation, filtration, precipitation, sorption, plant uptake, macrophyte degradation, microbial decomposition, and nitrogen transformations take place (Rousseau et al., 2004). These processes are often active in parallel and mutually influence each other (Langergraber et al., 2009). Realistic descriptive models can produce reliable predictions of the performance of constructed treatment wetlands (Langergraber, 2007).

Looking back on the history of numerical wetland modeling, as reviewed by Rousseau et al. (2004), the primitive first-order models by Kadlec and Knight (1996), which simply regarded wetlands as a black box, used a first-order rate constant to evaluate the effects of autochthonous production, sedimentation, temperature, and retention time (Shepherd et al., 2001) on the variation of the concentrations of the contaminants in treated water. Owing to oversimplification by using the first-order rate assumption, further efforts using a variable order (known as Monod-type models) have been made to evaluate individual contaminant removal rates (Kemp and George, 1997; Mitchell and McNevin, 2001). However, these models do not recognize the influence of complex seasonal changes.

Multicomponent reactive transport models were proposed to model both the transport and reactions of the main constituents of municipal wastewater in the subsurface flow processes through wetlands (Langergraber et al., 2009). The reactive transport model highlighted the essential requirement for the description of the clogging process, which significantly influences the hydraulic conductivity of porous substrate beds and, consequently, the long-term behavior of wetland systems. Nevertheless, these clogging models do not recognize seasonal dynamics such as plant litter decay or the harvest of excessive plant growth.

Chapter 34 particularly provides the wetland modeling community with a statistically validated long-term data interpretation. This will allow modelers to define seasonal removal coefficients for individual water quality parameters and provide wetland managers with insight into seasonal removal processes, allowing them to revise wetland management plans accordingly.

The aim of Chapter 34 is to assess the long-term and seasonal treatment performance of various experimental vertical-flow constructed wetlands treating combined domestic wastewater and surface runoff. The key objectives are to critically evaluate (1) the inflow water quality, (2) the outflow water quality of different wetland filters, (3) the influence of design parameters such as aggregate size on seasonal water quality, (4) the influence of operational parameters such as contact time, resting time (i.e., filter temporarily without influent), and loading on seasonal water quality, and (5) the influence of seasonal water quality variations on filter clogging.

34.2 MATERIALS AND METHODS

34.2.1 Experimental Set-Up and Operation

The experimental vertical-flow constructed wetland system was located within a greenhouse on top of the roof of the Newton Building of The University of Salford, Greater Manchester, UK (Sani et al., 2013a). the set-up included two filters that are essentially controls receiving clean dechlorinated water.

Table 34.2.1.1 indicates an overview of the statistical experimental set-up used to test the impact of four variables. Filters 1 and 2 compared to Filters 3 and 4 test the influence of a larger aggregate diameter. Filters 5 and 6 compared to Filters 3 and 4 check the impact of a higher loading rate. The application of a lower contact rate is tested if Filter 7 is compared with Filters 3 and 4. Finally, a lower resting time is the difference between Filters 7 and 8.

TABLE 34.2.1.1 Comparison of the Statistical Experimental Set-Up of Wetland Filters (after Sani et al., 2013a)

Design and Operational Variables	Unit	Filters 1 and 2	Filters 3 and 4	Filters 5 and 6	Filter 7	Filter 8
Aggregate diameter	mm	20	10	10	10	10
Contact time	h	72	72	72	36	36
Resting time	h	48	48	48	48	24
Chemical oxygen demand	mg/l	138	138	285	138	138

Note: Annually treated approximate volumes of wastewater: Filters 1–6, 475 l/year; Filter 7, 680 l/year; Filter 8, 949 l/year. Filters 2, 4, and 6 are replicates for the most common operational scenarios.

Eight laboratory-scale vertical-flow constructed wetlands were constructed from Pyrex tubes with an inner diameter of 19.5 cm and a height of 120 cm. The filters were filled with siliceous (minimum of 30%) pea gravel up to a depth of 60 cm and planted with *Phragmites australis* (Cav.) Trin. ex Steud. (Common Reed). Dead macrophyte plant material was harvested in winter and returned to the corresponding wetland filters by placing it on top of the litter zone.

The main outlet valve is located at the bottom of each constructed wetland system. Eight further valves (used to test for clogging) are located on the sidewall of each wetland column. The sidewall valves are located at heights of 10, 20, 30, 40, 45, 50, 55, and 60 cm from the bottom of each column.

Wetland columns received 6.5 l of inflow water during the feeding mode, which was different between several filters (Table 34.2.1.1). Columns 1–6 were sampled after 72 h of contact time and then left to rest for 48 h, while columns 7 and 8 were sampled after 36 h contact time and left to rest for 48 and 24 h, respectively. All water quality parameters discussed in Chapter 34 were usually determined during or directly after sampling. The preliminary treated urban wastewater used for the inflow water was obtained from the Davyhulme Sewage Works (Greater Manchester), one of the largest wastewater treatment plants in Europe (http://en.wikipedia.org/wiki/Davyhulme), operated by the water utility company United Utilities. Fresh wastewater was collected approximately once per week and was stored and aerated by standard aquarium air pumps in a cold room before use. The wastewater quality was highly variable and comprised domestic and industrial wastewater, as well as surface water runoff.

Aqua Medic Titan chillers (Aquacadabra, Bexleyheath, UK) were used to maintain the root system and debris layer of all wetland systems at semi-natural below-surface temperatures of about 12 °C. This temperature simulates the temperature of the upper earth layer where the root system of the wetland plants of a real treatment system would be located.

The COD was used as the criterion to differentiate between low and high loads (Table 34.2.1.1). An inflow target COD of about 285 mg/l (usually between 122 and 620 mg/l) was set for wetlands with a high loading rate (Filters 5 and 6). The remaining filters, Filters 1, 2, 3, 4, 7, and 8, received wastewater diluted with dechlorinated tap water. The target inflow COD for these filters was approximately 138 mg/l (usually between 43 and 350 mg/l).

34.2.2 Water Quality Analysis

Routine obstruction observations and water quality sampling were carried out according to APHA (2005, unless stated otherwise) to monitor clogging evolution and long-term and seasonal treatment performance, respectively. The DR 2800 Hach Lange spectrophotometer (www.hach.com) was used for standard water quality analysis for variables including COD, ammonia-nitrogen, nitrite-nitrogen, nitrate–nitrogen, total phosphorus, ortho-phosphate-phosphorus, and suspended solids (SS).

The 5-day biochemical oxygen demand (BOD) was determined in all water samples with the OxiTop IS 12-6 system, a manometric measurement device, supplied by the Wissenschaftlich-Technische Werkstätten (WTW), Weilheim, Germany. Nitrification was suppressed by adding 0.05 ml of 5 g/l N-allylthiourea (WTW chemical solution no. NTH600) solution per 50 ml of sample water. Note that this device was not available during the first year of operation.

Turbidity was measured with a Turbicheck turbidity meter (Lovibond Water Testing, Tintometer Group, www.lovibond.com). Temperature data for the first year of operation were recorded outside and in the shade at an official weather station in Woodford, located southeast of Salford. The raw data were supplied by the UK MetOffice (www.metoffice.gov.uk). Concerning the second year of operation, inside temperature measurements were performed by project team members inside and outside the greenhouse.

34.2.3 Clogging Tests

Overall hydraulic conductivity measurements to assess the severity of clogging were performed. The columns were regularly filled with wastewater to the top of the debris layer and subsequently emptied after a resting time of 2 h to allow for air bubbles to escape the media. The time taken to drain each column and the associated water volume captured were noted. As an indirect measure of clogging, SS samples were taken from the inflow wastewaters, the layer of debris on top of each filter, the eight sampling ports of each column (see Section 34.2.1), and the outflows.

34.2.4 Statistical Tests

Microsoft Excel (www.microsoft.com) was used for the general data analysis. The nonparametric Mann–Whitney U test was computed using IBM SPSS Statistics Version 20 (www.ibm.com) and was used to compare the medians of two (unmatched) independent samples. This was required because virtually all sample data (even after data transformation) were not normally distributed so that an analysis of variance could not be applied.

34.3 RESULTS AND DISCUSSION

34.3.1 Inflow Water Quality Analysis

This section covers objective (a). The undiluted influent concentrations for COD, BOD, ammonia-nitrogen, nitrate–nitrogen, ortho-phosphate-phosphorus, SS, and turbidity were 285 mg/l, 71 mg/l, 47 mg/l, 5 mg/l, 15 mg/l, 157 mg/l, and 193 NTU, respectively. The seasonal data variability was relatively high for nitrate–nitrogen and SS.

34.3.2 Evaluation of Oxygen Demand Variables

This section addresses objectives (b)—(d), which are not covered individually and directly, but indirectly when different water quality parameters are being discussed. Figure 34.3.2.1(a) shows the seasonal variations of COD. The

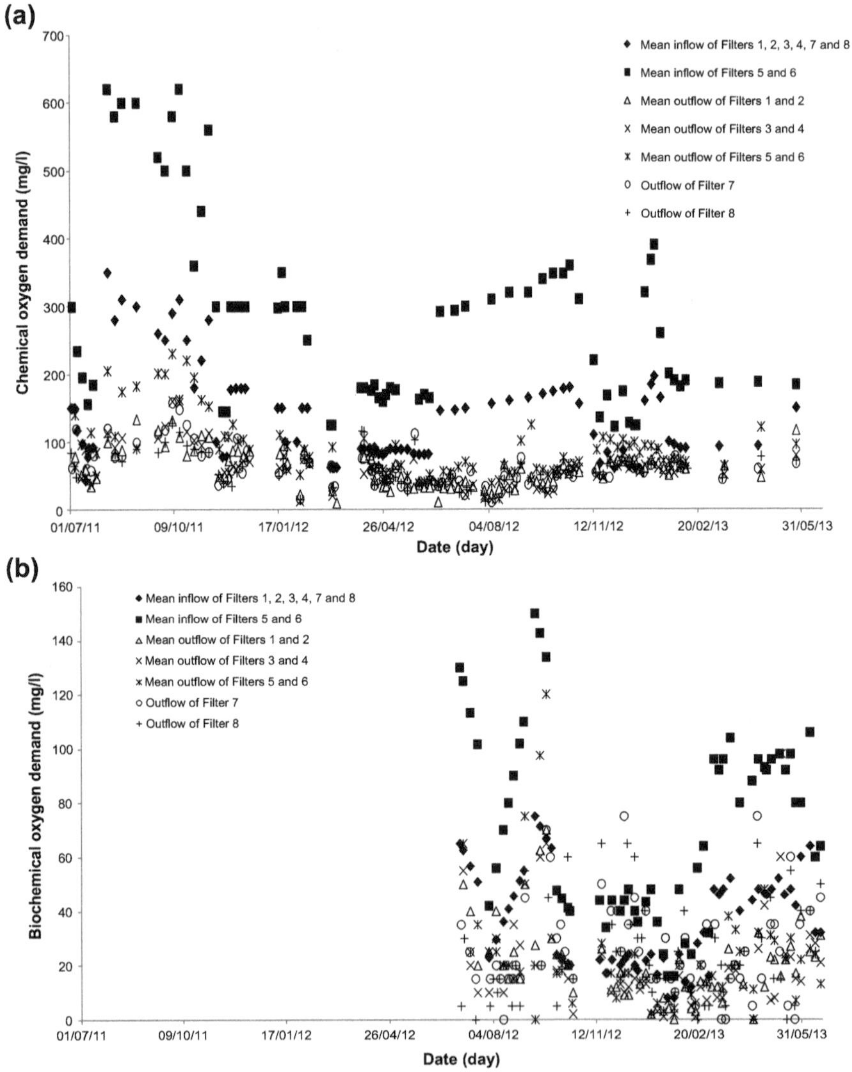

FIGURE 34.3.2.1 Seasonal variations in (a) chemical oxygen demand, (b) biochemical oxygen demand, (c) ammonia-nitrogen, (d) nitrate—nitrogen, and (e) ortho-phosphate-phosphorus. *After Sani et al. (2013a).*

(c)

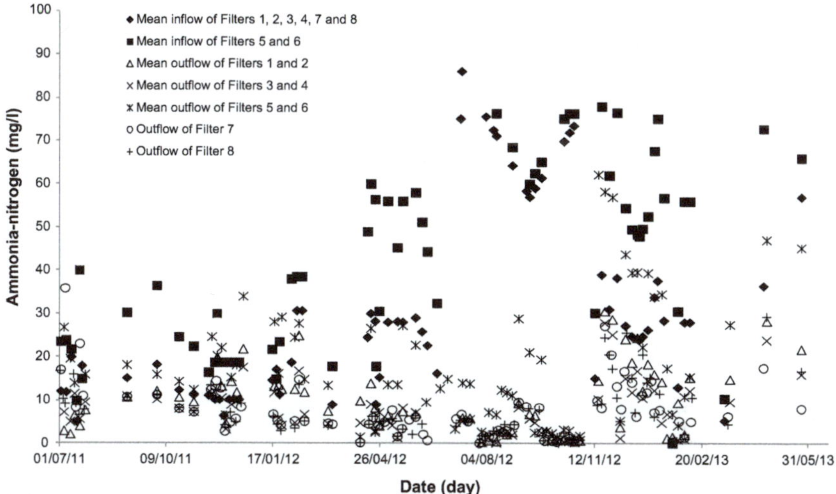

(d)

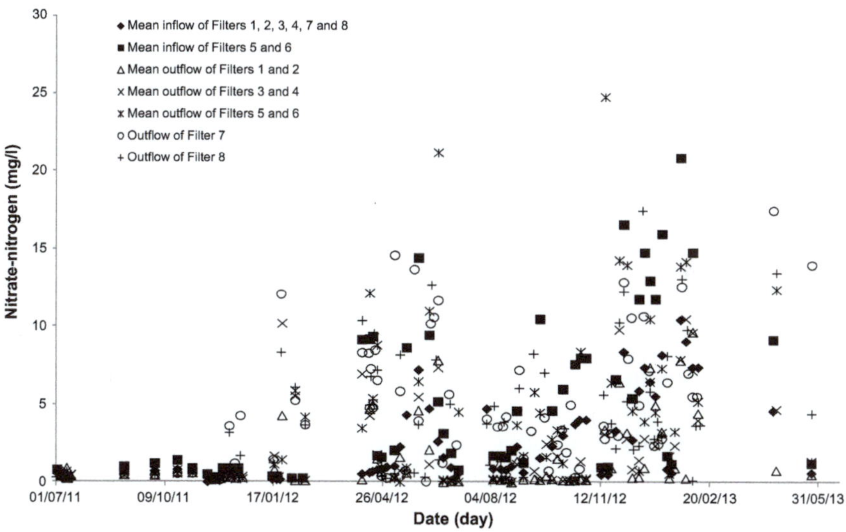

FIGURE 34.3.2.1 Cont'd.

Urban Waste Water Treatment (England and Wales) Regulations (UK Government, 1994), which implement Council Directive 91/271/EEC Concerning Urban Waste-water Treatment (European Community, 1991), set a threshold value of 125 mg/l for secondary wastewater treatment. Filters 1–8 were 3, 2, 2, 2, 15, 15, 3, and 2 times noncompliant, respectively. However, the experimental system tested in Chapter 34 may only be seen as one secondary wastewater treatment step that does not directly discharge to the natural

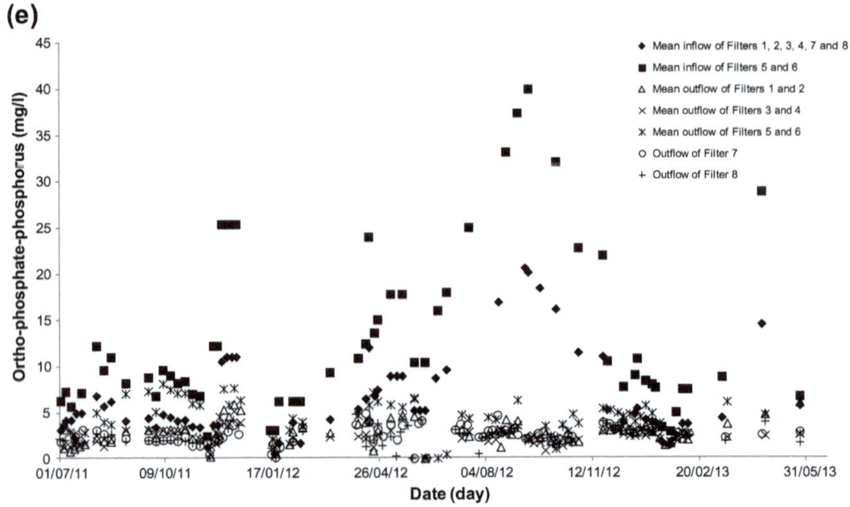

FIGURE 34.3.2.1 Cont'd.

environment. All filters demonstrated relatively good COD removal (excluding the time close to the start-up period). This can be explained by the fact that the biological activity necessary for microbial degradation takes time to develop and, as such, the treatment efficiency can be expected to improve after microbial acclimatization.

The overall mean COD removal efficiencies for Filters 5 and 6 (both with high loading rate) were higher than those for Filters 3 and 4 (both with low loading rate). This difference was statistically significant (based on the nonparametric Mann–Whitney U test). The opposite is the case for the equivalent COD loads. A comparison between Filters 3 and 4 with Filter 7 gives insight into the effect of contact time on the treatment performance. The overall removal efficiencies were rather similar. The COD removal for Filters 1–4 was also similar, indicating that aggregate size may not matter.

A clear seasonal trend with high COD values in fall and low COD values in summer was apparent, confirming previous findings by Song et al. (2006) but contradicting data by Merlin et al. (2002). Elevated COD inflow concentrations had a significantly positive impact on the treatment performance for COD. The low summer values in this study can be explained by increased biodegradation during warm summer months. A statistically significant seasonal variation for COD is only evident if filters with a low loading rate are compared to those with a high one. High-rate filters are likely to be overloaded, at least during fall, winter, and summer.

Figure 34.3.2.1(b) shows the distribution of the second and final oxygen demand variable BOD. The traditional UK standard for BOD removal from secondary wastewater is 20 mg/l and 25 mg/l for sensitive and less sensitive (e.g., many coastal discharges) areas, respectively (Royal Commission on Sewage Disposal, 1915). More recently, the Urban Waste Water Treatment (England and Wales) Regulations (UK Government, 1994) set a threshold value of 25 mg/l for secondary wastewater treatment. Nevertheless, authorities try to work toward the slightly more stringent traditional targets.

Regarding sensitive watercourses, Filters 1−8 were 31, 27, 23, 20, 27, 28, 12, and 14 times noncompliant, respectively. In comparison, Filters 1−8 were 19, 14, 13, 11, 20, 17, 7, and 6 times noncompliant for less sensitive areas, respectively. The relatively poor performance of Filters 5 and 6 can be explained by the high inflow loading rates. However, Filters 7 and 8 perform relatively well owing to a low loading rate. A seasonal trend with high values at the end of summer and low values in winter was apparent. This confirms findings by Tunçsiper et al. (2009) but contradicts those by Song et al. (2006). However, aggregate diameter, resting time, and loading rate had no clear influence on seasonal BOD removal trends.

34.3.3 Comparison of Nitrogen Compounds

The removal of nitrogen within vertical-flow constructed wetlands is usually mainly by microbial nitrification and denitrification. Nitrogen removal in many constructed wetland systems without adequate active or passive aeration is insufficient, mainly because of the lack of available oxygen used for aerobic biological degradation (Scholz, 2010).

The overall reduction rates of ammonia-nitrogen were relatively high, partially owing to temperatures usually above 15 °C during warm months (Tsihrintzis et al., 2007). The removal efficiencies regarding both concentrations and loads were low if undiluted wastewater was used. Inflow water high in COD was linked to a statistically significant ($P < 0.05$) difference between the overall mean daily values of Filters 5 and 6 in comparison to the overall mean daily values of Filters 3 and 4. However, aggregate size, resting time, and contact time were not important for the overall ammonia-nitrogen removal.

A clear seasonal trend for ammonia-nitrogen with high values at the end of spring and low values in summer was apparent, confirming earlier findings by Song et al. (2006). A statistically significant seasonal variation for ammonia-nitrogen is only evident if filters with a low loading rate are compared to those with a high one. High-rate filters are likely to be overloaded, at least during colder months.

The regulations (UK Government, 1994) set no threshold for ammonia-nitrogen that would relate to the treatment system discussed in Chapter 34. However, a realistic guideline threshold value for ammonia-nitrogen in the context of this experiment would be 20 mg/l. Filters 1−8 were 9, 10, 5, 5, 27, 29, 5, and 5 times noncompliant, respectively. Figure 34.3.2.1(c) shows the seasonal variations of ammonia-nitrogen.

The regulations (UK Government, 1994) also set no threshold for nitrate–nitrogen that would relate to the treatment system. Nevertheless, a realistic guideline threshold value for nitrate–nitrogen in the context of this experiment could be 50 mg/l. All filters were compliant.

Although the nitrate–nitrogen concentration in the inflow was relatively low, the outflow concentrations were relatively high for all filters. Only Filters 1 and 2 had positive removal efficiencies related to concentration reductions, but not to load reductions. In contrast, all other filters functioned as sources for nitrate–nitrogen. The negative removal efficiencies for nitrate–nitrogen indicated that denitrification was likely to be only a minor removal mechanism. However, the necessary conditions for denitrification to occur (e.g., anoxic environment and presence of easily biodegradable organic carbon) were not directly monitored within the entire small wetlands, because this would have been too destructive.

The nitrate–nitrogen reduction was higher for smaller compared to larger aggregate sizes (Figure 34.3.2.1(d)). Long contact and resting times and a low COD loading were also positive. The mean overall daily nitrate–nitrogen values of Filters 3 and 4 compared to Filter 7 were statistically significantly different from each other.

High nitrate–nitrogen values at the beginning of winter and low values in summer were recorded. Differences in contact time were significant ($P < 0.05$) during warm months. A longer contact time led to greater removal of nitrate–nitrogen. In comparison, differences in loading rate only result in significant variations of nitrate–nitrogen during warm months.

34.3.4 Evaluation of Ortho-Phosphate-Phosphorus

Figure 34.3.2.1(e) shows the seasonal variations of ortho-phosphate-phosphorus. In general, phosphorus is one of the most difficult pollutants to be removed by mature constructed wetlands (Pant et al., 2001). The ortho-phosphate-phosphorus removal efficiencies ranged between 68% and 73% for all filters regardless of the loading rate, which is surprising, considering that an overall statistically significant difference for COD and contact time has been noted.

Aggregate size and resting time were not crucial parameters in terms of overall ortho-phosphate-phosphorus removal. This can be explained by the fact that phosphorus is usually present in particulate form and does not dissolve well in filters that are not yet saturated by phosphorus or other compounds competing for adsorption sites (Scholz, 2006, 2010).

All filters performed insufficiently in terms of phosphorus removal compared to other key parameters such as COD (see Section 34.3.2) and ammonia-nitrogen (see Section 34.3.3). Findings confirm those by Smith et al. (2006) indicating that constructed wetlands were not efficient in removing phosphate in Nordic countries, especially during prolonged high loading periods. The regulations (UK Government, 1994) set a value of 2 mg/l for total phosphorus for communities of 10,000–100,000 inhabitants. A threshold for ortho-phosphate-phosphorus that would relate to the treatment system discussed in this chapter does not exist. However, a realistic guideline threshold value for ortho-phosphate-phosphorus could be 1 mg/l. Filters 1–8 would be 70, 70, 69, 68, 71, 70, 77, and 78 times noncompliant, respectively.

No clear seasonal trend for phosphorus removal was observed, contradicting findings by Merlin et al. (2002), who have noted low values in summer and high values in winter. High COD inflow concentrations had a significantly positive impact on the treatment efficiency for ortho-phosphate-phosphorus. Some significant seasonal ortho-phosphate-phosphorus differences were recorded for variations of contact time and loading rate for summer 2011, summer 2012, fall 2012, and winter 2012; and summer 2011, fall 2011, fall 2012, and winter 2012, respectively. Ortho-phosphate-phosphorus removal seems to be independent of temperature, if contact time and loading rate vary. This could be explained by the dominant physical and not biological removal processes (Kadlec and Knight, 1996; Scholz, 2010).

34.3.5 Comparison of Particles

The overall removal efficiencies for SS were high. However, the inflow concentrations decreased over time. A higher loading rate had a significantly ($P < 0.05$) negative impact on the overall treatment performance. Suspended solids accumulated in the top part of the filters, confirming findings by Hua et al. (2010) and Scholz (2010). The presence of different aggregates did not seem to have an influence on solids retention, at least in the early stages of operation.

The traditional UK standard for SS removal from secondary wastewater is 30 mg/l (Royal Commission on Sewage Disposal, 1915). Filters 1–8 were 7, 8, 4, 4, 12, 6, 7, and 8 times noncompliant, respectively. More recently, the regulations (UK Government, 1994) set a value of 35 mg/l. Filters 1–8 were 5, 6, 4, 4, 7, 6, 6, and 4 times noncompliant, respectively. However, authorities try to comply with the more stringent traditional guideline.

No clear seasonal trend for SS in the outflow of all filters was apparent. Relatively high COD inflow concentrations had a significantly positive impact on the treatment performance for SS. There are also only a few significant differences between filters subjected to different aggregate sizes (summer 2012) and loading rates (summer 2011, spring 2012, and summer 2012).

High turbidity outflow values for the end of spring and low values for the end of fall were recorded. There are no internationally recognized outflow compliance values for turbidity. Differences in loading rate influence turbidity removal significantly only in summer 2012 and spring 2013. Findings mirror those for SS during warm months, indicating a good relationship between turbidity and microbial activity leading to organic matter degradation and, subsequently, to an increase in particles.

34.3.6 Assessment of Filter Clogging Based on Water Quality Variables

Objective (e) is addressed in this section. A review discussing clogging of vertical-flow wetland systems has been published by Knowles et al. (2011). The key operational parameters in terms of clogging were hydraulic and solids loading rate. With respect to Chapter 34, no significant differences in hydraulic conductivity were usually recorded for all filters and seasons. For water quality variables with high (i.e., above 100 mg/l) overall mean inflow concentrations, an elevated COD inflow load and a smaller aggregate size made a significant difference in terms of overall COD and SS outflow concentrations and in terms of seasonal SS outflow concentrations, respectively.

The development of a litter zone on top of each filter was observed after the first year of operation in spring 2013. This was partly due to both the high strength and SS load of the wastewater, but mainly due to the dead macrophyte plant material that was harvested in winter and returned to the corresponding wetland filters. However, concerning filter bed clogging evolution measured in terms of SS accumulation, none of the systems have shown any signs of serious within-bed clogging after two years of operation, even for the high-rate Filters 5 and 6, which contrasts with the expectation of noticing clogging phenomena for this system as described by Sani et al. (2013b). However, a very modest but statistically significant breakthrough of turbidity has been noticed for high-rate filters.

34.4 CONCLUSIONS AND RECOMMENDATIONS

High removal of COD, ammonia-nitrogen, ortho-phosphate-phosphorus, and SS was noted. Inflow varied seasonally concerning nitrate—nitrogen and SS. Findings indicate usually good overall removal for small aggregates, long contact time, long resting time, and reduced loading. Different COD inflow concentrations had a significant impact on the removal of ammonia-nitrogen, ortho-phosphate-phosphorus, and SS. A small aggregate diameter, a short contact time, a long resting time, and a low inflow COD were beneficial in reducing SS. Seasonal outflow trends were recorded for COD, BOD, ammonia-nitrogen, nitrate—nitrogen, and turbidity. Ortho-phosphate-phosphorus and suspended solids removal were independent of season.

No clogging was observed. Further research on long-term (i.e., 10 or 20 years) clogging should be considered. However, such research is unlikely to receive funding and attract committed academic and industrial partners.

Chapter 35

Recycling of Domestic Wastewater Treated by Vertical-Flow Wetlands for Irrigation

35.1 INTRODUCTION

35.1.1 Background

Since water resources are limited, particularly in dry climatic regions, wastewater treatment and subsequent recycling is a viable alternative (Asano, 1994). Treated wastewater can be used for agricultural land irrigation, aquaculture, landscape irrigation, urban and industrial applications, recreational and ecosystem purposes, and artificial recharging of groundwater (Asano et al., 2007). Around 20 million ha of agricultural land is irrigated by both treated and untreated wastewater (Jiménez and Asano, 2008). Advantages associated with wastewater recycling include the supply of nutrients and trace minerals to plants, potentially leading to higher yields and a decrease in the demand for inorganic fertilizers (Bichai et al., 2012). For more detailed information on treated wastewater reuse and planning in warm countries, the reader may wish to consult Kalavrouziotis and Drakatos (2001), Kalavrouziotis et al. (2008, 2011, 2015), Pedrero et al. (2009), and Kalavrouziotis (2011).

Constructed treatment wetlands are implemented for environmental pollution control to treat a variety of wastewaters including industrial effluents, urban and agricultural runoff, animal wastewaters, sludge, and mine drainage (Sani et al., 2013a; Scholz, 2010). Recently, some large-scale wetland systems have also been successfully applied to treat domestic wastewater (Dong et al., 2011).

35.1.2 Wastewater Recycling for Crop Irrigation

The earliest documented sewage farms, where wastewater has been applied to land for disposal and for agricultural use, were operated in the sixteenth and

Wetlands for Water Pollution Control. http://dx.doi.org/10.1016/B978-0-444-63607-2.00035-6

seventeenth centuries in Bunzlau, Germany, and Edinburgh, Scotland (Shuval et al., 1986). Treated urban wastewater can be recycled and reused in arid regions that are confronting increasing water shortages. The evaluation of the effects of treated wastewater reuse on crops intended for human consumption is of particular interest (Aiello et al., 2007; Asano and Levine, 1996; Cirelli et al., 2012). Unfavorable concentrations of certain nutrients and trace elements are a challenge to the growth of plants fed by recycled pretreated wastewater. Moreover, traces of hydrocarbons from diesel spills associated with urban runoff or industrial effluent are a more recent challenge (García-Delgado et al., 2012; Scholz, 2010).

García-Delgado et al. (2012) undertook a greenhouse study in Spain to assess the effect of treated urban wastewater on soil and pepper quality. The wastewater application saved fertilizer (37% nitrogen, 66% phosphorus, and 12% potassium). Total polyaromatic hydrocarbons and the heavy metals cadmium, lead, and arsenic within the pepper fruits were low. The highest concentration (lower than proposed threshold concentrations for carcinogenicity) was recorded for phenathrene.

Boyden and Rababah (1996) assessed the recycling of nutrients from settled primary domestic wastewater (not disinfected) to produce value-added crops including capsicum and tomato. The crops grown in these systems considerably removed nitrogen and phosphorous from settled primary sewage and appeared healthy compared to the control using commercial nutrients.

Cheng et al. (2004) assessed an integrated system that recycles waste organics and treats wastewater from a swine farm to grow vegetables. An anaerobic digester with ambient temperature has been used to treat the swine wastewater and to produce biogas. A trickling nitrification biofilter has been developed to convert ammonium in the effluent into nitrate. The nitrified anaerobic effluent is used as both fertilizer and irrigation water.

Morari and Giardini (2009) assessed the treatment efficiency of pilot-scale vertical-flow constructed wetlands on municipal wastewaters and their suitability for irrigation reuse in Italy. Only water quality parameters with high removal efficiencies fulfilled the Italian guidelines for irrigation reuse, whereas parameters with lower efficiencies such as suspended solids and total phosphorus limited the potential water reuse.

Vertical-flow constructed wetlands treating septic tank effluent in Guangzhou (China) achieved removal rates for chemical oxygen demand (COD), 5-day biochemical oxygen demand (BOD), suspended solids, total nitrogen, and total phosphorus of 60, 80, 74, 49, and 79 mg/l, respectively (Cui et al., 2003). After that, the treated effluent was used for hydroponic cultivation of water spinach and romaine lettuce. The removal efficiencies of the whole system for COD, 5-day BOD, suspended solids, total nitrogen, and total phosphorus were 71%, 98%, 97%, 86%, and 87%, respectively. It was found that using treated effluent for hydroponic cultivation of vegetables could reduce the nitrate content in vegetables.

Lopez et al. (2006) assessed constructed wetlands treating municipal effluents to be reused in agriculture. Recorded average removal efficiencies for suspended solids, 5-day BOD, COD, total nitrogen, and total phosphorus were 85%, 65%, 75%, 42%, and 32%, respectively.

35.1.3 Plant Selection

Many vegetables have the potential to grow on recycled wastewater. However, there is the potential of some vegetables, such as lettuce and cabbage, to become contaminated by microbes because their edible leaves are too close to the ground receiving the treated wastewater. Therefore, it makes sense to select vegetables where the edible fruit is located far away from the ground. This may include peppers, tomatoes, maize, eggplants, beans, lentils, and peas. The next step in selecting suitable vegetables is to decide on easy-to-grow and relatively cost-effective plants with high nutritional value. Many vegetables may fit these conditions in particular geographical settings. However, the authors concentrated on two pepper types in this study because they can also be grown in greenhouses in the UK (Jones, 2013; Nickels, 2012).

Sweet Pepper (California Wonder; cultivar of *Capsicum annuum* Linnaeus Grossum Group) is described by the supplier, B&Q plc (see methodology section below), as a high-cropping large fruit growing from green to red that can be picked at either stage. The vegetable is usually used for salads and cooking.

Ideal growing conditions for Sweet Pepper are warm soil (21–29 °C), which should be kept moist but not waterlogged (Nickels, 2012). According to Haifa Chemicals (2014), the optimal temperature for Sweet Pepper during the germination stage is between 20 °C and 25 °C, and for the vegetative growth stage the corresponding range is between 20 °C and 25 °C through the day and between 16 °C and 18 °C through the night. Furthermore, for the flowering and fruiting stage, the recommended temperature should vary between 26 °C and 28 °C and between 18 °C and 20 °C during the day and night, respectively. It follows that commercial growing of Sweet Pepper in the UK is highly fragmented and restricted to greenhouses (or similar) due to low temperatures.

Sweet Pepper is sensitive to an abundance of moisture and excessive temperatures. They are known to be rich in vitamin C (Nickels, 2012) and sensitive to high levels of salinity, requiring salinity conditions below 1280 mg/l (FAO, 2003). Furthermore, the electronic conductivity for irrigation water should be less than 2000 μs/cm (Haifa Chemicals, 2014). Peppers prefer to grow in light and well-drained soil that should be rich in organic matter, such as sandy loam or loams with a pH value between 6.5 and 7.5 (Haifa Chemicals, 2014).

Chili (De Cayenne; *Capsicum annuum* (Linnaeus) Longum Group "De Cayenne") is described by B&Q as a good crop of slender and hot fruits ideal for growing in pots on the patio or balcony or in a greenhouse. It is also described as a perfect Chili for general cooking. This type of pepper needs approximately 100 days to mature.

Chilies prefer warm, moist, and nutrient-rich soil in a warm climate. The germination time is 5–14 days. The sowing to cropping time is approximately 18 weeks. Chilies are mostly perennial (often more than three years) in subtropical and tropical regions (Nickels, 2012). However, they are usually grown as annuals in temperate climates such as the UK. Commercial growing of Chilies in the UK is fragmented and also restricted to greenhouses due to low temperatures. Furthermore, Chilies prefer a loamy soil with a pH of between 7.0 and 8.5 (i.e., neutral to weakly alkaline soil), according to Nickels (2012).

35.1.4 Nutrients and Minerals

Jones (2013) discussed the positive and negative impacts of key nutrients and minerals on plants. The major minerals impaction on the growth of plants are nitrogen (predominantly ammonium), phosphorus, potassium, calcium, magnesium, and sulfur. Micronutrients that are beneficial in small amounts are (in no particular order) boron, copper, manganese, molybdenum, potassium, iron, zinc, iodine, and chlorine. Copper, manganese, molybdenum, iron, zinc, and aluminum (in no particular order) are often described as heavy metals. They may be toxic in high-inflow water concentrations for peppers. Under acid conditions (soil with a pH of less than 7), heavy metals could be a problem to sensitive plants (FAO, 2003).

Metals present in wastewater (cadmium, copper, molybdenum, nickel, and zinc) used for irrigation purposes can pose a health risk to humans and animals (Ebrazi Bakhshayesh et al., 2014; FAO, 2003). The Food and Agriculture Organization of the United Nations (FAO) (2003) states that the most toxic ions in treated wastewater are sodium, chloride, and boron. Boron exceeding 0.5 mg/l is toxic to sensitive plants. Sweet Pepper is semi-tolerant (about 2 mg/l) to boron. Pescod (1992) classified boron concentrations in irrigation water according to the degree of restriction on its use: there are no limitations for values of less than 0.7 mg/l, slight to moderate controls for values between 0.7 mg/l and 3.0 mg/l, and severe restrictions for measurements of more than 3.0 mg/l.

Pescod (1992) and FAO (2003) also recommend limits for trace minerals in reclaimed water used for irrigation. Long-term (for water used continuously on all soils) and short-term (for water used for a period of up to 20 years on fine-textured neutral alkaline soils) threshold values for 18 elements have been listed. Recommended maximum concentrations for the trace elements, which are often exceeded (see results below), are 5.0 mg/l for iron, 0.2 mg/l for manganese, and 2.0 mg/l for potassium.

FAO (1994) classified the suitability of treated wastewater for recycling in terms of pollutants. Acceptable ranges for ammonia-nitrogen, ortho-phosphate-phosphorous, and potassium were between 0 and 5, between 0 and 2, and between 0 and 2 mg/l, respectively. Furthermore, Pescod (1992) stated that for irrigation water there is no restriction for its reuse if nitrate–nitrogen values are $<$5.0 mg/l. Slight to moderate constraints exist for the range between 5 and 30 mg/l. Severe recycling restrictions are usually imposed for values of more than 30.0 mg/l.

Johnson and Decoteau (1996) reported that few reports are available concerning the required limits for nitrogen in soil for good growth of Chilies, which should not be grown when too much nitrogen ($>$280 kg/ha) is present in the soil, leading to excessive growth of foliage, apparently at the expense of fruit production.

FAO and the World Health Organization (FAO/WHO, 2001) recommended the following thresholds for metals in vegetables: cobalt (50.0 mg/kg), chromium (2.3 mg/kg), copper (73.3 mg/kg), nickel (66.9 mg/kg), lead (0.3 mg/kg), and zinc (9.4 mg/kg). In contrast, Chary et al. (2008) recommended a limit for copper in vegetables of 20 mg/kg, while for lead and zinc the corresponding values were 1 and 50 mg/kg, respectively.

The Ministry of Health of the People's Republic of China (2005) stated the following maximum levels of contaminants in food: arsenic (0.05 mg/kg), chromium (0.5 mg/kg), cadmium (0.05 mg/kg), and lead (0.1 mg/kg). EC (2001) has set maximum levels for certain contaminants in food: copper (20 mg/kg), lead (0.3 mg/kg), zinc (50 mg/kg), and cadmium (0.05 mg/kg).

Ciju (2013a) reports the following nutrition values for 100 g of fresh and raw green Sweet Peppers: phosphorus (20 mg), potassium (175 mg), calcium (10 mg), magnesium (10 mg), iron (0.34 mg), and zinc (0.13 mg). In comparison, Ciju (2013b) reports the following nutrition values for 100 g of sun-dried Chilies: phosphorus (159 mg), potassium (1870 mg), calcium (45 mg), magnesium (88 mg), iron (6.04 mg), and zinc (1.02 mg). Further beneficial elements may include silicon. Other elements such as aluminum should be present in low quantities within the irrigation water.

FAO (2003) stated the nutrient requirements for pepper which are appropriate for proper canopy formation: nitrogen (90 kg/ha), phosphorus (6 kg/ha), potassium (90 kg/ha), phosphorus pentoxide (14 kg/ha), and potassium oxide (108 kg/ha). The corresponding values for good fruit production are as follows: nitrogen (2.0 kg/ton), phosphorus (0.26 kg/ton), potassium (1.83 kg/ton), phosphorus pentoxide (0.6 kg/ton), and potassium oxide (2.2 kg/ton).

Haifa Chemicals (2014) stated the required rates of macro- and secondary plant nutrient uptake by pepper plants in greenhouses: nitrogen (390–920 kg/ha), phosphorus pentoxide (200–330 kg/ha), potassium oxide (640–1530 kg/ha), calcium oxide (100–210 kg/ha), magnesium oxide (60–150 kg/ha), and sulfur (40–50 kg/ha).

35.1.5 Motivation and Objectives

Effluent from different types of wetland systems treating domestic wastewater was selected to irrigate vegetables. Some of the wetlands received standard wastewater while the others received wastewater that was subject to a one-off diesel fuel spill. The treated wastewater from all wetland types was recycled for the irrigation of Sweet Peppers and Chilies. This experiment may provide the scientific justification for integrating treatment wetlands into agricultural food production.

The overall aim is to assess if vegetables can be grown successfully on recycled domestic wastewater treated by constructed wetlands. The corresponding key objectives related to the growing of Sweet Pepper and Chili are to assess (1) the suitability for growth when using recycled wastewater, (2) the impact of different treated wastewaters as a function of the wetland type, (3) the impact of the environmental conditions on growth, (4) the volume of treated wastewater for irrigation, (5) the suitability of different growth media, (6) the effect of a diesel oil spill on the suitability of the recycled wastewater, and (7) the economic viability of different experimental set-ups.

35.2 METHODOLOGY

35.2.1 Constructed Wetlands Set-Up and Operation

The vertical-flow constructed wetland system is located within a greenhouse (door left open) on top of the roof of the Newton Building of The University of Salford, Greater Manchester, UK (Sani et al., 2013a). The system was operated between June 27, 2011, and June 4, 2014. The set-up includes two filters that are essentially controls receiving clean dechlorinated water. Table 35.2.1.1 indicates an overview of the statistical experimental set-up used to test the impact of four variables. Filters 1 and 2 compared to Filters 3 and 4 test the influence of a larger aggregate diameter. Filters 5 and 6 compared to Filters 3 and 4 check the impact of a higher loading rate. The application of a lower contact rate is tested if Filter 7 is compared with Filters 3 and 4. Finally, a lower resting time is the difference between Filters 7 and 8.

Ten laboratory-scale vertical-flow constructed wetlands were constructed from Pyrex tubes with an inner diameter of 19.5 cm and a height of 120 cm. The filters were filled with siliceous (minimum of 30%) pea gravel up to a depth of 60 cm and planted with *Phragmites australis* (Cav.) Trin. ex Steud. (Common Reed). Dead macrophyte plant material was harvested in winter and returned to the corresponding wetland filters by placing it on top of the litter zone (Sani et al., 2013a).

The main outlet valve is located at the bottom of each constructed wetland system. Eight additional valves (used to test for clogging) are located on the sidewall of each wetland column. The sidewall valves were located at heights of 10, 20, 30, 40, 45, 50, 55, and 60 cm from the bottom of each column (Sani et al., 2013a).

Wetland columns received 6.5 l of inflow water during the feeding mode, which was different between several filters (Table 35.2.1.1). Columns 1–6

TABLE 35.2.1.1 Comparison of the Experimental Vertical-Flow Wetland Filter (F) and Control (C) Set-Up (after Almuktar et al., 2015)

Design or Operational Variable	Unit	F 1 and 2	F 3 and 4	F 5 and 6	F 7	F 8	C A	C B
Aggregate diameter	mm	20	10	10	10	10	10	10
Contact time	H	72	72	72	36	36	72	72
Resting time	H	48	48	48	48	24	48	48
Chemical oxygen demand	mg/l	122.8	122.8	243.8	122.8	122.8	2.3	2.3

Note: Annually treated volumes of wastewater: Filters 1–6, 470 l/year; Filter 7, 624 l/year; Filter 8, 858 l/year. On September 26, 2013, 130 g of diesel was added to Filters 1, 3, and 5 and Control A.

were sampled after 72 h contact time and then left to rest for 48 h, while columns 7 and 8 were sampled after 36 h contact time and left to rest for 48 and 24 h, respectively. All water quality parameters discussed in this paper were usually determined during or directly after sampling. The preliminary treated urban wastewater used for the inflow water was obtained from the Davyhulme Sewage Works, one of the largest wastewater treatment plants in Europe (http://en.wikipedia.org/wiki/Davyhulme), operated by the water company United Utilities in Greater Manchester. Fresh wastewater was collected approximately once per week and was stored and aerated by standard aquarium air pumps in a cold room before use. The wastewater quality was highly variable and comprised domestic and industrial wastewater as well as surface water runoff.

In order to simulate a one-off diesel fuel (100% pure; no additives) spill, 130 g (equivalent to an inflow concentration of 20 g/l) of diesel fuel (100% pure; no additives) was poured into Filters 1, 3, and 5 and into one of the two columns (Control A), on September 26, 2013 (Table 35.2.1.1). The fuel was obtained from a gas station operated by Tesco Extra (Pendleton Way, Salford, UK).

Aqua Medic Titan chillers (Aquacadabra, Bexleyheath, UK) were used to maintain the root system and debris layer of all wetland systems at semi-natural below-surface temperatures of about 12 °C. This temperature simulates the temperature of the upper earth layer where the root system of the wetland plants of a real treatment system would be located (Sani et al., 2013a).

The COD was used as the criterion to differentiate between low and high loads (Table 35.2.1.1). An inflow target COD of about 285 mg/l (usually between 122 and 620 mg/l) was set for wetlands with a high loading rate (Filters 5 and 6). The remaining Filters 1, 2, 3, 4, 7, and 8 received wastewater diluted with dechlorinated tap water. The target inflow COD for these filters was approximately 138 mg/l (usually between 43 and 350 mg/l).

35.2.2 Water Quality Analysis

Routine water quality sampling was carried out according to the American Public Health Association (APHA) (2005, unless stated otherwise) to monitor clogging evolution and long-term and seasonal treatment performance, respectively. A DR 2800 Hach Lange spectrophotometer (www.hach.com) was used for standard water quality analysis for variables including COD, ammonia-nitrogen, nitrate—nitrogen, ortho-phosphate-phosphorus, and suspended solids.

Total petroleum hydrocarbons (TPH) were determined by gas chromatography and flame ionization externally by Exova Health Sciences (Hillington Park, Glasgow, UK) according to their own TPH in Waters (with Aliphatic/Aromatic Splitting) Method (Exova Health Sciences, 2014), which is accredited to the British Standard (BS) method BS EN ISO IEC 17025 by the

United Kingdom Accreditation Service and is compatible with the International Organization for Standardization (ISO) standards (e.g., ISO17025), BS method BS DD 220 1994, and American Standard methods (United States Environmental Protection Agency (US EPA) Method 3510C and US EPA SW846 Method 8015).

The 5-day BOD was determined in all water samples with the OxiTop IS 12-6 system, a manometric measurement device, supplied by the Wissenschaftlich-Technische Werkstätten (WTW), Weilheim, Germany. Nitrification was suppressed by adding 0.05 ml of 5 g/l N-allylthiourea (WTW chemical solution No. NTH600) solution per 50 ml of sample water.

Turbidity was measured with a Turbicheck turbidity meter (Lovibond Water Testing, Tintometer Group, www.lovibond.com). The pH was measured with a sensION + Benchtop multi-parameter meter (Hach Lange, Düsseldorf, Germany). The redox potential for all water samples was measured using a VARIO pH meter (WTW, Weilheim, Germany). This meter was less reliable than the Hach Lange meter for measuring pH. Temperature data for the first year of operation were recorded outside and in the shade at an official weather station in Woodford, located southeast of Salford. The raw data were supplied by the UK MetOffice (www.metoffice.gov.uk). Concerning the second year of operation, inside temperature measurements were performed by project team members inside and outside the greenhouse.

35.2.3 Environmental Monitoring

Light measurement readings were undertaken using the LUX meter ATP-DT-1300 for the range between 200 and 50,000 lux (TIMSTAR, Winsford, Cheshire, UK) just above the top of the plants and showed values between 3855 and 12,316 lux (mean of 6921 lux) close to the plants.

Typical values for full-day light are between approximately 10,000 and 20,000 lux. The humidity and temperature were controlled with the support of a combined thermometer-hygrometer station provided by wetterladen24.de (JM Handelspunkt, Geschwend, Germany). The humidity measuring range was between 20% and 99%. The corresponding precision was ±4% between 35% and 75%.

The temperature was controlled by an electrical heater (Rhino H029400 TQ3 2.8 kW Thermo Quartz Infrared Heater 230 V) supplied by Express Tools Ltd (Bournemouth, UK). The humidity was artificially elevated by a varying number of humidity meters (Challenge 3.0L Ultrasonic Humidifier; Argos, Central Milton Keynes, England, UK).

35.2.4 Selected Vegetables

Sweet Pepper (California Wonder) and Chili (De Cayenne) were supplied by B&Q plc (Chandlers Ford, Hants, UK; www.diy.com) as part of their Verve brand. The Verve product codes were 311137 and 362387, respectively. All seeds were bought on September 14, 2013.

35.2.5 Replanting the Vegetables

In this experiment, the seeds (90 per plant type; see above) were sown thinly in a propagator (Verve; B&Q plc) into seed and cutting compost (Verve; B&Q plc) and covered with 6 mm of compost on September 16, 2013. Each propagator contained 72 planting cells with an average depth of 5 cm (only planted up to about 4 cm; measured before initial watering) and square sides of approximately 3.5 cm. The compost comprised 58% sustainably sourced (in terms of ecology, archaeology, and conservation) *Sphagnum* moss peat and unspecified amounts of composted bark, green compost, wood fiber, and coir (natural fiber extracted from the husk (outer shell) of coconuts), as well as oyster shells (optional), vermiculture (optional), perlite (optional), loam (optional), charcoal (optional), alcosorb (optional), sand (optional), grit (optional), wetting agent (to retain moisture better; between 200 and 400 ml/m^3), and essential nutrients and trace minerals lasting for approximately 6 weeks. The remaining 42% comprised, among other components, more than 48% nonpeat composted organic material, such as a mixture of composted green waste and spent brewery grains. The fertilizer content was between 0 and 3 kg/m^3. The dolomitic limestone content was between 0 and 7 kg/m^3. However, the exact combination of ingredients is commercially sensitive and was therefore not communicated to the authors.

The propagators were placed within a dark incubation room between September 17 and September 23, 2013. The transparent covers of the propagators were kept on top of the propagator bases. The temperature was maintained between 19.5 °C and 22.5 °C (mean of 20.8 °C). The recommended range according to the supplier is between 18 °C and 21 °C. The compost was kept moist until the seeds germinated.

After germination of some seeds was noticed on September 23, 2013, all seeds were relocated to a lab fitted with OSRAM HQL (MBF-U) high-pressure mercury lamp (400 W; Base E40) grow lights provided by OSRAM (Foshan, Guangdong, China) and supported by an H4000 gear unit, which was supplied by Philips (Croyden, UK). The bulbs were comparable to those used by Boyden and Rababah (1996). The temperature close to the plants ranged between 19.3 °C and 26.3 °C (mean of 24.2 °C). The lights were set on timers, simulating the sunrise and sunset times in Salford (http://www.timeanddate.com). The transparent covers of the propagators were kept on top of the propagator bases (gap of approximately 6.0 cm) until the first seedlings reached them on September 30, 2013.

The second planting (i.e., first replanting) of the strongest 70 Sweet Pepper and 76 Chili plants took place when most seedlings had at least two true leaves and were large enough to handle on October 11, 2013. The weakest 10 Sweet

Pepper and 1 Chili plants were not replanted. Moreover, 10 Sweet Pepper and 13 Chili plants either did not germinate or died before the first replanting.

Seedlings were transplanted into 60-mm-diameter (average) pots of moist multipurpose compost and grown for 3–4 weeks. Each tray contained 40 pots with a depth of 60 mm each. Multipurpose compost was filled up to a depth of 4.5 cm and covered with a 1.0-cm layer of bark (B&Q Verve range). The topping contained small chipped bark from mixed wood (responsibly sourced) and was applied to retain moisture and insulate soil.

Some vegetables were planted in pure sand to assess the impact of the organic grow substrate on plant growth. The product Play Pit Sand (silica), which is described by the supplier, Deko-Pak Limited (Hipperholme, Brighouse, UK), as nonstaining, nontoxic, safe, and clean, was used. Sand was filled up to a depth of 5.5 cm.

All vegetables were kept permanently indoors in the same heated laboratory fixed with grow lights. The environmental boundary conditions were essentially independent of a specific geographical region but were similar to those of a warm country.

The third planting (i.e., second and final replanting) took place 28 days after the first replanting on November 8, 2013. Table 35.2.5.1 outlines the experimental set-up. The Sweet Pepper and Chili plants were planted individually into 10 l plastic and round plant pot provided by Scotplants (Hedgehogs Nursery, Glenrothes, Scotland, UK). The plant pots dimensions were as follows: height of 22.0 cm, bottom diameter of 22.0 cm, and top diameter of 28.5 cm. The top 2 cm were left unplanted. Sand-based plants were planted to a depth of 20.0 cm. In comparison, soil-based plants were planted to a depth of 17.5 cm and covered by an additional 2.5 cm of bark (see above). Sufficient space between plants was always allowed for Sweet Pepper and Gardeners Delight. The plants remained indoors in the laboratory characterized above.

All plants were supported initially by small bamboo canes (diameter of approximately 0.3 cm; length of up to 30 cm) and later by bigger bamboo canes (diameter average of 0.8 cm; range between 0.6 and 1.2 cm; length of up to 150 cm), if and when necessary. When required, plants were supported using a string, which was loosely tied to the main stem against the cane.

35.2.6 Growth Conditions

Some plants were fed with a liquid and concentrated fruit and vegetable fertilizer from the B&Q Verve range. The fertilizer had a nitrogen to phosphorus to potassium ratio of 4:4:4, according to the European fertilizer solution for the UK. The total nitrogen component was 4%. Nitric nitrogen and ureic nitrogen parts were 1.1% and 2.1%, respectively. Phosphorus pentoxide (P_2O_5) and potassium oxide (K_2O) made up 4% each of the solution, but the corresponding P and K content were only 1.7% and 3.3%, respectively.

TABLE 35.2.5.1 Experimental Design in Terms of Plant Number Allocations after the Second Replanting (after Almuktar et al., 2015)

Inflow Source	Growth Media	Sweet Pepper	Chili
Filter 1 outflow	Compost with bark	P1; P2	C3; C4
Filter 2 outflow	Compost with bark	P5; P6	C6; C8; C9
Filter 3 outflow	Compost with bark	P8; P9; P10	C10; C11; C12
Filter 4 outflow	Compost with bark	P12; P16	C16; C17
Filter 5 outflow	Compost with bark	P18; P19; P20	C18; C19; C20
Filter 6 outflow	Compost with bark	P22; P23	C21
Filter 7 outflow	Compost with bark	P26; P28	C25; C26
Filter 8 outflow	Compost with bark	P31; P32; P33	C27; C28; C29
Control A outflow	Compost with bark	P35	C31; C33
Control B outflow	Compost with bark	P39	C37; C38
Deionized water	Compost with bark	P41	C41
Tap water (100%)	Compost with bark	P44	C42; C43
Tap water with fertilizer (0.7 ml/l)	Compost with bark	P45; P46	C45; C46
Wastewater (20%); tap water (80%)	Compost with bark	P47	C49
Wastewater (100%)	Compost with bark	P51; P54	C52; C54
Filter 1 outflow	Silica sand	P55; P56	C56
Filter 2 outflow	Silica sand	P57	C58
Filter 3 outflow	Silica sand	P59	C61
Filter 4 outflow	Silica sand	P61	C63; C64
Filter 5 outflow	Silica sand	P65	C66
Filter 6 outflow	Silica sand	P66; P67	C68
Filter 7 outflow	Silica sand	P17; P69	C71
Filter 8 outflow	Silica sand	P70; P71	C72; C73
Control A outflow	Silica sand	P73	C74
Control B outflow	Silica sand	P74	C76; C77
Deionized water	Silica sand	P80	C80

TABLE 35.2.5.1 Experimental Design in Terms of Plant Number Allocations after the Second Replanting (after Almuktar et al., 2015)—cont'd

Inflow Source	Growth Media	Sweet Pepper	Chili
Tap water (100%)	Silica sand	P81	C82
Tap water with fertilizer (0.7 ml/l)	Silica sand	P83; P84	C84; C85
Wastewater (20%); tap water (80%)	Silica sand	P86; P87	C87
Wastewater (100%)	Silica sand	P89; P90	C90

Note: Original seed planting reference numbers: Sweet Pepper (P1–P90) and Chili (C1–C90).

Moreover, the fertilizer also contained trace elements (names not listed) of unspecified quantities.

Domestic cultivars were selected to maximize self-fertility. In an outside setting, wind or insects provide sufficient motion to produce commercially viable crops (Jones, 2013). Therefore, mechanical movement of the plants and manual pollen transfer between plants were practiced in this study. Cross-pollination between Sweet Peppers and Chilies was prevented by separating the growing space with the help of temporary walls.

35.3 RESULTS AND DISCUSSION

35.3.1 Water Quality Analysis

The wetland effluent was used as the influent for the vegetable pots. Highly fluctuating values for TPH were observed in the outflow waters obtained from Filters 1, 3, 5, and 8 and Control A. The TPH concentrations followed this order: Control A > Filter 8 > Filter 1 > Filter 3 > Filter 5. Chemical oxygen demand values were the highest for Filter 5. In contrast, the lowest values were noted for Control B. Filters 1, 3, and 8 had relatively similar COD concentrations. Control A had higher COD values than Filter 6. No differences in COD values were noted for Filters 2, 4, and 7.

The 5-day BOD was high for raw wastewater, followed by wastewater samples that were diluted with 80% tap water. In comparison, the lowest 5-day BOD was observed for tap water with fertilizer, tap water, and deionized water.

High concentrations of ammonia-nitrogen, which exceeded the threshold of 5 mg/l (FAO, 1994), were noted for both Filters 5 and 6, followed by those for Filters 1 and 2. The nitrate–nitrogen for all filters' outflow waters was less than the maximum threshold value of 30 mg/l (Pescod, 1992). Based on the recommended threshold of 2 mg/l for ortho-phosphate-phosphorus (FAO, 1994), the outflow waters from all wetland filters (with the exception of Controls A and B) were associated with too high ortho-phosphate-phosphorus concentrations.

The highest value for suspended solids was noted for raw wastewater, followed by that for wastewater that was diluted with 80% tap water. In contrast, the lowest values were observed for Filter 7 outflow water and tap water with fertilizer. Turbidity was high for raw wastewater. Filter 7 had the lowest turbidity values. The pH values for all filter outflows were within the normal range, between 6.0 and 8.5 (Pescod, 1992). According to Haifa Chemicals (2014), temperature measurements for this experiment were within the recommended ranges for different growth stages of Chilies and Sweet Peppers.

35.3.2 Growth Comparisons

The statistical experimental set-up as specified in Table 35.2.5.1 was chosen for the second replanting stage. A surplus of nitrogen was noted for new plant parts of both Sweet Peppers and Chilies. A surplus of molybdenum was noted for the old plant parts for both Sweet Peppers and Chilies. However, this observation is ambiguous, because symptoms for some plants also indicate deficiencies.

In countries where wastewater is seen as a resource, low wastewater consumption by plants is an advantage. However, high wastewater use by plants is seen as an advantage in temperate regions. The productivity of plants in terms of harvest is, however, independent of the wastewater consumption. Nevertheless, a higher foliage production requires more water. A high yield commonly related to the most suitable provision of nutrients and trace elements.

The overall growth development of Sweet Peppers was rather disappointing, possibly due to the high concentrations of nutrients and trace minerals and the adverse environmental boundary conditions in the laboratory. In contrast, Chilies did reasonably well, but the growth of foliage was excessive and the harvest was delayed. The highest number of fruits is associated with tap water and an organic growth medium. Low fruit numbers correlate well with inorganic growth media. Findings indicate that nutrient concentrations supplied to the Chilies by a combination of compost and treated wastewater are usually too high to produce a good harvest. However, as the compost is depleted of nutrients after about eight months, the harvest increased for pots that received pretreated wastewater in comparison to those pots depending only on the nutrients associated with the compost.

35.3.3 Inductively Coupled Plasma Findings

High concentrations of iron, which exceeded the threshold of 5 mg/l (FAO, 2003; Pescod, 1992), were noted for both raw wastewater and tap water. Based on the recommended threshold of 2 mg/l for potassium (FAO, 1994), the outflow water from all wetland filters (with the exception of Controls A and B) was associated with too-high potassium concentrations. Furthermore, high concentrations were also observed for raw wastewater and for wastewater samples that were diluted with up to 80% tap water. Results for Filters 3 and 5 show relatively high manganese concentrations, which exceeded the threshold of 0.2 mg/l (FAO, 2003; Pescod, 1992).

35.3.4 Brief Cost–Benefit Analysis and Limitations

Sweet Pepper and Chili seed packets were purchased from B&Q plc for £1.48, or 148 pence, each. The corresponding seed numbers were 45 and 70, respectively. One seed of Sweet Pepper and Chili therefore costs 3.29 and 2.11 pence, respectively.

Considering the germination success rates of 78 and 74 for Sweet Pepper and Chili, respectively, each seedling costs 257 and 156 pence, in that order. However, only 50 seedlings of Sweet Pepper and Chili each reached maturity. This corresponds to a cost of 165 and 106 pence, respectively.

Sweet Pepper and Chili can be purchased in the UK for approximately 56 and 16 pence each or 362 and 1040 pence per kilogram. However, taking into account the costs of watering, fertilizer, and maintenance, the calculation becomes more complex.

The potential fear and disgust by consumers regarding eating microbially contaminated vegetables decrease considerably if vegetables are cooked for a long time at considerable heat. Menegaki et al. (2009) assessed the fear and disgust factors by comparing the effects of descriptive terms on farmers' willingness to use and willingness to pay for recycled water for irrigation and consumers' willingness to use and willingness to pay for products irrigated with recycled water. Treated effluent from wastewater treatment plants was described as "recycled water" for one experimental group and as "treated wastewater" for another. Although the two terms describe the same commodity, willingness to use the water was reliably higher with the "recycled water" descriptor for both farmers and consumers. However, the descriptor affected willingness to pay only in the consumer sample. Both farmers and consumers who were unwilling to use recycled water commodities cited feelings of disgust (32%) as the main cause of their rejection (Menegaki et al., 2009).

Sweet Peppers are eaten both raw and cooked. However, Chilies are usually cooked, and the risk of microbial contamination is therefore very low. Considering that Sweet Peppers in comparison to Chilies are more likely to be used in a salad than in a cooked dish, they are more difficult to sell because of the fear and disgust factors discussed above. Therefore, the likelihood of selling the selected plants at a fair price, taking the fear of contamination factor into account, is likely to be less for Sweet Peppers compared to Chilies.

Table 35.3.4.1 shows a proposed novel harvest classification scheme for Chilies. Only the higher classes are of great commercial interest. However, the estimated prices are dependent on the market. The highest numbers of fruits categorized as Class A were harvested from plants grown in organic media and watered with tap water. However, tap water was also associated with the highest fruit numbers categorized as Class E. The highest mean price of harvested fruits is also associated with tap water.

This study presents a partly incomplete picture of the reuse of treated wastewater for irrigation because microbiological parameters were not studied. However, microbial contamination of Chilies and Sweet Peppers is unlikely due to the relatively long distance between the fruits and the potentially contaminated soil (Cirelli et al., 2012). Moreover, vegetable pots receiving wastewater treated with wetlands can be considered as safer than those receiving only preliminary treated wastewater.

Municipal wastewater often lacks the required amount of potassium for growth of commercial crops (Boyden and Rababah, 1996). Therefore, potassium could have been added at an optimal dose for the growth of all tested plants. However, the results in this chapter indicate that potassium was sufficient in the outflow waters of most wetlands.

The humidity was relatively low throughout the experiment. However, Bakker (1989) found no significant effect of humidity on the *C. annuum* fruit shape, number of cavities per fruit, pericarp thickness, dry matter content, and fruit maturation rate. The relative humidity had no effect on the growth of Sweet Peppers between the range of 50% and 80% (Bakker, 1989, 1991).

35.4 CONCLUSIONS AND RECOMMENDATIONS

The experiment shows that Sweet Peppers and Chilies can be grown using wastewater treated by constructed wetlands. However, the yield of Sweet Peppers was insignificant in contrast to that of Chilies, possibly due to the high concentrations of nutrients (particularly nitrogen) and trace minerals and adverse environmental boundary conditions in the laboratory. Chilies did reasonably well, but the growth of foliage was excessive and the harvest was delayed due to high nitrogen concentrations in the inflow water. The highest number of fruits was associated with tap water and an organic growth medium. In contrast, plants associated with irrigation water contaminated by hydrocarbon were usually linked to a poor harvest. Standard wetland design parameters were only of secondary importance.

Findings also indicate that nutrient concentrations supplied to the Chilies by a combination of compost and treated wastewater are usually too high to produce a good harvest. A high yield was related to the most suitable provision of nutrients and trace elements. However, as the compost is depleted of nutrients such as nitrogen after about eight months, the harvest increased for

TABLE 35.3.4.1 Chili Harvest Classification Scheme (after Almuktar et al., 2015)

Variable	Class A	Class B	Class C	Class D	Class E
Quality class	Outstanding	Good	Good	Satisfactory	Unsatisfactory
Codex standard (2013)	"Extra" class	Class I	Class II	Not applicable	Not applicable
Price (pence per gram)	2.00	1.00	0.50	0.25	0.00
Target market	Top restaurant	National supermarket	Independent retailer	Vegetable industry	Waste company
Product	Fresh vegetable	Fresh vegetable	Fresh vegetable	Powder or canned	Waste
Contamination	Uncontaminated	Uncontaminated	Uncontaminated	Uncontaminated	Contaminated
Length (L, mm)	Very long ($L \geq 80$)	Long ($60 \leq L < 80$)	Medium ($40 \leq L < 60$)	Short ($20 \leq L < 40$)	Very short ($L < 20$)
Width (W, mm)	Very wide ($W \geq 20$)	Wide ($16 \leq W < 20$)	Medium ($12 \leq W < 16$)	Slim ($8 \leq W < 12$)	Very slim ($W < 8$)
Weight (w, g)	Very large ($w \geq 9$)	Large ($7 \leq w < 9$)	Medium ($5 \leq w \leq 7$)	Small ($3 \leq w < 5$)	Very small ($w < 3$)
Bending	$L/W \geq 3.5$	$L/W \geq 3.5$	$L/W \geq 3.5$	$L/W < 3.5$	$L/W < 3.5$
Color	Characteristically red	Characteristically red	Characteristically red	Not fully red or unripe	Not fully red or unripe

pots that received pretreated wastewater in comparison to those pots depending only on the nutrients associated with the compost.

The productivity of Chilies in terms of harvest was independent of the wastewater consumption. Nevertheless, higher foliage production due to excess nutrients and trace minerals required more water. The current research will be continued with the same plants to assess if further harvests are economical and to determine when the nutrients within the compost are fully depleted. Moreover, the accumulation of metals and their toxicity in the soil, as well as microbiological contamination, will also be studied. Further research with other vegetables receiving recycled treated wastewater from other wastewater treatment units should also be undertaken to select the best and most cost-effective technology in order to obtain the greatest crop yield.

Chapter 36

Piggery Wastewater Treatment with Integrated Constructed Wetlands

36.1 INTRODUCTION

36.1.1 Legislative Pressures

The lack of solutions for the sustainable and cost-effective treatment of agricultural wastewater is a widespread problem, particularly across the European Union (EU) (Martinez et al., 2009). Traditionally, solid-rich biological wastewaters are disposed of in great volumes by means of land-spreading, as the high nutrient concentrations are of great benefit to farmers as a cheap and effective source of fertilizer (Smith et al., 2000). Many farmers and farming industries do not have large-scale treatment facilities to deal with excessive amounts of wastewaters. Holding tanks are the most common storage solution, allowing for wastewaters to be collected and stored prior to disposal or land application (Meers et al., 2008).

In the European Union, Nitrates Directive 91/676/EEC (EEC, 1991) has traditionally put pressure on piggeries due to the restrictions placed on land-spreading of corresponding wastewaters at certain times of the year. Considering that suitable farmland for spreading is limited and other landowners are often unable to spread piggery wastewater as fertilizer, expensive and high-maintenance equipment, such as anaerobic digesters and membrane technology, is often purchased that can put great financial burdens on piggeries (Henkens and van Keulen, 2001).

The Nitrates Directive limits the amount of nitrogen that can be applied to land. This places pressure on farmers in the EU to manage their wastewaters effectively to prevent eutrophication of surface waters. The directive states "appropriate N [nitrogen] spreading calendars and sufficient manure storage, for availability only when the crop needs nutrients, and good spreading practices" (EEC, 1991) as types of actions that it promotes. The construction of alternative storage and treatment units or additions to traditional treatment and operational methods are traditionally required to abide with the Nitrates Directive (Henkens and van Keulen, 2001).

Wetlands for Water Pollution Control. http://dx.doi.org/10.1016/B978-0-444-63607-2.00036-8

36.1.2 Traditional Piggery Wastewater Management

Intensive animal production has resulted in a considerable overproduction of piggery wastewater. Using either tractor-mounted spreaders or fixed rotational spreaders, the wastewaters from yards, runoff, washings, manure, and slurry are spread on land as a form of cheap fertilizer utilizing the high nutrient concentrations of these wastewaters (Healy et al., 2007; Meers et al., 2008). Spreading the excess wastewater over arable land has resulted in contamination and eutrophication of groundwater and surface waters, respectively (Smith et al., 2000; Kumm, 2003; Meers et al., 2008; Martinez et al., 2009).

For agricultural industries such as dairy or cattle farming, landspreading is often a viable option due to the availability of large areas of farmland. In comparison, piggeries produce large volumes of waste with high nutrient concentrations and have a smaller available land area compared to dairy, cattle, or sheep farming due to the housed nature of large industrial units (Kumm, 2003; Healy et al., 2007; Meers et al., 2008).

Therefore, the use of anaerobic digesters for the treatment of pig manure and slurry has become more viable. This is partially due to the added benefit of biogas (methane) harvesting and subsequent recycling to power equipment on the farm. Moreover, a solids separator can also produce biosolids that can be used as a sustainable solid fuel source. However, these digesters primarily remove carbon from the manure and slurry, and the liquid fraction from the separator is still nutrient-rich and requires storage and subsequent treatment.

36.1.3 Alternative Piggery Wastewater Treatment with Wetlands

Wetlands can provide effective, low-cost, low-maintenance, and environmentally beneficial wastewater treatment alternatives to traditional treatment systems (Harrington and Scholz, 2010; Smith et al., 2000; Scholz, 2006). More recently, there is a move from treatment wetlands to more sustainable and integrated wetland systems. The integrated constructed wetland (ICW) concept is a holistic approach to the design of treatment wetlands that takes economic, social, environmental, and landscape aesthetic aspects into consideration in its design, construction, and operation. Moreover, the incorporation of wetlands into the landscape enhances biodiversity (Scholz et al., 2007; Harrington and McInnes, 2009). A comprehensive review of wetland systems for environmental pollution control in Ireland has been published by Babatunde et al. (2008).

Only recently, wetland systems have been used to treat agricultural wastewaters high in ammonia, which is toxic to a lot of wetland plants. For example, Meers et al. (2008) report on wetland systems that successfully treat piggery wastewater to the following standards: 15 mg/l for total nitrogen, 2 mg/l for total phosphorus, and 125 mg/l for chemical oxygen demand. Their pilot-scale system, which had a capacity of 1000 m^3/year, was subsequently used to design full-scale installations with annual capacities between 10,000 and 25,000 m^3/year.

Free water surface flow wetland systems are usually used for piggery wastewater treatment to avoid clogging. Chen et al. (2008) assessed a conventional three-stage treatment scheme followed by a modified free water surface wetland (with or without plants) with a two-day hydraulic retention time. This was a promising option to meet Chinese swine wastewater discharge limitations (chemical oxygen demand, 600 mg/l; biochemical oxygen demand, 80 mg/l; suspended solids, 150 mg/l).

The loading and flow rates in constructed wetlands vary greatly. For example, the Livestock Wastewater Treatment Database (Knight et al., 2000) holds data from 68 wetland sites in North America. The mean hydraulic loading rate was 4.7 cm/day. Mean system flows of 10 m^3/day have been calculated. Most of the systems used for livestock wastewater treatment are of small size with a mean area of only 0.6 ha. Constructed wetlands treating swine wastewater are slightly larger, with a mean area of about 1 ha (Knight et al., 2000).

Furthermore, the water depth has an important impact on the growth of macrophytes that are used in constructed wetland systems. Depending on the species that are planted, greater water depth can inhibit macrophyte growth and colonization of the wetland cells (Clarke and Baldwin, 2002; USEPA, 1988). High water levels also result in wetlands being more susceptible to the effects of high nutrient concentrations and may subsequently lead to the death of more sensitive species. Shallower systems also help to increase nitrification by increasing the aerobic conditions present in the cells (Carty et al., 2008; Scholz, 2006).

The dilution of wastewater with clean water to improve nutrient removal is common practice in constructed wetland operation, particularly if the wastewater is slightly toxic (e.g., piggery wastewater) and no separate pipes for individual wastewater streams exist (Scholz, 2006). Heavily polluted wastewater can also be diluted by less contaminated wastewaters such as roof and yard runoff. If the organic loading rates are excessive, this can result in decreased removal performances (Kantawanichkul et al., 2001) and an increase in the risk of ammonia toxicity to some constructed wetland plants (Hunt et al., 2002, 2004).

The most commonly used macrophytes in treatment wetlands are *Phragmites* and *Typha* (Scholz, 2006). More recently, *Glyceria* is being used because of its high tolerance to the toxicity of ammonia (Tylova-Munzarova et al., 2005). The most common macrophytes planted on floating mats in anaerobic lagoons treating swine wastewater have been assessed by Hubbard et al. (2004). The nutrient uptake rates were relatively high but variable.

Recirculation of partially treated wastewaters has been shown to have a positive effect on total nitrogen removal within wetlands. For example, Kantawanichkul et al. (2001) showed that the recirculation of effluent in a combined vertical-flow and horizontal-flow wetland system increased the total nitrogen removal rate from 71% to 85%. Moreover, Humenik (1999) reported that nitrified lagoon water added to constructed wetland microcosms led to nitrogen removal rates that were four to five times that of nonnitrified liquid. Pilot-scale plants in China (He et al., 2006) were used to assess different recirculation rates. Findings showed increased ammonia–nitrogen removal rates in comparison to nonrecirculated effluent. However, the researchers reported no significant increase in phosphorus removal. The partially nitrified state of recirculated effluent leads to a decrease of ammonia volatilization (Poach et al., 2004) and helps to abate ammonia toxicity with regard to the most commonly used macrophytes within wetland systems.

The most widespread method of enhancing denitrification is to recirculate the wastewater or to add partially nitrified water (He et al., 2006; Kantawanichkul et al., 2001; Poach et al., 2003). This is done by recycling the effluent back into the system or by the addition of partially nitrified storage or lagoon water. It follows that denitrification is promoted by supplying the system with greater concentrations of nitrate and nitrite throughout the treatment process, thus reducing the risk of volatilization. However, a complete removal of the volatilization process within open wetlands is unrealistic.

The aim of Chapter 36 is to test the applicability of the ICW concept for piggery wastewater treatment. The objectives are to assess:

- the effect of low and high flow rates on the wetland treatment efficiency;
- the effect of low and high nitrogen loading rates on the treatment performance;
- the potential benefits of recycling some of the partially treated effluent;
- boundary conditions such as temperature impacting on nutrient removal processes such as nitrification;
- operational modes in terms of their cost-to-benefit ratio; and
- if the proposed treatment options are good value for money compared to other treatment options and alternative methods.

36.2 SITE, MATERIALS, AND METHODS

36.2.1 Site Description, System Set-Up, and Test Conditions

Figure 36.2.1.1 provides an overview of the experimental ICW system comprising four operational modes with four replicates each. The influent is pumped from a control hut housing the storage tanks to the individual wetland systems. The flow between the cells is induced by gravity.

The overall mesoscale wetland system (Harrington and Scholz, 2010) was designed in the spirit of the ICW concept (Carty et al., 2008; Scholz et al., 2007; Harrington and McInnes, 2009). The experimental system was located at the Teagasc Centre in Moorepark (Fermoy, County Cork, Ireland). The mean annual temperature was 11.5 °C and the mean annual rainfall was 1150 mm.

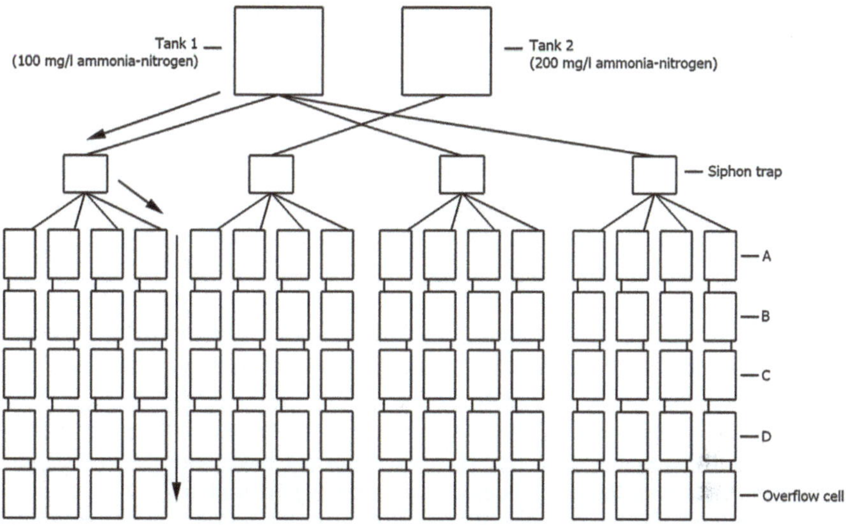

FIGURE 36.2.1.1 Sketch of the system set-up: Storage tank 1 (first from the left side) provides influent to the normal (wetland systems 1–4), recycling (wetland systems 5–8), and high flow rate (wetland systems 13–16) treatment systems (all cells: 100 mg/l ammonia–nitrogen). Storage tank 2 (second from the left side) supplies influent to the high nutrient (200 mg/l ammonia–nitrogen) loading wetland systems 9–12 (after Harrington and Scholz, 2010).

The approximate mean seasonal temperatures for the region were as follows: winter, 8.0 °C; spring, 10.5 °C; summer, 15.0 °C; and fall, 12.5 °C (Met Éireann, 2010).

Sixteen ICW systems were constructed in total, with each individual system having an overall surface area over four equally sized cells of 0.788 m^3. Each mesoscale system consisted of four wetland cells, which were constructed using polyethylene containers (length: 555 mm; width: 355 mm; depth: 300 mm) placed in series and buried within sloping trenches. Wastewater flowed sequentially from one cell to another cell via pipes with a diameter of 60 mm. A fifth cell was attached to the fourth cell of each ICW system to collect overflow, which represents the finally treated wastewater (Figure 36.2.1.1).

The experiment was conducted in two phases. During the first phase, a high flow rate was applied. Until April 7, 2009, the normal and the high flow rate operational modes had flow rates of 112 and 180 m^3/ha/day, respectively. These potentially too-high flow rates were dropped to 37 and 74 m^3/ha/day, respectively, during the second phase to allow for a longer filter run time and less sediment buildup and to avoid exposing the plants to a too-high level of ammonia toxicity.

Each ICW system received diluted anaerobic digester liquid from storage containers with a volume of 1 m^3. The dilution ratio (i.e., water to wastewater) for the ICW systems operated at normal loading rates was 1:16. In comparison, the systems operated at high nutrient loading rates had a corresponding ratio of 1:32. The influent was pumped using in-line submersible pumps set on timers to deliver approximately constant volumes to each system. The operation was constant except for periods when the wetlands were frozen (mostly between mid-December and mid-January).

The 16 ICW systems were divided into four sets with four replicates each (Figure 36.2.1.1). Operational differences were as follows: high and low hydraulic and nutrient loading and effluent recycling. ICW systems 1–4 received a standard treatment; i.e., low hydraulic load, low nutrient load (100 mg/l ammonia–nitrogen), and no effluent recycling. Wetlands 5–8 also had low hydraulic and nutrient loads (100 mg/l ammonia–nitrogen) but received recycled effluent to cell 1; i.e., 100% effluent recycled through the system weekly. Systems 9–12 were associated with a low hydraulic load but a high nutrient loading rate (200 mg/l ammonia–nitrogen) and no effluent recycling. Finally, a high flow rate (rates of 180 m^3/ha/day during phase 1 and 74 m^3/ha/day during phase 2) was used for wetland systems 13–16. However, nutrient loads were low (100 mg/l ammonia–nitrogen) and no effluent recycling was performed.

All wetland cells were densely planted (254 individual plants per m^2) with an approximately equal number and amount of mature and healthy stands of *Glyceria maxima* (Hartm.) Holmb. (syn. *G. aquatica* (L.) Wahlenb.), *G. spectabilis* Mert. & W.D.J. Koch, *Molinia maxima* Hartm. and *Poa aquatica* L. The plants were about two years old and were obtained from an existing semi-natural wetland near Kilmeaden, County Waterford, Ireland. All plants had not been treating any wastewater before replanting. *Glyceria maxima* is a highly competitive and invasive rhizomatous perennial mannagrass that is native to Europe and grows in wetlands such as riverbanks and ponds.

36.2.2 Analytical Methods

Sampling was performed weekly for the storage tanks containing the influents and the outflows of all wetland cells 1 and 4 of each system. Each sample was analyzed for ammonia–nitrogen, molybdate-reactive phosphorus, nitrate–nitrogen, nitrite–nitrogen, total organic nitrogen, and chloride. The 5-day at 20 °C N-allylthiourea biological oxygen demand (BOD) was analyzed for effluents fortnightly. The analyses were performed at the Teagasc (Moorepark) and Waterford County Council water laboratories using American Public Health Association standard methods (APHA, 1998).

36.3 RESULTS AND DISCUSSION

36.3.1 Raw and Diluted Wastewater

The influents to the normal, recycling, and high flow rate ICW treatment systems were associated with relatively high ammonia–nitrogen (approximately 100 mg/l) and nitrate–nitrogen (about 10 mg/l) concentrations. In comparison, the influent for the high nutrient loading treatment system had ammonia–nitrogen and nitrate–nitrogen concentrations that were considerably higher and lower, respectively.

The water quality of the raw liquid fraction of the separated anaerobic digestate was diluted with tap water to obtain the desired influents. This was necessary because of the very high concentration of ammonia–nitrogen (roughly between 1600 and 2500 mg/l), which cannot be treated effectively by wetland systems. Moreover, dilution was required to protect *G. maxima* from the toxicity of ammonia. The requirement for pretreatment of high-strength animal wastewater has been reviewed by Cronk (1996).

36.3.2 Standard Treatment

A direct comparison of the different treatment systems (Fig. 36.2.1.1) for the two flow phases is difficult. Seasons, indicated by temperature fluctuations, and the different lengths of the two flow phases are likely to have influenced the ICW system performances (Cronk, 1996; Kuschk et al., 2003). For example, the first phase (high flow rate) was conducted in winter and early spring, when the outside temperature was low, while the second phase (low flow) was tested at a higher temperature. This shortcoming makes a direct assessment of the absolute water quality values difficult. Moreover, a direct comparison with wetland systems operated by others is also impaired. However, considering that all systems with different test conditions were operated in parallel, it is justified to compare system performances relative to each other.

For the standard treatment at 100 mg/l ammonia–nitrogen, the outflow of cell 1 had particularly high ammonia–nitrogen concentrations for both the high flow rate (64 mg/l) and low flow rate (20 mg/l) ICW systems. The outflow of cell 4 indicates that these concentrations were reduced significantly ($P < 0.05$), to approximately 16 and 3 mg/l, respectively. Furthermore, it should be noted that the final outflow BOD concentration for the low flow rate system is relatively high (19.9 mg/l) and close to the threshold (20 mg/l) of some international secondary wastewater treatment guidelines that could be used for comparative purposes (Scholz, 2006).

Appropriate standards for piggery wastewater treatment by wetland systems are currently being discussed. A reasonable threshold for treated piggery wastewater with ICW is 50 mg/l BOD. This threshold is only exceeded once by the standard treatment operational mode. All other operational modes performed exceptionally well. However, other international standards might be as stringent as 20 mg/l (Irish Urban Waste Water Treatment Regulations, 2001; Scholz, 2006).

36.3.3 Effluent Recycling Treatment

For the outflows of the recycling treatment mode at 100 mg/l ammonia–nitrogen, the entire effluent from cell 3 was used as additional inflow to cell 1. The outflow of cell 1 had particularly high ammonia–nitrogen concentrations for the high flow rate (62 mg/l) systems. The outflow of cell 4 indicates that this concentration was reduced significantly ($P < 0.05$) to approximately 14 mg/l.

Figure 36.3.3.1 indicates the temporal distributions of nitrogen species for the outflows of all ICW systems. Figure 36.3.3.1(b) shows ammonia–nitrogen concentrations for the overall outflows of all ICW systems operated in normal, recycling, high nutrient load, and high flow rate modes. If someone is interested in protecting receiving freshwaters, a threshold of 0.5 mg/l should be considered as a guide value (Scholz, 2006). All systems failed this target during the initial phase of the project, where the flow rates were high. With the exception of the recycling operational mode during the second phase of the project, when the flow rates were relatively low, all ICW systems experienced some ammonia–nitrogen removal challenges even during the second winter. The observed improved nitrogen removal for the recycling treatment mode supports similar findings reported by Kantawanichkul et al. (2001), Poach et al. (2003), and He et al. (2006) for other wetland systems.

Figure 36.3.3.2 shows a comparison of the outflow nutrient concentrations associated with the ICW systems operated in the recycling treatment mode. The graphs for the outflows of cell 1 (Figure 36.3.3.2(a)) and cell 4 (Figure 36.3.3.2(b)) are relatively similar. However, the ammonia–nitrogen concentrations are lower than the nitrate–nitrogen concentrations for the outflows of cell 4 in comparison to the outflows of cell 1 in summer 2009. This indicates that nitrification, in comparison to denitrification, performs relatively better in summer than in winter (Figure 36.3.3.2). This confirms findings by Kuschk et al. (2003) highlighting that higher temperatures in summer (August) compared to winter (January) have led to up to five times higher nitrogen removal.

Stone et al. (2004) and Dong and Reddy (2010) discussed the successful treatment of swine wastewater by marsh–pond–marsh constructed wetlands. Moreover, Dong and Reddy (2010) detected organisms responsible for anaerobic ammonium oxidation, which might contribute considerably to the overall nutrient removal. The potential presence of these organisms within the ICW discussed in this chapter might explain the relatively good removal efficiency of ammonia. However, marsh–pond–marsh systems are different from ICW, which usually do not have (unplanted) pond-like cells, except for occasional polishing purposes at the end of a treatment train of four (planted) wetland cells (Scholz et al., 2007).

Seasonal and temperature effects on denitrification have previously been examined (Cronk, 1996; Reddy et al., 2001; Kuschk et al., 2003; Trias et al., 2004). There is a moderate positive correlation between temperature and both nutrient and suspended solids removal rates in wastewaters. Trias et al. (2004) reported variable findings for swine wastewater treatment with respect to total suspended solids, ranging from 77% at moderate temperatures to 42% during the

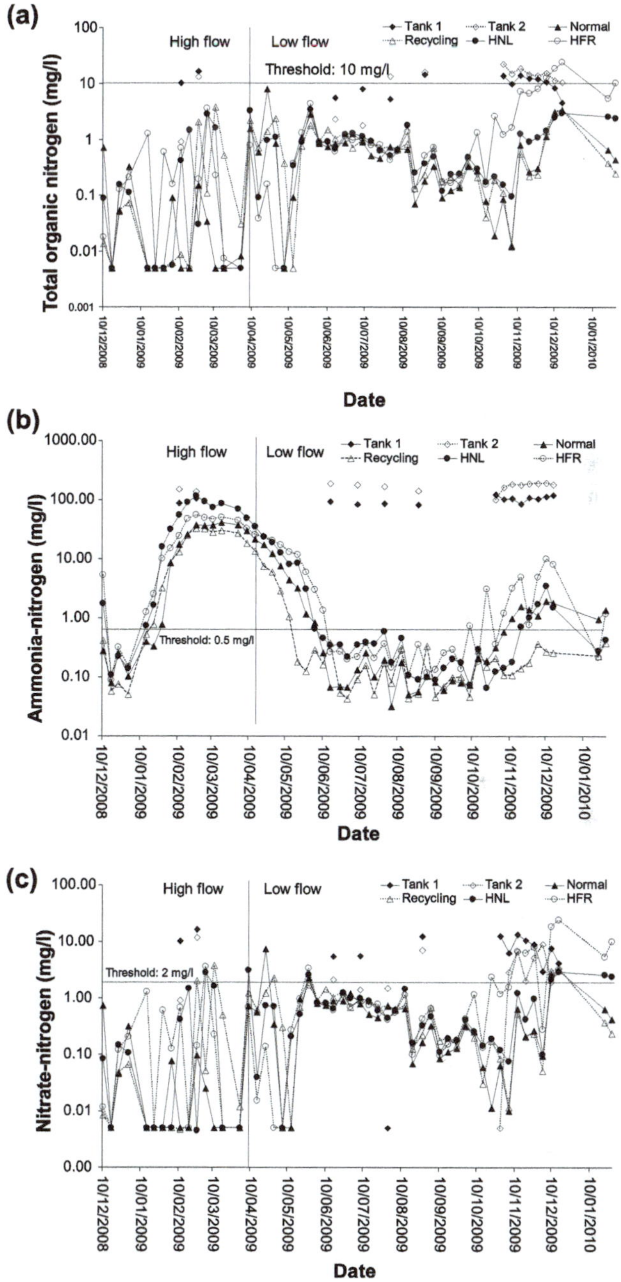

FIGURE 36.3.3.1 (a) total organic nitrogen, (b) ammonia–nitrogen, and (c) nitrate–nitrogen for the overall outflows of all integrated constructed wetland systems operated in normal, recycling, high nutrient load (HNL), and high flow rate (HFR) modes. Each data point usually represents four replicates according to Harrington and Scholz (2010).

warmest period. However, suspended solids were not determined in this study, because surface-flow systems such as ICW usually do not clog (Scholz, 2006).

36.3.4 High Nutrient Load Treatment

Concerning the high nutrient load treatment (37 m^3/ha/day at 200 mg/l ammonia–nitrogen), the outflow of cell 1 had exceptionally high ammonia–nitrogen concentrations for both the high flow rate (125 mg/l) and low flow rate (62 mg/l) systems. The outflow of cell 4 indicates that these concentrations were reduced significantly ($P < 0.05$) to 43 and 5 mg/l, respectively. Furthermore, it should be noted that the final outflow BOD concentrations for both the high flow rate and low flow rate systems were relatively high (19 and 22 mg/l) and close to the threshold (20 mg/l) of some international secondary wastewater treatment guidelines (Scholz, 2006).

36.3.5 High Flow Rate Treatment

The outflow of the high flow rate treatment systems operated at 100 mg/l ammonia–nitrogen for both phases was analyzed. Until April 7, 2009 (end of the first phase), a high flow rate of 180 m^3/ha/day was applied. This potentially too high flow rate was subsequently dropped to 74 m^3/ha/day during the second phase to allow for a longer filter run time, less sediment buildup, and a reduced risk of ammonia toxicity for wetland plants.

Considering that the high flow rate was followed by a low flow rate, it is likely that the negative aspects of the high nutrient load, such as ammonia toxicity, have also impacted on plant growth at the beginning of the second phase (Scholz, 2006). Figure 36.3.3.2(b) shows that ammonia–nitrogen values were still relatively high for about one month after the phase change. However, ammonia toxicity on plant growth was not assessed, and a detailed investigation was beyond the scope of this chapter.

The outflow of cell 1 had particularly high ammonia–nitrogen concentrations for both the high flow rate (80 mg/l) and low flow rate (38 mg/l) systems. The outflow of cell 4 indicates that these concentrations were reduced significantly ($P < 0.05$) to approximately 24 and 5 mg/l, respectively. It was surprising that the final BOD outflow concentrations were relatively low (14 mg/l) despite the higher flow rate.

Figure 36.3.3.2(a) indicates the distribution of the total organic nitrogen concentration for the outflows of cell 4 with respect to the normal, recycling, high nutrient load, and high flow rate operational modes. The usually accepted standard for total nitrogen is either 10 or 15 mg/l (Irish Urban Waste Water Treatment Regulations, 2001; Meers et al., 2008; Scholz, 2006). Concerning piggery wastewater treatment with ICW, there is currently no internationally accepted guide threshold value for total organic nitrogen. Nevertheless, all systems, with the exception of the one operated at a high flow

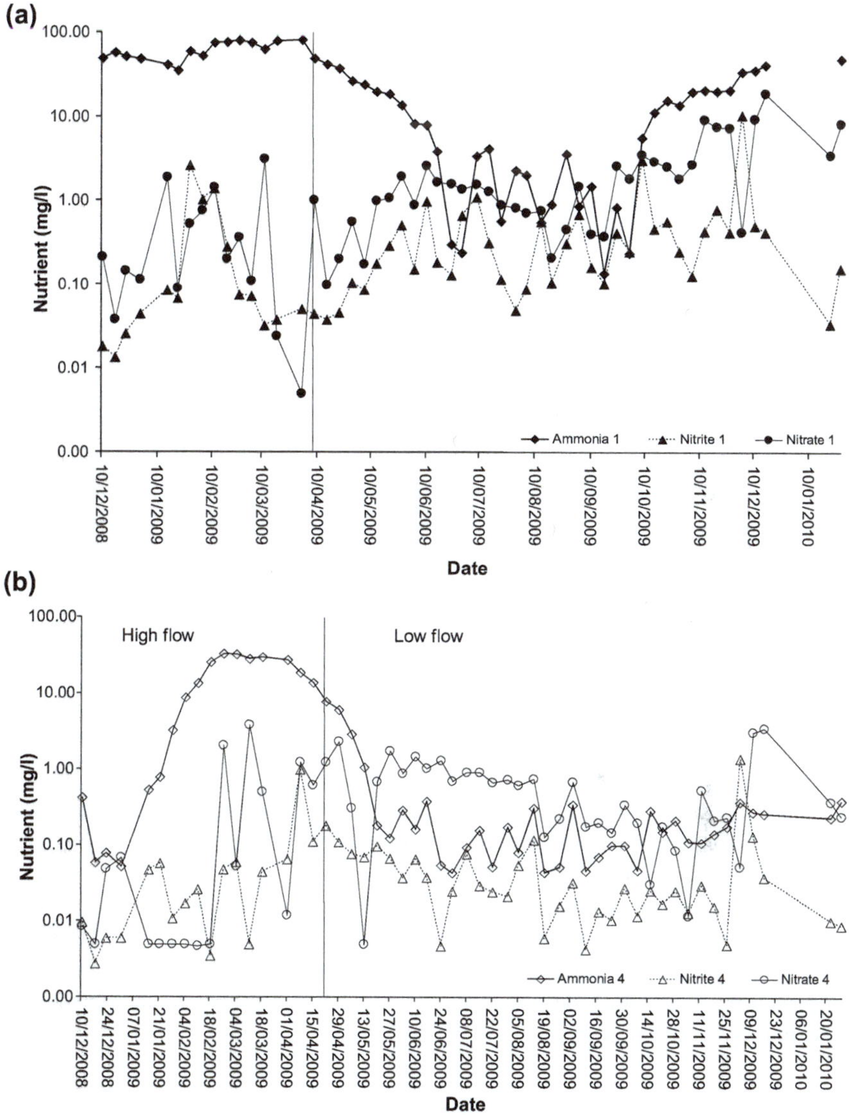

FIGURE 36.3.3.2 Outflow nutrient concentrations associated with recycling treatment (37 m^3/ha/day at 100 mg/l ammonia–nitrogen, 100% of the effluent from cell 3 was used as additional inflow to cell 1) of the integrated constructed wetland system for (a) cell 1 and (b) cell 4 (after Harrington and Scholz, 2010).

rate (74 m^3/ha/day) during the second winter, performed very well. This indicates that the high flow rate treatment has led to the ICW system's being at least partially overloaded, which subsequently leads to water and air pollution (Martinez et al., 2009).

The nitrate–nitrogen concentrations for the overall outflows of all ICW systems operated in normal, recycling, high nutrient load, and high flow rate modes are shown in Figure 36.3.3.2(c). A reasonable guide threshold value for nitrate–nitrogen in piggery effluent associated with ICW treatment is 2 mg/l. Regarding the second low flow phase of the project, only the systems operated at high flow rates (74 m^3/ha/day) had serious problems staying below this threshold during the second winter. These findings also suggest that the high flow rate treatment has led to the wetland system being partially overloaded.

Lee et al. (2004) assessed wetland systems treating pretreated swine effluent at three different hydraulic retention times. Results showed that the systems responded well to the changing hydraulic loads in removing total phosphorus (47%–59%) and total nitrogen (10%–24%). While physical mechanisms were dominant in removing pollutants, the contributions of microbial mechanisms increased with the duration of wetland use, achieving 16% of total nitrogen removed during the longest hydraulic retention time. These findings indicate that a change of flow rate has a measurable impact on nutrient removal. However, the absolute nutrient concentration values are relatively similar, if compared with each other for different flow rates. These findings have been confirmed in the present study, particularly for phosphorus removal and to a lesser extent for nitrogen removal (Figure 36.3.3.2).

A suitable threshold for molybdate-reactive phosphorus is 1 mg/l, which has been set, for example, in the Irish Phosphorus Regulations (1998). Only the system operated at high flow rate had an occasional problem with exceeding this threshold. Overall, molybdate-reactive phosphorus was rather the limiting nutrient in all ICW systems regardless of the operational modes. This contrasts findings by Stone et al. (2004) showing that their marsh–pond–marsh constructed wetland system had problems in effectively removing phosphorus from swine lagoon effluent.

The removal of nutrients such as phosphorus is frequently influenced by temperature. A higher temperature leads to more phosphorus removal (Cronk, 1996). However, the findings presented in this study do not confirm this observation. This could be explained by the relatively low concentrations of phosphorus in the first place.

36.3.6 Simplified Cost–Benefit Analysis and Limitations

The benefit of the ICW system may be expressed in ammonia–nitrogen removal, which is by far the most problematic parameter to be removed in piggery wastewater treatment with ICW. After mid-June 2009, the removal rates of ammonia–nitrogen were relatively stable. However, some entries in fall and winter for the high flow rate operational mode were an exception.

The removal rates observed in this study were relatively high (about 99.8% for the high nutrient load operation) if compared against similar studies. Similar free water surface wetlands in China have shown lower ammonia removal rates for agricultural wastewaters (47.9%–69.7%; Lu et al., 2009). Alternative wetland designs using subsurface flow have also shown lower ammonia–nitrogen removal rates of between 63% and 75% (Gonzalez et al., 2009).

For a cost–benefit analysis to be of universal value, some key assumptions need to be made. The overall pumping costs (e.g., pump hut, pumps, pipes, maintenance, and electricity) for the standard treatment system are assumed to be x cost units. There might be differences in summer and winter electricity prices distorting the calculation.

The additional pumping costs for the recycling of treated effluent are assumed to be $0.9 \times x$. Finally, the costs associated with doubling the inflow volume are estimated to be $2 \times x$. All other capital and operational costs for a wetland system of 1 ha that is in operation for 10 years are similar for all treatment systems and are assumed to be $20 \times x$. The estimated multipliers of x are based on the assumption that the farmer is heavily involved in designing and constructing the wetland system to reduce the overall capital costs. Moreover, the site is assumed to be optimal for ICW construction in terms of its topography and hydrogeology.

Based on the assumptions discussed above, the analysis shows that for ammonia–nitrogen removal, the operational modes normal, recycling, high nutrient load, and high flow rate have removal ratios of 0.52:0.53:0.51:1.00. This indicates that all systems with the standard flow rate perform virtually similarly. There is no statistically significant difference between them. Moreover, taking into account that systems with the high flow rate mode have about twice the flow rate compared to the other systems, there is also no benefit in terms of ammonia–nitrogen removal when operating the ICW at twice the flow rate.

The cost ratio for the operational modes normal, recycling, high nutrient load, and high flow rate was 0.56:0.55:0.54:1.00, which reflects that the pumping costs are relatively small in comparison to other costs such as wetland construction. By dividing the cost by the removal (i.e., the benefit) ratios, one can obtain the cost–benefit ratio, which is 1.08:1.04:1.06:1.00. It follows that the high flow rate system is marginally better than the other operational modes. However, this difference is not statistically significant and comes with the drawback of several water quality parameters being either close to or above likely water quality parameter threshold values.

The research study has its clear limitations. For example, an industrial-size wetland unit would be by a factor of between 5000 and 40,000 times larger (Carty et al., 2008; Meers et al., 2008). The larger the system would be, the less important would be the relative pumping costs, making the recycling mode option financially more viable.

It is likely that an industrial-size wetland would also be deeper by a factor between 2 and 3, taking the accumulation of debris and litter into account (Carty et al., 2008; Scholz, 2006). However, a deeper experimental wetland would have performed similarly, considering that shallow-rooting *G. maxima* was used and that most microbial activity takes place within the litter zone.

36.3.7 Impact on Future Piggery Farming

The wider benefits of the ICW concept have been documented by Scholz et al. (2007) and Harrington and McInnes (2009). Wetlands treating piggery wastewater are sustainable and inexpensive but require relatively high footprints in terms of land requirement. Moreover, dilution of wastewater and effluent recycling may be considered as unsustainable in countries where water is scarce.

Alternative treatment methods include standard wastewater treatment systems such as trickling filters, activated sludge processes, industrial-scale membrane technology, and digestion units (O'Brien et al., 2005). The main alternative to wetlands, particularly in Ireland, is anaerobic digestion, which is, however, expensive and maintenance-intensive but provides additional benefits such as methane gas used to power farms. Only piggeries operating on an industrial scale may be able to justify the purchase of anaerobic digesters. The ICW concept therefore has the potential to provide a low-cost and in situ treatment system allowing farmers to process excess pig wastewater themselves without the requirement of expensive ex situ treatment based on industrial-scale technologies.

36.4 CONCLUSIONS AND RECOMMENDATIONS

The findings indicate that effluent recycling within ICW treating piggery wastewater is likely to lead to better outflow concentrations of most key pollutants such as ammonia–nitrogen. High flow rate treatment led to the wetland systems being partially overloaded, particularly by nitrogen species.

Nitrification performed better in summer compared to winter, which made direct performance comparisons between the low and high flow rate phases difficult. However, considering that all systems with different test conditions were operated in parallel, it is justified to compare system performances relative to each other. In general, low, in comparison to high, flow rates led to better system performance.

A cost–benefit analysis based on ammonia–nitrogen loads retained indicated that all ICW systems performed similarly, regardless of treatment mode (i.e., nutrient and hydraulic loads and effluent recycling). However, doubling of the flow rate is likely to be most economical, even if outflow concentrations for ammonia–nitrogen are slightly higher than for operational modes with a standard flow rate.

For small piggeries with sufficient land, the proposed ICW system is a viable sustainable and cost-effective alternative to anaerobic digestion, biomembranes, and other standard wastewater treatment methods such as the activated sludge process. Further research on field scale and application in different climates is recommended.

Chapter 37

Wetland Systems as Part of the Sustainable Flood Retention Basin Concept

37.1 INTRODUCTION

37.1.1 Rationale for Rapid Survey and Classification Method

The European Union (EU) Flood Directive (2007/60/EC) is driving a move to sustainable flood risk management planning in Europe (EU, 2007). The directive requires that flood risk planning be completed on a catchment scale and aligns this with the preexisting Water Framework Directive (2000/60/EC) catchments and River Basin Districts (EU, 2007). In particular, climate change is likely to increase the severity and frequency of flood events, thus increasing the associated hazard. This may threaten some existing flood defenses, which were designed and built prior to climate change being identified as an issue; therefore, they may require modification to ensure their sustainability.

The EU recognizes that member states may face significant challenges in implementing the Flood Directive, which also requires the survey of water bodies with flood control potential, and has responded with funding programs such as the Strategic Alliance for Water Management Actions (SAWA, 2010), which aimed to provide tools and guidance to aid the member states in implementation. In particular, SAWA is aiding North Sea area member states in the implementation of the Flood Directive, providing guidance on the characterization of adaptive measures such as sustainable flood retention basins (SFRB).

Existing survey methods for water bodies and catchment assessments are based on the ecology, chemistry, and hydrology of a catchment. The methods to determine these characteristics are time consuming and expensive (Watzin, 1999). As these methodologies are predominantly ecological, their outputs tend to overemphasize the ecological status of a water body, and this can give rise to conflict in the case of flood defense impoundments with high flood return periods because these basins become overgrown and often achieve a high biodiversity.

In some cases in Europe, this has resulted in expensive flood defense structures that cannot be used because flooding would damage the ecology.

Wetlands for Water Pollution Control. http://dx.doi.org/10.1016/B978-0-444-63607-2.00037-X

Such conflicts need to be resolved through impartial debate and discussion with an objective assessment of the structure, its design purpose, and its current status. Many existing hydrological models do not consider the flood control potential of existing dams and impoundments to contribute to hydraulic management, though reservoir release from drinking water reservoirs to maintain river ecology is an established management practice (Montaldo et al., 2004).

A further aspect of sustainability in flood risk management should be to consider the existing flood defense infrastructure and impoundments that already exist within a catchment. Considering that many agencies will have to undertake assessments of their areas and objectively classify the flood defense potential of the existing infrastructure (SEPA, 2007), a detailed, expensive investigation is not always going to be practical. The system outlined in the SFRB guidance manual (Scholz and Yang, 2010) has proven to be inexpensive, rapid, and reliable as an assessment tool for existing flood retention basin infrastructure, such as most SFRB. The SFRB concept is based on the views of diverse international groups of engineers, landscape planners, and environmental scientists and has withstood detailed scientific scrutiny.

37.1.2 Sustainable Flood Retention Basin Typology

The suggested typology for SFRB (Table 37.1.2.1) is based on the views of a multidisciplinary and multinational team of environmentalists, scientists, engineers, and landscape planners (Scholz, 2007a,b) and has been refined subsequently (Scholz and Sadowski, 2009). This collaborative approach has resulted in identification of six types of SFRB (Table 37.1.2.1) as the minimum practical number to accommodate the variety of roles and modes of operation of this diverse group of structures and semi-natural water bodies. Most SFRB are used for the collection of river flow and runoff, which is slowed down and later released downstream, resulting in discharge waves being flattened and discharge periods extended, mitigating potential flooding (Scholz, 2007a,b).

Many retention basins perform additional tasks, such as infiltration of runoff for groundwater recharge; drinking water supply; diffuse pollution mitigation; enhancement of recreational benefits such as water skiing, bird watching, and fishing; and green space provision (Scholz, 2006). In fact, some SFRB have even become sites of special scientific interest (SSSI; Rowell, 2009) after years of neglect in terms of engineering management, resulting in high biodiversity. The current multifunctionality of SFRB is largely what makes these structures sustainable. However, it is the resulting diversity of stakeholders that can lead to conflicts over the status and function of an SFRB and hinder the successful implementation of sustainable flood risk management planning.

The proposed survey method combines hard scientific and engineering data such as the dimensions of dams and structures and catchment land types with softer, more holistic landscape, environmental, and ecological variables characterizing the potential for diffuse pollution mitigation. This provides an

TABLE 37.1.2.1 Definitions of Sustainable Flood Retention Basin (SFRB) Types (after Scholz and Yang, 2010)

Type	Name	Definition of SFRB Type	Typical Examples	Basin Characteristics
1	Hydraulic flood retention basin (HFRB)	Managed, traditional SFRB that is hydraulically optimized and captures sediment	Drinking water reservoir (in operation); highly engineered and large flood retention basin	Very high rainfall and seasonal impact; high site elevation; very highly engineered or even automated with high outlet flexibility; fully managed and tidy in appearance; very high floodwater volume; very deep flooding depth; potentially high basin gradient; very large flood surface area; very long wetted perimeter and dam; usually very problematic terrestrial and aquatic animal passage; often high levels of algae in spring and summer; can be permanently wet (hydropower supply) or virtually permanently dry (flood control); inorganic sediment; if purpose-built SFRB, often very high pollution
2	Traditional flood retention basin (TFRB)	Aesthetically pleasing retention basin used for flood protection adhering to sustainable drainage and best management practice	Former drinking water reservoir; traditional flood retention basin	Very high rainfall and high seasonal impact; high site elevation; managed, highly engineered, or even automated with high outlet flexibility; quite high floodwater volume and deep flooding depth; potentially high basin gradient; large flood surface area and long wetted perimeter; high and long dam; problematic aquatic and terrestrial animal passage; frequent algal bloom in summer (if eutrophic); mostly inorganic sediment; low vegetation cover; not excessively polluted in uplands; little groundwater infiltration; mixed catchment, usually with forestry and farming

Continued

TABLE 37.1.2.1 Definitions of Sustainable Flood Retention Basin (SFRB) Types (after Scholz and Yang, 2010)—cont'd

Type	Name	Definition of SFRB Type	Typical Examples	Basin Characteristics
3	Sustainable flood retention wetland (SFRW)	Aesthetically pleasing retention and treatment wetland used for passive flood protection adhering to sustainable drainage and best management practice	Sustainable drainage systems or best management practices such as some retention basins, detention basins, large ponds, or wetlands	High rainfall and clearly recognizable seasonal impact; relatively low score for engineered appearance; usually some outlet flexibility; acceptable aquatic and terrestrial animal passage; small to medium floodwater volume and typically shallow flooding depth; usually highly polluted if wet; partly wet; mainly inorganic substrate (often remains from construction); if matured and unmanaged, the sediment becomes increasingly organic; substantial vegetation cover (often macrophytes present); usually highly urbanized catchment
4	Aesthetic flood treatment wetland (AFTW)	Treatment wetland for the retention and treatment of contaminated runoff, which is aesthetically pleasing and is well integrated into the landscape and has some social and recreational benefits	Some modern constructed treatment wetlands; integrated constructed wetland	Fairly low rainfall; highly engineered with emphasis on water treatment; high floodwater volume and shallow flooding depth; acceptable aquatic and terrestrial animal passage; flat and short dam; short wetted perimeter; large flood surface area; usually highly polluted and often wet; substantial and often lush vegetation; variable catchment size with mixed uses
5	Integrated flood retention wetland (IFRW)	Integrated flood retention wetland for passive treatment of runoff; flood retention and enhancement of recreational benefits	Some artificial water bodies within parks or near motorways that have a clear multipurpose function such as water sport and fishing	Semi-natural; flat and short dam; low floodwater volume and very shallow flooding depth; small floodwater surface area and short wetted perimeter; easy animal passage if not in urban areas; usually highly polluted with high organic proportion within sediments; usually substantially wet; very high, dense, and lush vegetation cover; small catchment (often located within a park); substantial recreational activities

6	Natural flood retention wetland (NFRW)	Passive, natural flood retention wetland that may have become a site of special scientific interest (SSSI) requiring protection from adverse human impacts	Natural or semi-natural lakes and large ponds, potentially with restricted access	Very natural (like a lake) and often an SSSI or at least with a high potential for becoming an SSSI; dam typically absent and no outlet flexibility; very low floodwater volume and very shallow flooding depth; often very small flood surface area (unless a large managed lake) and short wetted perimeter; easy aquatic and terrestrial animal passage; usually very wet or permanently wet; usually deep; natural organic sediment (originating predominantly from basin vegetation if the SFRB is mature); little pollution; very high proportion of vegetation cover; very small catchment with dominant pasture cover; high groundwater infiltration; possibly neglected for decades; high, rather natural catchment proportions

impartial assessment of SFRB characteristics, which recognizes both its design purpose and the current uses of the structure. The SFRB categorization methodology is therefore relevant to a wide range of stakeholders, including flood risk management planners, engineers, local authorities, and community groups. It provides an impartial, quantifiable, and consistent assessment and should be used to identify infrastructure with the potential to contribute to flood risk management planning.

37.1.3 Survey Template

The SFRB survey method is based on completing a site survey template, which contains a total of 40 variables. Details of these are provided in the subsequent section with practical guidance on how to determine each of them. A vital aspect of the classification system is to assign an estimated confidence level to each of the variables as they are determined. The confidence value (%) is an estimate by the assessors of how accurately each variable has been determined and the confidence that they have in the determination.

The confidence value has been banded into high, medium, and low confidence levels. A high confidence value is typically linked to a variable that has been measured or can be estimated with a very high degree of confidence based on knowledge and experience. The confidence value then assigned is between 61% and 100%. In cases where the confidence value is between 31% and 60%, additional investigations should be conducted to improve the confidence value. In cases where the confidence value assigned to a variable is $\leq$30%, the variable should be treated as missing. It has been found that assessors who use the system tend to assign confidence values in 5% increments. This is undesirable if sufficient expertise to undertake the assessment is available, as it changes a 100-point to a 20-point scale system.

The proposed survey methodology has been tested for case studies in temperate and oceanic climates such as Baden (Germany), Scotland, and England. It requires modification to be applied effectively in other climatic zones and to accommodate various scales of infrastructure. In particular, descriptions for variables such as annual rainfall and seasonal impact should be adjusted to the application area. Moreover, it is recommended that national weather and mapping data should be used, where available, to decide on landscape and climatic variables and the appropriate ranges for these parameters. An assessment with the proposed methodology should provide a relatively objective and consistent output, which can be used to facilitate stakeholder discussions and identify infrastructure.

37.1.4 Manpower and Equipment

The proposed methodology to characterize water bodies including SFRB is rapid and inexpensive to apply within a catchment and therefore should not require expensive equipment or detailed measurements. The variable characterization is a two-stage process of combining a desk study and a field visit. The desk study can provide an estimate of most variables using a standard personal computer with an internet connection in typically less than 40 min. The site visit involves locating the water body, recording SFRB (if identified) and catchment details using a survey form, and assessing the water body characterization variables visually. This typically requires 20 min per site.

The basic equipment and materials thus required for the entire process are a personal computer with an internet connection to carry out research, a digital camera to record details of the catchment and the SFRB, and a 1:25,000 to 1:50,000 scale map of the survey area. A global positioning system (GPS) unit with <5-m error is a useful additional tool to allow geographical locations to be recorded, which can subsequently be used in digital modeling.

37.1.5 Rationale and Objectives

Chapter 37 contains all the required information and guidance to allow for a comprehensive water body (particularly potential SFRB) assessment in a survey case study area. It is intended that the method should be applied at a catchment scale where the catchment boundaries ought to be defined and the SFRB characteristics within the catchment have to be identified.

Once the locations of potential SFRB in a study area have been identified, a desk study for each basin is undertaken, using all available sources of information. Variables are then estimated or measured, and their numerical values are subsequently recorded on the survey form. The next step is to perform a site visit to confirm or correct the measured and estimated variables from the desk study, and then the survey form is completed and finalized. A short description of each of the variables is supplied in Section 37.2 with information on how to determine each of the parameters. The entire process can be completed within approximately 1 h for most sites.

Scholz (2007a) outlined an empirical tool for the categorization of SFRB. As the SFRB concept developed further, a more statistical tool was developed for estimating SFRB types (Table 37.1.2.1). This tool has been discussed by Scholz and Sadowski (2009).

This chapter aims to provide a concise guidance document for the characterization of SFRB, which are water bodies that have a flood control and diffuse pollution control purpose. The objectives are as follows:

- to justify the need for an SFRB typology;
- to describe in detail the most relevant rapid SFRB characterization variables; and
- to discuss and justify new characterization variables.

A discussion of the SFRB classification system (Scholz and Sadowski, 2009) based on the variables outlined in detail within this chapter is beyond the scope of this guidance document. Moreover, the applicability of the guidance document (Scholz and Yang, 2010) might be restricted to variables characterizing water bodies located in temperate and oceanic climates, considering that the experience of the authors is restricted to countries such as Germany and the United Kingdom.

37.2 ASSESSMENT OF CLASSIFICATION VARIABLES

This section provides a detailed overview of each of the water body characterization variables and how to assess them. Definitions and advice are given to ensure consistency of application of the SFRB determination and classification method. Nevertheless, the author accepts that there may be specific case study sites where the information given below might be insufficient to determine a variable with high confidence.

37.2.1 Engineered

A body of water can either be formed naturally or it can be created by man. Man-made structures can be highly diverse and range from very large water supply and hydropower dams of earth or concrete to small-scale structures often built to supply water to industrial processes. The Engineered (%) variable determines whether the SFRB is natural or man-made and how pronounced these tendencies are.

Dam structures with full engineering control, such as a modern drinking water supply reservoir or a purpose-built hydraulic flood retention basin (Table 37.1.2.1), are examples of potentially highly engineered structures. These types of basin would typically receive values of between 70% and 95%.

A structure that is natural is one that has not been extensively modified by man. The basin is typically a natural landscape feature and there is little (e.g., only a formal outlet or protected embankment present) or no evidence of human interference. This type of basin will normally receive a value of <20%.

37.2.2 Dam Height

Dam Height (m) is defined as the height of the man-made structure that creates the impoundment. The height of the structure is taken from the highest to the lowest point, usually just below the bottom outlet.

For a rapid assessment, the Dam Height can be visually estimated, and a valuable tool is a photograph with the face of the dam and a team member on top of it. Accurate GPS can also be used to record spatial coordinates at the top and base of the SFRB, though these can have high levels of error associated with them as well. Typically, the height has to be estimated from the front face of the dam, as this is usually clearly visible.

37.2.3 Dam Length

Dam Length (m) is defined as the span of the structure creating the impoundment. The structure may span a valley or, in rare cases, a dam may surround virtually the entire water body. Note that a natural lake restricted only by the topography of a valley attracts a Dam Length value of 0 m.

If written documentation or local displays of dam details are available, then Dam Length can be determined from this information. Otherwise, Dam Length can be paced or measured during the site visit, or for structures that are >500 m long, they can be measured from maps with a scale of ≤1:50,000.

37.2.4 Outlet Arrangement

The Outlet Arrangement (%) variable describes how and where water leaves the SFRB. In the case of small natural water bodies, there is usually a single river or stream leaving the water body. This arrangement can be considered as a single, independent, simple, and uncontrolled outlet. The river outlet will not have any form of dam, weir, or sluice to control water levels and would generally receive a value of between 0% and 8% (with zero applied where there are no control structures). An entry of 8% might be given if there was a simple control or measurement structure present, but which does not significantly impact on water levels.

In the case of natural flood retention wetlands (Table 37.1.2.1), the river outlet is typically natural and would receive a value of zero or close to it. The outlets in many shallow wetland systems can be choked, blocked, or overgrown by reeds and other vegetation, and these may slow down water release and treat it at the same time. The additional benefit of water treatment is one of the reasons why these natural systems are considered to be sustainable basins.

Many relatively small dams and impoundments have a minimum of two types of outlets. Typically, a dam will have an overflow (spillway) and pipes within the dam structure for various purposes; e.g., bottom outlet to drain the reservoir, outlet to control the river base flow, and outlet to convey the water for subsequent treatment. A complex outlet structure is usually considered to be a combined system (typically with manual operation and a single fixed spillway) and attracts values between 15% and 75%, higher values being awarded to highly engineered systems with potentially fully automated and remotely controlled structures.

Large and modern drinking water reservoirs or hydropower dam constructions typically have combined outlets with fixed spillways and one or more water extraction points. These larger dams are often fully automated and typically receive a value higher than 75%.

37.2.5 Aquatic Animal Passage

Key to the movement of aquatic animals expressed by the variable Aquatic Animal Passage (%) is the flow of water from the basin and the geomorphology of the water course or overflow. In the case of many dams as part of wet basins, the only possible route of movement to upstream areas for fish and other aquatic organisms would be up the spillway and/or through outlet pipes. Typically, these are long concrete or masonry channels with significant drops in height and very thin layers of water, making aquatic animal passage almost impossible (Larinier, 2000). In the absence of a fish pass or fish ladder, these should be considered significant barriers to aquatic animal passage, typically receiving values between 0% and 10%.

In the case of some smaller SFRB, steps may or may not have been taken to adequately facilitate aquatic animal passage. Generally, those SFRB with an adequate flow of water, no significant drops in height, and an insignificant barrier such as a small dam would score between 10% and 40%. Most SFRB that have been designed with a small fish pass or bypass stream would score between 41% and 69%.

Semi-natural water bodies and dams with a modern fish ladder are considered to allow for adequate aquatic animal movement from the impoundments to the wider environment, and typically these would receive scores >70%. Fish ladder design and passage has been a controversial issue for some time, and peer-reviewed information (in the regional context) on the effectiveness of fish ladders is a valuable aid in determining this variable.

37.2.6 Land Animal Passage

The Land Animal Passage (%) variable is intended to provide information on how easily terrestrial animals such as deer, squirrels, and birds can navigate across a dam or around a water body. This variable requires consideration of the structure of the SFRB and the wider landscape context along with any natural or man-made barriers such as the dam itself. The basin location, Dam Height and Dam Length, fencing, gates, bridges, paths, and thickness and type of fringing vegetation can be important factors in this assessment.

Sustainable flood retention basins located in the remote and/or upper reaches of catchments where there are generally sparse population and infrastructure typically represent areas where terrestrial animal passage is good and attract values between 70% and 100%. It is possible for very large dams, often crossing steep valleys, to pose a significant barrier to animal movement due to the high dam structure and the large size of the impoundment it creates. Spillways often create a break in the dam wall, which can be difficult for animals to cross. A lot of SFRB located in urban areas and near main roads may pose a significant barrier to terrestrial animal movements (Shepherd et al., 2009); typically, these circumstances result in values between 0% and 20%. Some large natural water bodies can be barriers to animal movement, and in these circumstances a value between 21% and 69% is typically awarded.

37.2.7 Flood Plain Elevation

Flood Plain Elevation (m) is defined as the maximum additional height that the water rises above the normal water level of the basin to reach the flood plain. It is usually possible to estimate the normal water level of an impoundment or lake by the distribution of debris and water marks around the edges. These can be submerged, hidden, or washed away during periods of high water levels, which can make this variable difficult to estimate. It should, however, be noted that the grass level is often a good indicator of maximum flood plain elevation, as this plant cannot tolerate long periods of submergence. A clear line that is close to the shore where healthy grass ends is often a good indicator of maximum floodwater level.

In the case of many reservoirs, there is a dam with a spillway present either in a lowland area or in a valley. In virtually all such cases, the site has a 0 m flood plain elevation, as the spillway ultimately sets the impoundment capacity.

37.2.8 Basin Channel Connectivity

The Basin Channel Connectivity (m) considers how an SFRB is connected to its water inlets and outlets and whether it provides a direct path for water flow during flood events. It is an estimate of how directly connected the SFRB is to its water supply and main drainage route; that is, whether the basin is online or off-line. For an online structure the entire inflow water stream flows through the basin easily, while for an off-line basin most of the floodwater bypasses the impoundment via an additional channel. The shortest distance between the bypass channel usually incorporating the SFRB and the main (often original) stream bed is named the basin channel connectivity. However, there are also some more complex cases (see below).

An online SFRB may have several water inlets and outlets that are virtually part of the river system; i.e., the river effectively flows straight through the SFRB or water body. Such SFRB and natural water bodies receive a basin channel connectivity value of 0 m.

In the case of some purpose-built off-line SFRB, they may only receive water when the river reaches significant flood volumes. These basins are typically built for long-return-period flood events and are often located directly adjacent to the river prone to flooding. Such basins are also considered to be off-line and the distance of this offset is also recorded as basin channel connectivity, as offset basins typically have a lower negative ecological impact on the river than those that are built online (Colin et al., 2000).

37.2.9 Wetness

The variable Wetness (%) has been added to aid in distinguishing between permanently flooded basins such as some drinking water supply reservoirs, industrial impoundments, and purpose-built flood retention basins, which are predominantly dry and are only flooded occasionally in response to major storm events, when they fulfill their flood control function.

Many current and former drinking water supply reservoirs, which have also a flood control and diffuse pollution control potential, effectively run at their maximum design capacity, discharging down their spillways. Such SFRB typically receive a value of >90%, depending on how much of the basin is covered by water. Deep natural water bodies typically receive similar values.

Some shallow SFRB are silting up due to a range of factors, including natural landscape processes such as siltation accelerated by eutrophication. For example, in some Scottish locations, this process is stepped up by the removal of arable farming restrictions in the catchment once a drinking water supply reservoir is no longer used for its design purpose. These basins are typically shallow and boggy, with extensive fringes of dense reeds and other macrophytes, and they may therefore have very little open water. Such sites typically receive a value between 10% and 74%, depending on the proportion of open water present.

Purpose-built SFRB designed for long-term flood events may only be partly flooded once or twice per decade. Such sites are typically dry basins with a high density of vegetation, and some may be used predominantly for other purposes such as recreation or farming. These dry basins typically receive a value of between 0% and 9%.

37.2.10 Proportion of Flow within the Channel

The variable Proportion of Flow within the Channel (%) determines to which fraction the stream water will flow directly through an SFRB and is linked to the variable Basin Channel Connectivity (see Section 37.2.8). If an off-line SFRB is present, the mean proportions of flow through the main channel and the bypass need to be considered during the estimation. An off-line SFRB receives a value <100% for the main or original river channel, while an online SFRB always receives a corresponding value of 100%.

In the case of disused online impoundments, which are often found at full capacity, with water leaving via the dam spillway, the majority of the flow will be within the spillway channel and typically a score of 100% is awarded. Natural water bodies and wetlands typically have a single main channel or river draining them and these also receive a score of 100%.

37.2.11 Mean Flooding Depth

The Mean Flooding Depth (m) is a combined parameter that is determined by adding the mean additional depth due to an average flood to the mean depth of the basin (see Section 37.2.38). Many small impoundments, natural water bodies, and SFRB are very shallow basins typically with an average depth <3 m. Such SFRB have relatively low flooding depths and are often well vegetated with dense stands of macrophytes such as reeds (Bayley and Guimond, 2008).

Large basins such as some drinking water reservoirs, purpose-built SFRB, and some large industrial impoundments are designed to supply water at quantities of $>100,000$ m^3/day. Typically, the impoundment's, overall capacity is many times this volume, and such SFRB and reservoirs can have depths >10 m. This is typically reflected in a very high dam. For such basins, flooding depths can reach a value >10 m. For some hydraulic electric power stations and for some large natural lakes, flooding depths can be >30 m due to steep sides, constrained outlets, and/or deep basins.

37.2.12 Typical Wetness Duration

Typical Wetness Duration (day/year) is an estimate of the mean number of days during which the basin is wet within a given year. This variable has been added to distinguish between permanently flooded types of SFRB such as some drinking water reservoirs and industrial impoundments (Table 37.1.2.1) and purpose-built SFRB, which may only be flooded very occasionally and are therefore predominantly dry adaptive measures to control flooding.

Natural water bodies, drinking water supply reservoirs, and other forms of large-scale impoundments are often permanently wet, typically receiving a value close to 365 day/year. It is recognized that dams are periodically drained for maintenance and inspection. However, this does not detract from the predominantly permanently wet nature of these structures.

Many SFRB designed for long flood return events may receive a value as low as 1 day/year (or even less). Typically, such impoundments are designed to deal with long-return-period flooding events such as 20, 50, or even 100 years.

37.2.13 Flood Duration

Flood Duration (days/year) is different from the variable Typical Wetness Duration (see Section 37.2.12), as it only considers the mean number of days in a given year that the SFRB is actually flooded rather than being wet. Typically, this variable is estimated from information based on rainfall patterns, flood event return periods, and the variable seasonal influence (see Section 37.2.22) to arrive at a probable number. If information is available on water levels for an SFRB, this can be used to accurately determine the number of days of flooding.

In areas of high annual rainfall (≥ 2 months/year) and with even a moderate seasonal influence, there can be relatively frequent flooding events, and sites may be flooded as often as 20–30 days/year (e.g., west coast of Scotland). In areas of low rainfall (≤ 0.4 months/year) with a low seasonal influence, such as Mediterranean climates, flooding is likely to have a very low frequency of ≤ 2 days/year.

37.2.14 Basin Bed Gradient

The Basin Bed Gradient (%) indicates to which degree the depth changes along the flow from the main inlet to the main outlet points. The Basin Bed Gradient can be calculated by dividing the mean depth of the basin (see Section 37.2.38) by the length of the basin (see Section 37.2.39).

37.2.15 Mean Basin Flood Velocity

The Mean Basin Flood Velocity (cm/s) is defined as the average speed of the water traveling through the entire basin from the inlet to the outlet during a flood. An "educated guess" is usually used to estimate this value, with the support of other variables such as the slope of the basin, expressed by the mean Basin Bed Gradient (see Section 37.2.14). Other means of investigation, particularly for permanently wet basins, are too expensive and time consuming for a brief investigation.

A high value for this variable is associated with a purpose-built SFRB located in upland areas where heavy flooding occurs and where a high basin gradient is an apparent landscape feature. The value for such basins can be as high as 150 cm/s. In comparison, a typical value for basins in lowland areas is below 15 cm/s.

37.2.16 Wetted Perimeter

The Wetted Perimeter (m) is the length of land and solid material that the water in the basin comes into contact with. Such components that are included in the total length are the entire perimeter of the basin, any islands that are within the basin, and any vegetation (e.g., tree trunks and reed stems) that is protruding through the surface of the water. The use of ordinance survey maps can help to roughly determine the wetted perimeter of large, deep, and geometrically simple water bodies. A brief experiment can be undertaken to estimate the perimeter of reeds; e.g., three small square frames of 10×10 cm should be placed around a representative section of reeds to obtain a composite estimate of the reed stem density (Sutherland, 2006).

37.2.17 Maximum Floodwater Volume

The Maximum Floodwater Volume (m^3) is reached when a basin is flooded to its maximum capacity and can retain no more water without its spilling over into another basin or catchment. The two main variables on which one should focus when calculating the maximum floodwater volume are the Mean Flooding Depth (see Section 37.2.11) of the basin and the Flood water Surface Area (see Section 37.2.18).

The numerical value for this variable depends predominantly on the size of the water body. If the surface area is small, then the volume will also be small in comparison to a water body that has a much larger surface area. For upland areas, the water depth of an SFRB is relatively more important than the surface area, while the opposite is the case for lowland areas.

37.2.18 Floodwater Surface Area

The variable Floodwater Surface Area (m^2) determines the mean area of the water surface when the basin has been flooded. This information cannot be found on a map, as the water surface on the map is often based on the maximum or mean depth. Therefore, an estimate of the flood surface area has to be drawn onto a map and the surface area should subsequently be calculated from this drawing.

Depending on the steepness of the surrounding landscape, some floodwater surface areas can be much larger than the existing mean surface area. This is particularly the case for lowland areas. In contrast, for many upland locations within steep valleys, the flood surface area remains fairly similar to that of the actual surface area of the water (see Section 37.2.17).

37.2.19 Mean Annual Rainfall

The Mean Annual Rainfall (mm) is the long-term average of the depth of rain that falls within the catchment area within a given year (Bronstert et al., 2007). It is highly related to the basin capacity and flood event. This information cannot be gathered from a site visit alone. However, the value can be obtained from a database (e.g., Meteorological Office) or local weather station. For the UK, the Flood Estimation Handbook CD-ROM contains exact rainfall data and may be obtained from the Centre for Ecology and Hydrology.

37.2.20 Drainage

The variable Drainage (cm/day) represents how efficiently water moves through the unsaturated zone of the soil and away from the basin. It estimates the mean distance at which water can drain through the unsaturated zone of soil within a typical day.

If the soil series around the basin can be identified, the drainage should be characterized from its known drainage properties. Groundwater vulnerability maps are widely available, for example, for most EU countries. Areas where groundwater is considered vulnerable to pollution typically have excellent drainage, and there are few flow delaying reactions in the overlying soil. Therefore, these areas can be considered to have high drainage values. Equally, areas of low groundwater vulnerability are typically those with poor drainage properties or where a layer of clay or other impermeable material underlies the soil (Ramchunder et al., 2009).

There are exceptions to these general conditions, such as cases of extensive and deep organic soils, which protect groundwater from potential pollution, or thin soils underpinned by hard igneous rocks. It is important to assess the soil while on site to confirm or correct the initial desk study assessment.

37.2.21 Impermeable Soil Proportion

Permeability is the ease with which a soil allows water to pass through it and is largely determined by the soil type surrounding a basin. Highly permeable strata such as sand and gravel easily allow water to pass through with little retardation. Such highly permeable strata typically contain <2% clay, which is relatively impermeable. In areas dominated by other soil types, there will be variable and significantly higher proportions of clay, with the most impermeable soils being those with a high proportion of clay. Heavy clay soils generally have very poor drainage characteristics (Bah et al., 2009).

The Impermeable Soil Proportion (%) can be estimated in the field, and there is a wide range of national guidelines to estimate soil types. The clay content is often used in conjunction with the amount of rock present in the soil to arrive at an estimate of the overall impermeable proportion of the soil surrounding an SFRB. Alternatively, soil series maps can be used to determine the soil type and the proportions of sand, silt, and clay within that soil. These findings need to be compared to the soil in the field to determine the proportion of rock present.

37.2.22 Seasonal Influence

The variable Seasonal Influence (%) is a sum parameter, which is easy to estimate if general knowledge of regional weather conditions and the landscape topography is available. Climatic conditions have a pronounced effect on whether a basin is wet or dry and the frequency of flooding. For example, in the UK, winter periods are wetter than those encountered during summer. Most lowland areas around the Mediterranean can be considered to have little seasonal variation, and therefore SFRB located in these areas are considered to be subject to low Seasonal Influence (<20%). In contrast, for SFRB located in mountainous areas such as the Alps or in regions that are subject to wet climates such as the west coasts of Scotland and Norway, there may be a highly pronounced Seasonal Influence (>70%). Consequently, SFRB identified in these areas are considered to be subject to high seasonal variations (Dawson et al., 2008).

37.2.23 Altitude

The variable Altitude (m) is a measure of the height above mean sea level at which the site is located and is simply determined from a spot height on a map or by using GPS equipment. The site elevation for a reservoir with a dam is usually taken at the bottom of the side of the dam facing upstream, assuming that this is the deepest point within the basin.

37.2.24 Vegetation Cover

The variable Vegetation Cover (%) refers to the SFRB and not to the corresponding catchment. Dry SFRB can be completely vegetated and, if maintained, are typically grass-covered and achieve a Vegetation Cover value between 20% and 70%. Mature vegetation such as a full basin covered with trees is associated with a value close to 100%. In contrast, no vegetation or tarmac will result in values close to zero. It is worth noting whether the SFRB is maintained and vegetated with short grass, which will not impede water flow through the basin, or if the basin is fully covered with mature vegetation, which may reduce its capacity and slow down the movement of floodwater.

In the case of wet basins, the area of the basin occupied by emergent and floating plants is estimated during the site visit. SFRB are highly diverse and some may contain values as low as 1%, in the case of steep-sided maintained drinking water reservoirs, or >90%, for basins that are silted up naturally and that are fully covered by mature reed stands.

37.2.25 Algal Cover in Summer

The variable Algal Cover in Summer (%) provides an estimate of water surface cover with undesirable phytoplankton growth that occurs in a wet SFRB, and it is easiest to estimate accurately during a summer site visit. In contrast, dry SFRB used solely for flood control purposes may have no potential for phytoplankton growth.

Water bodies that are rich in nutrients often undergo one or more extensive algal blooms in summer. In countries with nutrient-poor waters such as most upland areas of Scotland and Norway, it is unproblematic to estimate the likely potential for pollution-related blooms. However, it can be more challenging in some lowland areas in central Europe where waters are typically higher in nutrient content and support more extensive algal and surface macrophyte communities (Anderson et al., 2002).

37.2.26 Relative Total Pollution

The variable Relative Total Pollution (%) is a measure of how impacted an SFRB is by predominantly diffuse agricultural pollution, urban runoff, and municipal wastewater (Scholz, 2006), and it is largely a function of the breakdown of land types within a catchment and the way that the land is used (see Sections 37.2.31–37.2.36).

Diffuse agricultural pollution arises as a consequence of the normal arable and livestock application of fertilizers and agrochemicals as management tools (Stevens and Quinton, 2009). Some of these chemicals are transported by surface and groundwater movement and can be captured within water bodies including SFRB, particularly those designed to trap sediment. The degree of runoff and impact depends on farming practices, chemical application rates, and tillage practices. Livestock farming can result in significant inputs of nitrates and phosphates from the animals, and corresponding microbiological contamination can be a problem during storm events (Edwards and Withers, 2008).

Old mines and associated spoil heaps and processing areas can be significant sources of pollution of the water environment. There is a substantial history of mining for most minerals and metals throughout the world. It is therefore valuable to consider whether there may be any water contamination from this source that may affect the pollution status of an SFRB (Shepherd et al., 2009).

Industrial processes can be a source of pollution within catchments. In Europe and North America such facilities are closely regulated and monitored so that they do not exceed strict consent conditions (IPCC, 2007). The potential presence of these facilities would be considered in the assessment, and a low value of between 3% and 5% pollution would be associated with such a site. Old, derelict industrial facilities can be associated with significant land contamination problems, and such sites should therefore be considered to have a high pollution potential (Loures and Panagopoulos, 2007).

37.2.27 Mean Sediment Depth

The Mean Sediment Depth (cm) is the average depth of the sediment within an SFRB structure, whether wet or dry. The mean sediment depth can be simply determined in dry basins by digging a shallow hole and subsequently assessing the sediment profile.

In permanently flooded SFRB, the sediment is not easily accessible, and an estimate of the sediment depth is therefore derived. Freshwater sedimentation rates are highly variable and depend on a wide variety of factors. It is standard practice in limnological investigations to express sedimentation rates as grams of sediment per meter per year. If the bulk density of the sediment is known, it is a simple matter to convert this to the depth of sediment (Ramos-Scharron and MacDonald, 2007).

37.2.28 Organic Sediment Proportion

The Organic Sediment Proportion (%) within basin sediment is determined by complex interactions in the water environment of a catchment. Organic material is provided by decaying terrestrial plants and animals, combined with the production from aquatic algae, plants, and animals within a wet SFRB, to establish the overall production of the water body. This organic input is then metabolized by bacteria and sediment-dwelling invertebrates, which utilize a large proportion of the organic carbon as an energy source. The organic proportion of the sediment is also based on the interaction of these metabolic processes and the deposition of gravel, sand, silt, and clay from within the catchment (Kuhn and Diekmann, 2003).

In upland areas with high rainfall, there are usually relatively low proportions of organic matter present in the sediment. This type of catchment tends to host oligotrophic water bodies with limited primary and secondary production, and the majority of the organic matter within the basin is cycled through the biota. Most upland catchments therefore tend to have low proportions of organic matter present within the basin sediment and would typically receive Organic Sediment Proportion values between 1% and 3%. An exception to this general rule can be where dense conifer plantations have acidified the water in an upland catchment. In these cases, refractory conifer leaf litter is a feature of all visible sediments (Heal, 2001), and such sites should be considered to have a high Organic Sediment Proportion of >7%.

In lowland areas and for areas where the waters are oligotrophic to mesotrophic, sediments are considered to have relatively high organic matter content due to the increased productivity of such waters, and these are typically assigned values between 7% and 15%. Where waters are eutrophic and highly productive, there is often significant deposition of organic matter at the bottom of a permanently wet SFRB (Chung et al., 2009). These sites are considered to have a relatively high organic matter content of between 16% and 30%.

Hypereutrophic water bodies have a high primary production and regularly suffer from algal blooms. When dead algae sink to the bottom of a water body, they tend to cause anoxic conditions, reducing the breakdown rate of organic materials in the sediments. If nitrate, phosphate, and ammonia are released into a water body, it is possible to find organic matter at between 30% and 60% of the sediment (Zhang et al., 2009).

Values >60% of organic sediment should not routinely be assigned to SFRB unless there is clear and compelling evidence that this is the case. For example, an exception is where an SFRB is surrounded by peat bogs. Peat bogs often contain decayed and water-logged *sphagnum* moss and other plant materials that have partially decomposed. This type of soil comprises virtually 100% organic material (Ukonmaanaho et al., 2006).

The ideal solution to determine the organic carbon content of sediment is by direct measurement of a homogenized sample, typically achieved using an organic carbon analyzer, which heats the sample and converts the organic matter present to carbon dioxide that is then measured (Schumacher, 2002). This measurement could be used directly in the SFRB classification system. However, such accurate determinations are relatively costly and time consuming.

Any survey team should consider whether the boundaries proposed above are suitable for the corresponding survey area. Information on areas with sedimentation problems can be found in river basin management plans prepared, for example, by the European environmental agencies.

37.2.29 Flotsam Cover

Flotsam is defined as debris and waste that is floating on the surface of a water body. It can include items such as debris from trees, rubbish thrown into the water by humans, or even abandoned boats or drowned cars. The principal objective of the variable Flotsam Cover (%) is to determine if there is flotsam present that currently restricts or could restrict the flow of water out of an outlet structure. An estimate of the proportion of outflow area covered by flotsam needs to be made by the survey team.

The value awarded for Flotsam Cover should reflect the mean proportion of cover of the average outlet area by flotsam. This variable is an indirect measure of flow restriction. Therefore, basins with outlets that have little Flotsam Cover receive low values (typically <5%), whereas basins that are full of flotsam, such as some invasive plant species or a lot of human rubbish, would receive high values if the outlet structure is severely covered with flotsam.

37.2.30 Catchment Size

The variable Catchment Size (km^2) is the area of land from which water feeds into an SFRB. In the absence of a national register of dams and reservoirs, for example, the European environmental agencies have defined river basin districts and sub-catchments within their corresponding larger units.

There are some cases where one SFRB is fed by another one or by a series of SFRB. It follows that the corresponding catchments are linked with each other. For example, a downstream SFRB has a catchment that comprises those of SFRB located upstream.

37.2.31 Urban Catchment Proportion

The urbanized proportion of a catchment is simply the area occupied by man-made structures such as roads, farms, and towns. The variable Urban Catchment Proportion (%) is likely to be an important indicator of diffuse urban pollutants to the water environment.

In the case of SFRB that have a very high Urban Catchment Proportion (>90%), it may be appropriate to replace the rather natural land use categories below with different types of urban development. Such a division could include light or heavy industry, industrial estates, and retail and residential areas, as appropriate.

37.2.32 Arable Catchment Proportion

Arable Catchment Proportion (%) is defined as the area within a catchment where crops are grown, either for commercial agricultural purposes or for subsistence farming. The type of crop is likely to vary with climatic conditions and weather, and the variable can be used for any type of farming where rows of crops are interspersed with bare soil. In some parts of the world, separate categories for rice farming and aquaculture will be needed. A high proportion of arable area generally contributes greatly to the diffuse pollution of water bodies, so the Relative Total Pollution (see Section 37.2.26) of the basin and the Organic Sediment Proportion (see Section 37.2.28) might be high.

37.2.33 Pasture Catchment Proportion

Pasture Catchment Proportion (%) is the land within a catchment where animals are taken to graze. In temperate climatic zones such as Northern Europe, pasture land is typically managed and consists of relatively short and dense grass that looks like a monoculture. It is given a separate category as it can be a significant source of specific nutrient and microbiological contamination of the water environment.

37.2.34 Viniculture Catchment Proportion

Viniculture is the practice of growing vines to produce grapes, which are later made into wine. Many vineyards are on steeply sloping land and have relatively bare soils. These conditions can result in significant runoff and soil erosion if badly managed (Casali et al., 2009). In many areas of the world, viniculture is a common practice. However, the variable Viniculture Catchment Proportion (%) is not relevant to cool temperate climates such as Scotland and Norway.

37.2.35 Forest Catchment Proportion

Forest Catchment Proportion (%) is simply the fraction of the catchment that is covered by predominantly managed forest and can easily be estimated or measured using maps and should be ground-truthed during a site visit. In heavily forested catchments, it may be desirable to distinguish between natural woodlands and forestry plantations, particularly where forestry plantations can be a significant source of diffuse pollution and acidification (Nisbet, 2002).

37.2.36 Natural Catchment Proportion

Natural Catchment Proportion (%) is the final category of land use that is intended to cover the remaining part of the catchment and is considered to be the land where there is no or minimal interference from man. It may therefore constitute potentially remote grassland, scrubland, moor, and similar types of land.

Generally, basins located in the upper lands or in the deep valleys have high natural catchment proportions. In contrast, basins located in urban areas always have very little or no natural features. The variable Natural Catchment Proportion can be assigned a value during the site visit when comparing its characteristics with those of the corresponding urban and forest catchment proportions, all of which can be obtained from maps. A high fraction of natural catchment also correlates positively to a low Organic Sediment Proportion (see Section 37.2.28).

37.2.37 Groundwater Infiltration

Groundwater is considered to be the water that lies beneath the saturated zone of the soil. It is composed principally of surface waters and rainfall that has percolated through the soil and into the underlying rocks and typically intersects with water bodies such as SFRB and natural lakes. Groundwater Infiltration (%) indicates the proportion of the water within an SFRB that comes from groundwater, and it can be a significant source of water for some basins located in the lowlands.

In the case of purpose-built dry SFRB, there is no groundwater infiltration. Moreover, former industrial impoundments and drinking water supply reservoirs are typically lined with an impermeable layer of clay and are therefore isolated from the surrounding groundwater. Both types of SFRB receive a value of usually well under 5%.

Some wet SFRB and natural lakes may receive a fair proportion of their water from groundwater. Such basins are typically shallow and, due to the groundwater flow, contain a small lake or pond within the deeper parts of the basin. This is generally most apparent during the drier period of the year, when springs can become visible. Such basins typically receive a value of between 5% and 10%, with around 10% considered to be a typical value for a natural lake or pond.

Basins may be encountered in some regions where groundwater composes between 10% and 40% of the flow of water from the SFRB, and these are considered to have a high dependence on groundwater. Very high values for groundwater infiltration to a basin would be between 41% and 50%.

37.2.38 Mean Depth of Basin

The Mean Depth of Basin (m) is simply the average depth of the dry or wet impoundment. In the vast majority of small and wet SFRB, the mean depth of the basin will need to be estimated during the site visit. An estimate can be derived based on the likely basin bed gradient and the maximum height of the dam, which is often close to the maximum depth of the impoundment by assessing the surrounding landscape.

37.2.39 Length of Basin

The Length of Basin (m) is the distance from the two points of the basin perimeter that are farthest away from each other at normal environmental conditions (e.g., no flooding). In an ideal case, this is the distance between the main inlet and the main outlet.

37.2.40 Width of Basin

The Width of Basin (m) is the distance across the basin at normal environmental conditions (e.g., no flooding) and is ideally at right angles to the basin length (see Section 37.2.39).

37.3 CONCLUSIONS AND RECOMMENDATIONS FOR FURTHER WORK

This chapter provides the scientific basis for the determination of 40 variables characterizing water bodies located in temperate and oceanic climates. The main practical application of the proposed methodology is the characterization of SFRB used for flood control and diffuse pollution control purposes. This chapter is likely to contribute to an increase in the uptake of the SFRB methodology by practitioners working as urban and landscape planners and by civil and environmental engineers involved in SFRB design and construction.

The research describes in detail all key characterization variables, including novel ones such as engineered, outlet arrangement, aquatic animal passage, land animal passage, basin and channel connectivity, seasonal influence, relative total pollution, and flotsam cover. Traditional variables that are intuitively clear, such as altitude and length of basin, received less coverage. Most of these variables can be used for a rapid characterization of water bodies. Moreover, inexpensive sum parameters such as Relative Total Pollution were discussed and justified.

Further work is encouraged to expand the list of variables and their definitions for other climatic and geographical regions. It would also be beneficial to reduce the time spent in the field to a minimum if labor costs are high. Finally, a quality assurance scheme should be developed to allow for an increased numerical accuracy when repeating variable determinations with a different team.

Chapter 38

Classification of Sustainable Flood Retention Basins

38.1 INTRODUCTION TO TRADITIONAL CLASSIFICATION

38.1.1 Flood Risk Management

The most recent climate change projections for the United Kingdom (UK) for 2080 indicate a 10%–20% increase in winter precipitation for the east coast and a corresponding increase of between 10% and 30% for the west coast (Department of Food and Rural Affairs, 2009). Applying these estimates to current UK rainfall patterns, a potential increase in precipitation of between 40 and 160 mm per annum for the east coast and between 285 and 1200 mm for the west coast of Scotland can be predicted based on the data provided by the Meteorological Office (2007). Moreover, the remaining regions of the UK are expected to face similar changes and challenges (Department of Food and Rural Affairs, 2009). The UK government has recognized the adverse impacts of flooding as a policy priority and has therefore doubled its flood defense spending since 1997 (Department of Food and Rural Affairs, 2009).

The European Union has introduced the Flood Directive 2007/60 EC (European Union, 2007), which requires each member state to develop flood risk management plans. The concept of sustainability has been applied to flood defenses, and this implies that the use of all available infrastructure and resources contribute toward this process for society's gain (Shih and Nicholls, 2007).

Environmental classification and assessment required by this directive is a resource-intensive process requiring additional expert staff and laboratory facilities. Flood risk management is often an equally resource-intensive process; for example, the development of detailed flood risk maps available to the public on a user-friendly geographical information system is costly (SEPA, 2009). It follows that a new rapid assessment methodology supporting the work associated with the directive would help to reduce costs.

A logical assessment system to classify and assess flood defense structures, which includes consideration of their sustainability, is of value for the flood risk management process. A standard classification system providing

Wetlands for Water Pollution Control. http://dx.doi.org/10.1016/B978-0-444-63607-2.00038-1

definitions and examples for all sustainable flood retention basin (SFRB) types, which is based on clear and relevant characterization variables, will support communication between stakeholders, including practitioners and the general public (Scholz and Sadowski, 2009; McMinn et al., 2010; Yang et al., 2012). To be widely accepted, such a system should allow for rapid screening of existing infrastructure and should ideally be neither labor nor resource intensive. The classification of SFRB to optimize flood control is a novel approach addressing this problem (see also Chapter 37). The SFRB methodology moves away from traditional engineering solutions, which rely predominantly on hard engineering control variables, toward soft variables (Scholz, 2007a). This approach readily lends itself to solving multidisciplinary challenges.

38.1.2 Traditional and Adaptive Structural Measures and Their Classification

There are a wide range of traditional engineering solutions that can be applied to provide flood defenses in urban and rural areas. These traditional approaches predominantly use hard engineering solutions such as barriers and dykes to protect the public from the economic and social costs of flooding (Kendrick, 1988). More recently, sustainable drainage systems (SuDS), which generally operate by absorbing water and slowing the rate of runoff from urban areas, are used in Europe. SuDS can also assist in diffuse pollution control by providing large surface areas for biochemical reactions to take place and retaining the first foul flush of pollutants released during significant rainfall events (ATV-DVWK, 2001; Scholz, 2006).

Most natural and constructed retention basins retain runoff for subsequent release, thus reducing downstream flooding problems. Some basins, such as wetlands, perform other tangible, albeit less "visible" roles, including diffuse pollution control and infiltration of treated runoff, promoting groundwater recharge. The diversity of retention basin types is therefore high and is further complicated by often multiple and competing functions that these structures fulfill.

A classification system is therefore beneficial in allowing clear communication between stakeholders such as politicians, planners, public interest groups, engineers, and environmental scientists. The absence of a universal classification scheme for retention basins results in potential confusion about the status of individual structures and their functions. This can lead to conflicts between stakeholders concerning the management of retention basins including wetlands, as discussed by Scholz and Sadowski (2009), who therefore proposed a conceptual classification model based on 141 SFRB located in the River Rhine Valley, Baden, Germany. Six SFRB types were defined based on the expert judgment of engineers, scientists, and environmentalists.

The European Union has acknowledged that member states may face significant challenges in complying with the flood directive. Therefore, member states have financed programs such as the Strategic Alliance for Water Management Actions (SAWA, 2010) to develop guidance on adaptive measures such as SFRB to assist the member states in developing flood risk management plans.

The aim of Sections 38.2 and 38.3 is to characterize types (i.e., subclasses) of SFRB in Scotland with the help of a revised rapid conceptual classification model, originally proposed by Scholz and Sadowski (2009). The key objectives are as follows:

- to aid stakeholder communication by avoiding misunderstandings with respect to planning and legal matters concerning the purpose of different Scottish SFRB types;
- to determine and characterize all relevant and, particularly, the key independent classification variables using a principal component analysis (PCA) and a sensitivity analysis using the Wilcoxon test;
- to develop a conceptual classification methodology with the support of a large and detailed example case study data set;
- to compare the Scottish with the German data set (Scholz and Sadowski, 2009) to verify the validity of the methodology and to improve the research approach; and
- to illustrate and discuss examples of the most dominant Scottish SFRB types that are also highly relevant for water and environmental engineers and landscape planners.

Furthermore, the aim of Sections 38.4–38.7 is to introduce multilabel classification to SFRB data sets to analyze and predict multiple functions of SFRB and thus to better understand and manage them. The importance lies in the application of machine learning techniques to solve problems in a different and new subject area. The key objectives are as follows:

- to predict the types of unlabeled SFRB allowing one SFRB to belong to more than one type; and
- to verify the efficiency and applicability of multilabel classification techniques with the help of different case studies showing a high variability.

38.2 TRADITIONAL METHODOLOGY

38.2.1 Identification of Sites and Definitions

Previous work by Scholz and Sadowski (2009) has established a draft classification system for SFRB based on a database of approximately 180 sites in Baden. A revised version of this method has been applied to a wide range of water bodies in the wider central Scotland region. The database combines engineering features with landscape and catchment attributes to produce a holistic classification.

A mathematically sound methodology has been developed to justify the very concept of classifying SFRB. Six SFRB subclasses were defined based on expert judgment provided by groups of British, German, Swedish, French, Irish, and American professional engineers, scientists, environmentalists, and landscape and urban planners during informal consultation workshops held in Germany (2006) and Scotland (2007, 2009).

Moreover, a set of water body characterization parameters that relates to real SFRB was required, and this had to capture the multinational groups' conceptual classification. The variables were obtained during informal brainstorming and consultation sessions and were compared with those obtained via a previous literature review (Scholz, 2007a) and site visits in Germany, UK, Sweden, Ireland, and Denmark held between 2006 and 2010. These parameters have been named "classification variables" and are intended to be a reasonable compromise between the accuracy and rapidity required in their assessment.

The methodology also makes use of two powerful statistical techniques, notably cluster analysis and PCA. The latter can help to discover the relationships between variables and identify the most important variables. Cluster analysis involves the forceful grouping of the results of the classification surveys of all SFRB into distinct clusters. Then, based on the clustered groups and expert judgment, it is possible to identify which cluster relates best to which SFRB definition.

One hundred and sixty-seven sites were selected for classification using the 1:50 000 scale Ordnance Survey Maps of central Scotland (Figure 38.2.1.1). In the context of this investigation, the sites of interest are those that may be able to play a role in either flood management or diffuse pollution control. Structures that may be able to play a role in flood control are considered to be those where the water level can be controlled either manually or automatically and are typically former or current engineered water supply reservoirs. Sites with the potential to contribute to diffuse pollution control are typically more natural and relatively small water bodies.

The most important classification variables for various types of SFRB in Scotland were identified and subsequently grouped. Any site investigation is a two-stage process comprising the estimation of variables during a desk study and site visit. The desk study was supported by information obtained from the Scottish Flood Defence Asset Database (Bassett et al., 2007) and relevant other regional references such as Foster et al. (2008), Meteorological Office (2007), and SEPA (2009).

The characteristics of each SFRB type are based on the interpretation of findings obtained from the statistical evaluation (see below). The six subclasses are the following: hydraulic flood retention basin (type 1), traditional flood retention basin (type 2), sustainable flood retention wetland (type 3), aesthetic flood retention wetland (type 4), integrated flood retention wetland

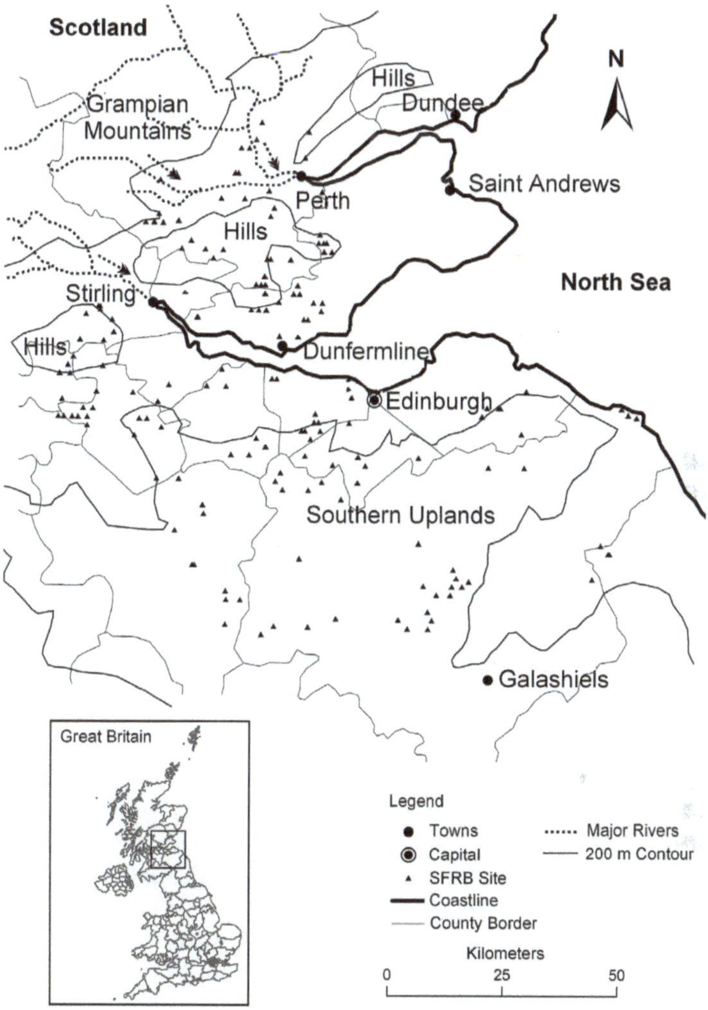

FIGURE 38.2.1.1 Study area, administrative boundaries, and key sustainable flood retention basins (SFRB) in the wider central Scotland area (after McMinn et al., 2010).

(type 5), and natural flood retention wetland (type 6). The revised definitions of SFRB subclasses are independent of all statistical analyses and were formulated based on expert judgment.

38.2.2 Recognition of Classification Variables

Most variables characterizing water bodies in Scotland were adopted from those initially proposed by Scholz and Sadowski (2009): (1) Engineered (%); (2) Dam Height (m); (3) Dam Length (m); (4) Outlet Arrangement and

Operation (%); (5) Aquatic Animal Passage (%); (6) Land Animal Passage (%); (7) Floodplain Elevation (m); (8) Basin and Channel Connectivity (m); (9) Wetness (%); (10) Proportion of Flow within Channel (%); (11) Mean Flooding Depth (m); (12) Typical Wetness Duration (day/annum); (13) Estimated Flood Duration (1/a); (14) Basin Bed Gradient (%); (15) Mean Basin Flood Velocity (cm/s); (16) Wetted Perimeter (m); (17) Maximum Floodwater Volume (m^3); (18) Floodwater Surface Area (m^2); (19) Mean Annual Rainfall (mm); (20) Drainage (cm/day); (21) Impermeable Soil Proportion (%); (22) Seasonal Influence (%); (23) Site Elevation (m); (24) Vegetation Cover (%); (25) Algal Cover in Summer (%); (26) Relative Total Pollution (%); (27) Mean Sediment Depth (cm); (28) Organic Sediment Proportion (%); (29) Flotsam Cover (%); (30) Catchment Size (km^2); (31) Urban Catchment Proportion (%); (32) Arable Catchment Proportion (%); (33) Pasture Catchment Proportion (%); (34) Viniculture Catchment Proportion (%); (35) Forest Catchment Proportion (%); (36) Natural Catchment Proportion (%); (37) Groundwater Infiltration (%); (38) Mean Depth of the Basin (m); (39) Length of Basin (m); (40) Width of Basin (m).

Variables such as Engineered, Floodplain Elevation, Basin and Channel Connectivity, Mean Flooding Depth, Flood Duration, and Relative Total Pollution were refined and clarified to fit within the Scottish context. It has been appreciated that there are differences in the built environment and landscape. For example, the variable Mean Flooding Depth recognizes high slope values for the Scottish landscape and deep flooding depths of some rather natural lakes.

The methodology has been updated by including the new variables Mean Depth of the Basin, Length of the Basin, and Width of the Basin in the classification template. The previous variable Aquatic and Land Animal Passage was divided into the following separate variables: Aquatic Animal Passage and Land Animal Passage. This accounts for fundamentally different obstacles concerning the freedom of unrestricted movement for animals.

Similarly, the previous variable Forest and Natural Catchment Proportion was split into Forest Catchment Proportion and Natural Catchment Proportion. The former variable Viniculture Catchment Proportion was not suitable for Scotland, so it was removed from the classification template.

The variable Wetness was further refined to make a strong distinction between permanently wet systems such as reservoirs and lakes (Scottish data set) and SFRB, which may be dry and become wet only occasionally (German data set). Typical Wetness Duration became more important because it distinguishes between permanently flooded features such as reservoirs and lakes and SFRB designed for occasional flood control. The variable Flood Frequency is very difficult to determine with high certainty. Moreover, this variable is obsolete if no flood frequency or rainfall-runoff data are available.

The variable Wetted Perimeter, previously used by Scholz and Sadowski (2009), is highly important for the Scottish data set, which predominantly comprises wet basins in contrast to dry basins. However, dry basins dominate the German data set. A high wetted perimeter value is likely to indicate a higher diffuse pollution control potential, considering that the stems of partly submerged reeds also contribute to the perimeter. Vegetation Cover has been further specified, considering that the vegetation within a predominantly dry basin is completely different from the aquatic vegetation within a wet basin. Furthermore, the following additional purposes for Scottish SFRB were identified: industrial production (mainly of historic importance) and drinking water reservoir. These are in addition to the purposes flood retention, sustainable drainage, environmental protection, recreation, and landscape enhancement identified for the German data set.

38.2.3 Rationale for the Elimination of Less Relevant Variables

The application of the PCA analysis with Matlab Version 7.1 (Pratap, 2002) provided a better overview of the underlying data structure. On the basis of the loading plot it is possible, where several variables are grouped closely together, to extract one single variable, which may then replace the entire group. The main point of the PCA is to remove redundant variables, hence reducing the risk of multicollinearity.

The cluster analysis and classification was performed twice: (1) using all 39 variables and (2) using only 18 key variables identified by the PCA. The final classification system was intended to be based on variables that are accurate and easy to obtain and associated with a high confidence value assigned to them during their determination.

Dominant variables were retained from the group and used for a subsequent cluster analysis. The remaining variables within each group were discarded. This procedure has been followed because too many variables may overcomplicate the decision-making tool, making the end product rather user-unfriendly. Furthermore, variables with similar principal components were effectively related in terms of what phenomena they are describing. By keeping one variable representing a specific group, the other variables within this group naturally become redundant.

Another technique to assess the suitability of a characterization variable is to evaluate its repeatability. Therefore, three independent research groups comprising two final-year project engineering students each (all briefed and supervised by the authors) assessed the same set of variables for 17 randomly selected case study sites independently from each other during four field visits each in 2009. The ranked reliability of water body characterization variables and assessed purpose according to the Wilcoxon signed rank test was calculated.

The Wilcoxon signed rank test, also known as the Wilcoxon matched pairs test, is a nonparametric test used to assess the median difference in paired data. The test avoids the distributional assumption because it is based on the rank order of the differences rather than the actual value of the differences.

Nondirectional hypotheses were made that there would be a significant difference between paired data (the initial site visit and the revisits). The statistical analysis was carried out using SPSS software (SPSS, 2009) based on a two-tailed hypothesis.

38.2.4 Assignment of Basin Types with the Help of Cluster Analyses

The statistical software package Matlab 7.1 (Pratap, 2002) was used to perform cluster analyses on the standardized example data set. The clustering technique used was an agglomerative method (otherwise known as a "bottom-up approach"). The results are displayed on a dendrogram, which allows an unambiguous appreciation of the cluster properties of the data. The cluster analysis technique "Ward's linkage," which effectively forced the data into a predefined number of clusters, thus eliminating outliers, was applied (Kaufman and Rousseauw, 1990). In this case, the objective was to obtain as many clusters as there are SFRB subclasses, of which there are six.

After the Ward cluster analysis had forced the data points (one point corresponds to all 39 variable values per site) into seven groups (six groups of SFRB and one group of non-SFRB), the general statistics of each cluster were found. The objective was to determine which SFRB type corresponded best to which newfound cluster, and this was done on the basis of expert judgment, supported by the case study information obtained during the site investigation. The dominant basin purpose greatly influences the selection of the most likely SFRB type; e.g., a modern drinking water reservoir is likely to be a hydraulic flood retention basin.

The classification methodology for SFRB developed by Scholz and Sadowski (2009) was applied to the Scottish database. The results indicate that the system could only identify three types of SFRB for the wider central Scotland region: type 2 (traditional flood retention basin), type 5 (integrated flood retention wetland), and type 6 (natural flood retention wetland). It follows that type 1 (hydraulic flood retention basin), type 3 (sustainable flood retention wetland), and type 4 (aesthetic flood treatment wetland) were not well represented in Scotland. This modeling output was verified by the authors and their research team during field visits (supported by desk studies) to all sites between 2007 and 2009.

The clustering approach forces the case study sites into groups that correspond to the different SFRB types previously identified. The clustering outcome shows that the main types of SFRB in central Scotland are types 2 and 6. Case study sites that were forced by the German prediction template (Scholz and Sadowski, 2009) into type 5 contain some sites that have a tendency to belong to types 2 and 6, while sites forced into type 2 consist of some sites that had characteristics of type 1. This can be explained by the multipurpose function of most SFRB.

38.3 DISCUSSION OF THE TRADITIONAL METHOD

38.3.1 Reduction Exercise for the Classification Variables

The total number of variables was reduced based on the results of the PCA and sensitivity analysis. The loading plot allowed for the identification of seven definite independent variables (3, 6, 14, 18, 27, 30, and 39) and 11 groups of dependent variables to be identified. The key variables representing each group were as follows: 2, 3, 6, 7, 9, 14, 16, 18, 19, 23, 24, 27, 29, 30, 31, 34, 37, and 39. For Scotland, Wetted Perimeter, Maximum Floodwater Volume, Floodwater Surface Area, Engineered, Catchment Size, Outlet Arrangement and Operation, Dam Height, Land Animal Passage, Impermeable Soil Proportion, and Mean Sediment Depth were the most important SFRB characterization variables, which greatly contributed to the variability expressed by the first and second component.

The results of the sensitivity analysis demonstrated that variables such as Annual Rainfall are most reliable. This parameter, for example, is determined from high-quality and reliable data sets (e.g., UK Meteorological Office maps) and is consistent between different assessors, regardless of age and experience. In contrast, the least reliable variables are those that are estimated during site visits. The most surprising finding was the variability between assessors when estimating Dam Height, which might be an artifact. The variables Mean Sediment Depth and Impermeable Soil Proportion are important when considering the PCA findings and the international literature but difficult to estimate in the field. A detailed desk study (wherever possible) is therefore required to support their determination. If insufficient information is available and fieldwork studies are too time consuming and/or expensive, the variables Mean Sediment Depth and Impermeable Soil Proportion should be removed from the group of selected variables.

38.3.2 Cluster Analyses

The cluster analysis was performed twice for the Scottish database: based on all (39) variables and based on the reduced set of variables (18). The analysis performed on the reduced set of variables (listed above) indicated seven clusters containing the six SFRB types and a group comprising non-SFRB sites (predominantly unmanaged natural lakes). Dendrograms for six SFRB based on 39 and 18 variables were computed.

Concerning the complete set of 39 variables, the clusters from left to right correspond to type 5 (group A; 16 sites), type 2 (group B; 57 sites), type 1 (group C; 4 sites), type 6 (group D; 68 sites), type 4 (group E; 9 sites), and type 3 (group F; 5 sites). Moreover, eight sites were identified as non-SFRB. With respect to the analysis based on the reduced set of variables, the clusters correspond to SFRB type 6 (group A; 61 sites), type 3 (group B; 11 sites), type 5 (group C; 13 sites), type 4 (group D; 12 sites), type 2 (group E; 52 sites), and type 1 (group F; 5 sites). In addition, 13 sites were identified as non-SFRB.

38.3.3 Groupings Based on Cluster Analysis and Case Study Application

Each cluster can be directly linked to an SFRB type. The distribution of cluster entries in the corresponding SFRB types was both explainable and expected. For example, some SFRB were originally built for drinking water supply purposes. However, the main purposes have often changed, and the types have "shifted" from the original purely drinking water supply function to something more sustainable, aesthetic, and/or natural and have become more integrated into the landscape. Moreover, some sites have become so polluted that they cannot be used for drinking water supply purposes and have instead become fishing ponds or even nature reserves, some even protected by law (usually type 6). The conceptual model provides clear definitions for the past and current (i.e., after aging) status of SFRB, aiding communication between different stakeholders.

The number of classification variables was reduced with the help of a PCA. With respect to flood control, Dam Length, Basin Bed Gradient, Floodwater Surface Area, Catchment Size, and Width of the Basin were the most important independent SFRB characterization variables, which greatly contributed to the variability expressed by the first and second component.

A cluster analysis was performed with a reduced set of variables, which had been identified as independent, easy, and reliable to determine. Seven clusters containing the six SFRB types and a group comprising non-SFRB sites (predominantly unmanaged natural lakes) were identified. The largest groups are natural flood retention wetlands (61 sites) and traditional flood retention basins (52 sites). The former includes passive natural flood retention wetlands characterized by a relatively high wetted perimeter and the latter comprises managed traditional reservoirs that are hydraulically optimized (partly automated). The relatively small groups represent SFRB, which could also be classed as wetlands with strong flood and diffuse pollution control functions. Findings indicate that Scotland has a lower diversity of SFRB types than, for example, Baden, where six clear SFRB groups were identified.

This finding may have two principal reasons. The first reason may be bias in the selection of water bodies for investigation. However, this was not the case during this study, which was undertaken by a large and diverse team of experts over three years. The second reason is that there is a lower diversity of SFRB types in Scotland due to a simple or underdeveloped flood infrastructure, which lacks retention structures overall.

Traditional flood retention basins (SFRB type 2) and natural flood retention wetlands (SFRB type 6) dominate the wider central Scotland area. There is a great underutilized potential in using former and less important current potable supply reservoirs for flood control purposes. There is also a great missed opportunity in integrating semi-natural water bodies into diffuse pollution strategies.

The fieldwork program has identified a large number of water supply reservoirs that are currently surplus to requirements. These reservoirs were predominantly identified as SFRB type 2 with the help of the cluster analyses. In the vast majority of cases, these structures now fulfill multiple roles, providing opportunities for recreation, nature conservation, and angling, with many former drinking water or industrial water supplies being managed as fisheries.

A feature of these sites, based on the majority of current (SFRB type 1) and former (SFRB type 2) drinking water supply reservoirs surveyed, is that they are maintained at their maximum volumes, and the spillways are continually in operation. In this mode of operation, this extensive infrastructure is making very little contribution to water retention in the upper reaches of catchments.

It follows that a change in management practice of these structures by Scottish Water could assist in sustainable flood risk management and result in more sustainable reservoirs (SFRB type 3). Effectively, this would require some water to be released from the reservoirs prior to expected heavy rainfalls. As the vast majority of former drinking water reservoirs have manual level control, this would require someone to visit the sites and open the valves to release the water, returning prior to the main rainfall event to close the valves. This simple operation would create capacity to enhance water storage in the upper reaches of the catchments and retard the peak flows from the upper catchment, which has the potential to reduce the chances of flooding downstream. Combining this approach with conventional solutions such as SuDS, barriers, and dykes will help to reduce the size, cost, and land take of other flood defenses. It is critical to the success of such an approach that appropriate compensation is provided to the owners of the structures to reflect the value of this service and the mild inconvenience it may cause. However, it is beyond the scope of Chapter 38 to discuss how to evaluate the value of the service in practice.

Critical issues to be addressed in this approach are the needs of the owners and operators of the reservoirs. In particular, many of these reservoirs (SFRB type 5) are leased to fishing clubs, typically for *Salmo trutta* (Brown Trout) angling. A balance between the requirements of anglers and water quality targets needs to be found. The majority of these sites only operate as fisheries between March 15 and October 15, which is the traditional trout fishing season in Scotland.

As the most severe rainfall and storm events are predicted for the winter months (Department of Food and Rural Affairs, 2009), the reservoirs could be used for flood control purposes outside the fishing season (i.e., shift from SFRB type 5 to SFRB type 3). A major concern of the fisheries owners will be the retention of the fish within the reservoirs during periods of water release, and this may require the fitting of fine screens onto the valve-controlled outlets of a reservoir. Equally, water supply organizations such as Scottish Water will need to be reassured that the change of management practice will not negatively impact the water quality (required for SFRB type 1) within the basin and any management action would need to ensure that all the SFRB purposes and uses were maintained.

38.4 INTRODUCTION TO MULTILABEL CLASSIFICATION

38.4.1 Basins and Wetland Systems

A SuDS can be regarded as one type of adaptive structure to mitigate flood risk and diffuse pollution on a local scale. On a large scale, flood retention basins can play important roles in adapting catchments to flooding. In comparison to a general flood retention basin, which is typically used for flood attenuation, an SFRB can also be used for direct economic purposes such as a drinking water supply or recreational activities (Scholz, 2007a; see also Chapter 37).

The study of multifunctional flood retention basins is important, considering that it is becoming good practice to prepare risk assessments of river basins and coastal areas on a global scale. For example, the European Union has responded to an increase in the perceived severity of flooding by introducing Flood Directive 2007/60/EC for the assessment and management of flood risks (European Union, 2007). The European Union member states are required to assess river basins, which will be used as the basis for flood risk management. Sustainable flood risk management plans will focus on prevention, protection, and preparedness.

Different risk assessment and classification strategies for flood retention basins and other structures have been reviewed previously by Scholz (2007b), who subsequently proposed the SFRB concept together with his international research team (Scholz, 2007a). One aim of the multifunctional SFRB concept was to aid communication between stakeholders by providing a simple classification scheme for any natural or artificial structures that have a potential flood control purpose.

Various techniques have been proposed to distinguish between different SFRB types; e.g., correlation analysis among characteristic variables (Scholz, 2007a), cluster analysis (McMinn et al., 2010), and traditional classification (Yang et al., 2011) has been developed to distinguish SFRB types by engineers and scientists. Classification is a primary technique used to predict class membership for data instances involving two steps: training and testing. In the training step, the classifier learns the different class concepts on training data with known class labels. In the test step, the classifier predicts the class label of unlabeled instances based on the learned information. However, all these traditional methods are limited to assigning only one label (SFRB type) to one SFRB site, which means that one SFRB is predicted as one specific (i.e., the only one) SFRB type. In fact, one basin usually has multiple functions and thus should fall into more than one SFRB category.

For instance, Harlaw Reservoir (55.87°N, 3.32°W) not only supplies water for Edinburgh (SFRB type 2) but also plays an important role in flood control (SFRB type 1) after a change of the management strategy in 2010. During the periods of high rainfall, sufficient space in the reservoir is left to collect unexpected floodwater. The reservoir holds back water running off the Pentland Hills (southwest of Edinburgh) and releases the runoff gradually, avoiding floods downstream.

Many existing drinking water reservoirs perform flood control functions. However, previous classifications of these reservoirs only focused on the primary function of drinking water supply and ignored the secondary potential function of flood control. Thus, previous SFRB classification methods cannot characterize the functions of SFRB accurately and comprehensively. Exploring the potential functions of SFRB is very valuable for planners and engineers to better manage SFRB. For instance, if it is known that an SFRB has a potential function in terms of flood control, this may help to reduce the size, cost, and land take of other new flood defenses. If an SFRB has a potential for recreational activities to take place, this information may guide landscape planners to change the SFRB type or to integrate recreation into future landscape planning, in agreement with the recent Natural Environment White Paper (Defra, 2011).

The ambiguity of traditional classification (one basin is associated with only one SFRB type) is likely to cause conflicts and confusion regarding the understanding of "real" basin functions; e.g., by ignoring aesthetic and recreational values of an SFRB, planners might only focus on the flood control purpose, which may result in the building of more highly engineered purpose-built concrete structures. In this case, the basin would consequently lose its valuable social benefits. With the help of multilabel classification, planners can clearly understand the potential functions of SFRB.

Another challenge of traditional SFRB classification is data training. In reality, each SFRB is likely to perform different functions. However, traditional classifiers cannot learn multiple functions of SFRB, but only the predominant function of an SFRB. This leads to the incomplete learning of the pattern of the corresponding data set. Therefore, it is impossible to predict SFRB types effectively. To better understand and capture the functions of SFRB and improve the development and management of SFRB in the context of flood risk management planning, the multilabel classification model is thus introduced in this study.

38.4.2 Environmental Modeling and Multilabel Classification

Environmental modeling is becoming more relevant for environmental engineers and scientists. Therefore, this chapter helps to make environmental modeling more accessible and relevant to environmentalists by attempting to demonstrate how a generic modeling solution in the area of classification can solve real practical problems for environmental engineers and scientists. This chapter is examining an example of a wider set of environmental problems.

There are plenty of other relevant methods and applications in different environmental sectors that should help the reader. Some recent and relevant examples are as follows: application of machine learning methods to spatial interpolation of environmental variables (Gibert and Sànchez Marrè, 2011); artificial neural networks and support vector machines for application in environmental modeling (Li et al., 2011); and self-organizing map analysis of environmental data (Hedmark et al., 2009).

Traditional classification is concerned with learning from a set of examples that is associated with a single label l from a set of labels L, $|L| > 1$. If examples are associated with a set of labels, this is called multilabel classification, which refers to the classification problems where an instance can belong to more than one category simultaneously (Tsoumakas and Katakis, 2007).

Various multilabel classification approaches have been extensively studied (Yang et al., 2012). Initially, this approach originated from an investigation of text categorization problems, where each document may belong to several labels (topics) simultaneously (McCallum, 1999). Joachims (1998) constructed a set of binary support vector machine (SVM) classifiers, training each possible class versus the remaining ones and assigning a real value to each class to indicate the class relationship. However, this method did not address multilabel training models and specific testing criteria. McCallum (1999) proposed a generative approach in which a model was trained by expectation maximization; i.e., selection of the most probable set of labels from the set of possible classes. However, this generative model was limited to specific text applications. Schapire and Singer (2000) proposed BoosTexter, extending AdaBoost to handle multilabel text categorization. The algorithm was somewhat inefficient due to a high space complexity and time-per-round complexity.

Multilabel learning has been widely applied in many complex real-world problems ranging from bioinformatics (Clare and King, 2001; Elisseeff and Weston, 2002; Zhang and Zhou, 2006; Zhu et al., 2009), music (Li and Ogihara, 2006; Trohidis et al., 2008), directed marketing (Zhang et al., 2006), and video annotation (Snoek et al., 2006; Qi et al., 2007) to semantic scene classification (Boutell et al., 2004). For bioinformatics, the C4.5 algorithm was adapted by Clare and King (2001), who modified the definition of entropy to include multilabel data of gene expression. Choosing a decision tree as the baseline algorithm, they only aimed to learn the symbolic rules. However, this was not a complete classification. Elisseeff and Weston (2002) proposed a multilabel kernel method for yeast gene functional classification, minimizing the ranking loss (see also below). Zhang and Zhou (2006) extended the back-propagation neural network algorithm for multilabel learning (BP-MLL). They introduced a new error function that takes multilabel learning into account.

In addition, Boutell et al. (2004) proposed a multilabel support vector machine (MLSVM) framework by decomposing the multilabel learning problems into multiple independent binary classification problems. They presented an effective cross-training strategy, the C-Criterion, in testing and two new evaluation metrics. Zhang and Zhou (2007) proposed a lazy learning approach to multilabel learning named multilabel k-nearest neighbor (MLKNN), extending the k-nearest neighbor (KNN) algorithm with a Bayesian approach. It uses the maximum a posteriori principle to determine the label set of the test instance, based on prior and posterior probabilities for the frequency of each label within the k-nearest neighbors.

38.5 DATA AND METHODOLOGY

38.5.1 Data Acquisition

A guidance manual on how to determine the variables characterizing water bodies including SFRB has been published by Scholz and Yang (2010). Data acquisition includes conventional hard engineering variables such as Dam Length and Dam Height, along with more holistic variables such as how Engineered the structure appears and variables such as Aquatic Animal Passage and Land Animal Passage (Scholz and Sadowski, 2009, McMinn et al., 2010). This combination of hard and soft variables readily lends itself to solving multidisciplinary problems such as working toward a sustainable water management strategy. Sites situated in the wider central Scotland area and in Baden have been surveyed. Each site obtains data for 43 variables along with a confidence level for each data point.

Each SFRB was characterized by 43 variables. During data acquisition, the functions of each SFRB were assigned different scores by an international expert team to obtain the corresponding class labels of SFRB. If the SFRB performs multiple functions, it is thus associated with multiple labels simultaneously. In comparison, for traditional classification approaches, the dominant type of each SFRB, which had the highest score, was considered as the class label of the SFRB. The distribution of SFRB types in the two data sets indicates that each instance (SFRB) was associated with about two labels on average in the Scottish data set and 1.5 labels on average in the German data set.

In traditional supervised learning, an instance is associated with a class label. In contrast, for multilabel problems, each instance is associated with multiple class labels. Currently, various multilabel classification algorithms have been proposed. Three popular algorithms, MLSVM, MLKNN, and BP-MLL, were applied for multilabel classification on the data sets from Scotland and Baden. There are two reasons for selecting these three multilabel classification algorithms. One reason lies in the properties of the SFRB data sets. For the Scottish and German data sets, their sample sizes are relatively small and each instance is associated with multiple labels. The learning models of these three algorithms fit the SFRB data sets well. Specifically, SVM is known for its high classification accuracy regarding small data sets and its generalization capability (Joachims, 1998; Li and Shawe-Taylor, 2003). The lazy learning algorithm KNN is well suited for multimodal classes (Sudhakaran and Vasantha, 2011). The Naive Bayes algorithm also only requires a small amount of training data and has good performance results (Sudheep et al., 2011). The second reason is to investigate whether a multilabel model is suitable for the classification of SFRB, as these three algorithms are the representatives of various multilabel classification paradigms (Boutell et al., 2004; Zhang and Zhou, 2006, 2007).

38.5.2 Multilabel Support Vector Machine

The SVM model is a popular and promising tool for data classification (Chen and Lin, 2006). Its basic idea is to construct a separating hyperplane between the training instances of both classes. Among all possible hyperplanes, that one with the maximum margin between classes is selected (Chen and Lin, 2006). Given that training vectors are in two classes, then a linear SVM solves a quadratic optimization problem. In this study, to reduce the impact of the penalty parameter on the training error, five settings of this parameter were used during the experiment. The results are averaged over all different parameters.

Boutell et al. (2004) proposed multilabel self-organizing map learning techniques (MLSVM) for classification purposes. Cross-training (also called MODEL-x) is used where multilabel data are trained more than once. If a basin belongs, for example, to SFRB type 1 and 3 simultaneously, for learning the concept of type 1, the SFRB is selected as the positive instance. It is also further treated as the positive instance when learning the concept of type 3. For handling multiclass problems, the one-versus-all strategy is applied, where each class is compared with all other classes. Although the findings indicate the effectiveness of this strategy, its performance may also be affected by other factors, which is not consider here, such as small sample size and unbalanced data training. For more detail, the reader is referred to Rifkin and Klautau (2004) regarding experimental studies on different multiclass classification algorithms, although the problem of which strategy can yield better practical performance is still unresolved.

After the classifier learns each concept of the classes during the testing phase, given a test instance, it outputs a score for each class. The magnitude of the score indicates the degree of membership in the corresponding class. To generalize the one-versus-all approach with respect to multilabel classification, a labeling criterion (i.e., the C-Criterion) is used to determine the labels of an instance in the testing phase.

38.5.3 Multilabel k-Nearest Neighbor

The KNN algorithm is a lazy learning algorithm that predicts the membership of objects based on closest training examples in the feature space. Given an instance, its k closest neighbors are found in terms of the Pearson correlation coefficient, and then its label value is determined by these k neighbors using the majority vote manner principle. Derived from the KNN algorithm, a multilabel learning algorithm named Multilabel k-Nearest Neighbor has been proposed (Zhang and Zhou, 2007). Like KNN, for each test instance, the number of nearest neighbors should first be identified. To mitigate the impact of classification results in terms of a different number of neighbors, in this study, $k = \{6, 8, 10, 12, 14\}$ were used. After the search of k-nearest neighbors, the prior and posterior probabilities for the frequency of each label are estimated from the training set based on frequency counting.

38.5.4 Back-Propagation for Multilabel Learning

An artificial neural network (ANN), also called neural network, is a mathematical model that simulates the system of biological neural networks (Kohonen, 1988). Usually, a neural network consists of three layers of neurons: input layer, hidden layer, and output layer. The input layer receives the data and passes them on to the hidden layer, and finally the output layer generates the predicted values. All neurons in their layers are fully weighted and connected and the weights are adjusted during the training phase. Currently, back-propagation (BP) is the most popular algorithm for training ANN, where the prediction error is fed back into the neural network for updating the connection weights so as to minimize the total error between the prediction and the figure in the training procedure.

Zhang and Zhou (2006) extended the traditional multilayer feed-forward neural network to the multilabel classification problem. They arranged neurons in layers, of which the first layer takes inputs, the last layer produces the outputs, and the middle one is called the hidden layer. As the number of neurons in the hidden layer is hard to determine in this study, the heuristic method where the number of the hidden neurons $= \gamma \times$ input dimensionality was used. The input dimensionality equates to the number of variables (43 in this study) and γ is a bias parameter of the feed-forward networks. To reduce the sensitivity of the parameter γ, the authors set it as 0.5, 1.0, 1.5, 2.0, and 2.5 in the experiments, according to work undertaken by Zhang and Zhou (2006). To capture the characteristics of multilabel learning, a classical back-propagation algorithm (Rumelhart et al., 1986) was employed to minimize the global error function.

In this global error function, the summation in the i-th error term takes account of the accumulated difference between the outputs of any pair of labels with one belonging to x_i and another not belonging to x_i, which is then normalized by the total number of possible pairs (Zhang and Zhou, 2006). In this way, the labels belonging to an instance should be ranked higher than those not belonging to that instance.

In the learning phase, all training instances are fed to the network one by one. With the weight updating functions, the weights (and biases) are modified. The training process is terminated when the global error E does not decrease any more or the number of training epochs increases to a threshold.

After the network is trained, for each test instance, the actual outputs can be obtained. The predicted labels for the instance are finally determined by a threshold function (Elisseeff and Weston, 2002; Zhang and Zhou, 2006).

38.5.5 Evaluation Measures

In the single-label system, a class prediction is either correct or incorrect and the standard evaluation metrics include precision, recall, accuracy, and F-measure (Sebastiani, 2002). Here, for comparison, accuracy is used, which is defined in Eq. (38.5.5.1):

$$\text{Accuracy} = \frac{n}{m} \tag{38.5.5.1}$$

where n denotes the number of instances that are correctly predicted; m means the number of instances in the evaluated data set.

For the multilabel data, various evaluation measures have been studied (Tsoumakas and Katakis, 2007; Tsoumakas et al., 2010). In this work, the following evaluation metrics proposed by Schapire and Singer (2000) are applied to evaluate some learning algorithms, including Average Precision, Coverage, One-error, Ranking-Loss, and Harming-Loss.

Average Precision (Avg. Pre.; Eq. 38.5.5.2) measures the average fraction of labels ranked above a particular label. It derives from the performance measure used in information retrieval systems (Salton, 1991). It is used to evaluate the document ranking performance for query retrieval. When average precision equals 1, it means that the ranking runs perfectly.

$$\text{Avg. Pre.} = \frac{1}{m}\sum_{i=1}^{m}\frac{1}{|Y_i|}\sum_{\lambda\in Y_i}\frac{|\{\lambda'\in Y_i : r_i(\lambda')\le r_i(\lambda)\}|}{r_i(\lambda)} \tag{38.5.5.2}$$

where m is the number of the instances in the evaluated data set; Y_i is the set of actual labels of a given instance x_i; $r_i\ (\lambda)$ represents the rank of a label λ predicted by the ranking method; and λ' indicates the class label where the rank is higher than that of class label λ.

Coverage (Eq. 38.5.5.3) is designed to assess how far one must go, on average, down the list of labels to cover all the possible labels that the instances actually belong to. The smaller the value of coverage is, the better the performance.

$$\text{Coverage} = \frac{1}{m}\sum_{i=1}^{m}\max_{\lambda \in = Y_i} r_i(\lambda) - 1 \tag{38.5.5.3}$$

where m is the number of the instances in the evaluated data set; $r_i(\lambda)$ represents the rank of a label λ predicted by the ranking method; Y_i is the set of actual labels of a given instance x_i.

One-error (Eq. 38.5.5.4) evaluates how many times the top-ranked label does not appear in the set of ground truth labels. The smaller the value of one-error is, the better the performance.

$$\text{One-error} = \frac{1}{m}\sum_{i=1}^{m}\delta\left(\arg\max_{\lambda \in L} r_i(\lambda)\right) \tag{38.5.5.4}$$

where m is the number of the instances in the evaluated data set; $r_i(\lambda)$ represents the rank of a label λ predicted by the ranking method; Y stands for the set of labels; $\delta(\lambda)$ equals one, if $\lambda \notin Y_i$, otherwise it equals zero.

Ranking-Loss (Ran.-Loss) expresses how many times the relevant labels are reversely ordered. The performance is perfect when the ranking loss equals zero. It is defined in Eq. (38.5.5.5).

$$\text{Ran.-Loss} = \frac{1}{m}\sum_{i=1}^{m}\frac{1}{|Y_i||\overline{Y_i}|}\left|\left\{(\lambda_a,\lambda_b): r_i(\lambda_a) > r_i(\lambda_b), (\lambda_a,\lambda_b) \in Y_i \times \overline{Y_i}\right\}\right| \tag{38.5.5.5}$$

where m is the number of the instances in the evaluated data set λ_a and λ_b indicate class labels; $r_i(\lambda)$ represents the rank of a label λ predicted by the ranking method; Y_i is the set of actual labels of a given instance x_i, and $\overline{Y_i}$ denotes the complementary set of Y_i in Y.

Hamming-Loss (Ham.-L.^{ml}) evaluates how many times an instance–label pair is classified incorrectly (Schapire and Singer, 2000). It is defined in Eq. (38.5.5.6).

$$\text{Ham-L.}^{ml} = \frac{1}{m}\sum_{i=1}^{m}\frac{|Y_i \Delta Z_i|}{|Q|} \tag{38.5.5.6}$$

where m is the number of the instances in the evaluated data set; Q is the size of the set of labels. Given an instance x_i, Y_i and Z_i stand for the actual set of labels and the predicted set of labels, respectively; Δ stands for the symmetric difference of two sets.

To further evaluate the performance of the predicted results, the evaluation matrices Precision (Pre.^{ml}), Recall (Rec.^{ml}), and F_1 (F_1^{ml}) are extended into multilabel classification tasks (Godbole and Sarawagi, 2004). These measures are based on "per instance," calculating the average differences of the actual and the predicted sets of labels over all instances of the evaluation data set. Specifically, they are defined in Eqs. (38.5.5.7)–(36.5.5.9). The higher the values are, the better the multilabel classifiers perform.

$$\text{Pre.}^{\text{ml}} = \frac{1}{m}\sum_{i=1}^{m}\frac{|Y_i \cap Z_i|}{|Z_i|} \tag{38.5.5.7}$$

where m is the number of the instances in the evaluated data set; given an instance x_i, Y_i and Z_i stand for the actual set of labels and the predicted set of labels, respectively; $|Y_i \cap Z_i|$ means the number of labels that are correctly predicted; $|Z_i|$ represents the number of labels in the predicted set of labels.

$$\text{Rec.}^{\text{ml}} = \frac{1}{m}\sum_{i=1}^{m}\frac{|Y_i \cap Z_i|}{|Y_i|} \tag{38.5.5.8}$$

where m is the number of the instances in the evaluated data set; given an instance x_i, Y_i and Z_i stand for the actual set of labels and the predicted set of labels, respectively; $|Y_i|$ means the number of labels in the actual set of labels.

$$\text{F}_1^{\text{ml}} = \frac{1}{m}\sum_{i=1}^{m}\frac{2|Y_i \cap Z_i|}{|Z_i| + |Y_i|} \tag{38.5.5.9}$$

where m is the number of the instances in the evaluated data set; given an instance x_i, Y_i and Z_i stand for the actual set of labels and the predicted set of labels, respectively; $|Y_i|$ and $|Z_i|$ represent the number of labels in the two sets, respectively; $|Y_i \cap Z_i|$ stands for the number of labels that are correctly predicted.

Two data sets have been used to evaluate the multilabel classification approaches for analyzing the multiple functions of SFRB. For each method and all data, during the training phase, one-leave-out cross-validation was performed. In comparison to the traditional method of splitting data into only two parts (i.e., training data and test data), cross-validation is performed using different partitions on multiple rounds, and the validation results are averaged over the rounds. Therefore, in this study, to reduce the impact of variability of the data splitting for classification, one-leave-out cross-validation has been applied. Specifically, each data instance from the data set is selected as the test data, and then the remaining data instances are used as the training data. In addition, five different values for each parameter were selected in a heuristic way suggested by the algorithms.

For comparison, the corresponding traditional classifiers SVM, KNN, and BP were also applied in experiments. Multilabel classifiers MLSVM, MLKNN, and BP-MLL were implemented in Matlab. The corresponding traditional classifiers used in this chapter are available in WEKA (http://www.cs.waikato.ac.nz/ml/weka/).

38.6 EXPERIMENTAL RESULTS AND DISCUSSION

The multilabel algorithms (MLSVM, MLKNN, and BP-MLL) and the traditional classification techniques (SVM, KNN, and BP) were performed on the same SFRB data sets. The performances of these classifiers were evaluated by using the metrics discussed above.

Table 38.6.1 summarizes the classification results of each classifier for both data sets. For the Scottish data set, all multilabel classifiers achieved good prediction results: The precisions of the MLSVM, MLKNN, and BP-MLL techniques were 89.0% ± 3.2%, 87.4% ± 6.9%, and 78.1% ± 30.2%, respectively. All three multilabel classifiers obtained high recall and F_1 values; particularly, BP-MLL achieved the highest recall at 90.9% ± 24.0%. Other criteria such as One-error, Ranking-Loss, and Hamming-Loss were associated with relatively low values. Average Precision values were all above 90%, which confirms the good performance of the learning algorithm. In comparison, the corresponding precisions for the German data were slightly higher; i.e., 97.3% ± 1.4%, 96.0% ± 3.7%, and 93.5% ± 17.5% for MLSVM, MLKNN, and BP-MLL, respectively. The recall and F_1 values were as high as 82.3% and 85.9% on average, respectively. Furthermore, the Hamming-Loss and One-error values for the data set from Baden were all smaller than those for the Scottish data set. Although multilabel learning is more complicated, the good classification results indicated that all three multilabel classifiers can predict the types of SFRB effectively, among which MLSVM slightly outperforms the other two methods. Nevertheless, the proposed multilabel classifiers have about 10% false hits regarding SFRB classification. The impact of this error on planning decisions is likely to be small and considerably less than human error. However, no data evidence for this claim is currently available.

Furthermore, Table 38.6.1 displays the classification results for the three traditional learning algorithms SVM, KNN, and BP applied on both data sets in terms of accuracy. However, since multilabel classification and traditional classification use different evaluation metrics, it is difficult to compare the classification results directly with each other. The reason is that Accuracy is defined to evaluate traditional (single-label) classification in which the predicted class is either correct or not, while Precision is used to assess whether the labels predicted by the multilabel classifiers are either fully correct, partly correct, or fully incorrect. Only if all types of an SFRB are predicted perfectly is the classification regarded as fully correct. Although the evaluation matrices for traditional classification algorithms and multilabel classification algorithms have different definitions, it is interesting to note that most of the performances of traditional classifications were weaker than those of the multilabel methods based on the experiments. For instance, the traditional classifiers SVM, KNN, and BP obtained accuracies of 86.1% ± 2.0%, 83.0% ± 0.6%, and 88.3% ± 0.2%, respectively, for the Scottish data. In comparison, the precision values of MLSVM and MLKNN increased by about 4%. For the German SFRB data,

TABLE 38.6.1 Experimental Results Based on Multilabel Learning Algorithms and Traditional Learning Algorithms (Mean ± Standard Deviation) (after Yang et al., 2012)

	SFRB in Scotland		
Evaluation Criteria	**MLSVM**	**MLKNN**	**BP-MLL**
Avg. Pre.	0.918 ± 0.011	0.921 ± 0.033	0.901 ± 0.198
Coverage	0.878 ± 0.050	0.833 ± 0.149	0.839 ± 1.039
One-error	0.105 ± 0.018	0.102 ± 0.044	0.151 ± 0.359
Ran.-Loss	0.066 ± 0.009	0.062 ± 0.025	0.069 ± 0.150
Ham.-L.^{ml}	0.111 ± 0.017	0.109 ± 0.036	0.163 ± 0.151
Pre.^{ml}	0.890 ± 0.032	0.874 ± 0.069	0.781 ± 0.302
Rec.^{ml}	0.742 ± 0.039	0.777 ± 0.062	0.909 ± 0.240
$F_1{}^{ml}$	0.783 ± 0.038	0.796 ± 0.069	0.743 ± 0.253
	SVM	KNN	BP
Accuracy	0.861 ± 0.020	0.830 ± 0.006	0.883 ± 0.002
	SFRB in Baden		
Evaluation Criteria	**MLSVM**	**MLKNN**	**BP-MLL**
Avg. Pre.	0.919 ± 0.010	0.959 ± 0.013	0.942 ± 0.119
Coverage	1.083 ± 0.063	0.721 ± 0.108	0.856 ± 1.232
One-error	0.014 ± 0.005	0.020 ± 0.000	0.020 ± 0.140
Ran.-Loss	0.084 ± 0.010	0.035 ± 0.014	0.054 ± 0.109
Ham.-L.^{ml}	0.077 ± 0.009	0.082 ± 0.021	0.083 ± 0.114
Pre.^{ml}	0.973 ± 0.014	0.960 ± 0.037	0.935 ± 0.175
Rec.^{ml}	0.816 ± 0.021	0.817 ± 0.034	0.837 ± 0.249
$F_1{}^{ml}$	0.863 ± 0.018	0.858 ± 0.039	0.856 ± 0.196
	SVM	KNN	BP
Accuracy	0.900 ± 0.002	0.905 ± 0.000	0.897 ± 0.010

Note: MLSVM, multilabel support vector system; MLKNN, multilabel k-nearest neighbor; BP-MLL, back-propagation for multilabel learning; Avg. Pre., Average Precision; Ran.-Loss, Ranking-Loss; Ham.-L.^{ml}, Hamming-Loss; Pre.^{ml}, Precision; Rec.^{ml}, Recall; $F_1{}^{ml}$, F_1.

the classifiers SVM, KNN, and BP achieved accuracies of 90.0% ± 0.2%, 90.5% ± 0.0%, and 89.7% ± 1.0%, respectively, compared with the precision of multilabel classifiers ranging from 93.5% to 97.3%. From the experimental results, it is obvious that multilabel classification algorithms yield better results. The benefits of the multilabel classification models over traditional methods will be further illustrated by three representative case studies.

The comparative findings show that multilabel classification provides a robust and efficient tool to better classify and understand SFRB than the traditional methods. This tool allows one SFRB to belong to several types simultaneously. Multiple types associated with a single SFRB verify what the authors observed in practice; i.e., one SFRB often performs multiple functions. Thus, multilabel classification better represents the complex reality than do traditional approaches. Furthermore, this tool encourages people to gain new insights into the true status and multiple functions of SFRB in a comprehensive way, avoiding conflicts and confusion about SFRB assessment between engineers, stakeholders, and planners. For example, for a natural water body partially developed for water sports, some people may state it belongs to SFRB type 5 due to the recreational activities, while others insist it belongs to SFRB type 6 because of its natural character. Conflicts may be avoided when it is simultaneously judged by planners and decision makers as both type 5 and 6 by multilabel classifiers, as the public recognizes that the SFRB has multiple and justified functions at the same time.

When making plans for the development and management of SFRB, planners may realize the benefits of multiple functions of SFRB. Similarly, some designers may promote more people-friendly, rather than highly engineered, facilities. An SFRB originally used for sustainable drainage (SFRB type 3) could have a great landscape value, which needs to be addressed in maintenance strategies. Moreover, multilabel classifiers may estimate desirable functions of SFRB that are currently not noticed. For example, a large number of drinking water reservoirs (SFRB type 2) located in central Scotland are currently used (and only used) for water supply purposes. Multilabel classifiers identified them also as type 1, giving planners a wider range of opportunities. It follows that multilabel classification can guide people in identifying and understanding new functions of SFRB.

38.7 REPRESENTATIVE CASE STUDIES

The relationship between multilabel classification and SFRB functionality will be demonstrated with the help of three relevant examples in this section, which also shows the need for multilabel classification and its application value in practice. The multilabel classification of SFRB performed very well under the scheme of multilabel learning. It allows one SFRB to belong to multiple types simultaneously, which helps people consider the development of SFRB from a multidisciplinary and holistic perspective.

Johnston Loch (55.89°N, 4.09°W) is located in North Lanarkshire, Scotland. It drains the runoff from the surrounding residential area called Gartcosh. Additionally, it is an ideal place for walking and relaxing because of its peaceful and beautiful natural sights. The overall structure of the loch is entirely natural; i.e., there are currently no engineered structures such as a dam or a spillway. Thus, it can be intuitively characterized as an SFRB of types 3 (mainly used for sustainable drainage), 5 (largely used for aesthetic recreation), and 6 (predominantly natural water bodies). Johnston Loch was predicted as SFRB types 3, 5, and 6 by both classifiers MLKNN and BP-MLL, while MLSVM predicted it as SFRB types 5 and 6. However, Johnston Loch was judged as SFRB type 6 only by the traditional classifiers SVM, KNN, and BP. Therefore, the predictions of the multilabel classifiers represented the diverse functions of Johnston Loch well.

The results of the traditional classifications were too limited. The findings may mislead people, which could lead to a neglect of the loch's drainage and landscape aesthetic functions. Realizing the fact that Johnston Loch has multiple functions, planners and designers can develop the basin more sustainably and cost-effectively.

Murg Ausgleichsbecken (48.40°N, 8.21°W) located in Forbach (Baden, Germany) is taken as another random example. It is a purpose-built SFRB used for flood protection. However, the integrative functions of enhancing the landscape aesthetics and recreational activities have been growing in importance after its construction. Therefore, Murg Ausgleichsbecken is characterized not only as an SFRB type 1 but also as an SFRB type 5. All three multilabel classifiers MLSVM, MLKNN, and BP-MLL predicted the basin as types 1 and 5 simultaneously. This confirms the findings of experts who visited this SFRB in 2006 and 2010. However, Murg Ausgleichsbecken is only predicted as type 1 using the traditional classifiers. Realizing that the Murg Ausgleichsbecken actually belongs also to type 5 using the new classification method, planners might enhance its aesthetic and recreational attributes by providing wider public access. Without the help of multilabel classification, planners might only focus on the flood control purpose and thus lose many valuable societal benefits of the basin.

The previously introduced Harlaw Reservoir was also correctly predicted as SFRB types 1 and 2 by the multilabel classifiers. The prediction results verified the fact that the reservoir plays an important role in flood defense by adjusting runoff release quantities. More importantly, realizing that reservoirs and other basins can be adapted to contribute to flood control will save huge sums of money that can be spent elsewhere on new flood defense structures. If one can clearly identify an SFRB as being purely of type 1 and if it is subsequently managed properly, the structure is likely to contribute substantially to sustainable flood risk management planning, supporting the implementation of legislation (Defra, 2010).

38.8 CONCLUSIONS AND RECOMMENDATIONS

In comparison to six clearly defined SFRB types identified for Baden in Germany, the Scottish data set contained only two main SFRB types. Traditional flood retention basins consisting predominantly of former drinking water reservoirs are a clearly noticeable component of the Scottish landscape. These structures could be used for low-cost flood control purposes if their water level were actively controlled on a seasonal basis.

Natural flood retention wetlands also dominate the case study area and could make a significant contribution to diffuse pollution control if they are managed appropriately. The most important independent and accurately determined SFRB variables that resemble wetland systems with a high diffuse pollution treatment function were Wetted Perimeter, Floodwater Surface Area, Engineered, Catchment Size, Outlet Arrangement and Operation, and Mean Sediment Depth.

The proposed slightly revised conceptual classification model may be used elsewhere by stakeholders for decision-making processes. The rapid classification system appears to be robust and has the potential, with minor modifications, to be applied across temperate oceanic and temperate continental regions in Europe.

Chapter 38 also applies multilabel classification algorithms for SFRB case studies. To better understand, assess, and manage SFRB, a multilabel classification model to predict the multiple types of SFRB has been introduced. Highly variable case studies in central Scotland and southern Baden were successfully used to validate the effectiveness and efficiency of the multilabel classification of SFRB.

The multilabel algorithms MLSVM, MLKNN, and BP-MLL were successfully applied to address the problem of potential ambiguity in classifying SFRB into different types. The evaluation measures Average Precision, Coverage, One-error, Ranking-Loss, Hamming-Loss, Precision, Recall, and F_1 were calculated to verify the effectiveness of the classification model.

The findings showed that all multilabel classifiers achieved good results, which obtained precisions ranging from 78.1% to 89.0 % for Scottish data and varying from 93.5% to 97.3% for German data. Furthermore, all three multilabel classifiers achieved high recall and F_1 values. The corresponding traditional classifiers (SVM, KNN, and BP) yielded classification accuracies between 83.0% and 88.3% for Scottish SFRB and ranging from 89.7% to 90.5% for SFRB in Baden. In addition, three SFRB case studies were further investigated to verify the benefits of the multilabel learning algorithms in practice.

The findings showed that the multilabel classifiers, which predicted multiple types of SFRB, would bring a deeper and more comprehensive insight into the status and functions of SFRB than the traditional classification methods. The predicted multiple functions of SFRB thus help to reduce and/or avoid confusion and misunderstandings concerning SFRB assessment and management among planners, engineers, and authorities. And especially, it provides planners and practitioners with an efficient, unbiased, and scientific tool to distinguish existing SFRB and manage them properly in terms of flood risk management, water resources management, and social and recreational benefit development. The SFRB concept is a practical tool for planners, helping them with the implementation of the Water and Flood Management Act and addressing challenges of the Natural Environment White Paper.

The SFRB methodology should be used to develop a decision support model for the identification of the most suitable SFRB type based on boundary input conditions characterizing a case study site requiring an adaptive measure to control flooding and diffuse pollution. The value for each characteristic SFRB variable was associated with an uncertainty value when it was estimated or measured. However, the current multilabel classification framework does not take uncertainty issues into account. The focus should be on classification or clustering of SFRB with uncertain data in the future.

References

Aamodt, A., Plaza, E., 1994. Case-based reasoning: foundational issues, methodological variations and system approaches. AI Commun. 7 (1), 39–59.

Abbott, C.L., Comino-Mateos, L., 2001. In situ performance monitoring of an infiltration drainage system and field testing of current design procedures. J. Chart. Inst. Water Environ. Manage. 15 (3), 198–202.

Ahn, C., Mitsch, W.J., Wolfe, W.E., 2001. Effects of recycled FGD liner material on water quality and macrophytes of constructed wetlands: a mesocosm experiment. Water Res. 35 (3), 633–642.

Aguilera, P.A., Frenich, A.G., Torres, J.A., Castro, H., Vidal, J.L.M., Canton, M., 2001. Application of the Kohonen neural network in coastal water management: methodological development for the assessment and prediction of water quality. Water Res. 35 (17), 4053–4062.

Aiello, R., Cirelli, G.L., Consoli, S., 2007. Effects of reclaimed wastewater irrigation on soil and tomato fruits: a case study in Sicily (Italy). Agric. Water Manage. 93 (1–2), 65–72.

Allen, S.E., 1974. Chemical Analysis of Ecological Materials. Blackwell, Oxford, UK.

Alloway, B.J., 1995. Heavy Metals in Soils, second ed. Chapman and Hall, Suffolk, UK.

Almuktar, S.A.A.-A.N., Scholz, M., Al-Isawi, R.H.K., Sani, A., 2015. Recycling of domestic wastewater treated by vertical-flow wetlands for irrigating chillies and sweet peppers. Agric. Water Manage. 149, 1–22.

Anderson, D.M., Gilbert, P.M., Burkholder, J.M., 2002. Harmful algal blooms and eutrophication: nutrient sources, composition, and consequences. Estuaries 25 (4B), 704–726.

Andoh, R.Y.G., Declerck, C., 1997. A cost effective approach to stormwater management? Source control and distributed storage. Water Sci. Technol. 36 (8–9), 307–311.

Ann, Y., Reddy, K.R., Delfino, J.J., 1999a. Influence of chemical amendments on phosphorus immobilization in soils from a constructed wetland. Ecol. Eng. 14 (1–2), 157–167.

Ann, Y., Reddy, K.R., Delfino, J.J., 1999b. Influence of redox potential on phosphorus solubility in chemically amended wetland organic soils. Ecol. Eng. 14 (1–2), 169–180.

APHA, 1998. Standard Methods for the Examination of Water and Wastewater, twentieth ed. American Public Health Association (APHA), American Water Works Association and Water and Environmental Federation, Washington D.C., USA.

APHA, 2005. Standard Methods for the Examination of Water and Wastewater, twenty-first ed. American Public Health Association (APHA), American Water Works Association, and Water and Environment Federation, Washington D.C., USA.

Arditi, D., Tokdemir, O.B., 1999. Comparison of case-based reasoning and artificial neural networks. J. Comp. Civ. Eng. 13 (3), 162–168.

Arheimer, B., Brandt, M., 1998. Modelling nitrogen transport and retention in the catchments of southern Sweden. Ambio 27 (6), 471–480.

Asano, T., 1994. Reusing urban wastewater – an alternative and a reliable water resource. Water Int. 19 (1), 36–42.

Asano, T., Burton, F.L., Leverenz, H., Tsuchihashi, R., Tchobanoglous, G., 2007. Water Reuse: Issues, Technologies, and Applications. McGraw Hill, New York, USA.

Asano, T., Levine, A.D., 1996. Wastewater reclamation, recycling and reuse: past, present, and future. Water Sci. Technol. 33 (10–11), 1–14.

Asare, S.N., Rudra, R.P., Dickinson, W.T., Wall, G.J., 1993. Seasonal variability of hydraulic conductivity. Trans. ASAE 36 (2), 451–457.

Astebøla, S.O., Hvitved-Jacobsenb, T., Simonsen, O., 2004. Sustainable stormwater management at Fornebu: from an airport to an industrial and residential area of the city of Oslo, Norway. Sci. Total Environ. 334–335, 239–249.

ATV-DVWK, 2001. Hochwasserrückhaltebecken – Probleme und Anforderungen aus wasserwirtschaftlicher und ökologischer Sicht (Deutsche Vereinigung für Wasserwirtschaft, Abwaser und Abfall e.V. (ATV-DVWK), Gesellschaft zur Förderung der Abwassertechnik e.V, Hennef, Germany (in German).

ATV-DVWK-Arbeitsgruppe, 2002. Planung, Bau und Betrieb von Anlagen zur Versickerung von Niederschlagswasser (Design, Construction and Operation of Rainwater Drainage Systems), Regelwerk A-138, ES-41. ATV-DTWK (German Association for Water, Wastewater and Waste), Gesellschaft zur Förderung der Abwassertechnik e.V, Hennef, Germany, ISBN 3-935669-83-6 (in German).

van der Aa, M., Trepel, M., van Gaans, P.F.M., Bleuten, W., Kluge, W., 2001. Modelling water flow and fluxes of a valley mire for use in restoration. Landnutzung und Landentwicklung 42, 72–78 (in German).

Babatunde, A.O., Zhao, Y.Q., O'Neill, M., O'Sullivan, B., 2008. Constructed wetlands for environmental pollution control: a review of developments, research and practice in Ireland. Environ. Int. 34 (1), 116–126.

Bachand, P.A.M., Horne, A.J., 1999. Denitrification in constructed free-water surface wetlands: I. Very high nitrate removal rates in a macrocosm study. Ecol. Eng. 14 (1–2), 9–15.

Backstrom, M., Karlsson, S., Backman, L., Folkeson, L., Lind, B., 2004. Mobilisation of heavy metals by deicing salts in a roadside environment. Water Res. 38 (3), 720–732.

Bah, A.R., Kravchuk, O., Kirchhof, G., 2009. Sensitivity of drainage to rainfall, vegetation and soil characteristics. Comput. Electron Agric. 68 (1), 1–8.

Bakker, B.H., 1989. The effects of air humidity on flowering, fruit set, seed set and fruit growth of glasshouse sweet pepper (*Capsicum annuum* L.). Scientia Horticulturae 40 (1), 1–8.

Bakker, B.H., 1991. Analysis of Humidity Effects on Growth and Production of Glasshouse Fruit Vegetables (PhD thesis). Wageningen University, Wageningen, The Netherlands.

Ball, A.S., William, M., Vincent, D., Robinson, J., 2001. Algae growth control by barley straw extract. Biores. Technol. 77 (2), 177–181.

Barrington, S.F., Madramootoo, C.A., 1989. Investigating seal formation from manure infiltration into soils. Trans. ASAE 32, 851–856.

Barrington, S.F., Jutras, P., Broughton, R.S., 1987. The sealing of soils by manures. Can. Agric. Eng. 29, 99–108.

Bassett, D., Pettit, A., Anderson, C., Grace, P., 2007. Scottish Flood Defence Asset Database – Final Report. JBA Consulting for the Scottish Government, Edinburgh, Scotland, UK. http://www.scotland.gov.uk/Publications/2007/08/20111904/0.

Bastviken, S.K., Eriksson, P.G., Premrov, A., Tonderski, K., 2005. Potential denitrification in wetland sediments with different plant species detritus. Ecol. Eng. 25 (2), 183–190.

Bayley, S.E., Guimond, J.K., 2008. Effects of river connectivity on marsh vegetation community structure and species richness in montane floodplain wetlands in Jasper National Park, Alberta, Canada. Ecoscience 15 (3), 377–388.

Belanche, L., Valdes, J.J., Comas, J., Roda, I.R., Poch, M., 2000. Prediction of the bulking phenomenon in wastewater treatment plants. Artif. Intell. Eng. 14 (4), 307–317.

Bettess, R., 1996. Infiltration Drainage manual of Good Practice (R156). Construction Industry Research and Information Association (CIRIA), CIRIA, London, UK, ISBN 0-86017-457-3. Report 156.

Bichai, F., Polo-Lopez, M.I., Ibanez, P.F., 2012. Solar disinfection of wastewater to reduce contamination of lettuce crops by *Escherichia coli* in reclaimed water irrigation. Water Res. 46 (18), 6040–6050.

Blockley, D., 2005. The New Penguin Dictionary of Civil Engineering. Penguin Books, Michigan, USA.

Boar, R.R., Crook, C.E., Moss, B., 1989. Regression of *Phragmites australis* reedswamps and recent changes of water chemistry in the Norfolk Broadland, England. Aquat. Bot. 35 (1), 41–55.

Bockholt, R., Koch, G., Ebert, W., Fedderwitz, E., 1992. Nutrient contents of drainwater and drainage ditchwater on agriculturelly used areas in the drinking water territory of the River Kösterbeck. Zeitschrift für Kulturtechnik und Landentwicklung 33, 178–185 (in German).

Boon, P.I., Mitchell, A., 1995. Methanogenesis in the sediments of an Australian freshwater wetland: comparison with aerobic decay and factors controlling methanogenesis. FEMS Microbiol. Ecol. 18 (3), 174–190.

Boutell, M.R., Luo, J., Shen, X., Brown, C.M., 2004. Learning multi-label scene classification. Pattern Recog. 37 (9), 1757–1771.

Boyden, B.H., Rababah, A.A., 1996. Recycling nutrients from municipal wastewater. Desalination 106 (1–3), 241–246.

Braskerud, B.C., 2002. Factors affecting phosphorous retention in small constructed wetlands treating agricultural non-point source pollution. Ecol. Eng. 1 (19), 41–61.

Braun-Blanquet, J., 1932. Plant sociology. In: Fuller, G.D., Conrad, H.C. (Eds.), The Study of Plant Communities. McGraw-Hill, New York, USA.

Brix, H., 1997. Do macrophytes play a role in constructed treatment wetlands? Water Sci. Technol. 35 (5), 11–17.

Brix, H., 1999. Functions of macrophytes in constructed wetlands. Water Sci. Technol. 29 (4), 71–78.

Brix, H., Arias, C.A., del Bubba, M., 2001. Media selection for sustainable phosphorus reduction in subsurface flow constructed wetlands. Water Sci. Technol. 44 (11–12), 47–54.

Bronstert, A., Baardossy, A., Bismuth, C., Buiteveld, H., Disse, M., Engel, H., Fritsch, U., Hundecha, Y., Lammersen, R., Niehoff, D., Ritter, N., 2007. Multi-scale modelling of land-use change and river training effects on floods in the Rhine basin. River Res. Appl. 23 (10), 1102–1125.

Brown, T.C., Bergstrom, J.C., Loomis, J.B., 2007. Defining, valuing and providing ecosystem services. Nat. Resour. J. 47 (2), 330–376.

Building Research Establishment, 1991. Soakaway Design, Building Research Establishment (BRE) Digest 365, Which Replaced BRE Digest 151. BRE Bookshop, Watford, UK.

Bulc, T., Slak, A.S., 2003. Performance of constructed wetland for highway runoff treatment. Water Sci. Technol. 48 (2), 315–322.

Burt, T.P., 2003. Monitoring change in hydrological systems. Sci. Total Environ. 310 (1–3), 9–16.

Busch, M., Notte, A.L., Laporte, V., Erhard, M., 2012. Potentials of quantitative and qualitative approaches to assessing ecosystem services. Ecol. Indic. 21, 89–103.

Butler, D., Davies, J.W., 2004. Urban Drainage, second ed. Spon Press, London, UK.

Butler, D., Davies, J.W., 2000. Urban Drainage. E & FN Spon, London, UK.

Butler, D., Parkinson, J., 1997. Towards sustainable urban drainage. Water Sci. Technol. 35 (9), 53–63.

Butler, D., Xiao, Y., Karunaratne, S.H.P.G., Thedchananamoorthy, S., 1995. The gully pot as a physical, chemical and biological reactor. Water Sci. Technol. 31 (7), 219–228.

Buss, S.R., Rivett, M.O., Morgan, P., Bemment, C.D., 2005. Attenuation of Nitrate in the Subsurface Environment. Environment Agency, Bristol, UK.

Campbell, C.S., Ogden, M.H., 1999. Constructed Wetlands in the Sustainable Landscape. John Wiley & Sons, New York, USA.

Cao, L.J., Tay, F.E.H., 2003. Support vector machine with adaptive parameters in financial time series forecasting. IEEE Trans. Neural Networks 14 (6), 1506–1518.

Carty, A., Scholz, M., Heal, K., Gouriveau, F., Mustafa, A., 2008. The universal design, operation and maintenance guidelines for farm constructed wetlands (FCW) in temperate climates. Bioresour. Technol. 99 (15), 6780–6792.

Caquet, T., Lagadic, L., Jonot, O., Baturo, W., Kilanda, M., Simon, P., LeBras, S., Echaubard, M., Ramade, F., 1996. Outdoor experimental ponds (mesocosms) designed for long-term ecotoxicological studies in the aquatic environment. Ecotoxicol. Environ. Saf. 34 (2), 125–133.

Carignan, R., Kaill, J., 1980. Phosphorus sources for aquatic weeds: water or sediments? Science 207 (4434), 987–989.

Carpenter, G.A., Markuzon, N., 1998. ARTMAP-IC and medical diagnosis: instance counting and inconsistent cases. Neural Networks 11 (2), 323–336.

Carroll, P., Harrington, R., Keohane, J., Ryder, C., 2005. Water treatment performance and environmental impact of integrated constructed wetlands in the Anne Valley Watershed, Ireland. In: Dunne, E.J., Reddy, K.R., Carton, O.T. (Eds.), Nutrient Management in Agricultural Watersheds: A Wetlands Solution. Wageningen Academic Publishers, Wageningen, USA, pp. 207–217.

Casali, J., Gimenez, R., De Santisteban, L., Alvarez-Mozos, J., Mena, J., Del Valle de Lersundi, J., 2009. Determination of long-term erosion rates in vineyards of Navarre (Spain) using botanical benchmarks. Catena 78 (1), 12–19.

Cedergren, H.R., 1997. Seepage, Drainage, and Flow Nets, third ed. John Wiley and Sons, Inc., New York, USA.

Cereghino, R., Giraudel, J.L., Compin, A., 2001. Spatial analysis of stream invertebrates distribution in the Adour-Garonne drainage basin (France) using Kohonen self-organizing maps. Ecol. Model. 146 (1–3), 167–180.

Chang, C.-C., Lin, C.-J., 2005. LIBSVM: A Library for Support Vector Machines. Software available at http://www.csie.ntu.edu.tw/~cjlin/libsvm/.

Chang, A.C., Olmstead, W.R., Johanson, J.B., Yamashita, G., 1974. The sealing mechanism of wastewater ponds. J. Water Pollut. Control Fed. 46, 1715–1721.

Chary, N., Kamala, C., Raj, D., 2008. Assessing risk of heavy metals from consuming food grown on sewage irrigated soil and food chain transfer. Ecotoxicol. Environ. Saf. 69 (3), 513–524.

Chaudhry, M.H., 1996. Principles of flow of water. In: Mays, L.W. (Ed.), Resources Handbook. McGraw-Hill, New York, USA.

Chen, Y.W., Lin, C.J., 2006. Combining SVMs with various feature selection strategies. In: Guyon, I., Gunn, S., Nikravesh, M., Zadeh, L. (Eds.), Feature Extraction, Foundations and Applications. Springer, London, UK, pp. 315–323.

Chen, S.W., Kao, C.M., Jou, C.R., Fu, Y.T., Chang, Y.I., 2008. Use of a constructed wetland for post-treatment of swine wastewater. Environ. Eng. Sci. 25 (3), 407–417.

Cheng, S.P., Grosse, W., Karrenbrock, F., Thoennessen, M., 2002. Efficiency of constructed wetlands in decontamination of water polluted by heavy metals. Ecol. Eng. 18 (3), 317–325.

Cheng, J., Shearin, T.E., Peet, M.M., Willits, D.H., 2004. Utilization of treated swine wastewater for greenhouse tomato production. Water Sci. Technol. 50 (2), 77–82.

Chin, D.A., 2000. Water Resources Engineering. Prentice Hall, Upper Saddle River, New Jersey, USA.

Chung, E.G., Bombardelli, F.A., Schladow, S.G., 2009. Modeling linkages between sediment resuspension and water quality in a shallow, eutrophic, wind-exposed lake. Ecol. Model. 220 (9–10), 1251–1265.

Ciju, R.J., 2013a. Bell Peppers – Growing Practices and Nutritional Value. Amazon, Ridgmont, England, UK.

Ciju, R.J., 2013b. Chile Peppers. Amazon, Ridgmont, England, UK.

Cirelli, G.L., Consoli, S., Licciardello, F., Aiello, R., Giuffrida, F., Leonardi, C., 2012. Treated municipal wastewater reuse in vegetable production. Agric. Water Manage. 104, 163–170.

CIRIA, 2000. Sustainable Urban Drainage Systems: Design Manual for Scotland and Northern Ireland, Construction Industry Research and Information Association (CIRIA) Report C521. Cromwell Press, London, UK.

CIRIA, 2004. Sustainable Drainage Systems. Hydraulic, Structural and Water Quality Advice, Report C609. Construction Industry Research and Information Association, London, UK.

CIRIA, 2007. The SuDS Manual. Construction Industry Research and Information Association, London.

CIRIA, 2010. Planning for SuDS – Making It Happen. Construction Industry Research and Information Association, London, UK.

CIRIA, 2012. The Benefits of Large Species Trees in Urban Landscapes: A Costing, Design and Management Guide. Construction Industry Research and Information Association, London.

City of Edinburgh Council, 1999. Flood Assessment Report – Edinburgh Flood Assessment Study. Edinburgh City Development Department, Edinburgh, Scotland, UK.

Clare, A., King, R.D., 2001. Knowledge discovery in multi-label phenotype data. In: De Raedt, L., Siebes, A. (Eds.), Proceedings of the 5th European Conference on Principles of Data Mining and Knowledge Discovery. Springer, London, England, UK, pp. 42–53.

Clarke, E., Baldwin, A.H., 2002. Responses of wetland plants to ammonia and water level. Ecol. Eng. 18 (3), 257–264.

Clesceri, L.S., Greenberg, A.E., Eaton, A.D., 1998. Standard Methods for the Examination of Water and Wastewater, twentieth ed. American Public Health Association, American Water Works Association and Water Environment Federation, Washington, D.C., USA.

Colin, F., Puech, C., de Marsily, G., 2000. Relations between triazine flux, catchment topography and distance between maize fields and the drainage network. J. Hydrol. 236 (3–4), 139–152.

Costanza, R., d'Arge, R., de Groot, R., Farber, S., Grasso, M., Hannon, B., Limburg, K., Naeem, S., O'Neill, R.V., Paruelo, J., Raskin, R.G., Sutton, P., van den Belt, M., 1997. The value of the world's ecosystem services and natural capital. Nature 387 (6630), 253–260.

Convention on Wetlands of International Importance Especially as Waterfowl Habitat, 1971. Ramsar, Iran.

Cooper, P.F., Job, G.D., Green, M.B., Shutes, R.B.E., 1996. Reed Beds and Constructed Wetlands for Wastewater Treatment. WRc, Swindon, UK.

Council of European Communities, 2000. Directive of 23 October 2000 establishing a framework for community action in the field of water policy (2000/60/EC). Official J. L327, 0001–0073.

Cronk, J.K., 1996. Constructed wetlands to treat wastewater from dairy and swine operations: a review. Agric. Ecosyst. Environ. 58 (2–3), 97–114.

Cui, L.-H., Luo, S.-M., Zhu, X.-Z., Liu, Y.-H., 2003. Treatment and utilization of septic tank effluent using vertical-flow constructed wetlands and vegetable hydroponics. J. Environ. Sci. 15 (1), 75–82.

Culley, J.L.B., Phillips, P.A., 1982. Sealing of soils by liquid cattle manure. Can. Agric. Eng. 24, 87–89.

Culley, J.L.B., Phillips, P.A., 1989. Retention and loss of nitrogen and solids from unlined earthen manure storages. Trans. ASAE 32 (2), 677–683.

Curds, C.R., 1992. Protozoa in the Water Industry. Cambridge University Press, Cambridge, UK.

Dawson, J.J.C., Soulsby, C., Tetzlaff, D., Hrachowitz, M., Dunn, S.M., Malcolm, I.A., 2008. Influence of hydrology and seasonality on DOC exports from three contrasting upland catchments. Biogeochemistry 90 (1), 93–113.

Danso-Amoako, E., Kalimeris, N., Scholz, M., Yang, Q., Shao, J., 2012. Predicting dam failure risk for sustainable flood retention basins: a generic case study for the wider Greater Manchester area. Comput. Environ. Urb. Syst. 36 (5), 423–433.

D'Arcy, B.J., Frost, A., 2001. The role of best management practices in alleviating water quality problems associated with diffuse pollution. Sci. Total Environ. 265 (1), 359–367.

Davidson, T.E., Trepel, M., Schrautzer, J., 2002. Denitrification in drained and rewetted minerotrophic peat soils in Northern Germany (Pohnsdorfer Stauung). J. Plant Nutr. Soil Sci. 165, 199–204.

Davis, L., 1995. A Handbook of Constructed Wetlands: A Guide to Creating Wetlands for Agricultural Wastewater, Domestic Wastewater, Coal Mine Drainage, and Stormwater in the Mid-Atlantic Region, vol. 3. Agricultural wastewater (United States Department of Agriculture (USDA), Natural Resources Conservation Service, and the Environmental Protection Agency (EPA) Region III in cooperation with the Pennsylvania Department of Environmental Resources, Washington, D.C.

Davis, S., Fairbank, W., Weisheit, H., 1973. Dairy waste ponds effectively self-sealing. Trans. ASAE 16 (6), 69–71.

Debo, T.N., Reese, A.J., 2003. Municipal Stormwater Management, second ed. Lewis Publishers, New York, USA.

DEFRA, 2004. Development and Flood Risk – Joint Report to the Department for Environment, Food and Rural Affairs (DEFRA) and the Office of the Deputy Prime Minister. Environment Agency, London, UK.

DEFRA, 2010. Flood and Water Management Act. Department of Environment, Food and Rural Affairs (Defra). The Stationary Office, London, UK. http://www.legislation.gov.uk/ukpga/2010/29/pdfs/ukpga_20100029_en.pdf.

DEFRA, 2011a. The Natural Choice: Securing the Value for Nature, Natural Environment White Paper, Department of Environment, Food and Rural Affairs (Defra). The Stationary Office, London, England, UK. http://www.official-documents.gov.uk/document/cm80/8082/8082.pdf.

DEFRA, 2011b. Overarching Impact Assessment for the Natural Environment White Paper. UK Department of the Environment, Food and Rural Affairs (Defra), London, UK. http://www.archive.defra.gov.uk/environment/natural/documents/newp-ia-110607.pdf.

Demmy, G.G., Bottcher, A.B., Nordstedt, R.A., 1993. Measurement of Leakage from Dairy Waste Holding Ponds. ASAE paper No. 93–4017. The American Society of Agricultural Engineers, St. Joseph, Michigan, USA.

Department of Food and Rural Affairs, 2009. Adapting to Climate Change, Climate Projections. Department of Food and Rural Affairs, London, England, UK, p. 45. http://www.defra.gov.uk/environment/climatechange/adapt/pdf/uk-climate-projections.pdf.

DeTar, W.R., 1979. Infiltration of liquid dairy manure into soil. Trans. ASAE 22 (3), 520–528.

Deutsches Institut für Normung, 1996a. DIN EN ISO 13395: Water Quality – Determination of Nitrite Nitrogen and Nitrate Nitrogen and the Sum of Both by Flow Analysis (CFA and FIA) and Spectrometric Detection. Deutsches Institut für Normung e.V., Beuth Verlag, Berlin, Germany.

Deutsches Institut für Normung, 1996b. DIN EN 1189: Wasserbeschaffenheit - Bestimmung von Phosphor - Photometrisches Verfahren mittels Ammoniummolybdat. Deutsches Institut für Normung e.V., Beuth Verlag, Berlin, Germany (in German).

Deutsches Institut für Normung, 2003. DIN EN ISO 11732: Wasserbeschaffenheit -Bestimmung von Ammoniumstickstoff -Verfahren mittels Fließanalyse (CFA und FIA) und spektrometrischer Detektion (ISO/DIS 11732:2003). Deutsches Institut für Normung e.V., Beuth Verlag, Berlin, Germany (in German).

Dickey, E.C., Vanderholm, D.H., 1977. Feedlot runoff holding ponds: nutrient levels and related management aspects. J. Environ. Qual. 6 (3), 307–312.

Dobbs, C., Escobedo, F.J., Zipperer, W.C., 2011. A framework for developing urban forest ecosystem services and goods indicators. Landscape Urb. Plann. 99 (3–4), 196–206.

DOEHLG, 2010. Integrated Constructed Wetlands. Guidance Document for Farmyard Soiled Water and Domestic Wastewater Applications. Department of Environment, Heritage and Local Government, Dublin, Ireland.

Dong, X., Reddy, G.B., 2010. Soil bacterial communities in constructed wetlands treated with swine wastewater using PCR-DGGE technique. Biores. Technol. 101 (4), 1175–1182.

Dong, B., Cao, C., Lee, S.E., 2005a. Applying support vector machines to predict building energy consumption in tropical regions. Energy Build. 37 (5), 545–553.

Dong, Y., Wiliński, P.R., Dzakpasu, M., Scholz, M., 2011. Impact of hydraulic loading rate and season on water contaminant reductions within integrated constructed wetlands. Wetland 31 (3), 499–509.

Dong, N., Lu, W.C., Chen, N.Y., Zhu, Y.C., Chen, K.X., 2005b. Using support vector classification for SAR of fentanyl derivatives. Acta Pharmacol. Sin. 26 (1), 107–112.

Donkor, V.A., Häder, D.-P., 1996. Effects of ultraviolet radiation on photosynthetic pigments in some filamentous cyanobacteria. Aquat. Microb. Ecol. 11 (2), 143–149.

Dubey, A., Realff, M.J., Lee, J.H., Bommarius, A.S., 2005. Support vector machines for learning to identify the critical position of a protein. J. Theor. Biol. 234 (3), 351–361.

Dubois, D., Prade, H., 1998. An introduction to fuzzy systems. Clin. Chim. Acta 270 (1), 3–29.

Duch, W., Grudzinski, K., June 1999. Weighting and selection of features. In: Proceedings of Intelligent Information Systems VIII Workshop, Ustron, Poland, pp. 32–36.

Dunne, E.J., Culleton, N., O'Donovan, G., Harrington, R., Daly, K., 2005a. Phosphorous retention and sorption by constructed wetland soils in Southeast Ireland. Water Res. 39 (18), 4355–4362.

Dunne, E.J., Culleton, N., O'Donovan, G., Harrington, R., Olsen, A.E., 2005b. An integrated constructed wetland to treat contaminants and nutrients from diary farmyard dirty water. Ecol. Eng. 24 (3), 221–234.

Dunne, T., Leopold, L.B., 1978. Water in Environmental Planning. W.H. Freeman and Company, New York, USA.

Dzakpasu, M., Scholz, M., Harrington, R., Jordan, S.N., McCarthy, V., 2012. Characterising infiltration and contaminant migration beneath earthern-lined Integrated Constructed Wetlands. Ecol. Eng. 41, 41–51.

Dzakpasu, M., Hofmann, O., Scholz, M., Harrington, R., Jordan, S.N., McCarthy, V., 2011. Nitrogen removal in an integrated constructed wetland treating domestic wastewater. J. Environ. Sci. Health A Tox. Hazard. Subst. Environ. Eng. 46 (7), 742–750.

Ebrazi Bakhshayesh, B., Delkash, M., Scholz, M., 2014. Response of vegetables to cadmium-enriched soil. Water 6 (5), 1246–1256.

EC, March 2001. Setting maximum levels for certain contaminants in food stuffs. Commission Regulation No. 466/2001. Off. J. Eur. Comm. 16, L77/1 (2001).

Edwards, A.C., Withers, P.J.A., 2008. Transport and delivery of suspended solids, nitrogen and phosphorus from various sources to freshwaters in the UK. J. Hydrol. 350 (3–4), 144–153.

EEC, 1991. European economic community (EEC) Council Directive 91/676/EEC of 12 December 1991, concerning the protection of waters against pollution caused by nitrates from agricultural sources. Off. J. L 375, 1–8.

Eigenbrod, F., Bell, V.A., Davies, H.N., Heinemeyer, A., Armsworth, P.R., Gaston, K.J., 2011. The impact of projected increases in urbanization on ecosystem services. Proc. Royal Soc. B Biolog. Sci. 278 (1722), 3201–3208.

Ellis, J.B., 1995. Integrated approaches for achieving sustainable development of urban storm drainage. Water Sci. Technol. 32 (1), 1–6.

Ellis, J.B., D'Arcy, B.J., Chatfield, P.R., 2002. Sustainable urban drainage systems and catchment planning. J. Chart. Inst. Water Environ. Manage. 16 (4), 286–291.

Ellis, J.B., Deutsch, J.-C., Mouchel, J.-M., Scholes, L., Revitt, M.D., 2004. Multicriteria decision approaches to support sustainable drainage options for the treatment of highway and urban runoff. Sci. Total Environ. 334–335, 251–260.

Ellis, J.B., Revitt, M.D., Shutes, R.B.E., Langley, J.M., 1994. The performance of vegetated biofilters for highway runoff control. Sci. Total Environ. 146–147 (5), 543–550.

Ellis, J.B., Shutes, R.B.E., Revitt, M.D., 2003. Constructed Wetlands and Links with Sustainable Drainage Systems. Technical Report P2–159/TR1. Environment Agency, Bristol, UK.

Elisseeff, A., Weston, J., 2002. A kernel method for multi-labelled classification. In: Dietterich, T.G., Becker, S., Ghahramani, Z. (Eds.), Advances in Neural Information Processing Systems 14. MIT Press, Cambridge, Massachusetts, USA, pp. 681–687.

EPA, 1999. Storm Water. Technology Fact Sheet. Wet Detention Pond. 832-F-99–048. United States Environmental Protection Agency (EPA), Office of Water, Washington, D.C., USA.

EPA, 2008. The Provision and Quality of Drinking Water in Ireland – A Report for the Years 2006-2007. Environmental Protection Agency, Wexford, Ireland.

Eriksson, P.G., Weisner, S.E.B., 1996. Functional differences in epiphytic microbial communities in nutrient-rich freshwater ecosystems: an assay of denitrifying capacity. Freshwater Biol. 36, 552–562.

Eriksson, P.G., Weisner, S.E.B., 1999. An experimental study on effects of submersed macrophytes on nitrification and denitrification in ammonium-rich aquatic systems. Limnol. Oceanogr. 44 (8), 1993–1999.

Eriksson, E., Baun, A., Scholes, L., Ledin, A., Ahlman, S., Revitt, M., Noutsopoulos, C., Mikkelsen, P.S., 2007. Selected stormwater priority pollutants – a European perspective. Sci. Total Environ. 383, 41–51.

Essex County Council, 2012. Street Materials Guide – Design and Good Practice. Essex County Council, Chelmsford, UK.

Etherington, J.R., 1983. Wetland Ecology. Edward Arnold, London, UK.

EU, 2007. Directive 2007/60/EC of the European parliament and of the Council of 23 October 2007 on the assessment and management of flood risks. Off. J. Eur. Union (Eur Lex). Ref. L 288, 27–34.

Euliss, N.H., Smith, L.M., Liu, S., Duffy, W.G., Faulkner, S.P., Gleason, R.A., Eckles, S.D., 2011. Integrating estimates of ecosystem services from conservation programs and practices into models for decision makers. Ecol. Appl. 21, 128–134.

European Community, 1991. Council Directive 91/271/EEC of 21 May 1991 Concerning urban waste-water treatment. Off. J. Eur. Comm. L 135, 40–52.

European Community, 2006. Good Agricultural Practice for the Protection of Waters, Regulations. Irish Government, Dublin, Ireland.

European Environment Agency, 1999. Environment in the European Union at the Turn of the Century, Environmental Assessment Report 2. European Environment Agency, Copenhagen, Denmark.

Everall, N.C., Lees, D.R., 1997. The identification and significance of chemicals released from decomposing barley straw during reservoir algal control. Water Res. 31 (3), 614–620.

Exova Health Sciences, 2014. Total petroleum hydrocarbons in waters (with aliphatic/aromatic splitting) method. Exova Health Sciences, Glasgow, Scotland, UK.

Fan, J., Wang, W., Zhang, B., Guo, Y., Ngo, N.H., Guo, W., Zhang, J., Wu, H., 2013. Nitrogen removal in intermittently aerated vertical flow constructed wetlands: impact of influent COD/N ratios. Biores. Technol. 143, 461–466.

FAO, 1994. The State of Food and Agriculture Organization (FAO) of the United Nations. FAO Agriculture Series Number 27, Rome, Italy.

FAO, 2003. Users Manual for Irrigation with Treated Wastewater (Food and Agriculture Organization (FAO) of the United Nations. FAO Regional Office of the Near East, Cairo, Egypt.

FAO/WHO, 2001. Report on the 32nd Session of the Codex Committee on Food Additives and Contaminants. Joint Food and Agriculture Organization (FAO) and World Health Organization (WHO), Geneva, Switzerland.

Faulkner, S.P., Richardson, C.J., 1989. Physical and chemical characteristics of freshwater wetland soils. In: Hammer, D.A. (Ed.), Constructed Wetlands for Wastewater Treatment. Lewis Publishers, Chelsea, MI, USA, pp. 41–72.

Fay, D., Kramers, G., Zhang, C., McGrath, D., Grennan, E., 2007. Soil Geochemical Atlas of Ireland. Teagasc and Environmental Protection Agency, Ireland.

Fennessy, M.S., Brueske, C.C., Mitsch, W.J., 1994. Sediment deposition patterns in restored freshwater wetlands using sediment traps. Ecol. Eng. 3 (4), 409–428.

Ferguson, E., 1992. Engineering and the minds eye. Massachusetts Institute of Technology Press, Cambridge, USA.

Fetter, C.W., 2001. Applied Hydrogeology, fourth ed. Prentice Hall, Upper Saddle River, New Jersey, USA.

Field, R., Sullivan, D., 2003. Wet-weather Flow in the Urban Watershed: Technology and Management. Lewis Publishers, New York, USA.

Foster, G.C., Chiverrell, R.C., Harvey, A.M., Dearing, J.A., Dunsford, H., 2008. Catchment hydrogeomorphological responses to environmental change in the Southern Uplands of Scotland. Holocene 18 (6), 935–950.

Fowler, J., Cohen, L., 1998. Practical Statistics for Field Biology. John Wiley and Sons, West Sussex, UK.

Foy, C., Chaney, R., White, M., 1978. The physiology of metal toxicity in plants. Ann. Rev. Plant Physol. 29, 511–566.

Frede, H.-G., Dabbert, S., 1998. Handbuch zum Gewässerschutz in der Landwirtschaft. Ecomed, Landsberg, Germany (in German).

Fredrickson, L.H., Reid, F.A., 1990. Impacts of hydrologic alteration on management of freshwater wetlands. In: Sweeney, J.M. (Ed.), Management of Dynamic Ecosystems. North Central Section, Wildlife Society, West Lafayette, Indiana, USA, pp. 71–90.

Freeze, R.A., Cherry, J.A., 1979. Groundwater. Prentice Hall, Englewood Cliffs, New Jersey, USA.

Galuzzi, M.R., Pflaum, J.M., 1996. Integrating drainage, water quality, wetlands, and habitat in a planned community development. J. Urban Plann. Dev. 122 (3), 101–108.

Gambrell, R.P., Patrick, Jr., W.H., 1978. Chemical and microbiological properties of anaerobic soils and sediments. In: Hook, D.D., Crawford, R.M.M. (Eds.), Plant Life in Anaerobic Environments. Ann Arbor Science, Ann Arbor, pp. 375–423.

Garcia, H.L., Gonzalez, I.M., 2004. Self-organizing map and clustering for wastewater treatment monitoring. Eng. Appl. Artif. Intel. 17 (3), 215–225.

García-Delgado, C., Eymar, E., Contreras, J.I., Segura, M.L., 2012. Effects of fertigation with purified urban wastewater on soil and pepper plant (*Capsicum annuum* L.) production, fruit quality and pollutant contents. Span. J. Agric. Res. 10 (1), 209–221.

Gächter, R., Meyer, J.S., 1993. The role of micro-organisms in mobilization and fixation of phosphorus in sediments. Hydrobiology 253 (6), 103–121.

Gächter, R., Meyer, J.S., Mares, A., 1988. Contribution of bacteria to release and fixation of phosphorus in lake sediments. Limnol. Oceanogr. 1542–1558, 33 (6), part 2.

Gernaey, K.V., van Loosdrecht, M.C.M., Henze, M., Lind, M., Jørgensen, S.B., 2004. Activated sludge wastewater treatment plant modeling and simulation: state of the art. Environ. Model. Software 19 (9), 763–783.

Gervin, L., Brix, H., 2001. Reduction of nutrients from combined sewer overflows and lake water in a vertical-flow constructed wetland system. Water Sci. Technol. 44 (11–12), 171–176.

Gessner, M.O., 2000. Breakdown and nutrient dynamics of submerged *Phragmites* shoots in the littoral zone of a temperate hardwater lake. Aquat. Bot. 66 (1), 9–20.

Gevrey, M., Rimet, F., Park, Y.S., Giraudel, J.L., Ector, L., Lek, S., 2004. Water quality assessment using diatom assemblages and advanced modelling techniques. Freshwater Biol. 49 (2), 208–220.

Gibert, K., Sànchez-Marrè, M., 2011. Application of machine learning methods to spatial interpolation of environmental variables. Environ. Model. Software 26 (7), 983–985.

Gill, S.E., Handley, J.F., Ennos, A.R., Pauleit, S., 2007. Adapting cities for climate change: the role of the green infrastructure. Built Environ. 33 (1), 115–133.

Gleeson, C., Gray, N., 1997. The Coliform Index and Waterborne Disease. E & FN Spon, London, UK.

Godbole, S., Sarawagi, S., 2004. Discriminative methods for multi-labeled classification. In: Dai, H., Srikant, R., Zhang, C. (Eds.), Proceedings of the 8th Pacific-Asia Conference on Knowledge Discovery and Data Mining. PAKDD, Sydney, Australia, pp. 22–30.

González, F.T., Vallejos, G.G., Silveira, J.H., Franco, C.Q., García, J., Puigagut, J., 2009. Treatment of swine wastewater with subsurface-flow constructed wetlands in Yucatán, Mexico: Influence of plant species and contact time. Water SA 35 (3), 335–342.

Green, M., Friedler, E., Safrai, I., 1998. Enhancing nitrification in vertical-flow constructed wetlands utilizing a passive air pump. Water Res. 32 (12), 3513–3520.

Gregory, S.V., Swanson, F.J., McKee, W.A., Cummins, K.W., 1991. An ecosystem perspective of riparian zones. Bioscience 41 (8), 540–551.

Grieu, S., Traore, A., Polit, M., Colprim, J., 2005. Prediction of parameters characterizing the state of a pollution removal biological process. Eng. Appl. Artif. Intel. 18 (5), 559–573.

Groffman, P.M., Gold, A.J., Addy, K., 2000. Nitrous oxide production in riparian zones and its importance to national emission inventories. Chemosphere Global Change Sci. 2 (3–4), 291–299.

Grohmann, A., Hässelbarth, U., Schwerdtfeger, W., 2002. Die Trinkwasserverordnung, Einführung und Erläuterungen für Wasserversorgungsunternehmen und Überwachungsbehörden, fourth ed. Erich Schmidt Verlag, Berlin, Germany (in German).

de Groot, R.S., Wilson, M.A., Boumans, R.M.J., 2002. A typology for the classification, description and valuation of ecosystem functions, goods and services. Ecol. Econ. 41 (3), 393–408.

Guo, Y., 2001. Hydrologic design of urban flood control detention ponds. J. Hydrol. Eng. ASCE 6 (6), 472–479.

Haifa Chemicals, 2014. Nutritional Recommendations for Pepper. http://www.Haifa-group.com/knowledge_centre/crop_guide/pepper.

Hallin, S., Lindgren, P.E., 1999. PCR detection of genes encoding nitrite reductase in denitrifying bacteria. Appl. Environ. Microbiol. 65 (4), 1652–1657.

Ham, J.M., 2002. Seepage losses from animal waste lagoons: a summary of a four-year investigation in Kansas. Trans. ASAE 45 (4), 983–992.

Ham, J.M., DeSutter, T.M., 1999. Seepage losses and nitrogen export from swine-waste lagoons: a water balance study. J. Environ. Qual. 28 (4), 1090–1099.

Hamed, M.M., Khalafallah, M.G., Hassanien, E.A., 2004. Prediction of wastewater treatment plant performance using artificial neural networks. Environ. Model. Software 19 (10), 919–928.

Harrington, R., McInnes, R., 2009. Integrated Constructed Wetlands (ICW) for livestock wastewater management. Biores. Technol. 100 (22), 5498–5505.

Hathaway, J.M., Cook, M.J., Evans, R.O., 2010. Nutrient removal capability of a constructed wetland receiving groundwater contaminated by swine lagoon seepage. Trans. ASABE 53 (3), 741–749.

Hammer, D.A., 1989. Constructed Wetlands for Wastewater Treatment – Municipal, Industrial and Agricultural. Lewis Publishers, Chelsea, Michigan, USA.

Hammer, D.A., Pullin, B.P., McCaskey, T.A., Eason, J., Payne, V.W.W., 1993. Treating livestock wastewaters with constructed wetlands. In: Moshiri, A. (Ed.), Constructed Wetlands for Water Quality Improvement. CRC Press, Boca Raton, Florida, USA, pp. 343–347.

Hanssen, G., 1842. Das Amt Bordesholm im Herzogtum Holstein: eine statistische Monographie auf historischer Grundlage. Amt Bordesholm, Kiel, Germany (in German).

Harrington, R., Carroll, P., Carty, A.H., Keohane, J., Ryder, C., 2007. Integrated constructed wetlands: concept, design, site evaluation and performance. Int. J. Water 3 (3), 243–256.

Harrington, C., Scholz, M., 2010. Assessment of pre-digested piggery wastewater treatment operations with surface flow Integrated Constructed Wetland systems. Biores. Technol. 101 (20), 7713–7723.

Hawke, C.J., José, P.V., 1996. Reed Bed Management for Commercial and Wildlife Interests. The Royal Society for the Protection of Birds, Sandy, UK.

He, L.S., Liu, H.L., Xi, B.D., Zhu, Y.B., 2006. Enhancing treatment efficiency of swine wastewater by effluent recirculation in vertical-flow constructed wetland. J. Environ. Sci. China 18 (2), 221–226.

Heal, K.A.V., 2001. Manganese and land-use in upland catchments in Scotland. Sci. Total Environ. 265 (1–3), 169–179.

Healy, M.G., Rodgers, M., Mulqueen, J., 2007. Treatment of dairy wastewater using constructed wetlands and intermittent sand filters. Biores. Technol. 98 (12), 2268–2281.

Hedmark, Å., Zhang, L., Scholz, M., Aronsson, P., Elowson, T., 2009. Self-organizing map analysis of planted soil infiltration systems for treatment of log yard runoff. For. Sci. 55 (2), 183–188.

Henkens, P.L.C.M., van Keulen, H., 2001. Mineral policy in the Netherlands and nitrate policy within the European Community. Neth. J. Agric. Sci. 49 (2–3), 117–134.

Heimburg, K., 1984. Hydrology of north-central Florida cypress domes. In: Ewel, K.C., Odum, H.T. (Eds.), Cypress Swamps. University Presses of Florida, Gainesville, Florida, USA.

Herschy, R.W., 1998. Hydrometry – Principles and Practices, second ed. John Wiley and Sons, Chichester, UK.

Hibberd, B.G., 1989. Urban Forestry Practice, Forestry Commission Handbook 5. Her Majesty's Stationery Office Publications Centre, London.

Hills, D.J., 1976. Infiltration characteristics from anaerobic lagoons. J. Water Pollut. Control Fed. 48 (4), 695–709.

Holdren, G.C., Armstrong, D.E., 1980. Factors affecting phosphorus release from intact lake sediment cores. Environ. Sci. Technol. 14 (1), 79–87.

Holt, J.E., 1997. The designer's judgement. Des. Stud. 18 (1), 113–123.

Hong, Y.-S.T., Rosen, M.R., Bhamidimarri, R., 2003. Analysis of a municipal wastewater treatment plant using a neural network-based pattern analysis. Water Res. 37 (7), 1608–1618.

Hua, G.F., Zhu, W., Zhao, L.F., Huang, J.Y., 2010. Clogging pattern in vertical-flow constructed wetlands: insight from a laboratory study. J. Hazard. Mater. 180 (1–3), 668–674.

Hubbard, R.K., Gascho, G.J., Newton, G.L., 2004. Use of floating vegetation to remove nutrients from swine lagoon wastewater. Trans. Am. Soc. Agric. Eng. 47 (6), 1963–1972.

Humenik, F.J., Szogi, A.A., Hunt, P.G., Broome, S., Rice, M., 1999. Wastewater utilization: a place for managed wetlands – Review. Asian Australas J. Anim. Sci. 12 (4), 629–632.

Hunt, P.G., Matheny, T.A., Stone, K.C., 2004. Denitrification in a coastal plain riparian zone contiguous to a heavily loaded swine wastewater spray field. J. Environ. Qual. 33 (6), 2367–2374.

Hunt, P.G., Szogi, A.A., Humenik, F.J., Rice, J.M., Matheny, T.A., Stone, K.C., 2002. Constructed wetlands for treatment of swine wastewater from an anaerobic lagoon. Trans. Am. Soc. Agric. Eng. 45 (3), 639–647.

Hyndman, R.J., Koehler, A.B., 5 May, 2005. In: Another Look at Measures of Forecast Accuracy, Monash Econometrics and Business Statistics Working Papers.

Ice, G., 2004. History of innovative best management practice development and its role in addressing water quality limited waterbodies. J. Envir. Eng. 130 (6), 684–689.

International Institute for Land Reclamation and Improvement, 1964. Code of Practice for the Design of Open Watercourses and Ancillary Structures. Royal van Gorcum, Assen, The Netherlands.

IPCC, 2007. Contribution of working group II to the Fourth assessment report of the Intergovernmental Panel on climate change. In: Parry, M.L., Canziani, O.F., Palutikof, J.P., van der Linden, P.J., Hanson, C.E. (Eds.), Climate Change 2007: Impacts, Adaptation and Vulnerability. Cambridge University Press, Cambridge, UK.

Irish Phosphorus Regulations, 1998. Water Quality Standards for Phosphorus, Local Government (Water Pollution) Act 1977. S.I. No. 258 1998. Irish Government, Dublin, Ireland.

Irish Urban Waste Water Treatment Regulations, 2001. Urban Waste Water Treatment Regulations. S.I. No. 252 2001. Irish Government, Dublin, Ireland.

IWA, 2000. Constructed Wetlands for Pollution Control, International Water Association (IWA) Specialist Group 'Use of Macrophytes in Water Pollution Control'. IWA Publishing, London, UK.

Jefferies, C., Aitken, A., McLean, N., MacDonald, K., McKissock, G., 1999. Assessing the performance of urban BMPs in Scotland. Water Sci. Technol. 39 (12), 123–131.

Jensen, K., Granke, O., Hoppe, B., Kieckbusch, J., Trepel, M., Leiner, U., 2001. Weidelandschaft Eidertal – Naturschutz durch extensive Beweidung und Wiedervernässung. Petermanns Geographische Mitteilungen 145 (1), 38–49 (in German).

Jim, C.J., 1996. Roadside trees in urban Hong Kong: part II species composition. Arboricult. J. 20 (3), 279–298.

Jiménez, B., Asano, T., 2008. Water reclamation and reuse around the World. In: Jiménez, B., Asano, T. (Eds.), Water Reuse: An International Survey of Current Practice, Issues and Needs. International Water Association Publishing, London, UK, pp. 3–26.

Joachims, T., 1998. Text Categorization with Support Vector Machines: Learning with Many Relevant Features. In: Proc. Eur. Conf. Mach. Learn. Springer, Berlin, Germany 1398 (7), 137–142.

Joachims, T., 1999. Making large-scale SVM learning practical. In: Scholkopf, B., Burges, C., Smola, A. (Eds.), Advances in Kernel Methods – Support Vector Learning. MIT-Press, Cambridge, USA. http://svmlight.joachims.org.

Johnson, C., Decoteau, D., 1996. Nitrogen and potassium fertility affect jalapeno pepper plant growth, pod yield, and pungency. Hortscience 31 (7), 1119–1123.

Jones Jr., J.B., 2013. Instructions for Growing Tomatoes in the Garden and Greenhouse. Gro-Systems, Anderson, South Carolina, USA.

Junk, W.J., Bayley, P.B., Sparks, R.E., 1989. The food pulse concept in river-foodplain systems. In: Dodge, D.P. (Ed.), Proceedings of the International Large River Symposium, Canadian Journal of Fisheries and Aquatic Sciences, 106 (special issue), 11–127

Jurdo, G.B., Johnson, J., Feeley, H., Harrington, R., Kelly-Quinn, M., 2010. The potential of integrated constructed wetlands (ICWs) to enhance macroinvertebrate diversity in agricultural landscapes. Wetlands 30 (3), 393–404.

Jury, W.A., Horton, R., 2004. Soil Physics, sixth ed. John Wiley and Sons, Inc., Hoboken, New Jersey, USA.

Kadlec, R.H., 1999. Chemical, physical and biological cycles in treatment wetlands. Water Sci. Technol. 40 (2), 37–44.

Kadlec, R.H., 2002. Effects of Pollutant Speciation in Treatment Wetlands Design. Wetland Management Services, Chelsea, Michigan, USA.

Kadlec, R.H., Knight, R.L., 1996. Treatment Wetlands. CRC Press, Boca Raton, Florida, USA.

Kadlec, R., Knight, R.L., Vymazal, J., Brix, H., Cooper, P.F., Haberl, R., 2000. Constructed Wetlands for Pollution Control, International Water Association (IWA) Specialist Group 'Use of Macrophytes for Water Pollution Control,' Scientific and Technical Report Number 8. IWA Publishing, London, UK.

Kalavrouziotis, I.K., 2011. Basic principles of treated wastewater reuse planning in ecologically sensitive areas. Water Air Soil Pollut. 221 (1), 159–168.

Kalavrouziotis, I.K., Drakatos, P.A., 2001. The future of irrigation by using processed wastewater in Greece. J. Environ. Waste Manage. 4 (2), 107–110.

Kalavrouziotis, I.K., Arambatzis, C., Kalfountzos, D., Varnavas, S.P., 2011. Wastewater reuse planning in agriculture: the case of Aitoloakarnania, Western Greece. Water 3 (4), 988–1004.

Kalavrouziotis, I.K., Robolas, P., Koukoulakis, P.H., Papadopoulos, A.H., 2008. Effects of municipal reclaimed wastewater on the macro and microelements status of soil and of *Brassica oleracea* var. *italica*, and *Brassica oleracea* var. *gemmifera*. Agric. Water Manage. 95 (4), 419–426.

Kalavrouziotis, I.K., Kokkinos, P., Oron, G., Fatone, F., Bolzonella, D., Vatyliotou, M., Fatta-Kassinos, D., Koukoulakis, P.H., Varnavas, S.P., 2015. Current status in wastewater treatment, reuse and research in some Mediterranean countries. Desalin. Water Treat. 53 (8), 2015–2030.

Kamp-Nielson, L., 1974. Mud-water exchange of phosphate and other ions in undisturbed sediment cores and factors affecting exchange rates. Arch. Hydrobiol. 73 (2), 218–237.

Kantawanichkul, S., Somprasert, S., 2005. Using a compact combined constructed wetland system to treat agricultural wastewater with high nitrogen. Water Sci. Technol. 51 (9), 47–53.

Kantawanichkul, S., Neamkam, P., Shutes, R.B.E., 2001. Nitrogen removal in a combined system: vertical vegetated bed over horizontal sand bed. Water Sci. Technol. 44 (11–12), 137–142.

Karathanasis, A.D., Potter, C.L., Coyne, M.S., 2003. Vegetation effects on fecal bacteria, biochemical oxygen demand and suspended solids removal in constructed wetlands treating domestic wastewater. Ecol. Eng. 20 (2), 157–169.

Kaster, D.S., Medeirosand, C.B., Rocha, H.V., 2005. Supporting modeling and problem solving from precedent experiences: the role of workflows and case-based reasoning. Environ. Model. Software 20 (6), 689–704.

Kaufman, L., Rousseeuw, P.J., 1990. Finding Groups in Data – an Introduction to Cluster Analysis. John Wiley and Sons, Hoboken, New York, USA.

Kayranli, B., Scholz, M., Mustafa, A., Hofmann, O., Harrington, R., 2010. Performance evaluation of integrated constructed wetlands treating domestic wastewater. Water Air Soil Pollut. 210 (1–4), 435–451.

Keffala, C., Ghrabi, A., 2005. Nitrogen and bacterial removal in constructed wetlands treating domestic waste water. Desalination 185 (1–3), 383–389.

Kemp, M.C., George, D.B., 1997. Subsurface flow constructed wetlands treating municipal wastewater for nitrogen transformation and removal. Water Environ. Res. 69 (7), 1254–1262.

Kendrick, M., 1988. The Thames barrier. Landscape Urb. Plann. 16, 57–68.

Kiely, G., 1997. Environmental Engineering. McGraw-Hill International, Maidenhead, UK.

Klove, B., 2001. Characteristics of nitrogen and phosphorus loads in peat mining wastewater. Water Res. 35 (10), 2353–2362.

Knight, R.L., Kadlec, R.H., Ohlendorf, H.M., 1999. The use of treatment wetland for petroleum industry effluents. Environ. Sci. Technol. 33 (7), 973–980.

Knight, R.L., Payne, V.W.E., Borer, R.E., Clarke, R.A., Pries, J.H., 2000. Constructed wetlands for livestock wastewater management. Ecol. Eng. 15 (1–2), 41–55.

Knowles, P.R., Dotro, G., Nivala, J., Garcia, J., 2011. Clogging in subsurface-flow treatment wetlands: occurrence and contributing factors. Ecol. Eng. 37 (2), 99–112.

Kohavi, R., 2005. A Study of cross-validation and bootstrap for accuracy estimation and model selection. In: Proceedings of the 14th International Joint Conference on Artificial Intelligence, pp. 1137–1143.

Kohonen, T., 1988. An introduction to neural computing. Neural Networks 1 (1), 3–16.

Kohonen, T., 2001. Self-organizing Maps, third ed. Springer Verlag, Berlin, Germany.

Korboulewsky, N., Wang, R., Baldy, V., 2012. Purification processes involved in sludge treatment by a vertical flow wetland system: Focus on the role of the substrate and plants on N and P removal. Biores. Technol. 105, 9–14.

Kouki, S., M'hiri, F., Saidi, N., Belaïd, S., Hassen, A., 2009. Performances of a constructed wetland treating domestic wastewaters during a macrophytes life cycle. Desalination 246 (1–3), 252–267.

Kuehn, K.A., Gessner, M.O., Wetzel, R.G., Suberkropp, K., 1999. Standing litter decomposition of the emergent macrophyte *Erianthus giganteus*. Microb. Ecol. 38 (1), 50–57.

Kuehn, E., Moore, J.A., 1995. Variability of treatment performance in constructed wetlands. Water Sci. Technol. 32 (3), 241–250.

Kuhn, G., Diekmann, B., 2003. Data report: bulk sediment composition, grain-size, clay, and silt mineralogy of Pleistocene sediments from ODP Leg 177 Sites 1089 and 1090. Proc. ODP Sci. Results 177, 1–10.

Kumm, K.I., 2003. Ways to reduce nitrogen pollution from Swedish pork production. Nutri. Cycl. Agroecosyst. 66 (3), 285–293.

Kuschk, P., Wiessner, A., Kappelmeyer, U., Weissbrodt, E., Kastner, M., Stottmeister, U., 2003. Annual cycle of nitrogen removal by a pilot-scale subsurface horizontal flow in a constructed wetland under moderate climate. Water Res. 37 (17), 4236–4242.

Kvet, J., Husak, S., 1978. Primary data on biomass and production estimates in typical stands of fish pond littoral plant communities. In: Dykyjová, D., Kvet, J. (Eds.), Pond Littoral Ecosystems. Springer Verlag, Berlin, Germany, pp. 211–216.

Larinier, M., 2000. Dams and Fish Migration, World Commission on Dams. FAO Fisheries Technical Paper 419. Food and Agriculture Organization of the United Nations, Rome, Italy, 45–89.

Larson, A.C., Gentry, L.E., David, M.B., Cooke, R.A., Kovacic, D.A., 2000. The role of seepage in constructed wetlands receiving agricultural tile drainage. Ecol. Eng. 15 (1), 91–104.

Lawton, J.H., Brotherton, P.N.M., Brown, V.K., Elphick, C., Fitter, A.H., Forshaw, J., Haddow, R.W., Hilborne, S., Leafe, R.N., Mace, G.M., Southgate, M.P., Sutherland, W.L., Tew, T.E., Varley, J., Wynne, G.R., 2010. Making Space for Nature: A Review of England's Wildlife Sites and Ecological Network, Report. UK Department of the Environment, Food and Rural Affairs, London, UK. http://archive.defra.gov.uk/environment/biodiversity/documents/201009space-for-nature.pdf.

Langergraber, G., 2007. Simulation of the treatment performance of outdoor subsurface flow constructed wetlands in temperate climates. Sci. Total Environ. 380 (1–3), 210–219.

Langergraber, G., Giraldi, D., Mena, J., Meyer, D., Pena, M., Toscano, A., Brovelli, A., Korkusuz, E.A., 2009. Recent developments in numerical modelling of subsurface flow constructed wetlands. Sci. Total Environ. 407, 3931–3943.

Länderarbeitsgemeinschaft Wasser, 1998. Beurteilung der Wasserbeschaffenheit von Fließgewässern in der Bundesrepublik Deutschland – Chemische Gewässergüteklassifikation. Kulturbuchverlag, Berlin, Germany (in German).

Lee, K., Booth, D., Alam, P., 2005a. A comparison of supervised and unsupervised neural networks in predicting bankruptcy of Korean firms. Expert Syst. Appl. 29 (1), 1–16.

Lee, C.Y., Lee, C.C., Lee, F.Y., Tseng, S.K., Liao, C.J., 2004. Performance of subsurface flow constructed wetland taking pretreated swine effluent under heavy loads. Biores. Technol. 92 (2), 173–179.

Lee, B.-H., Scholz, M., Horn, A., Furber, A., 2005b. Constructed wetlands: prediction of performance with case-based reasoning (Part B). Environ. Eng. Sci. 23 (2), 203–211.

Leflaive, J., Cereghino, R., Danger, M., Lacroix, G., Ten-Hage, L., 2005. Assessment of self-organizing maps to analyze sole-carbon source utilization profiles. J. Microbiol. Methods 62 (1), 89–102.

Li, J., Heap, A.D., Potter, A., Daniell, J.J., 2011. A new library to combine artificial neural networks and support vector machines with statistics and a database engine for application in environmental modelling. Environ. Model. Software 26 (12), 1647–1659.

Li, T., Ogihara, M., 2006. Toward intelligent music information retrieval. IEEE T. Multimedia 8 (3), 564–574.

Li, Y., Shawe-Taylor, J., 2003. The SVM with uneven margins and Chinese document categorization. In: Proceedings of the 17th Pacific Asia Conference on Language, Information and Computation. PACLIC17, Singapore, pp. 216–227.

Lim, P.E., Tay, M.G., Mak, K.Y., Mohamed, N., 2003. The effect of heavy metals on nitrogen and oxygen demand reduction in constructed wetlands. Sci. Total Environ. 301 (1–3), 13–21.

Liu, H.-X., Zhang, R.-S., Yao, X.-H., Liu, M.-C., Hu, Z.-D., Fan, B.-T., 2004. Prediction of electrophoretic mobility of substituted aromatic acids in different aqueous-alcoholic solvents by capillary zone electrophoresis based on support vector machine. Anal. Chim. Acta 525 (1), 31–41.

Lopez, A., Pollice, A., Lonigro, A., Masi, S., Palese, A.M., Cirelli, G.L., Toscano, A., Passino, R., 2006. Integrated concepts in water recycling. Desalination 187 (1–3), 323–334.

Loures, L., Panagopoulos, T., 2007. From derelict industrial areas towards multifunctional landscapes and urban renaissance. WSEAS Trans. Environ. Dev. 3 (10), 181–188.

Lowrance, R., Altier, L.S., Newbold, J.D., Schnabel, R.R., Groffman, P.M., Denver, J.M., Correll, D.L., Gilliam, J.W., Robinson, J.L., Brinsfield, R.B., Staver, K.W., Lucas, W., Todd, A.H., 1979. Water quality functions of riparian forest buffers in Chesapeake Bay watersheds. Environ. Manage. 21 (5), 687–712.

Lu, R.-S., Lo, S.-L., 2004. Diagnosing reservoir water quality using self-organizing maps and fuzzy theory. Water Res. 36 (9), 2265–2274.

Lu, W.-Z., Wang, W.-J., 2005. Potential assessment of the support vector machine method in forecasting ambient air pollutant trends. Chemosphere 59 (5), 693–701.

Lu, S., Zhang, P., Jin, X., Xiang, C., Gui, M., Zhang, J., Li, F., 2009. Nitrogen removal from agricultural runoff by full-scale constructed wetland in China. Hydrobiologia 621 (1), 115–126.

Luederits, V., Eckert, E., Lange-Weber, M., Lange, A., 2001. Nutrient removal efficiency and resource economics of vertical-flow and horizontal-flow constructed wetlands. Ecol. Eng. 18 (2), 157–171.

Lund, L.J., Horne, A.J., Williams, A.E., 2000. Estimating denitrification in a large constructed wetland using stable nitrogen isotope ratios. Ecol. Eng. 14 (1–2), 67–76.

Lundy, L., Wade, R., 2011. Integrating sciences to sustain urban ecosystem services. Prog. Phys. Geogr. 35 (5), 653–669.

Mander, Ü., Kuusemets, V., Ivask, M., 1995. Nutrient dynamics of riparian ecotones: a case study from the Porijogi River catchment, Estonia. Landscape Urb. Plann. 31 (1–3), 333–348.

Maier, H.R., Morgan, N., Chow, C.W.K., 2004. Use of artificial neural networks for predicting optimal alum doses and treated water quality parameters. Environ. Model. Software 19 (5), 485–494.

Mandernack, K.W., Lynch, L., Krouse, H.R., Morgan, M.D., 2000. Sulfur cycling in wetland peat of the New Jersey Pinelands and its effect on stream water chemistry. Geochimica et Cosmochimica Acta 22 (5), 3949–3964.

Mark, O., Weesakul, S., Apirumanekul, C., Aroonnet, S.B., Djordjevic, S., 2004. Potential and limitations of 1D modelling of urban flooding. J. Hydrobiol. 299 (3–4), 284–299.

Martin, E.C., Pegelow, E.J., Watson, J., 1999. Using drainage lysimeters to evaluate irrigation and nitrogen interactions in cotton production. In: Silvertooth, J. (Ed.), Cotton: A College of Agriculture Report. College of Agriculture, The University of Arizona, Tucson, Arizona, USA, pp. 204–210.

Martin, P., Turner, B., Waddington, K., Dell, J., Pratt, C., Campbell, N., Payne, J., Reed, B., 2000. Sustainable Urban Drainage Systems – Design Manual for England and Wales. Construction Industry Research and Information Association (CIRIA), London, UK. CIRIA Report Number C522.

Martinez, J., Dabert, P., Barrington, S., Burton, C., 2009. Livestock waste treatment systems for environmental quality, food safety, and sustainability. Biores. Technol. 100 (22), 5527–5536.

Maulé, C.P., Fonstad, T.A., 1996. Ion Movement Immediately beneath Earthen Hog Manure Storages: How Much, How Deep, How Fast. ASAE paper No. 96–2049. The American Society of Agricultural Engineers, St. Joseph, Michigan, USA.

Maulé, C.P., Fonstad, T.A., Vanpalli, S.K., Majumdar, G., 2000. Hydraulic conductivity reduction due to ponded hog manure. Can. Agric. Eng. 42 (4), 157–163.

Mayer, T., Kramer, J.R., 1986. Effect of lake acidification on the adsorption of phosphorus by sediments. Water Air Soil Pollut. 31 (3–4), 949–958.

Mays, P.A., Edwards, G.S., 2001. Comparison of heavy metal accumulation in a natural wetland and constructed wetlands receiving acid mine drainage. Ecol. Eng. 16 (4), 487–500.

McCallum, A.K., 1999. Multi-label Text Classification with a Mixture Model Trained by EM. In: Proceedings of the AAAI' 99 Workshop on Text Learning. AAAI, Orlando, Florida, USA, pp. 1–7.

McMinn, W.R., Yang, Q., Scholz, M., 2010. Classification and assessment of water bodies as adaptive structural measures for flood risk management planning. J. Environ. Manage. 91 (9), 1855–1863.

McNaughton, S.J., 1966. Ecotype function in the *Typha* community-type. Ecol. Monogr. 36 (4), 297–325.

McNeill, A., Olley, S., 1998. The effects of motorway runoff on watercourses in south-west Scotland. J. Chart. Inst. Water Environ. Manage. 12 (6), 433–439.

McPherson, E.G., Simpson, J.R., Peper, P.J., Qingfu, X., 1999. Benefit-cost analysis of Modesto's municipal urban forests. J. Aboricult. 25 (5), 235–248.

Meers, E., Tack, F.M.G., Tolpe, I., Michels, E., 2008. Application of a full-scale constructed wetland for tertiary treatment of piggery manure: monitoring results. Water Air Soil Pollut. 193 (1–4), 15–24.

Menegaki, A.N., Mellon, R.C., Vrentzou, A., Koumakis, G., Tsagarakis, K.P., 2009. What's in a name: framing treated wastewater as recycled water increases willingness to use and willingness to pay. J. Econ. Psychol. 30 (3), 285–292.

Merlin, G., Pajean, J.L., Lissolo, T., 2002. Performances of constructed wetlands for municipal wastewater treatment in rural mountainous area. Hydrobiologia 469 (1–3), 87–98.

Memon, F.A., Butler, D., 2002. Assessment of gully pot management strategies for runoff quality control using a dynamic model. Sci. Total Environ. 295 (1–3), 115–129.

Met Éireann, 2010. Various Data Sheets. http://www.met.ie/climate/climate-data-information.asp.

Meteorological Office, 2002. Bradford Monthly Long-term Average Rainfall. Online data between 1908 and 2001. http://www.metoffice.gov.uk/climate/uk/stationdata/bradforddata.txt.

Meteorological Office, 2007. Fact Sheet 4. Climate of the United Kingdom. UK Meteorological Office, Exeter, England, UK, p. 17.

Millennium Ecosystem Assessment, 2005. Ecosystems and Human Well-being. Island Press, Washington, D.C., USA.

Miller, M.H., Robinson, J.B., Gillham, R.W., 1985. Self-sealing of earthen liquid manure storage ponds: I. A case study. J. Environ. Qual. 14 (4), 533–538.

Ministry of Health of the People's Republic of China, 2005. The Maximum Levels of Contaminants in Foods. Ministry of Health of the People's Republic of China, Beijing, China (in Mandarin).

Minitab Inc, 2003. Minitab Statistical Software, Release 14 for Windows. State College, Pennsylvania, USA.

Misstear, B.D.R., Daly, D., 2000. Groundwater in the Celtic regions: studies in hard rock and quaternary hydrogeology. Geol. Soc. Spec. Publ. 182, 53–65.

Mitchell, G., 2005. Mapping hazard from urban non-point pollution: a screening model to support sustainable urban drainage planning. J. Environ. Manage. 74 (1), 1–9.

Mitchell, C., McNevin, D., 2001. Alternative analysis of BOD removal in subsurface flow constructed wetlands employing Monod kinetics. Water Res. 35 (5), 1295–1303.

Mitsch, W.J., Gosselink, J.G., 2000. Wetlands, third ed. John Wiley & Sons, New York, USA.

Mohamed, M.A.A., Terao, H., Suzuki, R., Babiker, I.S., Ohta, K., Kaori, K., Kato, K., 2003. Natural denitrification in the Kakamigahara groundwater basin, Gifu prefecture, central Japan. Sci. Total Environ. 307 (1–3), 191–201.

Montaldo, N., Mancini, M., Rosso, R., 2004. Flood hydrograph attenuation induced by a reservoir system: analysis with a distributed rainfall-runoff model. Hydrology Proc. 18 (3), 545–563.

Moore, T.L.C., Hunt, W.F., 2012. Ecosystem service provision by stormwater wetlands and ponds – a means for evaluation? Water Res. 46 (20), 6811–6823.

Moore, P.A., Reddy, K.R., 1994. Role of Eh and pH on phosphorus geochemistry in sediments of Lake Okeechobee, Florida. J. Environ. Qual. 23 (5), 955–964.

Morari, F., Giardini, L., 2009. Municipal wastewater treatment with vertical flow constructed wetlands for irrigation reuse. Ecol. Eng. 35 (5), 643–653.

Morgan, M.D., Mandernack, K.W., 1996. Biogeochemistry of sulfur in wetland peat following 3.5 Y of artificial acidification (HUMEX). Environ. Intern. 22 (5), 605–610.

Moss, B., 1998. Ecology of Fresh Waters, third ed. Blackwell Science, Oxford, UK.

Mungur, A.S., Shutes, R.B.E., Revitt, D.M., House, M.A., 1997. An assessment of metal reduction by laboratory-scale wetlands. Water Sci. Technol. 35 (5), 125–133.

Munoz-Pedreros, A., 2004. Landscape evaluation: an environmental management. Revista Chilena de Historia Natural 77, 139–156.

Mustafa, A., Scholz, M., Harrington, R., Carroll, P., 2009. Long-term performance of a representative integrated constructed wetland treating farmyard runoff. Ecol. Eng. 35 (5), 779–790.

Nisbet, T.R., Welch, D., Doughty, R., 2002. The role of forest management in controlling diffuse pollution from the afforestation and clearfelling of two public water supply catchments in Argyll, West Scotland. For. Ecol. Manage. 158 (1–3), 141–154.

Nairn, R.W., Mitsch, W.J., 2000. Phosphorus removal in created wetland ponds receiving river overflow. Ecol. Eng. 14 (1–2), 107–126.

Newman, J.M., Clausen, J.C., Neafsey, J.A., 2000. Seasonal performance of a wetland constructed to process dairy milkhouse wastewater in Connecticut. Ecol. Eng. 14 (1–2), 181–198.

Nickels, J., 2012. Growing Chillies – a Guide to the Domestic Cultivation of Chilli Plants. Jason Nickels, UK.

Norrstrom, A.C., Jacks, G., 1998. Concentration and fractionation of heavy metals in roadside soils receiving de-icing salts. Sci. Total Environ. 218 (2–3), 161–174.

Nowak, D.J., Hoehn, R.E., Crane, D.E., Stevens, J.C., Fisher, C.L., 2010. Assessing Urban Forest Effects and Values, Chicago's Urban Forest, Resource Bulletin NRS-37. United States Department of Agriculture, Forest Service, Pennsylvania, USA.

Nuttal, P.M., Boom, A.G., Rowell, M.R., 1997. Review of the Design and Management of Constructed Wetlands. Construction Industry Research and Information Association (CIRIA). CIRIA Report 180. CIRIA, London, UK.

Nygaard, B., Ejrnæs, R., 2009. The impact of hydrology and nutrients on species composition and richness: evidence from a microcosm experiment. Wetlands 29 (1), 187–195.

Obarska-Pempkowiak, H., Klimkowska, K., 1999. Distribution of nutrients and heavy metals in a constructed wetland system. Chemosphere 39 (2), 303–312.

O'Brien, C.A., Scholz, M., McConnachie, G.L., 2005. Membrane bioreactors and constructed wetlands for treatment of rendering plant wastewater. Water Environ. J. 19 (3), 189–198.

O'Connor, M., 1989. Models of human behaviour and confidence in judgment: a review. Int. J. Forecast. 5 (2), 159–169.

Onkal-Engin, G., Demir, I., Engin, S.N., 2005. Determination of the relationship between sewage odor and BOD by neural networks. Environ. Model. Software 20 (7), 843–850.

Ostendorp, W., 1989. 'Die-back' of reeds in Europe – a critical review of literature. Aquat. Bot. 35 (1), 5–26.

Overbeck, J., 1988. Qualitative and quantitative assessment of the problem. In: Jørgensen, S.E., Vollenweider, R.A. (Eds.), Guidelines for Lake Management: Principles of Lake Management, vol. 1. International Lake Environment Committee, United Nations Environment Programme, pp. 19–36.

Pai, P.-F., Hong, W.-C., 2005. Support vector machines with simulated annealing algorithms in electricity load forecasting. Energy Convers. Manage. 46 (17), 2669–2688.

Pant, H.K., Reddy, K.R., Lemon, E., 2001. Phosphorus retention capacity of root bed media of subsurface flow constructed wetlands. Ecol. Eng. 17 (4), 345–355.

Park, S.J., Yoon, T.I., 2003. Weighted coagulation with glass and diatomite for stormwater treatment and sludge disposal. Environ. Eng. Sci. 20 (4), 307–317.

Parkin, T.B., 1987. Soil microsites as a source of soil denitrification variability. Soil Sci. Soc. Am. J. 51 (5), 1194–1199.

Parr, T.W., 1990. Factors affecting reed (*Phragmites australis*) growth in UK reed bed treatment systems. In: Cooper, P.F., Findlater, B.C. (Eds.), Constructed Wetlands in Water Pollution Control. Pergamon Press, Oxford, UK, pp. 67–76.

Pauleit, S., Jones, N., Garcia-Martin, G., Garciavaldecantos, J.L., Rivère, L.M., Vidal-Beaudet, L., Bodson, M., Randrup, T.B., 2002. Tree establishment practice in towns and cities – results from a European survey. Urb. For. Urb. Green 1 (2), 83–86.

Pedrero, F., Kalavrouziotis, I.K., José Alarcón, J., Koukoulakis, P., Asano, T., 2009. Use of treated municipal wastewater in irrigated agriculture – review of the practices in Spain and Greece. Agric. Water Manage. 97 (9), 1233–1241.

Perkin Elmer, 1982. Perkin Elmer 1100 Series Atomic Absorption Spectrophotometer. Perkin Elmer, London, UK.

Perkins, J., Hunter, C., 2000. Removal of enteric bacteria in a surface flow constructed wetland in Yorkshire, England. Water Res. 34 (6), 1941–1947.

Perschke, S.P., Wright, P.E., 1989. Evaluation of Soil under an Earthen Manure Storage Pond. ASAE paper No. 98–2145. The American Society of Agricultural Engineers, St. Joseph, Michigan, USA.

Pescod, M.B., 1992. Wastewater Treatment and Use in Agriculture, FAO Irrigation and Drainage Paper Number 47. http://www.fao.org/docrep/T0551E/T0551E00.htm.

Pettersson, T.J.R., 1997. FEM-modelling of open stormwater detention ponds. Nord. Hydrol. 28 (4–5), 339–350.

Picard, C.R., Fraser, L.H., Steer, D., 2005. The interacting effects of temperature and plant community type on nutrient removal in wetland microcosms. Biores. Technol. 96 (9), 1039–1047.

Pinney, M.L., Westerhoff, P.K., Baker, L., 2000. Transformations in dissolved organic carbon through constructed wetlands. Water Res. 34 (6), 1897–1911.

Poach, M.E., Hunt, P.G., Reddy, G.B., Stone, K.C., Matheny, T.A., Johnson, M.H., Sadler, E.J., 2004. Ammonia volatilization from marsh-pond-marsh constructed wetland treating swine wastewater. J. Environ. Qual. 33 (3), 844–851.

Poach, M.E., Hunt, P.G., Vanotti, M.B., Stone, K.C., Matheny, T.A., Johnson, M.H., Sadler, E.J., 2003. Improved nitrogen treatment by constructed wetlands receiving partially nitrified liquid swine manure. Ecol. Eng. 20 (2), 183–197.

Pratap, R., 2002. Getting Started with MATLAB: A Quick Introduction for Scientists and Engineers. Oxford University Press, Oxford, England, UK.

Preston, C.D., Croft, J.M., 1997. Aquatic plants in Britain and Ireland. Harley Books, Colchester, England.

Qi, G.J., Hua, X.S., Rui, Y., Tang, J., Mei, T., Zhang, H.J., 2007. Correlative multi-label video annotation. In: Proceedings of the 15th International Conference on Multimedia. ACM Press, New York, USA, pp. 17–26.

Reddy, G.B., Hunt, P.G., Phillips, R., Stone, K., Grubbs, A., 2001. Treatment of swine wastewater in marsh-pond-marsh constructed wetlands. Water Sci. Technol. 44 (11–12), 545–550.

Ramchunder, S.J., Brown, L.E., Holden, J., 2009. Environmental effects of drainage, drain-blocking and prescribed vegetation burning in UK upland peatlands. Prog. Phys. Geog. 33 (1), 49–79.

Ramos-Scharron, C.E., MacDonald, L.H., 2007. Measurement and prediction of natural and anthropogenic sediment sources, St. John, US Virgin Islands. Catena 71 (2), 250–266.

Reddy, K.R., Kadlec, R.H., Flaig, E., Gale, P.M., 1999. Phosphorus retention in streams and wetlands: a review. Crit. Rev. Environ. Sci. Technol. 29 (1), 83–146.

Reed, S.C., Crites, R.W., Middlebrooks, E.J., 1995. Natural Systems for Waste Management and Treatment, second ed. McGraw-Hill, New York, USA.

Reiche, E.W., 1994. Modelling water and nitrogen dynamics on catchment scale. Ecol. Model. 75–76, 371–384.

Richardson, M.J., Whoriskey, F.G., 1992. Factors influencing the production of turbidity by goldfish (*Carassius auratus*). Can. J. Zool. 70 (8), 1585–1589.

Rifkin, R., Klautau, A., 2004. In defence of one-versus-all classification. J. Mach. Learn. Res. 5, 101–141.

Ritter, W.F., Walpole, E.W., Eastburn, R.P., 1984. Effect of an anaerobic swine lagoon on groundwater quality in Sussex County, Delaware. Agric. Wastes 10 (4), 267–284.

Robertson, W.D., 2003. Enhanced attenuation of septic system phosphate in non-calcareous sediments. Ground Water 41 (1), 48–56.

Robertson, W.D., 2008. Irreversible phosphorus sorption in septic tank plumes? Ground Water 46 (1), 51–60.

Ross, S.M., 1994. Toxic Metals in Soil-plant Systems. John Wiley and Sons, Sussex, UK.

Rousseau, D.P.L., Vanrolleghem, P.A., De Pauw, N., 2004. Model-based design of horizontal subsurface flow constructed treatment wetlands: a review. Water Res. 38 (6), 1484–1493.

Rowell, T.A., 2009. Management planning guidance for protected sites in the UK; a comparison of decision-making processes in nine guides. J. Nat. Conserv. 17 (3), 168–180.

Rowsell, J.G., Miller, M.H., Groenevelt, P.H., 1985. Self-sealing of earthen liquid manure storage ponds: II. Rate and mechanism of sealing. J. Environ. Qual. 14 (4), 539–542.

Royal Commission on Sewage Disposal, 1915. Final Report of the Commissioners Appointed to Inquire and Report what Methods of Treating and Disposing of Sewage May Properly Adopted. His Majesty's Stationary Office, London, UK.

Rudd, J.W.M., Kelly, C.A., St Louis, V., Hesslein, R.H., Furutani, A., Holoka, M.H., 1986. The role of sulfate reduction in long term accumulation of organic and inorganic sulfur in lake sediments. Limnol. Oceanogr. 31 (6), 1281–1291.

Ruiz-Jimenez, J., Priego-Capote, F., Garcia-Olmo, J., Luque de Castro, M.D., 2004. Use of chemometrics and mid infrared spectroscopy for the selection of extraction alternatives to reference analytical methods for total fat isolation. Anal. Chim. Acta 525 (2), 159–169.

Rumelhart, D.E., Hinton, G.E., Williams, R.J., 1986. Learning internal representations by error propagation. In: Rumelhart, D.E., McClelland, J.L. (Eds.), Parallel Distributed Processing: Explorations in the Microstructure of Cognition. MIT Press, Cambridge, Massachusetts, USA, pp. 318–362.

Salton, G., 1991. Developments in automatic text retrieval. Science 253 (5023), 974–980.

Sander, H.A., Haight, R.G., 2012. Estimating the economic value of a cultural ecosystem services in an urbanizing area using hedonic pricing. J. Environ. Manage. 113, 194–205.

Sani, A., Scholz, M., Bouillon, L., 2013a. Seasonal assessment of experimental vertical-flow constructed wetlands treating domestic wastewater. Biores. Technol. 147, 585–596.

Sani, A., Scholz, M., Babatunde, A., Wang, Y., 2013b. Impact of clogging on the outflow water quality of vertical-flow constructed wetlands treating urban wastewater. Water Air Soil Pollut. 224 (3), 1488–1506.

Sansalone, J.J., 1999. Adsorptive infiltration of metals in urban drainage – media characteristics. Sci. Total Environ. 235 (1–3), 179–188.

Sasaki, K., Ogino, T., Endo, Y., Kurosawa, K., 2003. Field studies on heavy metal accumulation in a natural wetland receiving acid mine drainage. Mater. Trans. 44 (9), 1877–1884.

SAWA, 2010. SAWA – a Strategic Alliance in the North Sea Region, Strategic Alliance for Water Management Actions (SAWA) Project Information. http://www.sawa-project.eu.

Schapire, R.E., Singer, Y., 2000. Boostexter: a boosting-based system for text categorization. Mach. Learn. 39 (2/3), 135–168.

Schlesinger, W.H., 1991. Biogeochemistry: An Analysis of Global Change. Academic Press, San Diego, USA.

Schmitt, T.G., Thomas, M., Ettrich, N., 2004. Analysis and modelling of flooding in urban drainage systems. J. Hydro. 299 (3–4), 300–311.

Scholz, M., 2003. Performance predictions of mature experimental constructed wetlands, which treat urban water receiving high loads of lead and copper. Wat. Res. 37 (6), 1270–1277.

Scholz, M., 2006. Wetland Systems to Control Urban Runoff. Elsevier, Amsterdam, The Netherlands.

Scholz, M., 2010. Wetland Systems – Storm Water Management Control. Springer, Berlin, Germany.

Scholz, M., 2004a. Treatment of gully pot effluent containing nickel and copper with constructed wetlands in a cold climate. J. Chem. Tech. Biotechnol. 79 (2), 153–162.

Scholz, M., 2004b. Stormwater quality associated with a silt trap (empty and full) discharging into an urban watercourse in Scotland. Int. J. Environ. Stud. 61 (4), 471–483.

Scholz, M., 2004c. Case study: design operation, maintenance and water quality management of sustainable storm water ponds for roof run-off. Biores. Technol. 95 (3), 269–279.

Scholz, M., 2007a. Classification methodology for sustainable flood retention basins. Landscape Urb. Plann. 81 (3), 246–256.

Scholz, M., 2007b. Ecological effects of water retention in the River Rhine Valley: a review assisting future retention basin classification. Int. J. Environ. Stud. 64 (2), 171–187.

Scholz, M., Lee, B.-H., 2005. Constructed wetlands: a review. Int. J. Environ. Stud. 62 (4), 421–447.

Scholz, M., Martin, R.J., 1998a. Control of bio-regenerated granular activated carbon by spreadsheet modelling. J. Chem. Tech. Biotechnol. 71 (3), 253–261.

Scholz, M., Martin, R.J., 1998b. Biological control in granular activated carbon beds. Int. Rev. Hydrobiol. 83 (special issue), 657–664.

Scholz, M., Sadowski, A.J., 2009. Conceptual classification model for sustainable flood retention basins. J. Environ. Manage. 90 (1), 624–633.

Scholz, M., Trepel, M., 2004. Water quality characteristics of vegetated groundwater-fed ditches in a riparian peatland. Sci. Total Environ. 332 (1–3), 109–122.

Scholz, M., Uzomah, V., 2013. Rapid decision support tool based on novel ecosystem service variables for retrofitting of permeable pavement systems in the presence of trees. Sci. Total Environ. 458–460, 486–498.

Scholz, M., Xu, J., 2002. Performance comparison of experimental constructed wetlands with different filter media and macrophytes treating industrial wastewater contaminated with lead and copper. Biores. Technol. 83 (2), 71–79.

Scholz, M., Yang, Q., 2010. Guidance on variables characterising water bodies including sustainable flood retention basins. Landscape Urb. Plann. 98 (3–4), 190–199.

Scholz, M., Zettel, S., 2004. Stormwater quality associated with a full silt trap discharging into urban watercourses. J. Chart. Inst. Water Environ. Manage. 18 (4), 226–229.

Scholz, M., Corrigan, N.L., Yazdi, S.K., 2006. The Glasgow sustainable urban drainage system management project: case studies (Belvidere hospital and Celtic FC stadium areas). Environm. Eng. Sci. 23 (6), 908–922.

Scholz, M., Hedmark, Å., Hartley, W., 2012. Recent advances in sustainable multifunctional land and urban management in Europe: a review. J. Environ. Plann. Manage. 55 (7), 833–854.

Scholz, M., Höhn, P., Minall, R., 2002. Mature experimental constructed wetlands treating urban water receiving high metal loads. Biotechnol. Prog. 18 (6), 1257–1264.

Scholz, M., Morgan, R., Picher, A., 2005. Stormwater resources development and management in Glasgow: two case studies. Int. J. Environ. Stud. 62 (3), 263–282.

Scholz, M., Harrington, R., Carroll, P., Mustafa, A., 2007. The integrated constructed wetlands (ICW) concept. Wetlands 27 (2), 337–354.

Scholz, M., Harrington, R., Carroll, P., Mustafa, A., 2010. Monitoring of nutrient removal within Integrated Constructed Wetlands (ICW). Desalination 250 (1), 356–360.

Schultze, T., Mojzisch, A., Schulz-Hardt, S., 2012. Why groups perform better than individuals at quantitative judgment tasks: group to individual transfer as an alternative to differential weighting. Organ. Behav. Hum. Decis. Process. 118 (1), 24–36.

Schumacher, B.A., 2002. Methods for the Determination of Total Organic Carbon (TOC) in Soils and Sediments. Ecological Risk Assessment Support Center, Office of Research and Development, US Environmental Protection Agency, Washington, D.C., USA.

Seaman, E.A., 1979. Observations on *Carassius auratus* Linneaus harvesting *Potamogeton foliosus* Raf. in a small pond in north Virginia. Fisheries 4 (3), 24–25.

SEPA, 1999. Protecting the Quality of Our Environment – Sustainable Urban Drainage: An Introduction, Scottish Environmental Protection Agency (SEPA). Stationary Office, London, UK.

Seattle Department of Transportation, 2012. Approved Street Tree Lists. http://www.seattle.gov/trees/docs/2011-Street_Tree_List.pdf.

Sebastiani, F., 2002. Machine learning in automated text categorization. ACM Comput. Surv. 34 (1), 1–47.

SEPA, 2003. SUDS Advice Note – Brownfield Sites. Scottish Environmental Protection Agency (SEPA), Stirling, UK.

SEPA, 2007. Scotland's WFD Aquatic Monitoring Strategy. http://www.sepa.org.uk/water.

SEPA, 2009. Indicative River and Coastal Flooding Map. http://www.multimap.com/clients/browse.cgi?client=sepa&X=312000&Y=723000&width=550&height=450&scale=25000&coordsys=gb&nosnap=true&overlay=layer1.

Seppelt, R., Dormann, C.F., Eppink, F.V., Lautenbach, S., Schmidt, S., 2011. A quantitative review of ecosystem service studies: approaches, shortcomings and the road ahead. J. Appl. Ecol. 48 (3), 630–636.

Shepherd, H.L., Tchobanoglous, G., Grismer, M.E., 2001. Time-dependent retardation model for chemical oxygen demand removal in a subsurface-flow constructed wetland for winery wastewater treatment. Water Environ. Res. 73 (5), 597–606.

Shepherd, T.J., Chenery, S.R.N., Pashley, V., Lord, R.A., Ander, L.E., Breward, N., Hobbs, S.F., Horstwood, M., Klinck, B.A., Worrall, F., 2009. Regional lead isotope study of a polluted river catchment: river Wear, Northern England, UK. Sci. Total Environ. 407 (17), 4882–4893.

Shih, S.C.W., Nicholls, R.J., 2007. Urban managed realignment: application to the Thames estuary, London. J. Coast. Res. 23 (6), 1525–1534.

Shutes, R.E., Revitt, D.M., Scholes, L.N.L., Forshaw, M., Winter, B., 2001. An experimental constructed wetland system for the treatment of highway runoff in the United Kingdom. Water Sci. Technol. 44 (11–12), 571–578.

Shuval, H.I., Adin, A., Fattal, B., Rawitz, E., Yekutiel, P., 1986. Wastewater Irrigation in Developing Countries: Health Effects and Technical Solutions, Technical Paper No. 51. World Bank, Washington, D.C., USA.

Siegrist, R.L., Tyler, E.J., Jenssen, P.D., 2000. Design and performance of onsite wastewater soil absorption systems. In: Proceedings of National Research Needs Conference: Risk-based Decision Making for Onsite Wastewater Treatment. Washington University, St. Louis, Missouri, USA.

Simpson, F., Chapman, M., 1999. Comparison of urban governance and planning policy – East looking West. Cities 16 (5), 353–364.

Smith, R.L., 1980. Ecology and Field Biology, third ed. Harper and Row, New York, USA.

Smith, C.S., Adams, M.S., 1986. Phosphorus transfer from sediments to *Myriophyllum spicatum*. Limnol. Oceanogr. 31 (6), 1312–1321.

Smith, K.A., Charles, D.R., Moorhouse, D., 2000. Nitrogen excretion by farm livestock with respect to land spreading requirements and controlling nitrogen losses to ground and surface waters. Part 2: pigs and poultry. Biores. Technol. 71 (2), 183–194.

Smith, E., Gordon, R., Madani, A., Stratton, G., 2006. Year-round treatment of diary wastewater by constructed wetlands in Atlantic Canada. Wetlands 26 (2), 349–357.

Snoek, C.G.M., Worring, M., van Gemert, J.C., Geusebroek, J.M., Smeulders, A.W.M., 2006. The challenge problem for automated detection of 101 semantic concepts in multimedia. In: Proceedings of the 14th Annual ACM International Conference on Multimedia. ACM Press, New York, USA, pp. 421–430.

Song, Z., Zheng, Z., Li, J., Sun, X., Han, X., Wang, W., Xu, M., 2006. Seasonal and annual performance of a full-scale constructed wetland system for sewage treatment in China. Ecolog. Eng. 26 (3), 272–282.

Spieles, D.J., Mitsch, W.J., 2000. The effects of season and hydrologic and chemical loading on nitrate retention in constructed wetlands: a comparison of low- and high-nutrient riverine systems. Ecol. Eng. 14 (1–2), 77–91.

Spratt, H.G., Morgan, M.D., 1990. Sulfur cycling in a cedar dominated freshwater wetland. Limnol. Oceanogr. 35 (7), 1586–1593.

SPSS, 2009. Statistical Package for the Social Sciences (SPSS); Recently Re-branded as Predictive Analytics SoftWare (PASW). http://www.spss.com/uk/statistics/?gclid=CNvH2-XFuZsCFY4U4wodMkPQEA.

Stevens, C.J., Quinton, J.N., 2009. Diffuse pollution swapping in arable agricultural systems. Crit. Rev. Environ. Sci. Technol. 39 (6), 478–520.

Stottmeister, U., Wießner, A., Kuschk, P., Kappelmeyer, U., Kästner, M., Bederski, O., Müller, R.A., Moormann, H., 2003. Effects of plants and microorganisms in constructed wetlands for wastewater treatment. Biotechnol. Adv. 22 (1–2), 93–117.

Stumm, W., Morgan, J.J., 1996. Aquatic Chemistry: Chemical Equilibria and Rates in Natural Waters, third ed. John Wiley and Sons, New York.

Stone, K.C., Poach, M.E., Hunt, P.G., Reddy, G.B., 2004. Marsh-pond-marsh constructed wetland design analysis for swine lagoon wastewater treatment. Ecol. Eng. 23 (2), 127–133.

Sudhakaran, J.K., Vasantha, R., 2011. Mixed method approach for efficient component retrieval from a component repository. J. Software Eng. Appl. 4 (7), 442–445.

Sudheep, E.M., Sumam, M.I., Joseph, A., 2011. Design and performance analysis of data mining techniques based on decision trees and Naive Bayes classifier for employment chance prediction. J. Conver. Info. Tech. 6 (5), 89–98.

Sun, G., Ma, Y., Zhao, R., 2009. Study on purification efficiency of sewage in constructed wetlands with different plants. World Rural Obs. 1 (2), 35–39.

Sun, G., Gray, K.R., Biddlestone, A.J., Cooper, D.J., 1999. Treatment of agricultural wastewater in a combined tidal flow downflow reed bed system. Water Sci. Technol. 40 (3), 139–146.

Sundberg, C., Tonderski, K., Lindgren, P.-E., 2007. Potential nitrification and denitrification and the corresponding composition of the bacterial communities in a compact constructed wetland treating landfill leachates. Water Sci. Technol. 56 (3), 159–166.

Sundberg, C., Stendahl, J.S.K., Tonderski, K., Lindgren, P.E., 2006. Overland flow systems for treatment of landfill leachates — Potential nitrification and structure of the ammonia-oxidising bacterial community during a growing season. Soil Biol. Biochem. 39 (1), 127–138.

Sutherland, W.J., 2006. Ecological Census Techniques – a Handbook, second ed. Cambridge University Press, Cambridge, England, UK.

Sæbø, A., Benedidz, T., Randrup, T.B., 2003. Selection of trees for urban forestry in the Nordic countries. Urb. For. Urb. Green 2 (2), 101–114.

Tan, K.H., 1996. Soil Sampling, Preparation and Analysis. Marcel Dekker, New York, USA.

Tanner, C.C., 1996. Plants for constructed wetland treatment systems – a comparison of the growth and nutrient uptake of eight emergent species. Ecol. Eng. 7 (1), 59–83.

Tanner, C.C., Sukias, J.P.S., Upsdell, M.P., 1999. Substratum phosphorous accumulation during maturation of gravel-bed constructed wetlands. Wat. Sci. Technol. 40 (3), 147–154.

Tchobanoglous, G., Burton, F.L., Stensel, H.D., 2003. Wastewater Engineering: Treatment and Reuse, fourth ed. Metcalf and Eddy, McGraw Hill Companies, New York, USA.

TEEB, 2011. The Economics of Ecosystems and Biodiversity (TEEB) Manual for Cities: Ecosystem Services in Urban Management. http://www.teebweb.org.

Tota-Maharaj, K., Scholz, M., 2012. Artificial neural network simulation of combined permeable pavement and earth energy systems treating storm water. J. Environ. Eng. ASCE 138 (4), 499–509.

Trang, N.T.D., Konnerup, D., Schierup, H.H., Chiem, N.H., Tuan, L.A., Brix, H., 2010. Kinetics of pollutant removal from domestic wastewater in a tropical horizontal subsurface flow constructed wetland system: effects of hydraulic loading rate. Ecol. Eng. 36 (4), 527–535.

Trepel, M., 1999. Spatiotemporal simulation of water and nitrogen dynamics as a tool in fen restoration. Int. Peat J. 9, 45–52.

Trepel, M., 2001. Gedanken zur zukünftigen Nutzung Schleswig-Holsteinischer Niedermoore. Die Heimat 108 (11–12), 186–194 (in German).

Trepel, M., Kluge, W., 2002. Ecohydrological characterisation of a degenerated valley peatland in Northern Germany for use in restoration. J. Nat. Conserv. 10, 155–169.

Trepel, M., Kluge, W., 2004. WETTRANS: a flow-path-oriented decision support system for the assessment of water and nitrogen exchange in riparian wetlands. Hydrol. Proc. 18, 357–371.

Trepel, M., Palmeri, L., 2002. Quantifying nitrogen retention in surface flow wetlands for environmental planning at the landscape-scale. Ecol. Eng. 19 (2), 127–140.

Trepel, M., Holsten, B., Kieckbusch, J., Otten, I., Pieper, F., 2003. Influence of macrophytes on water level and flood dynamics in a riverine wetland in Northern Germany. In: Proceedings of the International Conference EcoFlood – Towards Natural Flood Reduction Strategies, Warshaw, 6–13 September 2003. Institute for Land Reclamation and Grassland Farming, Raszyn, Poland.

Trias, M., Hu, Z., Mortula, M.M., Gordon, R.J., Gagnon, G.A., 2004. Impact of seasonal variation on treatment of swine wastewater. Environ. Technol. 25 (7), 775–781.

Trohidis, K., Tsoumakas, G., Kalliris, G., Vlahavas, I., 2008. Multilabel classification of music into emotions. In: Proceedings of 9th International Conference on Music Information Retrieval. ISMIR, Philadelphia, Pennsylvania, USA, pp. 325–330.

Tsihrintzis, V.A., Akratos, C.S., Gikas, G.D., Karamouzis, D., Angelakis, A.N., 2007. Performance and cost comparison of a FWS and a VSF constructed wetland system. Environ. Technol. 28 (6), 621–628.

Tsoumakas, G., Katakis, I., 2007. Multi-label classification: an overview. Int. J. Data Warehous. Min. 3 (3), 1–13.

Tsoumakas, G., Katakis, I., Vlahavas, I., 2010. Mining multi-label data. In: Maimon, O., Rokach, L. (Eds.), Data Mining and Knowledge Discovery Handbook, second ed. Springer, Berlin, Germany, pp. 667–685.

Tunçsiper, B., Ayaz, S., Akça, L., Gunes, K., 2009. Performance of a pilot-scale, three-stage constructed wetland system for domestic wastewater treatment. Environ. Technol. 30 (11), 1187–1194.

Turner, M.K., 1995. Engineered reed-bed systems for wastewater treatment. Trends Biotechnol. 13 (7), 248–252.

Turok, I., Bailey, N., 2004. Twin track cities? Competitiveness and cohesion in Glasgow and Edinburgh. Progr. Plann. 62 (3), 135–204.

Tylova-Munzarova, E., Lorenzen, B., Brix, H., Votrubova, O., 2005. The effects of NH_4^+ and NO_3^- on growth, resource allocation and nitrogen uptake kinetics of *Phragmites australis* and *Glyceria maxima*. Aquat. Bot. 81 (4), 326–342.

Tyson, T.W., 2000. Best Management Practices for Animal Feeding Operations, ANR-1188. Alabama Cooperative Extension Systems, USA.

Tzoulas, K., Korpela, K., Venn, S., Yli-Pelkonen, V., Kaźmierczak, A., Niemela, J., James, J., 2007. Promoting ecosystem and human health in urban areas using green infrastructure: a literature review. Landscape Urb. Plann. 81 (3), 167–178.

Umweltbundesamt, 2001. Daten zur Umwelt: Der Zustand der Umwelt in Deutschland 2000. Erich Schmidt Verlag, Berlin, Germany (in German).

UK Government, 1994. Urban Waste Water Treatment (England and Wales) Regulations 1994 (SI 1994 No. 2841). UK Government, London, UK.

UK National Ecosystem Assessment, 2011. The UK National Ecosystem Assessment Technical Report. United Nations Environment Programme World Conservation Monitoring Centre, Cambridge, UK.

Ukonmaanaho, L., Nieminen, T.M., Rausch, N., Cheburkin, A., Le Roux, G., Shotyk, W., 2006. Recent organic matter accumulation in relation to some climatic factors in ombrotrophic peat bogs near heavy metal emission sources in Finland. Glob. Planet. Change 53 (4), 259–268.

United States Army Corps of Engineers, 2000. Wetlands Engineering Handbook – ERDC/EL TR-wrp-re-21. United States Army Engineer Research and Development Center Cataloging-in-Publication Data.

USBR, 1997. Water Measurement Manual, third ed. U.S. Department of the Interior, United States Bureau of Reclamation, U.S. Government Printing Office, Washington, D.C., USA.

USEPA, 1988. Constructed Wetlands and Aquatic Plant Systems for Municipal Wastewater Treatment. EPA/625/1–88/022. United States Environmental Protection Agency (USEPA), Office of Research and Development Cincinnati, Ohio, USA.

Uzomah, V., Scholz, M., Almuktar, S., 2014. Rapid expert tool for different professions based on estimated ecosystem variables for retrofitting of drainage systems. Comput. Environ. Urb. Syst. 44, 1–14.

van der Valk, A.G., Davis, C.B., 1978. Primary production of prairie glacial marshes. In: Good, R.E., Whigham, D.F., Simpson, R.L. (Eds.), Freshwater Wetlands: Ecological Processes and Management Potential. Academic Press, New York, USA, pp. 21–37.

Vapnik, V., 1995. The Nature of Statistics Learning Theory. Springer Verlag, New York, USA.

Verdenius, F., Broeze, J., 1999. Generalized and instance-specific modeling for biological systems. Environ. Model. Software 14 (5), 339–348.

Vesanto, J., Himberg, J., Alhoniemi, E., Parhankangas, J., November 1999. Self-organizing map in Matlab: the SOM Toolbox. In: Proceedings of the Matlab DSP Conference, Espoo, Finland, pp. 35–40. Software available at http://www.cis.hut.fi/projects/somtoolbox/.

Villarreal, E.L., Semadeni-Davies, A., Bengtsson, L., 2004. Inner city stormwater control using a combination of best management practices. Ecol. Eng. 22 (4–5), 279–298.

Vymazal, J., 2002. The use of sub-surface constructed wetlands for wastewater treatment in the Czech Republic: 10 years experience. Ecol. Eng. 18 (5), 633–646.

Vymazal, J., Krasa, P., 2003. Distribution of Mn, Al, Cu and Zn in a constructed wetland receiving municipal sewage. Water Sci. Technol. 48 (5), 299–305.

Vymazal, J., Brix, H., Cooper, P.F., Green, M.B., Haberle, R., 1998. Constructed Wetlands for Wastewater Treatment in Europe. Backhuys, Leiden, The Netherlands.

Wang, N., Mitsch, W.J., 2000. A detailed ecosystem model of phosphorus dynamics in created riparian wetlands. Ecol. Model. 126 (2–3), 101–130.

Walsh, C.J., Fletcher, T.D., Burns, M.J., 2012. Urban stormwater runoff: a new class of environmental flow problem. PLoS One 7 (9), e45814.

Watzin, M.C., McIntosh, A.W., 1999. Aquatic ecosystems in agricultural landscapes: a review of ecological indicators and achievable ecological outcomes. J. Soil Water Conserv. 54 (4), 636–644.

Werner, H., Obach, M., 2001. New neural network types estimating the accuracy of response for ecological modelling. Ecol. Model. 146 (1–3), 289–298.

White, I., Alarcon, A., 2009. Planning policy, sustainable drainage and surface water management: a case study of Greater Manchester. Built Environ. 35 (4), 516–530.

WHO, 2008. Guidelines for Drinking Water Quality (Incorporating the First and Second Addenda) In: Recommendations, third ed., vol. 1. World Health Organization, Geneva, Switzerland.

Wießner, A., Kappelmeyer, U., Kuschk, P., Kästner, M., 2004. Influence of the redox condition dynamics on the removal efficiency of a laboratory-scale constructed wetland. Water Res. 39 (1), 248–256.

Wild, A., 1993. Soils and the Environment. Cambridge University Press, Cambridge, UK.

Wilhelm, S.R., Schiff, S.L., Cherry, J.A., 1994. Biogeochemical evolution of domestic waste water in septic systems: 1. Conceptual model. Ground Water 32 (6), 905–916.

Willemen, L., Verburg, P.H., Hein, L., Mensvoort, M.E.F., 2008. Spatial characterization of landscape functions. Landscape Urb. Plann. 88 (1), 34–43.

Wilson, S., Bray, R., Cooper, P., 2004. Sustainable Drainage Systems. Hydraulic, Structural and Water Quality Advice. Construction Industry Research and Information Association. CIRIA, London, UK (CIRIA) Report C609.

Wisskirchen, R., Häupler, H., 1998. Standartliste der Farn- und Blütenpflanzen Deutschlands, Bundesamt für Naturschutz. Verlag Eugen Ulmer, Stuttgart (in German).

Wood, T.S., Shelley, M.L., 1999. A dynamic model of bioavailability of metals in constructed wetland sediments. Ecol. Eng. 12 (3–4), 231–252.

Woodland Trust, 2012. Tree Guide. http://www.british-trees.com/treeguide.

Wrubleski, D.A., Murkin, H.R., van der Valk, A.G., Nelson, J.W., 1997. Decomposition of emergent macrophyte roots and rhizomes in a northern prairie marsh. Aquat. Bot. 58 (2), 121–134.

Wu, X.Y., Mitsch, W.J., 1998. Spatial and temporal patterns of algae in newly constructed freshwater wetlands. Wetlands 18 (1), 9–20.

Yang, Q., Shao, J., Scholz, M., Plant, C., 2011. Feature selection methods for characterizing and classifying adaptive sustainable flood retention basins. Water Res. 45 (3), 993–1004.

Yang, Q., Shao, J., Scholz, M., Boehm, C., Plant, C., 2012. Multi-label classification models for sustainable flood retention basins. Environ. Model. Software 32, 27–36.

Zhang, M.J., Zhou, Z.H., 2006. Multi-label neural networks with applications to functional genomics and text categorization. IEEE Trans. Knowl. Data Eng. 18 (10), 1338–1351.

Zhang, M.L., Zhou, Z.H., 2007. ML-KNN: a lazy learning approach to multi-label learning. Pattern Recog. 40 (7), 2038–2048.

Zhang, Y., Burer, S., Street, W.N., 2006. Ensemble pruning via semi-definite programming. J. Mach. Learn. Res. 7, 1315–1338.

Zhang, Y., van Dijk, M.A., Liu, M., Zhu, G., Qin, B., 2009. The contribution of phytoplankton degradation to chromophoric dissolved organic matter (CDOM) in eutrophic shallow lakes: field and experimental evidence. Water Res. 43 (18), 4685–4697.

Zhao, Y.Q., Sun, G., Alen, S.J., 2004. Purification capacity of a highly loaded laboratory scale tidal flow reed bed system with effluent recirculation. Sci. Total Environ. 330 (1–3), 1–18.

Zheng, P.Q., Baetz, B.W., 1999. GIS-based analysis of development options from a hydrology perspective. J. Urb. Plann. Dev. 125 (4), 164–180.

Zhu, T., Sikora, F.J., 1995. Ammonium and nitrate removal in vegetated and unvegetated gravel bed microcosm wetlands. Water Sci. Technol. 32, 219–228.

Zhu, L., Yang, J., Shen, H.B., 2009. Multi label learning for prediction of human protein subcellular localizations. Protein J. 28 (9–10), 384–390.

Index

Note: Page numbers followed by f and t denote figures and tables, respectively.

A

Absorption, 169, 231
Acceptance, public, 192, 268, 317
Accuracy of prediction, 257
Acer pseudoplatanus (Sycamore), 287–289, 293–294
Activated sludge process, 91–93
  compared with percolating filtration, 93, 94t–95t
  compared with wetland system, 93, 94t–95t
  contact stabilization, 98
  conventional complete mix, 96
  deep shaft process, 99
  designs and kinetics, 99–103
  extended aeration, 97–98
  high-rate, 97
  oxidation ditches, 98–99
  series or plug flow system, 96
  step feed, 97
  tapered aeration, 96
Adaptation, sustainable flood retention basin, 433, 458–459
Adsorption, 42, 64
Aeration
  of activated sludge, 92
  diffused air, 99–100
  extended, 97–98
  mechanical, 100–101
  re-aeration, 17–18
  step, 104
  tapered, 96, 104
Aerobic digestion, 162
Aerobic ponds, 77–78
Aerobic self-purification, 77
Aerobic treatment, 17
Aesthetic and educational appreciation and inspiration for culture, art and design (AEAICAD), 273t–275t, 284
Aesthetic flood treatment wetland (AFTW), 435t–437t
Aesthetics, 435t–437t
Agarose gel electrophoresis, 365
Agglomeration, 10
Aggregates, 83, 155, 170, 173, 179–181, 190, 398
  fine, 155, 209
Agricultural water resources management, 401–418
Air
  blown, 55
  bubble, 58
  dissolved, 55–57
  drying, 164
Algae
  Activity. *See* Algal actions
  blue-green, 119
  control with goldfish, 267–268, 267f
  control with straw, 196
  growth, control of, 196–197
Algal actions, 66–67
Algal cover, in summer, 449
Alkalinity, 10–11
Alum (aluminum sulfate), 9, 35, 42
Aluminum hydroxide, 42, 64–65
Aluminum salt. *See* Alum (aluminum sulfate)
Ammonia, 132
  oxidation of, 5–6
  -nitrogen, 370t
Ammonia oxidizers, 370
Ammonification, 340
Analysis of variance (ANOVA), 224–225, 342, 381
Animal passage
  aquatic, 441–442
  land, 442
Annual outflow water qualities, comparison of, 217–223, 218t–221t, 222f
Apicomplexa, 118
Aquatic animal passage, 441–442
Aquatic plant management, 204–206, 204f–205f
Artificial neural network (ANN), 240, 473

Assimilation, 93
Attenuation pond, 191, 199t
Automated cleaning, 75

B

Back-propagation for multi-label learning (BP-MLL), 473–474, 478t
Backwashing, hydraulics, 75
Bacteria, 119
Basins
  bed gradient, 446
  infiltration, 191, 197t, 199t, 314
  length of, 454
  mean bed velocity, 446
  mean depth of, 454
  sustainable flood retention. *See* Sustainable flood retention basin (SFRB)
  width of, 455
Bearing failure, 89
Below-ground detention system, 227–238
  analytical methods, 229–232
  cost comparison, 232
  design and operation of, 228–229, 229t
  Ecosoil and turf, 236
  inflow water quality, 232–234, 233t
  outflow water quality, 234–236, 235t
Below-ground storage structure, 317
Best management practices (BMP), 279, 313
Biochemical oxygen demand (BOD), 5–6, 17, 19, 27, 37, 81, 83, 88, 104–105, 150–151, 217–225, 394f, 396–397
Biodegradation, 65–67
Biofilm, 82–83, 85, 89, 361, 386
Bioindicator, 170
Biological
  control (BC), 273t–275t, 284
  effects, of organic pollutants, 118–120
  growth, 5
  treatment, 77–80
Biological filtration, 81–86
  ecology of, 82–83
  process variants, 83–84
Biomass
  clogging, 81
  sloughing of, 82
Black box system, 239, 241, 256–257, 390
Blown air, flotation using, 55
Blue-green algae, 119
Brownian motion, 40–41
Bubbles, 56

C

Calcium hardness, 111
Capillary suction time (CST), 160
Capsicum annuum (Chilli or sweet pepper), 403–404
Carassius auratus (Goldfish), 267–268, 267f
Carbon, 142
  oxidation of, 77
  sequestration (CSS), 273t–275t, 283
Carcinogens, 136
Case-based reasoning (CBR) model, 332–333, 334t–335t, 337
Catalytic action, 108
Catchment, 362
  size, 452
Catchment proportion
  arable, 452–453
  forest, 453
  natural, 453–454
  pasture, 453
  urban, 452
  viniculture, 453
Centrifuge method, 165
Certainty estimations, 309
Channel
  connectivity, 443
  open, 345
  proportion of flow within, 444
Charge, electrical, 40–41, 44
Chemical oxygen demand (COD), 7, 387–388, 394–400, 394f
Chemistry
  water, 10–11
  wetlands, 140–147
Chilli (*Capsicum annuum*), 403–404
Chlorination, 129
  breakpoint, 133
  kinetics of, 133–134
Chlorine, 10
  addition, technology of, 134–135
  advantages and disadvantages of, 135–136
  applications of, 134
  as disinfectant, 131–133
Chlorine dioxide, as disinfectant, 131–133
Ciliophora, 118
Clarifier (sludge blanket), 47–54
Classification
  engineering, 14–15
  of settling behavior, 37–39
  sustainable flood retention basin (SFRB), 457–482

Clay, 41, 448
Cleaning
  automated, 23, 75
  manual, 23
Climate change, 433
Clogging
  filter, 81, 400
  process, 390
  tests, 393
Cluster analysis, 460, 464–468
C:N:P ratio, 7
Coagulants, 9–10, 35
Coagulation, 39
  chemicals, 42–44
  of colloidal solids, 77
  process, 41–42, 44–45
  sweep, 42
Colloidal suspensions, 40–41
Comminutors, 23–24
Common reed (*Phragmites australis*), 149–150, 210–212, 225
Community and environment (CE) variables, 279–280, 288t–289t, 300, 307–308
Composting, 163, 166–167
Concentration, residual, 133
Conditioning, chemical, 163
Constructed treatment wetland, 137–156
  metals, 155
  road runoff, 209–226
  roof runoff, 191–208
  storm water treatment, 154–155
  stream water treatment, 169–190
Constructed wetland performance, modeling of, 239–258
  comparison of applications, 251–256, 252t, 253f–254f, 254t, 255f–256f
  correlation analysis, 246, 247t
  experimental data and variables, 241–242
  input variables, optimization of, 246–251, 248f, 249t–250t
  K-nearest neighbors, 242–243
  performance evaluation, 246
  self-organizing map, 245–246
  support vector machine, 243–245
Contact stabilization, 98
Contaminant loading rates, 380–381
Conventional complete mix activated sludge process, 96
Copper, 169–170, 172–174, 175t–177t, 176, 179–182, 180t–181t, 190
Correlation
  analysis, 184–187, 185t–186t, 189f, 246, 247t
  coefficient, 247t
Corrosion, 107
Cost
  comparison (analysis), 232
  labor force, 68
  treatment, 93, 157, 227–228
Cost–benefit analysis, 415–416, 430–431
Crop irrigation, wastewater recycling for, 401–403
Crop yield, 418
Cryptosporidium, 2

D

Dam
  height, 440
  length, 441
Data acquisition, 471–472
Debris, 154–155, 211, 214, 392–393, 408, 431, 443, 451–452
Decision support matrix, for SuDS, 319–323, 324t–326t
Decision support system, 304
Decomposers, 117
Deep shaft process, 99
Degradation (biological), 397
Dendrogram, 464–465
Denitrification, 237, 340, 426–428
Dentrifiers, 370
Deoxyribonucleic acid (DNA) extraction, 364
Design
  of biological filters, 84–85
  circular, 57–59
  flow rate, 13–14
  geometry, 344–345
  loading criteria, 103
  rectangular, 57–59
Detention
  below-ground (system), 227–238
  cell, 234, 317
  tank, 107–108
Developer, 301t–302t
Development (urban), 268
Dewatering of sludge, 159–160
Diatoms, 119
Diesel, hydrocarbon, 401–402, 405–406, 408
Diffused air aeration, 99–100
Diffuse pollution control, 439, 444, 455

Digestion (aerobic), 162
Diseases
waterborne, 126–128
water-related, 123–124, 124t–125t
Disinfection, 129–136
chlorine, as disinfectant, 131–133
chlorine dioxide, as disinfectant, 131–133
by ozone, 130–131
requirements of, 129–130
traditional methods of, 130
Disk, 87, 89
Disposal, waste, 160–162, 161t
Dissolved air, flotation using, 55–57
Dissolved oxygen (DO), 17–19
Ditches, 343–345
disappearance due to vegetation growth, 353–354
drainage, 355
hydraulics, 343–344, 344f
oxidation, 98–99
vegetation, 343–345, 353f
Diversity index, 115–116
Dorr Clariflocculator, 35
Drainage, 447–448
engineer, 301t–302t, 308
engineering expert opinion, certainty estimations based on, 309
road runoff. *See* Road runoff, wetlands treating
roof runoff, 191
sustainable, 191
traditional, 317
Dry weather flow (DWF), 28, 32–34

E

Ecology
filter, 82–83
impact, 279–280, 300, 333–335, 443
Ecosoil (product), 236
Ecosystem mass balance, of wetlands, 147–148
Ecosystem services, 288t–289t, 295–312, 301t–302t
concept, 271–272
cultural, 273t–275t, 301t–302t
limitations of, 285–286, 306–307
linked, 272
methodology of, 278–279
provisioning, 273t–275t, 301t–302t
regulating, 273t–275t, 301t–302t
site assessment for, 297–299, 304–306
strength and weakness of, 284–285, 304–306
supporting, 273t–275t, 301t–302t
variable assessment, 282–284, 299, 299f
Edinburgh, SuDS management in, 318–319, 320f, 321t–322t, 333–337, 334t–336t
Effluent discharge standard principles, 19
Effluent recycling treatment, 425–428, 427f
Electroflotation, 57
Electrothermal atomic absorption spectrometer (ETAAS), 196
Elutriation, 164
End of pipe, traditional drainage, 191
Engineering methods, 194–195
Environmental monitoring, 409
Erosion prevention and maintenance of soil fertility (EPMSF), 273t–275t, 283
Eutrophication, 120–121, 121t–122t
Evapotranspiration, 379–380
Excess lime softening, 113–114
Expert system, 276, 282–283, 285–286, 295–312
Extended aeration, 97–98

F

Facultative ponds, 78–79
Farmyard, dirty water, 362, 368, 371, 386
Ferric chloride, 9
Ferrous sulfate, 9
Fertilizer, 178, 190
N–P–K, 173, 411–413
residual, 145
Filter, 10
clogging, 81, 400
media composition, 170, 171t
percolating. *See* Trickling filter
rapid gravity sand, 10
slow sand, 10
strips, 316
Filtration
biological, 81–86
hydraulics of, 74–75
percolating, 93, 94t–95t
physical, 128
pressure, 165
rapid, 69–76
slow, 61–68
vacuum, 165
First foul flush, 13–14

Fish experiment methodologies, 261–262
Flat-bottomed tank, 51
Floating solids, 23
Floc (structure), 50
Flocculation, 29, 39
  hopper-bottomed, 51
  mechanical, 35
  orthokinetic, 46–47
  perikinetic, 46–47
  process, 44–46
  zone, 47, 52
Flood
  control, 433–434, 439, 443–444, 449, 455
  defense, 433–434, 457–458, 469, 480
  duration, 139, 445
  frequency, 139
  plain evaluation, 443
  risk management, 433–438, 457–458
Flooding
  hydraulic changes due to summer, 358
  mean depth, 444–445
Flotation, 10
  electroflotation, 57
  system, 55–60
  units, 57–59
  using dissolved air, 55–57
Flotsam cover, 451–452
Flow
  dry weather, 28, 32–34
  rate, design, 13–14
  wet weather, 368
Fluoride, 10
Flush, first foul, 13–14
Food, 273t–275t
Food chain
  autotrophic, 116
  heterotrophic, 116–117
Free available chlorine, 132
Free water surface flow (FWS), 373–374, 421
Freezing, 163
Fresh water (FW), 273t–275t, 284
Fungus, sewage, 118–119

G

Glasgow, SuDS management, 318, 320f, 321t–322t, 333–337, 334t–336t
Goldfish (*Carassius auratus*), 267–268, 267f
Gravity settlement, 27
Gravity thickening, 164
Green roof, 315
Greenhouse gas, 152, 403, 406
  regulation, 273t–275t, 285, 306
  study, 402
Grit removal, 21, 24–25
Groundwater
  infiltration, 454
  level, 282, 298–299, 305, 324t–326t, 339–340
  oxygen-deficient, 108
Growth conditions (plants), 411–413
Guide level (GL), 4

H

Habitats for species (HS), 273t–275t, 283, 300–303
Hardness, 111
  carbonate hardness, 111–112
  magnesium hardness, 112–113
  noncarbonate hardness, 111–112
Heat, 130
Heavy metal
  copper, 169–170, 172–174, 175t–177t, 176, 179–182, 180t–181t, 190, 211, 214, 223–224
  determination, 173
  lead, 169–170, 172–174, 176, 179–183, 180t–181t, 190
  nickel, 211, 214, 223–224
  nitrates, 6, 17, 77, 83, 87–88, 88t, 340
  removal, 223–224
High flow rate treatment, 428–430
High nutrient load treatment, 428
High-rate activated sludge process, 97
Hopper-bottomed tank, 51
Horizontal velocity, 29
Hydraulic
  changes due to summer flooding, 358
  conductivity, 375–377, 377t, 379, 384
  ditch, 343–344, 344f
  filtration, 74–75
  head, 23, 383–384
  losses, 31
  open channels, 343–344, 344f
Hydraulic flood retention basin (HFRB), 435t–437t
Hydrocarbon
  diesel, 401–402
  total petroleum (TPH), 408–409
Hydrochloric acid (HCl), 132

Hydrology
methods, 261
precipitation, 139–140
of wetlands, 138–140
Hydro-period, 138–139
Hypochlorus acid (HOCl), 132–133

I

Imhoff flow tank, 27, 33–34
Impermeable soil proportion, 448
Impermeable surface, 278, 313, 317–319, 321t–322t
Incineration, 165–166
Inductively coupled plasma, 414–415
Infiltration
basin, 191, 197t, 199t, 314
groundwater, 454
ponds, design and operation of, 262–263, 262f
rate, 381–384, 382t
trench, 315
Infiltration wetland systems, 259–270
algae with goldfish, active control of, 267–268, 267f
fish experiment methodologies, 261–262
hydrological methods and quality analysis, 261
infiltration ponds, design and operation of, 262–263, 262f
rainfall, runoff and infiltration relationships, 263
study site, design of, 260–261, 260f
water quality assessment and management, 263–267
wetlands into urban and planning development, integration of, 268
Inlet distribution systems, 48
perforated-pipe, 48
perforated-plate, 48
point, 48
slot, 48
Instrumentation, 378
Integrated constructed wetland (ICW)
contaminant migration, 373–388
infiltration, 373–388
piggery wastewater treatment with, 419–432
for pollution control, 361–372, 362f, 363t
Integrated flood retention wetland (IFRW), 435t–437t
Interception, 139–140
Invertebrates (in main supplies), 126
Ion exchange, 108
Iron, 108
Iron compounds, 145
Iron removal, 107–109
Iron salt, 134, 163

J

Jar test, coagulation, 45

K

Kessener brush aerators, 101
Kinetics of chlorination, 133–134
Kingdom Protista, 117–118
K-nearest neighbors (KNN), 239, 242–243
multi-label, 471, 473

L

Lagoons, 162
Land
animal passage, 442
area requirement, 75
contamination, 324t–329t, 450
fragmentation of ownership, 324t–329t
value, 321t–322t, 324t–329t, 333
Landfill, sanitary, ecological, 161t, 167
Landscape planning, 434, 455
Lead, metal, 169–170, 172–174, 176, 179–183, 180t–181t, 190
Legislative pressures, piggery wastewater treatment with ICW, 419
Lime, 10
recovery, 114
softening, 113–114
Lime–soda softening, 112–113
Liner (soil-based), hydraulic conductivity of, 379
Litter
production, 206
removal, 232
zone, 209, 392, 400, 431
Loading
hydraulic, 380–381
rate methods, 28
rates, contaminant, 380–381
Local climate and air quality regulation (LCAR), 273t–275t, 283
Lysimeter, 378

M

Machine learning (technique), 239–241, 256–257
Macrophytes, in wetlands, 148–150, 173, 178–182, 268–269
Magnesium, 111–112
Maintenance of genetic diversity (MGD), 273t–275t
Manganese removal, 107–109
Mann–Whitney U test, 393, 396
Mastigophora, 117
Maximum admissible concentration (MAC), 4
Mean absolute scaled error (MASE), 246
Mean domestic water consumption, 13
Mechanical aeration, 100–101
Media breakage, 89
Medicinal resources (MR), 273t–275t
Mesosaprobic zones, 119–120
Metal
  determination, 173–174, 214–216
  nitrates, 214
  treatment and pH, link between, 224
Microbiology
  determination (examination), 174, 175t–176t
  microorganism, 7
  molecular, 362–364, 371
  statistics, 115–116
  water, 115–128
Microspora, 118
Mineralization, 93
Minerals, 404–405
Mixed liquor suspended solids (MLSS), 92, 96–97
Mixed liquor volatile suspended solids (MLVSS), 97, 99
Mixing
  hydraulic, 10
  mechanical, 10
  rapid, 10
Modeling (environmental), 469–471
Moderation of extreme events (MEE), 273t–275t, 283, 308–309
Molarity, 11
Molecular microbiology, 362–364, 371
Molybdate reactive phosphate (MRP), 365, 366t–367t, 368–369
Monitoring (environmental), 409
Multi-label classification, 469–471
Multi-label K-nearest neighbor (MLKNN), 471, 473, 478t
Multi-label support vector machine (MLSVP), 471–473, 478t
Multiple functions, of SFRB, 459, 468–469, 471, 476, 479–482

N

Natural flood retention wetland (NFRW), 435t–437t
Natural wetland treatment, 339–360
Newton's law, 37–38
Nitrate-nitrogen, 370t
Nitrates, 142
  metal, 214
Nitrification, 6, 17, 77, 83, 87–88, 88t, 340
Nitrogen, 7, 142–144
  ammonia, 5–6
  compounds, 397–398
  nitrate. *See* Nitrates
  nitrification, 6, 17, 77, 83, 87–88, 88t, 340
  organic, 7
  transformation processes, 369–370
Nutrient, 217, 340, 404–405
  removal process, 340
  requirements, 105
  transformation, 340

O

Odor
  natural, 122–123
  problems, 89
Oligosaprobic zone, 120
Opalinata, 117
Open channels, discharge determination for, 345
Operation
  problem, 89
  rate, 69–70, 74–75
Optimization (of input variables), 246–251, 248f, 249t–250t
Organic pollutants, biological effects of, 118–120
  saprobic system, 119–120
  sewage fungus, 118–119
Organic stream pollution, 17
  prediction of, 18–19
Ortho-phosphate-phosphorus, 398–399
Outlet arrangement, 441
Overflow rate, 39
Oxidation ditches, 98–99
Oxidation–reduction potential, 141

Oxygen, 140–141
demand, 104–105
dissolved, 17–19
Ozone, disinfection by, 130–131

P

Particle, colloidal, 35, 40–42
definition, 39
flocculation process, 46
producers, 116
rapid mixing, 45
van der Waals forces, 41
Particles, comparison of, 399
Pathogens, destroying, 129–130
Peat lands, 339, 341–342, 346
for cattle grazing purposes, 354–355
rehabilitation, 339–340
Percolating filtration, compared with activated sludge process, 93, 94t–95t
Permeable pavement, 272, 273t–275t, 276–294, 276f, 290t–292t, 316
pH, 10–11, 112–114
and metal treatment, link between, 224
Phosphate, 144–145
Phosphorus, 7, 144–145, 398–399
Photosynthesis, 116
Phragmites australis (common reed), 149–150, 210–212, 225
Pig, swine wastewater, 421, 426–428. *See also* Piggery wastewater treatment with ICW
Piggery wastewater treatment with ICW, 419–432
analytical methods, 424
cost–benefit analysis and limitations, 430–431
effluent recycling treatment, 425–428, 427f
high flow rate treatment, 428–430, 429f
high nutrient load treatment, 428
impact on future farming, 432
legislative pressures, 419–422
raw and diluted wastewater, 424–425
site description, 422–424
standard treatment, 425
system set-up, 422–424, 423f
test conditions, 422–424
traditional, 420
with wetlands, alternative, 420–422
Pista Grit Trap, 25
Planning
landscape, 295, 469
urban, 268
Plate settling, in sludge blanket clarifier, 53
Pollination (P), 273t–275t, 284
Pollution
control, diffuse, 439, 444, 455
control, integrated constructed wetlands for, 361–372
diffuse, 339–360
relative total, 449–450
stream, 17–20
Polyelectrolytes, 10
Polymerase chain reaction (PCR), 364–365
Polysaprobic zone, 119
Pond
aerobic, 77–78
attenuation, 191, 199t
dry, 200f
facultative, 78–79
infiltration, 191, 197t, 199t, 262–263, 262f, 314
lined, 336–337
premium, 279, 300, 301t–302t
retention, 313–314
stormwater, 313–314
waste stabilization, 77–79
wet, 191, 200f, 313–314
Prediction
accuracy of, 257
of organic stream pollution, 18–19
Preliminary treatment, 21–26
Pressure filtration, 165
Prevalence rating approach for sustainable drainage systems (PRAST), 330, 331t, 335–337, 336t
Primary productivity, 149
Primary treatment, 27–36
Principal component analysis (PCA), 460, 463
Protozoa, 116–118, 169–170, 174, 178, 182–183, 182f
ciliated, 119, 174, 190
flagellated, 119, 187
Kingdom Protista, 117–118
trophic structure, 116–117
Protozoology, 122
Public health, 123–128
Pulsator
plate, 53
super, 52–53
Pyrolysis, 166

Q

Quality, water, 345–346, 380, 408–409, 413–414
spring, 346–350
summer, 346–350
wetland, 365–368
Quiescent tanks, 33

R

Radial-flow tanks, 34
Rainfall, 263
mean annual, 447
Rapid mixing, 45–46
Rapid sand filter
compared with slow sand filter, 75
elements of, 69–70
Raw materials (RM), 273t–275t
Raw water, 9
Re-aeration, 17–18
Recreation and mental and physical health (RMPH), 273t–275t, 284
Rectangular horizontal-flow tanks, 33
Rectangular settling tanks
design details, 30–31
economics of construction, 30
Redox potential, 141, 145, 151
Reed bed (constructed wetland), 406
Reed Sweet-grass (*Glyceria maxima*), 424
Reedmace (*Typha latifolia*), 150, 194
Regeneration, town, 318
Regression analysis, 184–187, 188t, 189f, 189t
Relative total pollution, 449–450
Removal process, 107–108
advanced, 108–109
Residual concentration, 133
Resource, recovery, 17, 52–53
Respiration, 79
aerobic respiration, 140, 142
anaerobic respiration, 146
endogenous respiration, 5, 98
microbial respiration, 179
Retrofitting (SuDS), 271–294
Rhizome, of macrophyte, 142, 150, 169, 173, 190
Riparian wetlands, 139, 151–153
River Eider Valley, 343–345
characteristics of watercourses in, 346, 347f–348f
Road runoff, wetlands treating, 209–226
analysis of variance, 224–225
comparison of annual outflow water qualities, 217–223, 218t–221t, 222f
environmental conditions and operation, 212–214
filter design, 211–212, 212t
heavy metal removal, 223–224
inflow water quality, 217
limitations of, 211–212
media composition, 211–212
metal determinations, 214–216
metal nitrates, 214, 215t–216t, 218t–219t
nutrients, 217
organic strength, 217
pH and metal treatment, link between, 224
study site, 210, 210f
Roof runoff, wetland systems to control, 191–208
algal growth, control of, 196–197
aquatic plant management, 204–206, 204f–205f
engineering methods, 194–195
purpose of, 192
site description, 192
standard design considerations, 197–201, 197t–199t, 200f–201f
study site, design of, 192–194, 193f
sustainable roof runoff drainage, 191
system capacity, 197
system design comparisons, 201–202, 202t–203t
twenty-four-hour water quality monitoring, 202t–203t, 203–204
water quality analysis, 195–196
water quality management, 202–203
Rooting, macrophyte, 153
Rotating biological contactors (RBCs), 87–90
design and loading criteria, 87–88, 88t
operational problems, 89
principle elements, 89
principles of operation, 87
Roughing process, 84
Royal Commission, 1

Runoff
future, 324t–329t
quality, 321t–322t, 324t–329t
quantity, 321t–322t, 333
roof, 191–208
stormwater, 313–314
urban, 191

S

Safety. *See* Water quality standards
Sampling
twenty-four-hour, 202t–203t
water quality variable, 378–379
Sand bed, 70–71
Sand filter
rapid, 10, 69–70
slow, 10
Saprobic system, 119–120
Sarcodina, 117
Sarcomastigophora, 117
Schmutzdecke (layer of dirt), 64, 67
Screening (unit)
design details for, 23
design of, 21–23
Seasonal assessment, of vertical-flow wetlands, 389–400
clogging tests, 393
experimental set-up and operation, 391–392, 391t
filter clogging, 400
inflow water quality analysis, 393–400
nitrogen compounds, comparison of, 397–398
ortho-phosphate-phosphorus, evaluation of, 398–399
oxygen demand variables, evaluation of, 394–397, 394f
particles comparison, 399
statistical tests, 393
water quality analysis, 392–393
Sediment
mean depth, 450
organic proportion, 450–451
Sedimentation
aids, 35
efficiency, 64
Sedimentation tanks, 93
design, 28–29
design details, 30–31
economics of construction, 30
general details, 32
horizontal velocity, 29
hydraulic loss, 31
ideal, 39
primary, 33–34
secondary, 34
settling velocity, 29
time ratio, 30
Self-organizing map (SOM), 240–241, 245–246
Self-purification (aerobic), 77
Series or plug flow system, 96
Settling
behavior, classification of, 37–39
blanket or zone, 38
characteristics, 96
column, 39
hindered, 38–39
plate, 53
tanks. *See* Settling tanks
theory of, 37
velocity, 29
Settling tanks
design details, 30–31
economics of construction, 30
Sewage fungus, 118–119
Sewage treatment, 13–16
design flow rate, 13–14
principles of, 14
stages, engineering classification of, 14–15
Sewer, 13–14, 22
Shaft failure, 89
Shannon–Weaver Diversity Index, 115–116
Site
description, 192
identification, 459–461
Slope
bottom, 28
floor, 30, 34
internal, 162
ratio, 194
side, 24, 32, 34
Sloughing of biomass, 82
Slot distribution system, 48
Slow sand filter, 61–66
advantages and disadvantages of, 67–68
compared with rapid sand filter, 75
elements of, 61–63
mechanisms of, 63–66
Sludge, 32
aerobic digestion, 162
air drying, 164

bridging, 32
chemical, 158
chemical conditioning, 163
composting, 163, 166–167
deposition, 32
dewatering, 34, 159–160, 163–165
digested, 158, 161–162
disposal, 160–162, 161t
dumping on land, 167–168
elutriation, 164
excess activated, 158
freezing, 163
gravity thickening, 164
humus, 34, 158
incineration, 165–166
lagoons, 162
moisture and organic content of, 159t
primary, 34
pyrolysis, 166
raw or primary, 157
returned, 93
scrappers, 32–33
surplus, 34
thawing, 163
thickening, 57, 161t, 164
treatment, 32, 157–168
Sludge blanket clarifier, 38, 47–54
plate settling in, 53
types of, 51–53
Soak-away, 316
Soda
ash, 111
caustic, 43, 111
equivalents of, 112
softening. *See* Softening, lime–soda
Sodium aluminate, 43
Softening
excess lime, 113–114
lime recovery, 114
lime, 113
lime–soda, 112–113
water, 111–114
Soil
liner, hydraulic conductivity of, 379
proportion, impermeable, 448
Source control, sustainable drainage, 259, 313. *See also* Sustainable drainage system (SuDS)
Spiritual experience and sense of place (SESP), 273t–275t
Spring, water quality during, 346–350
Stabilization, contact, 98
Standards, 1–8
effluent discharge, 19
water quality, 1–8
Standard water treatment, 9–10
Statistics
analysis, 176
for applied microbiology, 115–116
mean, 115
median, 115
mode, 115
standard deviation, 115
honestly significant difference, 381
Stem-flow, 139–140
Step feed activated sludge process, 97
Stoke's law, 37–38
Storage (below-ground), 317
Storm runoff treatment (SRT), 273t–275t, 283, 308–309
Storm tanks, 32
Stormwater, runoff, 313–314
Straining, 9
Stream water treatment, 169–190
analytical procedures, 173–174
correlation analysis, 184–187, 185t–186t, 189f
efficiency, comparison of, 176–178, 177t
environmental conditions and operation, 172–173
experimental plan and limitations, 170
filter media composition, 170, 171t
macrophytes and water quality, 178–182, 178f–179f, 180t–181t
microbiological examinations, 174, 175t–176t
microbiology and water quality, 182–184, 182f–184f
regression analysis, 184–187, 188t, 189f, 189t
statistics, 176
Streeter–Phelps equation, 18–19
Sulfur, 146–147
Summer, water quality during, 346–350
Support vector machine (SVM), 240, 243–245, 470
multi-label, 471–473
Suspended solids (SS), 7, 19, 23, 37, 400
Suspension, colloidal, 40–41
Sustainable drainage system (SuDS), 15, 191–192, 227, 259–260, 262, 268–269, 313–317

Sustainable drainage system (SuDS) (*Continued*)
  decision support matrix (technique), 319–323, 324t–326t
  decision support model, 333–335, 334t–335t
  impact on water quality and quantity, 317–318
  model, 313–338
  retrofitting, 271–294
  shallow, 315
  standard, 314–315
  swale, 314–315
  variable determination approaches, 300–303
  weighting system, 319–323, 327t–329t
Sustainable flood retention basin (SFRB), 433–456
  aesthetic flood treatment wetland, 435t–437t
  algal cover in summer, 449
  altitude, 448
  aquatic animal passage, 441–442
  arable catchment proportion, 452–453
  basic channel connectivity, 443
  basin bed gradient, 446
  catchment size, 452
  classification of, 457–482
  classification variables, assessment of, 440
  dam height, 440
  dam length, 441
  drainage, 447–448
  engineered, 440
  flood duration, 445
  flood plain evaluation, 443
  flood risk management, 457–458
  flood water surface area, 447
  flotsam cover, 451–452
  forest catchment proportion, 453
  future work, recommendations for, 455
  groundwater infiltration, 454
  hydraulic flood retention basin, 435t–437t
  impermeable soil proportion, 448
  integrated flood retention wetland, 435t–437t
  land animal passage, 442
  length of basin, 454
  manpower and equipment, 438–439
  maximum flood water volume, 446–447
  mean annual rainfall, 447
  mean basin bed velocity, 446
  mean depth of basin, 454
  mean flooding depth, 444–445
  mean sediment depth, 450
  natural catchment proportion, 453–454
  natural flood retention wetland, 435t–437t
  organic sediment proportion, 450–451
  outlet arrangement, 441
  pasture catchment proportion, 453
  proportion of flow within the channel, 444
  rationale and objectives, 439–440
  rationale for, 433–434
  relative total pollution, 449–450
  seasonable influence, 448
  survey template, 438
  sustainable flood retention wetland, 435t–437t
  traditional flood retention basin, 435t–437t
  typical wetness duration, 445
  typology of, 434–438
  urban catchment proportion, 452
  vegetation cover, 449
  viniculture catchment proportion, 453
  wetness, 443–444
  wetted perimeter, 446
  width of basin, 455
Sustainable flood retention wetland (SFRW), 435t–437t
Swine (pig), water, 421, 426–428, 430
System capacity, 197

T

Tanks
  design, 28–29
  flat-bottomed, 51
  Imhoff flow, 27, 33–34
  quiescent, 33
  radial-flow, 34
  rectangular horizontal-flow, 33
  sedimentation. *See* Sedimentation tanks
  settling. *See* Settling tanks
  storm, 32
Tapered aeration, 96
Thawing, 163
Theory of settling, 37
Thickening
  gravity, 164
  sludge, 161t
Through-fall, 139–140
Time, ratio, 30
Tourism and area value (TAV), 273t–275t, 284

Toxins, natural, 122–123
Traditional flood retention basin (TFRB), 435t–437t
Treatment
biological, 77–80
chemical, 163
objectives, water quality standards and, 3–5
preliminary, 14, 21–26
primary, 14, 27–36
secondary, 14–15
standard, 425
tertiary, 14
Trickling filter, 81–82
Trophic structure of protozoa, 116–117
Turbidity, 64, 68, 151, 409
Turf (grass), 236
Twenty-four-hour sampling, 202t–203t, 203–204
Tyndall effect, 40
Typha latifolia (Reedmace), 150

U

Underdrain system, 71–74
Urban
area, 268, 271–272, 287–293
runoff, 191
tree, 272

V

Vacuum filtration, 165
Vegetables
replanting, 410–411, 412t–413t
selection of, 409–410
Vegetation
characterization, 346, 355–358, 356t–357t
control, 355–358
cover, 449
determination, 280–281
ditch, 343–345, 350–352, 353f
growth, ditches disappearance due to, 353–354
Velocity
downward, 56–57
horizontal, 29
settling, 29
Vertical-flow wetlands, seasonal assessment of, 389–400
Vertical settling tanks
design details, 31
economics of construction, 30

W

Ward's linkage, 464
Waste, organic, 17
Waste stabilization ponds, 77–79
aerobic ponds, 77–78
facultative ponds, 78–79
Wastewater
characterization of, 7
diluted, 424–425
domestic, 13
raw, 424–425
recycling, 401–403
Water
analysis, 195–196
budget, 138–139
chemistry, 10–11
drinking, 9–12
fluxes, 384
microbiology, 115–128
playground, 317
quality assessment, 263–267, 264t–266t
quality inflow, 217, 264t, 393–400
quality management, 202–203, 263–267
quality monitoring, 202t–203t, 203–204
quality outflow, 217–223, 218t–221t, 222f, 234–236, 235t
raw, 9
reclamation, 404
sampling, 378–379
softening, 111–114
sources of, 9
treatment, 3–5, 9–12
twenty-four-hour sampling, 202t–203t
water quality standards, 1–8, 19, 210, 350, 392, 408
Waterborne diseases, monitoring and prevention of, 126–128
Watercourse, waterway, 346, 349t
Water quality standards, 1–8
historical aspects of, 1–2
and treatment objectives, 3–5
Water-related diseases, 123–124, 124t–125t
Weighting system, SuDS, 319–323, 327t–329t
Wetland
chemistry, 140–147
constructed, 137–156
definitions of, 137–138
ecosystem mass balance, 147–148
hydrology of, 138–140
macrophytes in, 148–150
natural, 339–360
physical and chemical parameters, 150–151

Wetland (*Continued*)
reed bed, 406
riparian, 139, 151–153
sustainable drainage systems, 15
systems, 93, 94t–95t, 191–208, 468–469
vertical-flow, 213t
water quality, 365–368
Wildlife habitat, 279–280, 300–303, 301t–302t
Wilcoxon signed rank test, 463–464

X

X (or *x*), variable, 187, 188t–189t

Y

Y (or *y*), variable, 187, 188t–189t

Z

Zooplanktons, 183, 190, 206

Kolibri 360 GmbH
Ferdinand-Jühlke-Straße 7
99095 Erfurt, Deutschland
produktsicherheit@kolibri360.de